Grundlehren der mathematischen Wissenschaften 320

A Series of Comprehensive Studies in Mathematics

Springer
Berlin
Heidelberg
New York
Barcelona
Hong Kong
London
Milan
Paris
Singapore
Tokyo

Claude Kipnis
Claudio Landim

Scaling Limits of Interacting Particle Systems

Springer

Claude Kipnis†

Claudio Landim
Instituto de Matematica Pura e Aplicada
Estrada Dona Castorina 110
22460-320 Rio de Janeiro, Brasil
and
CNRS UPRES-A 6085
Université de Rouen
F-76128 Mont Saint Aignan Cedex, France

Library of Congress Cataloging-in-Publication Data

Kipnis, Claude, 1949-1993
Scaling limits of interacting particle systems / Claude Kipnis,
Claudio Landim.
p. cm. – (Grundlehren der mathematischen Wissenschaften ; 320)
Includes bibliographical references and index.
ISBN 3-540-64913-1 (Berlin : alk. paper)
1. Scaling laws (Statistical physics) 2. Probabilities.
3. Markov processes. 4. Hydrodynamics--Mathematics.
5. Mathematical physics. I. Landim, Claudio. II. Title.
III. Series.
QC174.85.S34L36 1999
532'.5--dc21 98-46051
CIP

Mathematics Subject Classification (1991): 60-02, 60K35, 82A05

ISSN 0072-7830

ISBN 3-540-64913-1 Springer-Verlag Berlin Heidelberg New York

Cover design: MetaDesign plus GmbH, Berlin
Typesetting by the author using a Springer TEX macro-package
SPIN: 10631667 41/3143-5 4 3 2 1 0 Printed on acid-free paper

A minha filha Anna

Aos meus pais Raul e Regina

Preface

The idea of writing up a book on the hydrodynamic behavior of interacting particle systems was born after a series of lectures Claude Kipnis gave at the University of Paris 7 in the spring of 1988. At this time Claude wrote some notes in French that covered Chapters 1 and 4, parts of Chapters 2, 5 and Appendix 1 of this book. His intention was to prepare a text that was as self-contained as possible. It would include, for instance, all tools from Markov process theory (cf. Appendix 1, Chaps. 2 and 4) necessary to enable mathematicians and mathematical physicists with some knowledge of probability, at the level of Chung (1974), to understand the techniques of the theory of hydrodynamic limits of interacting particle systems.

In the fall of 1991 Claude invited me to complete his notes with him and transform them into a book that would present to a large audience the latest developments of the theory in a simple and accessible form. To concentrate on the main ideas and to avoid unnecessary technical difficulties, we decided to consider systems evolving in finite lattice spaces and for which the equilibrium states are product measures. To illustrate the techniques we chose two well-known particle systems, the generalized exclusion processes and the zero-range processes. We also conceived the book in such a manner that most chapters can be read independently of the others.

Here are some comments that might help readers find their way. At the end of each chapter a list of references is given, possible or proved extensions are discussed, a brief survey of some related topics that could not be covered here is provided.

To present the main concepts and the principal goals of the theory in the simplest possible context, in Chapter 1 we prove the hydrodynamic behavior of a system with no interaction among particles, more precisely for a superposition of independent random walks. We recommend this chapter for someone who has never studied the subject before. We explain there the need for a microscopic and a macroscopic time scale, the natural time renormalization in order to observe non-trivial hydrodynamic phenomena for symmetric and asymmetric processes and the concepts of local equilibrium, conservation of local equilibrium, hydrodynamic behavior of interacting particle systems and hydrodynamic equations.

The purpose of Chapters 2 and 3 is to review the main properties of generalized exclusion processes and zero range processes and to introduce some weak formulations of local equilibrium. These two chapters can be skipped by someone familiar with interacting particle systems.

In Chapter 4 we start to present the main steps and the main tools in a general proof of the hydrodynamic behavior of interacting particle systems. To illustrate the approach we consider symmetric simple exclusion processes. The proof of the hydrodynamic behavior of these systems turns out to be very simple because for symmetric simple exclusion processes the density field is closed under the action of the generator. We discuss there the natural topological spaces to consider and we state some tightness results that are used throughout the book.

In Chapter 5 we explain the entropy method in the context of zero-range processes. Here the density field is no longer closed under the action of the generator and one needs to project local fields on the density field. This is made possible by the one and two block estimates that rely on bounds on the entropy and the Dirichlet form obtained through the examination of the time evolution of the entropy. The one and two block estimates are one of the most important results of this book. Most of the subsequent chapters depend on these estimates.

In Chapter 6 we present the relative entropy method, an alternative approach to prove the hydrodynamic behavior of systems whose hydrodynamic equation admits smooth solutions. This is a quite simple method to derive the hydrodynamic limit of an interacting particle system. Its development led to the investigation of the Navier–Stokes corrections to the hydrodynamic equations. This chapter can be read independently of the others, we just need the one block estimate.

In Chapter 7 we extend to reversible nongradient systems, i.e., to processes in which the instantaneous current cannot be written as the difference of a continuous function and its translation, the entropy method presented in Chapter 5. The proof of the hydrodynamic behavior of nongradient systems relies on a sharp estimate for the spectral gap of the generator of the process restricted to finite cubes and on the characterization of closed and exact forms of $\mathbb{N}^{\mathbb{Z}^d}$. The need for a sharp estimate on the spectral gap should be emphasized because it is completely hidden in the proof and because one of the main open questions in the field is that of obtaining a proof of the hydrodynamic behavior of nongradient systems that does not use any information on the size of the gap. The sharp estimate on the spectral gap and the characterization of closed and exact forms of $\mathbb{N}^{\mathbb{Z}^d}$ are contained in Appendix 3. There we also prove some estimates on the largest eigenvalue of small perturbations of Markov generators.

In Chapter 8 we prove the hydrodynamic behavior of asymmetric attractive processes in arbitrary dimensions. This chapter can also be read independently of the others, we just use the one block estimate. The proof relies on the theory of measure-valued entropy solutions of hyperbolic equations and on the entropy inequalities presented in Appendix 2.

In Chapter 9 we show how to derive the conservation of local equilibrium from a law of large numbers for local fields in the case of attractive systems. This result together with the methods presented in Chapter 8 for asymmetric processes and in Chapters 5, 6 and 7 for symmetric processes allow us to prove the conservation of local equilibrium for attractive systems. This chapter can be read independently of the previous ones.

In Chapter 10 we prove a large deviation principle for the density field by showing that the one and two block estimate are in fact superexponential. In Chapter 11 we investigate the equilibrium fluctuations of the density field. The nonequilibrium fluctuations remain a mainly open problem, though there exist some one-dimensional results.

In Appendix 1 we present all the tools of Markov processes used in the book. In Appendix 2 we derive, from the local central limit theorem for independent and identically distributed random variables, estimates on the distance between the finite marginals of canonical and grand canonical equilibrium measures. We also prove some results on large deviations needed in Chapter 10, fix the terminology and recall some well known results about weak solutions of hyperbolic and parabolic quasi-linear partial differential equations.

In September 1993, when Claude died, Chapters 1, 2, 4, 5 and parts of Appendix 1 were ready and I decided to conclude alone our original project. I just made one important modification: the language. All those who met Claude, certainly remember his incredible facility for learning languages. Even though he could speak and write in perfect English, the natural language in which to write a scientific book at the end of the twentieth century, Claude insisted on writing this book in French, in some sense our common native language. After he died in 1993, I decided, in order to maximise the potential readership, present and future, to switch to English. I hope Noemi will forgive me.

I take this opportunity to express my thanks to all our collaborators and colleagues from whom we learned most of the techniques presented here and to all those who read parts of the book for their comments and their encouragement. I particularly acknowledge the influence on us of Pablo Ferrari, Antonio Galves, Gianni Jona-Lasinio, Tom Liggett, Stefano Olla, Errico Presutti, Raghu Varadhan and Horng-Tzer Yau. I would like also to thank Ellen Saada for her careful reading of most of the chapters and for her uncountable suggestions and Olivier Benois for his helpful support with all the subtleties of TEX.

Campos dos Goytacazes, April 1998. Claudio Landim

Contents

Appendices

Frequently Used Notation

$\eta^{x,y}$ 26
ι_ε 69, 77
$\ll \cdot, \cdot \gg_\alpha$ 171
$\ll \cdot \gg_\alpha$ 162
$\ll g, h \gg_{\alpha,0}$ 161
$\ll g \gg_{\alpha,j}$ 161
$\mu \leq \nu$ 24
$\nabla\Psi$ 144
$\nabla\rho$ 15
$\nabla_{b_1,b_2} g$ 143
$\nu^N_{\rho_0(\cdot)}$ 42
$\nu^{\ell,j}$ 88
ν_α 26, 30
$\nu_{N,K}$ 27
$\nu_{\Lambda,K}$ 350
$\nu_{\ell,K}$ 159
$\nu^N_{\rho(\cdot)}$ 9
$\nu^N_{\rho_0(\cdot)}$ 12
$\nu_{\mathbb{T}^d_N,K}$ 11
$\partial^N_{u_i}$ 142, 300
$\pi^N(\eta, du)$ 43
$\pi^{N,\ell}$ 200
$\pi_j \bar{\mu}$ 238
$\pi^N_t(du)$ 47
$\sigma^{x,y}\eta$ 401
$\sigma_x \eta$ 281
$\tau_x \Psi$ 42
τ_x 13
$\underline{0}$ 214
φ^* 28
$a \leq b$ 44
$d(x, \Lambda)$ 163
$d_{i,j}(\alpha)$ 150
e_j 104
g^* 28
s_ψ 151
s_f 145
$w'_\mu(\gamma)$ 50
$w_\mu(\gamma)$ 51
$\mathbb{E}_\mu$ 9
$\mathbb{P}^N_{\mu^N}$ 69
$\mathbb{P}_{\mu^N}$ 141
$\mathbb{P}_\mu$ 9
$\mathbb{T}_N$ 7
$\mathcal{M}_+$ 47
$\mathcal{M}_+(\mathbb{T}^d)$ 47
$\mathcal{M}_1(\cdot)$ 13
$\mathfrak{A}^k$ 144, 161
$\mathfrak{D}(f)$ 343
$\mathfrak{D}(\nu_{\Lambda,K}, \cdot)$ 152
$\mathfrak{D}_b(\nu_{\Lambda,K}, \cdot)$ 152
$\mathfrak{F}$ 176
$\mathfrak{d}_x$ 106
[r] 12

Introduction

The problem we address in this book is to justify rigorously a method often used by physicists to establish the partial differential equations that describe the evolution of the thermodynamic characteristics of a fluid.

Suppose that we are interested in examining the evolution of a system constituted of a large number of components or degrees of freedom, e. g., a fluid or a gas. Since the total number of components is very large (typically of the order of 10^{23}), we are not interested nor able to give a precise description of the microscopic state of the system. Hence, following the statistical mechanics approach introduced by Boltzmann, we first examine the equilibrium states of the the system and characterize them by a small number of macroscopic quantities $\mathbf{p} = (p_1, \dots, p_a)$, called the thermodynamic characteristics, e. g. the temperature, the density, the pressure (which are often related through the equation of state).

Once the equilibrium states are known, we may investigate the evolution of a system out of equilibrium. Assume that the gas is confined in a volume V. For each point u in V, denote by V_u a small neighborhood of u, that is, small if compared to the total volume V, but large enough if compared to the intermolecular distance, in order to assume that each neighborhood contains an infinite number of particles. Due to the "strong" interaction among the molecules, it is natural to believe that the system reaches immediately a local equilibrium state, i.e., that in each neighborhood V_u the system is close to an equilibrium state characterized by the parameter $\mathbf{p}(u)$, that may depend on u.

As the system evolves, we expect this local equilibrium picture to evolve in a smooth way. More precisely, one would expect to observe at a later time t in a small neighborhood V_u a state close to a new equilibrium state characterized by a parameter $\mathbf{p}(t, u)$, which now depends on space and time. Moreover, it seems reasonable to believe that $\mathbf{p}(t, u)$ evolves smoothly in time according to a partial differential equation, called the hydrodynamic equation.

Despite many efforts, this program has not been completely achieved for Hamiltonian systems where particles evolve deterministically according to Newton's equations, mainly due to the lack of good ergodic properties of the system. Two simplifications have traditionally then been made. Either to assume the evolution of the microscopic system to be stochastic or to consider systems with low density of particles in such a way that the total number of collisions of each particles remains finite in finite intervals of time in order to avoid correlations.

Our main purpose in this book is to present general methods that permit to deduce the hydrodynamic equations of the thermodynamic characteristics of infinite systems assuming that the underlying microscopic dynamics is stochastic, i.e., to deduce the macroscopic behavior of the system from the microscopic interaction among particles. In all cases the microscopic dynamics will consist of random walks on the lattice submitted to some local interaction, the so-called interacting particle systems introduced by Spitzer (1970).

For simplicity, we shall assume that there is a unique conserved quantity: the total number of particles. Moreover, to highlight the main ideas and to avoid unnecessary technical difficulties, we shall present all results for processes whose equilibrium states are product measures. It should be emphasized, however, that all methods presented here depend on the *explicit* knowledge of the equilibrium states.

To illustrate the type of results we are going to prove, we conclude this introduction deriving the hydrodynamic equation of an interacting particle system under general assumptions on the microscopic evolution. To fix ideas, consider a system evolving on a one-dimensional torus, denoted by $\mathbb{T} = [0, 1)$. Fix a positive integer N, that represents the inverse of the distance between particles and that eventually will increase to infinity. Denote by $\eta(x)$ the total number of particles at time 0 in the interval $[x/N, (x+1)/N)$. The configuration $\eta = \{\eta(0), \dots, \eta(N-1)\}$ is therefore an element of $\mathbb{N}^{\mathbb{T}_N}$, provided $\mathbb{T}_N$ stands for the discrete one-dimensional torus with N points: $\mathbb{T}_N = \{0, \dots, N-1\}$. Two space scales appear naturally. The macroscopic scale $\mathbb{T}$ and the microscopic scale $\mathbb{T}_N$. Points of the macroscopic scale $\mathbb{T}$ are denoted by the letters u, v and correspond to the sites $[uN]$, $[vN]$ in the microscopic scale. Here, for a real number a, $[a]$ stands for the integer part of a. Sites in the microscopic scale $\mathbb{T}_N$ are denoted by the letters x, y, z and correspond to the points x/N, y/N, z/N in the macroscopic scale $\mathbb{T}$.

The time evolution of the system can be informally described as follows. We fix cylinder functions $\{c(x, \pm 1, \cdot), x \in \mathbb{T}_N\}$. If the state of the process is η, at rate $c(x, \pm 1, \eta)$ a particle jumps from x to $x \pm 1$. This dynamics corresponds to a Markov process whose generator L_N is given by

$$(L_N f)(\eta) = \sum_{\substack{x \in \mathbb{T}_N \\ |y|=1}} c(x, y, \eta)[f(\eta^{x,x+y}) - f(\eta)] .$$

In this formula $\eta^{x,x+y}$ is the configuration obtained from η by letting a particle jump from x to $x+y$:

$$\eta^{x,x+y}(z) = \begin{cases} \eta(z) & \text{if } z \neq x, \ x+y, \\ \eta(x) - 1 & \text{if } z = x, \\ \eta(x+y) + 1 & \text{if } z = x+y. \end{cases}$$

We shall assume that the process is translation invariant in the sense that the rate $c(x, y, \eta)$ is equal to $c(y, \tau_x \eta)$ for two cylinder functions $c(\pm 1, \eta)$. Here $\tau_x \eta$ stands for the configuration η translated by x so that $(\tau_x \eta)(y) = \eta(x+y)$. This group of translations is naturally extended to functions by the identity $(\tau_x f)(\eta) = f(\tau_x \eta)$.

To avoid degeneracies, we assume that $c(\pm 1, \eta) > 0$ if $\eta(0) > 0$ so that a jump from x to $x \pm 1$ is possible whenever there is at least one particle at x. Of course, when there are no particles at x a jump is impossible so that $c(\pm 1, \eta) = 0$ if $\eta(0) = 0$.

Denote by K the total number of particles at time 0: $K = \sum_{x \in \mathbb{T}_N} \eta(x)$. Notice that the total number of particles is conserved by the dynamics and that the process is irreducible for each fixed number of total particles. Hence, for each fixed K, $\{\eta_t,\, t \geq 0\}$ is an irreducible finite state Markov process. In particular, there exists a unique invariant measure, denoted by $\nu_{N,K}$. Since the process is translation invariant, so is the measure $\nu_{N,K}$. Therefore, since $\nu_{N,K}$ is concentrated on configurations with K particles, $E_{\nu_{N,K}}[\eta(x)] = K/N$ for all x in $\mathbb{T}_N$. Here, for a probability measure μ, $E_\mu[\cdot]$ stands for the expectation with respect to μ.

We shall assume that these invariant measures have nice local properties as $N \uparrow \infty$. More precisely, that there exists product probability measures $\{\nu_\alpha,\, \alpha \geq 0\}$ on $\mathbb{N}^{\mathbb{Z}}$ such that the finite marginals of $\nu_{N,K}$, as $N \uparrow \infty$ and $K/N \to \alpha$, converge to the finite marginals of ν_α:

$$\lim_{\substack{N \to \infty \\ K/N \to \alpha}} E_{\nu_{N,K}}[f] \;=\; E_{\nu_\alpha}[f]$$

for every bounded cylinder function f. The measures ν_α inherit all properties of the measure $\nu_{N,K}$. For each fixed $\alpha \geq 0$, ν_α is translation invariant, invariant for the infinite volume dynamics and parametrized by the density: $E_{\nu_\alpha}[\eta(0)] = \alpha$.

The requirement that the measures ν_α are product is not essential but simplifies the exposition and will systematically be satisfied by the examples considered here. This property permits to introduce in a simple way the idea of local equilibrium. Fix a continuous function $\rho\colon \mathbb{T} \to \mathbb{R}_+$. Denote by $\nu^N_{\rho(\cdot)}$ the product measure on $\mathbb{N}^{\mathbb{T}_N}$ with marginals given by

$$\nu^N_{\rho(\cdot)}\{\eta,\, \eta(x) = k\} \;=\; \nu_{\rho(x/N)}\{\eta,\, \eta(0) = k\}\;.$$

Thus, at site x (that corresponds to the macroscopic location x/N) we place particles according to the invariant distribution with density $\rho(x/N)$. It is easy to check that locally around the macroscopic point $u \in \mathbb{T}$ the measure $\nu^N_{\rho(\cdot)}$ is close to the invariant measure with density $\rho(u)$. Indeed, since the measure $\nu^N_{\rho(\cdot)}$ is product, for every positive integer ℓ and every sequence $(a_1, \dots, a_\ell)$,

$$\lim_{N \to \infty} E_{\nu^N_{\rho(\cdot)}}\Big[\prod_{x=1}^{\ell} \mathbf{1}\{\eta(x + [uN]) = a_x\}\Big] \;=\; E_{\nu_{\rho(u)}}\Big[\prod_{x=1}^{\ell} \mathbf{1}\{\eta(x) = a_x\}\Big]\;.$$

We shall now deduce the macroscopic evolution of the density profile $\rho(\cdot)$ assuming that the local equilibrium is conserved. More precisely, assume that at each macroscopic time t, that will correspond to a microscopic time $t\theta(N)$ for some scaling $\theta(N)$, the state of the process is still in local equilibrium, i.e., that around each u and at each time t the process is close to some equilibrium state.

Denote by $\rho(t,\cdot)$ the density profile at time t, so that at time t and in a small neighborhood of u the state of the process is close to $\nu_{\rho(t,u)}$.

Denote by $W_{x,x+1}$ the instantaneous current between x and $x+1$, i.e, the rate at which a particle jumps from x to $x+1$ minus the rate at which a particle jumps from $x+1$ to x. Since the process is translation invariant, $W_{x,x+1} = \tau_x W_{0,1}$ and $W_{0,1} = c(1,\eta) - c(-1,\tau_1\eta)$. There are two cases to be considered: Either in the stationary regime the current has mean zero with respect to all invariant measures or not. Assume first that $E_{\nu_\alpha}[W_{0,1}]$ does not vanish uniformly in α.

Fix a smooth function $H\colon \mathbb{T} \to \mathbb{R}$ and consider the martingale M_t^H defined by

$$M_t^H = N^{-1} \sum_{x\in\mathbb{T}_N} H(x/N)\eta_t(x) - N^{-1} \sum_{x\in\mathbb{T}_N} H(x/N)\eta(x) - \int_0^t ds\, N^{-1} \sum_{x\in\mathbb{T}_N} H(x/N) L_N \eta_s(x) .$$

By definition of the current, $L_N\eta(x) = W_{x-1,x} - W_{x,x+1}$. Therefore, after a summation by parts, the martingale M_t^H becomes

$$N^{-1} \sum_{x\in\mathbb{T}_N} H(x/N)\eta_t(x) - N^{-1} \sum_{x\in\mathbb{T}_N} H(x/N)\eta(x) - \int_0^t ds\, N^{-2} \sum_{x\in\mathbb{T}_N} (\partial^N H)(x/N) W_{x,x+1}(s) .$$

In this formula $(\partial^N H)$ stands for the discrete derivative of H: $(\partial^N H)(x/N) = N[H((x+1)/N) - H(x/N)]$. If we now rescale time by N, i.e., if we consider times of order N and change variables in the time integral, the martingale M_{Nt}^H turns out to be equal to

$$N^{-1} \sum_{x\in\mathbb{T}_N} H(x/N)\eta_{tN}(x) - N^{-1} \sum_{x\in\mathbb{T}_N} H(x/N)\eta(x) - \int_0^t ds\, N^{-1} \sum_{x\in\mathbb{T}_N} (\partial^N H)(x/N) W_{x,x+1}(sN) . \tag{0.1}$$

Since M_t^H is a martingale vanishing at time 0, its expectation is equal to 0 uniformly in time. Since we are assuming that the system is in local equilibrium at the macroscopic time t, which corresponds in this set up to the microscopic time tN, the expectation of $\eta_{tN}(x)$ is close to the expectation of $\eta(0)$ with respect to $\nu_{\rho(t,x/N)}$ because the site x in the microscopic scale corresponds to the point x/N in the macroscopic scale. By the same reasons, the expectation of $W_{x,x+1}(sN)$ is close to the expectation of $W_{0,1}$ with respect to $\nu_{\rho(s,x/N)}$. Therefore, taking expectation in (0.1), up to lower order terms, we have that

$$N^{-1}\sum_{x\in\mathbb{T}_N} H(x/N)\rho(t,x/N) \;-\; N^{-1}\sum_{x\in\mathbb{T}_N} H(x/N)\rho(x/N)$$
$$= \int_0^t ds\, N^{-1}\sum_{x\in\mathbb{T}_N} (\partial^N H)(x/N)\tilde{W}(\rho(s,x/N))\,,$$

provided $\tilde{W}(\alpha)$ stands for the expectation of the current $W_{0,1}$ with respect to the invariant measure with density α: $\tilde{W}_{0,1}(\alpha) = E_{\nu_\alpha}[W_{0,1}(\eta)]$. Since this expectation does not vanishes uniformly, integrating by parts, we obtain that the density $\rho(t,u)$ is a weak solution of the first order partial differential equation

$$\begin{cases} \partial_t\rho + \partial_u\tilde{W}(\rho) = 0\,, \\ \rho(0,\cdot) = \rho_0(\cdot)\,. \end{cases} \tag{0.2}$$

In the case where the current has mean zero with respect to all equilibrium measures, assume that it can be written as the difference of a continuous function and its translation: $W_{0,1} = h - \tau_1 h$. Interacting particle systems satisfying this assumption are called gradient processes. We discuss this property in Chapters 5 and 7. This assumption permits a second summation by parts in (0.1). The martingale M_t^H is now equal to

$$N^{-1}\sum_{x\in\mathbb{T}_N} H(x/N)\eta_t(x) \;-\; N^{-1}\sum_{x\in\mathbb{T}_N} H(x/N)\eta(x)$$
$$- \int_0^t ds\, N^{-3}\sum_{x\in\mathbb{T}_N} [\partial^N(\partial^N H)]((x-1)/N)\tau_x h(\eta_s)\,.$$

Rescaling now time by N^2, by the assumption of conservation of local equilibrium and since the expectation of $M_{tN^2}^H$ vanishes, up to lower order terms, we have that

$$N^{-1}\sum_{x\in\mathbb{T}_N} H(x/N)\rho(t,x/N) \;-\; N^{-1}\sum_{x\in\mathbb{T}_N} H(x/N)\rho(x/N)$$
$$= \int_0^t ds\, N^{-1}\sum_{x\in\mathbb{T}_N} [\partial^N(\partial^N H)]((x-1)/N)\tilde{h}(\rho(s,x/N))\,,$$

where $\tilde{h}(\alpha)$ is the expectation of the continuous function h with respect to the invariant measure with density α: $\tilde{h}(\alpha) = E_{\nu_\alpha}[h(\eta)]$. Therefore, $\rho(t,u)$ is a weak solution of the parabolic equation

$$\begin{cases} \partial_t\rho = \partial_u^2\tilde{h}(\rho)\,, \\ \rho(0,\cdot) = \rho_0(\cdot)\,. \end{cases} \tag{0.3}$$

It is therefore easy to derive the macroscopic evolution of the density profile $\rho(\cdot)$ under the assumption of local equilibrium. The purpose of this book is to present general methods that permit to prove the convergence in probability of the density field

$$N^{-1} \sum_{x \in \mathbb{T}_N} H(x/N) \eta_{t\theta(N)}(x)$$

to the integral $\int_{\mathbb{T}} du\, H(u)\rho(t,u)$, where ρ is the weak solution of a partial differential equation of type (0.2) or (0.3) and $\theta(N)$ a rescaling constant. In some cases we will be able to deduce from this convergence the conservation of local equilibrium (cf. Chap 9).

1. An Introductory Example: Independent Random Walks

The main purpose of this book is to present general methods that permit to deduce the hydrodynamic equations of interacting particle systems from the underlying stochastic dynamics, i.e., to deduce the macroscopic behavior of the system from the microscopic interaction among particles.

In order to present all main concepts that will appear throughout the book in a very simple setting, we consider in this chapter a system in which particles evolve according to independent continuous time random walks, a special case where there is no interaction among particles. In section 1 we describe all equilibrium states. In our stochastic context this corresponds to the specification of all invariant measures. We show further that these equilibrium states can be parametrized by the density of particles, which is also the unique quantity conserved by the stochastic dynamics. We then introduce in section 2 the concept of local equilibrium in our mathematical model. In section 3 we prove that the local equilibrium picture is conserved by the time evolution and deduce the (linear) hydrodynamic equations that describe the macroscopic evolution of the density. In section 4 we further discuss some properties of the equilibrium states.

1. Equilibrium States

In this chapter we investigate in detail the case of indistinguishable particles moving as independent random walks. Denote by $\mathbb{Z}$ the set of integers and by $\mathbb{Z}^d$ the d-dimensional lattice. For a positive integer N, denote by $\mathbb{T}_N$ the torus with N points: $\mathbb{T}_N = \mathbb{Z}/N\mathbb{Z}$ and let $\mathbb{T}_N^d = (\mathbb{T}_N)^d$. Here N represents the inverse of the distance between molecules. The points of $\mathbb{T}_N^d$, called sites, are represented by the last characters of the alphabet (x, y and z).

To describe the evolution of the system, we begin by distinguishing all particles. Let K denote the total number of particles at time 0 and let $x_1, \dots, x_K$ denote their initial positions. Particles evolve as independent translation invariant continuous time random walks on the torus. To be rigorous, fix a translation invariant transition probability $p(x, y)$ on $\mathbb{Z}^d$: $p(x, y) = p(0, y - x) =: p(y - x)$ for some probability $p(\cdot)$ on $\mathbb{Z}^d$, called the elementary transition probability of the system.

Let $p_t(x, y)$ represent the probability of being at time t on site y for a continuous time random walk with elementary transition probability $p(\cdot)$ starting from x. $p_t(\cdot\,,\cdot)$ inherits the translation invariance property from $p(\cdot\,,\cdot)$: $p_t(x, y) = p_t(0, y - x) =: p_t(y - x)$. Moreover, by the Markov property, $p_t(\cdot\,,\cdot)$ is the unique solution of the linear differential equations

$$\begin{cases} \partial_t p_t(x, y) = \sum_{z \in \mathbb{Z}^d} p(x, z) \Big[p_t(z, y) - p_t(x, y) \Big] , \\ p_0(x, y) \;=\; \mathbf{1}\{x = y\} \, . \end{cases}$$

In this formula $\partial_t p_t$ stands for the time derivative of p_t and $\mathbf{1}\{x = y\}$ for a function which is equal to 1 if $x = y$ and 0 otherwise.

We arc now in a position to describe the motion of each particle. Denote by $\{Z_t^i,\ 1 \le i \le K\}$ K independent copies of a continuous time random walk with elementary transition probability $p(\cdot)$ and initially at the origin. For $1 \le i \le K$, let X_t^i represent the position at time t of the i-th particle. We set

$$X_t^i \;=\; x_i \;+\; Z_t^i \quad \mathrm{mod}\ N \, .$$

Since particles are considered indistinguishable, we are not interested in the individual position of each particle but only in the total number of particles at each site. In particular the state space of the system, also called configuration space, is $\mathbb{N}^{\mathbb{T}_N^d}$. The configurations will be denoted by Greek letters η, ξ and ζ. In this way, for a site x of $\mathbb{T}_N^d$, $\eta(x)$ will represent the number of particles at site x for the configuration η. Therefore, if the initial positions are $x_1, \dots, x_K$, for every $x \in \mathbb{T}_N^d$:

$$\eta(x) \;=\; \sum_{i=1}^{K} \mathbf{1}\{x_i = x\} \, .$$

Inversely, given $\{\eta(x);\ x \in \mathbb{T}_N^d\}$, to define the evolution of the system, we can first label all particles and then let them evolve according to the stochastic dynamics described above.

Clearly, if we denote by η_t the configuration at time t, we have

$$\eta_t(x) \;=\; \sum_{i=1}^{K} \mathbf{1}\{X_t^i = x\} \, .$$

The process $(\eta_t)_{t \ge 0}$ inherits the Markov property from the random walks $\{X_t^i,\ 1 \le i \le K\}$ because all particles have the same elementary transition probability and they do not interact with each other.

The first question raised in the study of Markov processes is the characterization of all invariant measures. Since the state space is finite and since the total number of particles is the unique quantity conserved by the dynamics, for every positive integer K representing the total number of particles, there is only one invariant measure, as long as the support of the elementary transition probability $p(\cdot)$ generates $\mathbb{Z}^d$. The Poisson measures will, however, play a central role.

Recall that a Poisson distribution of parameter $\alpha \ge 0$ is the probability measure $\{p_{\alpha,k} = p_k,\ k \ge 1\}$ on $\mathbb{N}$ given by

$$p_k \;=\; e^{-\alpha}\,\frac{\alpha^k}{k!}, \qquad k \in \mathbb{N},$$

and its Laplace transform is equal to

$$e^{-\alpha}\sum_{k=0}^{\infty} e^{-\lambda k}\,\frac{\alpha^k}{k!} \;=\; e^{-\alpha}\, e^{\alpha e^{-\lambda}} \;=\; \exp\, \alpha(e^{-\lambda} - 1)$$

for all positive λ.

For a fixed positive function $\rho\colon \mathbb{T}_N^d \to \mathbb{R}_+$, we call Poisson measure on $\mathbb{T}_N^d$ associated to the function ρ a probability on the configuration space $\mathbb{N}^{\mathbb{T}_N^d}$, denoted by $\nu_{\rho(\cdot)}^N$, having the following two properties. Under $\nu_{\rho(\cdot)}^N$, the random variables $\{\eta(x),\ x \in \mathbb{T}_N^d\}$ representing the number of particles at each site are independent and, for every fixed site $x \in \mathbb{T}_N^d$, $\eta(x)$ is distributed according to a Poisson distribution of parameter $\rho(x)$. In the case where the function ρ is constant equal to α, we denote $\nu_{\rho(\cdot)}^N$ just by ν_α^N. Throughout this book, expectation with respect to a measure ν will be denoted by E_ν.

The measure $\nu_{\rho(\cdot)}^N$ is characterized by its multidimensional Laplace transform:

$$\begin{aligned} E_{\nu_{\rho(\cdot)}^N}\Big[\exp\Big\{-\sum_{x\in\mathbb{T}_N^d} \lambda(x)\,\eta(x)\Big\}\Big] &= \prod_{x\in\mathbb{T}_N^d} \exp\, \rho(x)\big(e^{-\lambda(x)} - 1\big) \\ &= \exp \sum_{x\in\mathbb{T}_N^d} \rho(x)\,\big(e^{-\lambda(x)} - 1\big) \end{aligned}$$

for all positive sequences $\{\lambda(x);\ x \in \mathbb{T}_N^d\}$ (cf. Feller (1966), Chap. VII).

The first result consists in proving that the Poisson measures associated to constant functions are invariant for a system of independent random walks.

Proposition 1.1 *If particles are initially distributed according to a Poisson measure associated to a constant function equal to α then the distribution at time t is exactly the same Poisson measure.*

Proof. Denote by $\mathbb{P}_{\nu_\alpha^N}$ the probability measure on the space $D(\mathbb{R}_+, \mathbb{N}^{\mathbb{T}_N^d})$ induced by the independent random walks dynamics and the initial measure ν_α^N. Expectation with respect to $\mathbb{P}_{\nu_\alpha^N}$ is denoted by $\mathbb{E}_{\nu_\alpha^N}$. Notice the difference between $E_{\nu_\alpha^N}$ and $\mathbb{E}_{\nu_\alpha^N}$. The first expectation is an expectation with respect to the probability measure ν_α^N defined on $\mathbb{N}^{\mathbb{T}_N^d}$, while the second is an expectation with respect to the probability measure $\mathbb{P}_{\nu_\alpha^N}$ defined on the path space $D(\mathbb{R}_+, \mathbb{N}^{\mathbb{T}_N^d})$. In particular, $\mathbb{E}_{\nu_\alpha^N}[F(\eta_0)] = E_{\nu_\alpha^N}[F(\eta)]$ for all bounded continuous function F on $\mathbb{N}^{\mathbb{T}_N^d}$.

Since a probability measure on $\mathbb{N}^{\mathbb{T}_N^d}$ is characterized by its multidimensional Laplace transform, we are naturally led to compute the expectation

$$\mathbb{E}_{\nu_\alpha^N}\left[\exp - \sum_{x\in\mathbb{T}_N^d} \lambda(x)\,\eta_t(x)\right]$$

for all positive sequences $\{\lambda(x);\ x\in\mathbb{T}_N^d\}$. For a site y in $\mathbb{T}_N^d$, we will denote by $X_t^{y,k}$ the position at time t of the k-th particle initially at y. In this way, the number of particles at site x at time t is equal to

$$\eta_t(x) = \sum_{y\in\mathbb{T}_N^d}\sum_{k=1}^{\eta_0(y)} \mathbf{1}\{X_t^{y,k}=x\}\,.$$

From this formula and inverting the order of summations we obtain the identity

$$\sum_{x\in\mathbb{T}_N^d}\lambda(x)\,\eta_t(x) = \sum_{y\in\mathbb{T}_N^d}\sum_{k=1}^{\eta_0(y)}\lambda(X_t^{y,k})\,.$$

Since each particle evolves independently and the total number of particles at each site at time 0 is distributed according to a Poisson distribution of parameter α,

$$\begin{aligned}
&\mathbb{E}_{\nu_\alpha^N}\left(\exp - \sum_{x\in\mathbb{T}_N^d}\lambda(x)\,\eta_t(x)\right)\\
&\quad = \prod_{y\in\mathbb{T}_N^d}\mathbb{E}_{\nu_\alpha^N}\left[\exp\left\{-\sum_{k=1}^{\eta_0(y)}\lambda(X_t^{y,k})\right\}\right]\\
&\quad = \prod_{y\in\mathbb{T}_N^d}\int \nu_\alpha^N(d\eta)\left(E\left[\exp\left\{-\lambda(X_t^{y,1})\right\}\right]\right)^{\eta_0(y)}\\
&\quad = \prod_{y\in\mathbb{T}_N^d}\exp\left\{\alpha\left(E\left[e^{-\lambda(y+X_t)}\right]-1\right)\right\}\,,
\end{aligned}$$

where X_t is the position at time t of a random walk on the torus $\mathbb{T}_N^d$ starting from the origin and with transition probability $p_t^N(\cdot)$ defined by

$$p_t^N(x,y) := \sum_{z\in\mathbb{Z}^d} p_t(x, y+Nz)$$

for x and y in $\mathbb{T}_N^d$. Since

$$E\left[e^{-\lambda(y+X_t)}\right] = \sum_{x\in\mathbb{T}_N^d} p_t^N(x-y)e^{-\lambda(x)},$$

inverting the order of summation, we obtain that

$$\mathbb{E}_{\nu_\alpha^N}\left(\exp - \sum_{x\in\mathbb{T}_N^d} \lambda(x)\,\eta_t(x)\right) = \exp\left\{\sum_{x\in\mathbb{T}_N^d} \alpha\left(e^{-\lambda(x)} - 1\right)\right\} . \quad \square$$

Remark 1.2 Since the total number of particles $\sum_{x\in\mathbb{T}_N^d} \eta(x)$ is conserved by the stochastic dynamics it might seem more natural to consider as reference probability measures the extremal invariant measures that are concentrated on the "hyper–planes" of all configurations with a fixed total number of particles. These measures are given by

$$\nu_{\mathbb{T}_N^d,K}(\,\cdot\,) := \nu_\alpha^N\Big(\,\cdot\,\Big|\, \sum_{x\in\mathbb{T}_N^d} \eta(x) = K\Big) .$$

Besides the fact that they enable easier computations, the Poisson distributions present other intrinsic advantages that will be seen in the forthcoming sections. We shall return to this discussion on extremal invariant measures in section 4.

Notice that only one quantity is conserved by the dynamics: the total number of particles. On the other hand, Poisson distributions are such that their expectation is equal to

$$\sum_{k\geq 0} e^{-\alpha}\frac{\alpha^k}{k!}\,k \;=\; \alpha .$$

The Poisson measures are in this way naturally parametrized by the density of particles. Furthermore, by the weak law of large numbers, if the number of sites of the set $\mathbb{T}_N^d$ is denoted by $|\mathbb{T}_N^d|$,

$$\lim_{N\to\infty} \frac{1}{|\mathbb{T}_N^d|} \sum_{x\in\mathbb{T}_N^d} \eta(x) \;=\; \alpha$$

in probability with respect to ν_α^N. The parameter α describes therefore the "mean" density of particles in a "large" box.

In conclusion, we obtained above in Proposition 1.1 a one–parameter family of invariant and translation invariant measures indexed by the density of particles, which is the unique quantity conserved by the time evolution.

2. Local Equilibrium

We announced that the passage from microscopic to macroscopic would be done performing a limit in which the distance between particles converges to zero. This point does not present any difficulty in formalization. We just have to consider the torus $\mathbb{T}_N^d$ as embedded in the d-dimensional torus $\mathbb{T}^d = [0,1)^d$, that is, taking the lattice $\mathbb{T}^d$ with "vertices" x/N, $x \in \mathbb{T}_N^d$. In this way the distances between molecules is $1/N$ and tends to zero as $N \uparrow \infty$.

We shall refer to $\mathbb{T}^d$ as the macroscopic space and to $\mathbb{T}^d_N$ as the microscopic space. In this way each macroscopic point u in $\mathbb{T}^d$ is associated to a microscopic site $x = [uN]$ in $\mathbb{T}^d_N$ and, reciprocally, each site x is associated to a macroscopic point x/N in $\mathbb{T}^d$. Here and below, for a d-dimensional real $r = (r_1, \dots, r_d)$, $[r]$ denotes the integer part of r: $[r] = ([r_1], \dots, [r_d])$.

On the other hand, since we have a one–parameter family of invariant measures, one way to describe a local equilibrium with density profile $\rho_0 \colon \mathbb{T}^d \to \mathbb{R}_+$ is the following. We distribute particles according to a Poisson measure with slowly varying parameter on $\mathbb{T}^d_N$, that is, for each positive N we fix the parameter of the Poisson distribution at site x to be equal to $\rho_0(x/N)$. Since this type of measure will appear frequently in the following chapters, we introduce the following terminology.

Definition 2.1 (Product measure with slowly varying parameter associated to a profile $\rho_0 \colon \mathbb{T}^d \to \mathbb{R}_+$). For each smooth function $\rho_0 \colon \mathbb{T}^d \to \mathbb{R}_+$, we represent by $\nu^N_{\rho_0(\cdot)}$ the measure on the state space $\Sigma_{\mathbb{T}^d_N} = \mathbb{N}^{\mathbb{T}^d_N}$ having the following two properties. Under $\nu^N_{\rho_0(\cdot)}$ the variables $\{\eta(x);\ x \in \mathbb{T}^d_N\}$ are independent and, for a site $x \in \mathbb{T}^d_N$, $\eta(x)$ is distributed according to a Poisson distribution of parameter $\rho_0(x/N)$:

$$\nu^N_{\rho_0(\cdot)}\{\eta;\ \eta(x) = k\} \;=\; \nu^N_{\rho_0(x/N)}\{\eta;\ \eta(0) = k\}$$

for all x in $\mathbb{T}^d_N$ and k in $\mathbb{N}$.

We have thus associated to each profile $\rho_0 \colon \mathbb{T}^d \to \mathbb{R}_+$ and each positive integer N a Poisson measure on the torus $\mathbb{T}^d_N$.

As the parameter N increases to infinity, the discrete torus $\mathbb{T}^d_N$ tends to the full lattice $\mathbb{Z}^d$. We can also define a Poisson measure on the space of configurations over $\mathbb{Z}^d$. For each $\alpha \geq 0$ we will denote by ν_α the probability on $\mathbb{N}^{\mathbb{Z}^d}$ that makes the variables $\{\eta(x);\ x \in \mathbb{Z}^d\}$ independent and under which, for every x in $\mathbb{Z}^d$, $\eta(x)$ is distributed according to a Poisson distribution of parameter α .

With the definition we have given of $\nu^N_{\rho_0(\cdot)}$, and since $\rho_0 : \mathbb{T}^d \to \mathbb{R}_+$ is assumed to be smooth, as $N \uparrow \infty$ and we look "close" to a point $u \in \mathbb{T}^d$ – that is "around" $x = [Nu]$ – we observe a Poisson measure of parameter (almost) constant equal to $\rho_0(u)$. In fact, since the function $\rho_0(\cdot)$ is smooth, for every positive integer ℓ and for every positive family of parameters $\{\lambda(x);\ |x| \leq \ell\}$,

$$\lim_{N\to\infty} E_{\nu^N_{\rho_0(\cdot)}}\left[e^{-\sum_{|x|\leq\ell}\lambda(x)\eta([uN]+x)}\right] \;=\; E_{\nu_{\rho_0(u)}}\left[e^{-\sum_{|x|\leq\ell}\lambda(x)\eta(x)}\right]. \tag{2.1}$$

In this formula and throughout this book, for $u = (u_1, \dots, u_d)$ in $\mathbb{R}^d$, $\|u\|$ denotes the Euclidean norm of u and $|u|$ the max norm:

$$\|u\|^2 \;=\; \sum_{1\leq i\leq d} u_i^2\,, \qquad |u| \;=\; \max_{1\leq i\leq d} |u_i|\,.$$

In this sense the sequence $\nu^N_{\rho_0(\cdot)}$ describes an example of local equilibrium. This definition of product measure with slowly varying parameter is of course too restrictive. To generalize it we introduce a concept of convergence.

In the configuration space $\mathbb{N}^{\mathbb{T}^d_N}$, endowed with its natural discrete topology, we denote by $\{\tau_x,\ x \in \mathbb{T}^d_N\}$ the group of translations. Thus, for a site x, $\tau_x \eta$ is the configuration that, at site y, has $\eta(x+y)$ particles:

$$(\tau_x \eta)(y) \;=\; \eta(y+x), \quad y \in \mathbb{T}^d_N .$$

The action of the translation group extends in a natural way to the space of functions and to the space of probability measures on $\mathbb{N}^{\mathbb{T}^d_N}$. In fact, for a site x and a probability measure μ, $(\tau_x \mu)$ is the measure such that

$$\int f(\eta)(\tau_x \mu)(d\eta) \;=\; \int f(\tau_x \eta)\mu(d\eta),$$

for every bounded continuous function f.

To perform the limit $N \uparrow \infty$ we embed the space $\mathbb{N}^{\mathbb{T}^d_N}$ in $\mathbb{N}^{\mathbb{Z}^d}$ identifying a configuration on the torus to a periodic configuration on the full lattice. We will endow the configuration space $\mathbb{N}^{\mathbb{Z}^d}$ with its natural topology, the product topology. By $\mathcal{M}_1(\mathbb{N}^{\mathbb{Z}^d})$ or simply by $\mathcal{M}_1$, we represent the space of probability measures on $\mathbb{N}^{\mathbb{Z}^d}$ endowed with the weak topology.

In this topological setting, formula (2.1) establishes that for all points u of $\mathbb{T}^d$ the sequence $\tau_{[uN]}\nu^N_{\rho_0(\cdot)}$ converges weakly to the measure $\nu_{\rho_0(u)}$. It is then natural to introduce the following definition.

Definition 2.2 (Local equilibrium) A sequence of probability measures $(\mu^N)_{N\geq 1}$ on $\mathbb{N}^{\mathbb{T}^d_N}$ is a local equilibrium of profile $\rho_0 \colon \mathbb{T}^d \to \mathbb{R}_+$ if

$$\lim_{N\to\infty} \tau_{[uN]}\mu^N \;=\; \nu_{\rho_0(u)}$$

for all continuity points u of $\rho_0(\cdot)$.

3. Hydrodynamic Equation

We turn now to the study of the distribution of particles at a later time t starting from a product measure with slowly varying parameter. Repeating the computations we did to prove Proposition 1.1 we see that if we start from a Poisson measure with slowly varying parameter then

$$
\begin{aligned}
\mathbb{E}_{\nu^N_{\rho_0(\cdot)}} & \left(\exp - \sum_{x \in \mathbb{T}_N^d} \lambda(x)\, \eta_t(x) \right) \\
&= \exp \sum_{x \in \mathbb{T}_N^d} \rho_0(x/N) \sum_{y \in \mathbb{T}_N^d} p_t^N(y - x) \left(e^{-\lambda(y)} - 1 \right) \\
&= \exp \sum_{y \in \mathbb{T}_N^d} \left(e^{-\lambda(y)} - 1 \right) \sum_{x \in \mathbb{T}_N^d} p_t^N(y - x)\, \rho_0(x/N) \\
&=: \exp \sum_{y \in \mathbb{T}_N^d} \left(e^{-\lambda(y)} - 1 \right) \psi_{N,t}(y).
\end{aligned}
$$

Therefore, at time t, we still have a Poisson measure with slowly varying parameter, which is now $\psi_{N,t}(\cdot)$ instead of $\rho_0(\cdot/N)$. Up to this point we have not used the particular form of $p_t(\cdot)$ besides the fact that it makes $p_t(\cdot,\cdot)$ translation invariant and thus bistochastic: $\sum_x p_t(x, y) = 1$ for every y. We shall now see what happens when t is fixed and N increases to infinity. In this case $p_t(\cdot)$ is a function with essentially finite support, that is, for all $\varepsilon > 0$, there exists $A = A(t, \varepsilon) > 0$ so that

$$
\sum_{|x| \le A} p_t(x) \;\ge\; 1 - \varepsilon \, .
$$

From the explicit form of $\psi_{N,t}$, we have that for every continuity point u of ρ_0,

$$
\lim_{N \to \infty} \psi_{N,t}([uN]) \;=\; \rho_0(u) \, .
$$

The profile has remained unchanged. The system did not have time to evolve and this reflects the fact that at the macroscopic scale particles did not move. Indeed, consider a test particle initially at the origin. Since it evolves as a continuous time random walk, if X_t denotes its position at time t, for every $\varepsilon > 0$, there exists $A = A(t, \varepsilon) > 0$ such that $P[\,|X_t| > A] \le \varepsilon$. Therefore, with probability close to 1, in the macroscopic scale, the test particle at time t is at distance of order N^{-1} from the origin. In a fluid, however, a "test" particle traverses a macroscopic distance in a macroscopic time, as do, for instance, particles in suspension.

We solve this problem distinguishing between two time scales (as we have two space scales: $\mathbb{T}^d$ and $N^{-1}\,\mathbb{T}_N^d$): a microscopic time t and a macroscopic time which is infinitely large with respect to t.

To introduce the macroscopic time scale, notice that the transition probabilities $p_t(\,\cdot\,)$ are equal to

$$
p_t(x) \;=\; \sum_{k=0}^{\infty} e^{-t} \, \frac{t^k}{k!} \, p^{*k}(x) \, ,
$$

where p^{*k} stands for the k-th convolution power of the elementary transition probability of each particle.

Assume that the elementary transition probability $p(\,\cdot\,)$ has finite expectation: $m := \sum x\, p(x) \in \mathbb{R}^d$. We say that the random walk is *asymmetric* if $m \neq 0$, that it is *mean–zero asymmetric* if $p(\cdot)$ is not symmetric but $m = 0$ and that it is *symmetric* if $p(\cdot)$ is symmetric. Recall that X_t stands for the position at time t of a continuous time random walk with transition probability $p(\cdot)$ and initially at the origin. By the law of large numbers for random walks, for all $\varepsilon > 0$,

$$\lim_{N\to\infty} \sum_{x;\, |x/N - mt| \le \varepsilon} p_{tN}(x) = \lim_{N\to\infty} P\Big[\Big|\frac{X_{tN}}{N} - mt\Big| \le \varepsilon\Big] = 1 .$$

In particular, from the explicit expression for $\psi_{N,tN}$ and since we assumed the initial profile to be smooth, we have that

$$\lim_{N\to\infty} \psi_{N,tN}([uN]) = \rho_0(u - mt) =: \rho(t, u)$$

for every u in $\mathbb{T}^d$.

We obtained in this way a new time scale, tN, in which we observe a new macroscopic profile: the original one translated by mt. More precisely, in this macroscopic scale tN we observe a local equilibrium profile that has been translated by mt since $\psi_{N,tN}$ is itself slowly varying in the macroscopic scale.

Of course, the profile $\rho(t, u)$ satisfies the partial differential equation

$$\partial_t \rho + m \cdot \nabla \rho = 0 \tag{3.1}$$

if $\nabla\rho$ denotes the gradient of ρ: $\nabla\rho = (\partial_{u_1}\rho, \dots, \partial_{u_d}\rho)$.

In conclusion, if we restrict ourselves to a particular class of initial measures, we have established the existence of a time and space scales in which the particles density evolves according to the linear partial differential equation (3.1). We have thus derived from the microscopic stochastic dynamics a macroscopic deterministic evolution for the unique conserved quantity.

An interacting particle system for which there exists a time and space macroscopic scales in which the conserved quantities evolve according to some partial differential equation is said to have a hydrodynamic description. Moreover, the P.D.E. is called the hydrodynamic equation associated to the system.

We summarize this result in the following proposition.

Proposition 3.1 *A system of particles evolving as independent asymmetric random walks with finite first moment on a d-dimensional torus has a hydrodynamic description. The evolution of the density profile is described by the solution of the differential equation*

$$\partial_t \rho + m \cdot \nabla \rho = 0 .$$

When the expectation m vanishes, the solution of this differential equation is constant, which means that the profile didn't change in the time scale tN. Nothing imposed, however, the choice of Nt as macroscopic time scale. In fact, when the mean displacement m vanishes, to observe an interesting time evolution, we need to consider a larger time scale, times of order N^2.

Assume that the elementary transition probability that describes the displacement of each particle has a second moment. Let $\sigma = (\sigma_{i,j})_{1\le i,j\le d}$ be the covariance matrix of this distribution:

$$\sigma_{i,j} = \sum_{x\in\mathbb{T}_N^d} x_i\, x_j\, p(x), \quad 1 \le i,j \le d\,.$$

By the central limit theorem for random walks, we see that

$$\begin{aligned}\lim_{N\to\infty} \psi_{N,N^2t}([Nu]) \;&=\; \lim_{N\to\infty} \sum_{x\in\mathbb{T}_N^d} p_t^N([Nu]-x)\rho_0(x/N)\\ &=\; \lim_{N\to\infty} E\Big[\rho_0(u - N^{-1}X_{tN^2})\Big] \;=\; \int_{\mathbb{R}^d} \overline{\rho_0}(\theta)G_t(u-\theta)\,d\theta\,,\end{aligned}$$

where $\overline{\rho_0}\colon \mathbb{R}^d \to \mathbb{R}_+$ is the periodic function, with period $\mathbb{T}^d$ and equal to ρ_0 on the torus $\mathbb{T}^d$ and G_t is the density of the Gaussian distribution with covariance matrix $t\,\sigma$.

Since the Gaussian distribution is the fundamental solution of the heat equation (which can be checked by a simple computation) we obtain the following result.

Proposition 3.2 *A system of particles evolving as independent mean–zero asymmetric random walks with finite second moment on a d-dimensional torus has a hydrodynamic description. The evolution of the density profile is described by the solution of the differential equation*

$$\begin{cases} \partial_t \rho \;=\; \displaystyle\sum_{1\le i,j\le d} \sigma_{i,j}\, \partial^2_{u_i,u_j}\rho \\ \rho(0,u) \;=\; \rho_0(u)\,. \end{cases}$$

Let $\{S^N(t),\, t\ge 0\}$ be the semigroup on $\mathcal{M}_1$ associated to the Markov process $(\eta_t)_{t\ge0}$. In Propositions 3.1 and 3.2, we have proved that there is a time renormalization θ_N such that

$$\lim_{N\to\infty} S^N(t\theta_N)\,\tau_{[uN]}\,\nu^N_{\rho_0(\cdot)} \;=\; \nu_{\rho(t,u)}\,,$$

for all $t\ge 0$ and all continuity points u of $\rho(t,\cdot)$. It is therefore natural to introduce the following definition.

Definition 3.3 (Conservation of local equilibrium) The local equilibrium $(\mu^N)_{N\ge1}$ of profile $\rho_0(\cdot)$ is conserved by the time renormalization θ_N if there exists a function $\rho\colon \mathbb{R}_+\times\mathbb{T}^d \to \mathbb{R}_+$ such that

$$\lim_{N\to\infty} S^N(t\theta_N)\,\tau_{[uN]}\,\mu^N \;=\; \nu_{\rho(t,u)}\,,$$

for all $t\ge 0$ and all continuity points u of $\rho(t,\cdot)$.

Usually $\rho(t,\cdot)$ is the solution of a Cauchy problem with initial condition $\rho_0(\cdot)$. As we said earlier, this differential equation is called the hydrodynamic equation of the interacting particle system. In this section we took advantage of several special features of the evolution of independent random walks to obtain an explicit formula for the profile $\rho(t,\cdot)$. The type of result, however, is characteristic of the subject. We have proved:

1. conservation of the local equilibrium in time evolution.

2. characterization at a later time of the new parameters describing the local equilibrium and derivation of a partial differential equation that determines how the parameters evolve in time.

The aim of the following chapters is to prove a weak version of the conservation of local equilibrium for a class of interacting particle systems. We would like in fact to prove a more general result, that is, one for initial states that are not product measures with slowly varying parameter – thus without assuming a strong form of local equilibrium at time 0 – but for initial states having a density profile and imposing that it is not too far, in a sense to be defined later, from a local equilibrium; the process establishing by itself a local equilibrium at later times.

4. Equivalence of Ensembles

In Section 2 we chose a class of invariant measures to describe the equilibrium states (the Poisson measures) when others would seem more appropriate. Indeed, since the "total number of particles" is conserved by the evolution it would have seemed more natural to choose the so-called canonical measures $\nu_\alpha^N\big(\,\cdot \mid \sum_{x\in\mathbb{T}_N^d}\eta(x)=K\big)$ (that do not depend on α).

However, since we want to describe the equilibrium state associated to a given density on the torus $\mathbb{T}^d$, we would be led to study the behavior, as $N\uparrow\infty$ and β is kept fixed, of

$$\nu_\alpha^N\left(\,\cdot\,\Big|\sum_{x\in\mathbb{T}_N^d}\eta(x)=N^d\beta\right)\,.$$

A simple computation presented below shows that for each fixed positive integer r, and for all sequences $(k_1,\dots,k_r)$ in $\mathbb{N}^r$ and $(x_1,\dots,x_r)$ in $\mathbb{Z}^r$,

$$\begin{aligned}\lim_{N\to\infty}\nu_\alpha^N&\left(\eta(x_1)=k_1,\dots,\eta(x_r)=k_r\,\Big|\sum_{x\in\mathbb{T}_N^d}\eta(x)=N^d\beta\right)\\&=\nu_\beta\big(\eta(x_1)=k_1,\dots,\eta(x_r)=k_r\big)\,.\end{aligned}$$

Indeed, the addition of independent Poisson distributions is still a Poisson distribution of parameter equal to the sum of the parameters, as we can see computing the Laplace transforms. Therefore, the left hand side of the above identity is equal to

$$\frac{\alpha^{k_1+\cdots+k_r}}{k_1!\ldots k_r!}e^{-N^d\alpha}\frac{[(N^d-r)\alpha]^{N^d\beta-(k_1+\cdots+k_r)}}{[N^d\beta-(k_1+\cdots+k_r)]!}\left[e^{-N^d\alpha}\frac{(\alpha N^d)^{N^d\beta}}{(N^d\beta)!}\right]^{-1}$$

$$=\frac{(N^d-r)^{N^d\beta-(k_1+\cdots+k_r)}}{k_1!\cdots k_r!(N^d)^{N^d\beta}}(N^d\beta)(N^d\beta-1)\cdots[N^d\beta-(k_1+\cdots+k_r)+1]$$

that, as $N\uparrow\infty$, converges to

$$\frac{1}{k_1!\ldots k_r!}e^{-r\beta}\beta^{k_1+\cdots+k_r}.$$

It is in this sense, known to the physicists as the "equivalence of ensembles", that the Poisson measures are indeed "natural" in our problem. One advantage that they present among others is that computations are much easier and that the definition of local equilibrium is expressed in a very simple and elegant way in terms of these measures.

In Appendix 2 we present in more detail and in a wide context some results connecting the canonical measures $\nu_\alpha^N\big(\,\cdot\mid\sum_{x\in\mathbb{T}_N^d}\eta(x)=K\big)$ to the grand canonical measures $\nu_\beta(\cdot)$. We obtain, for instance, some estimates of the total variation distance between these two measures.

5. Comments and References

The mathematical formulation of the derivation of macroscopic evolution equations from microscopic interactions goes back to Morrey (1955). The first rigorous results are due to Dobrushin and Siegmund–Schultze (1982), Rost (1981) and Galves, Kipnis, Marchioro and Presutti (1981). Dobrushin and Siegmund–Schultze (1982), that we essentially followed, consider the evolution of independent copies of stochastic processes on $\mathbb{R}^d$ and derive linear first order hydrodynamic equations. Rost (1981) obtains a first order quasi–linear hyperbolic equation for the one dimensional totally asymmetric simple exclusion process. Galves, Kipnis, Marchioro and Presutti (1981) deduce a non linear heat equation that describes the macroscopic evolution of a one dimensional non conservative spin system. De Masi, Ianiro and Presutti (1982) and Ferrari, Presutti and Vares (1987) prove the conservation of local equilibrium for nearest neighbor one dimensional symmetric exclusion processes and zero range processes with jump rate $g(k)=\mathbf{1}\{k\geq 1\}$.

Kipnis, Marchioro and Presutti (1982) consider a one dimensional system of harmonic oscillators in contact with reservoirs at different temperature. They obtain the stationary measure, the temperature profile and prove the local convergence to the Gibbs measure. De Masi, Ferrari, Ianiro and Presutti (1982) prove the same statement in the context of symmetric simple exclusion processes in contact with stochastic reservoirs at different temperature.

Presutti and Spohn (1983) derive the hydrodynamic behavior of the voter model. Malyshev, Manita, Petrova and Scacciatelli (1995) extend the result to weak perturbations of the voter model. They obtain equations of type

$$\partial_t \rho = m \cdot \nabla \rho - C\rho + \sum_{j=1}^{n_0} d_j \rho^j + e$$

$$\text{and} \quad \partial_t \rho = \sum_{i,j=1}^{d} B_{i,j} \partial^2_{u_i,u_j} \rho \; - \; C\rho + \sum_{j=1}^{n_0} d_j \rho^j + e \, .$$

Greven (1984) derives the hydrodynamic limit of a branching interacting particle system.

Conservation of local equilibrium for continuous spin systems or interacting diffusions were also considered. Rost (1984) derives the hydrodynamic behavior of Brownian hard spheres moving on $\mathbb{R}$. Fritz (1987a,b), (1989) proves the hydrodynamic limit of lattice Ginzburg–Landau processes by the method of resolvent. This method reduces the proof of conservation of local equilibrium to the verification of certain smoothness properties of the evolution as function of the initial condition. Funaki (1989a,b) extends the method to one dimensional Ginzburg–Landau models of diffusion processes evolving on the line.

Though we shall not consider this problem here, there exist an extended literature on the hydrodynamic behavior of deterministic or mechanical systems. The first results in this direction were obtained by Boldrighini, Dobrushin and Suhov (1983), Dobrushin, Pellegrinotti, Suhov and Triolo (1986), (1988), Dobrushin, Pellegrinotti and Suhov (1990) and Fritz (1982), (1985). We refer to Spohn (1991) for a clear and complete presentation of the subject. De Masi and Presutti (1991) is another excellent reference on hydrodynamic behavior of interacting particle systems. Some reviews have also been published or physical discussions on the derivation of the equations of motion. We mention Lebowitz and Spohn (1983), Presutti (1987), Boldrighini, De Masi, Pellegrinotti and Presutti (1987), Lebowitz, Presutti and Spohn (1988), Presutti (1997), Jensen and Yau (1997).

Applications. The methods and the results of the theory of hydrodynamic limit of interacting particle systems have been used to solve several different types of problems. Here are some examples.

Ulam's problem. For a positive integer N, denote by L_N the length of the largest increasing subsequence of a random permutation of $\{1, \dots, N\}$. Consider a Poisson point process on $\mathbb{R}^2$ and denote by R_N the maximal number of points in an increasing path contained in $[0, \sqrt{N})^2$. Hammersley (1972) pointed out the connection between the distribution of R_N and L_N. Proving the hydrodynamic behavior of a one dimensional continuous spin asymmetric process, Aldous and Diaconis (1995) deduce that $(1/\sqrt{N})E[L_N]$ converges to 2 (cf. also Seppäläinen (1996a) for another proof through hydrodynamic limit). Seppäläinen (1997b) solves the same problem on a planar lattice.

Stationary measures. Relying on the hydrodynamic behavior of a superposition of a speeded up exclusion process with a Glauber dynamics, derived by De Masi, Ferrari and Lebowitz (1986), Noble (1992) and Durrett and Neuhauser (1994) prove the existence of a non trivial stationary measure for a class of non conservative processes in the case where the stirring rate is large enough. Maes (1990) investigates the decay of correlations for the stationary measures of anisotropic lattice gases on $\mathbb{Z}^2$ through the behavior of the equilibrium fluctuations around the deterministic macroscopic evolution.

Occupation time large deviations. Landim (1992) and Benois (1996) deduce the occupation time large deviations for one dimensional symmetric exclusion processes and for independent one dimensional symmetric random walks from the large deviations from the hydrodynamic limit. Similar results were obtained previously by Cox and Griffeath (1984) for independent random walks and by Bramson, Cox and Griffeath (1988) for the voter model.

Random domino tillings. Jockush, Propp and Shor (1995) deduce from the hydrodynamic behavior of the totally asymmetric simple exclusion process obtained by Rost (1981), the asymptotic shape of a subregion in a random domino tillings of a square.

2. Some Interacting Particle Systems

We introduce in this chapter the interacting particle systems we consider throughout the book and present their main features. We shall refer constantly to Liggett (1985) for some proofs and some extensions of the results presented here.

In section 1 we review the basic properties of the product topology of $\mathbb{N}^{\mathbb{Z}^d}$. In section 2 we introduce the simple exclusion processes, in section 3 the zero range processes and in section 4 the generalized exclusion processes. In section 5 we present the main tools in the investigation of attractive systems and in section 6 we characterize the set of invariant and translation invariant measures of attractive zero range processes evolving on $\mathbb{N}^{\mathbb{Z}^d}$.

1. Some Remarks on the Topology of $\mathbb{N}^{\mathbb{Z}^d}$ and $\mathcal{M}_1(\mathbb{N}^{\mathbb{Z}^d})$

We briefly discuss in this section some aspects of the topology of $\mathbb{N}^{\mathbb{Z}^d}$ and of the weak convergence of probability measures. Recall that we denote the configurations of $\mathbb{N}^{\mathbb{Z}^d}$ by the Greek letters η, ξ and ζ so that, for each x in $\mathbb{Z}^d$, $\eta(x)$ stands for the total number of particles at x for the configuration η. We endow the space $\mathbb{N}^{\mathbb{Z}^d}$ with the product topology which is metrizable: Define, for instance, the distance $d(\cdot, \cdot)$ on $\mathbb{N}^{\mathbb{Z}^d}$ by

$$d(\eta, \xi) = \sum_{x \in \mathbb{Z}^d} \frac{1}{2^{|x|}} \frac{|\eta(x) - \xi(x)|}{1 + |\eta(x) - \xi(x)|} .$$

It is straightforward to check that d is a distance compatible with the product topology. Moreover, with this distance $\mathbb{N}^{\mathbb{Z}^d}$ is a complete separable metric space.

Since $\mathbb{N}$ is not a compact set, $\mathbb{N}^{\mathbb{Z}^d}$ is not itself compact. Nevertheless, the compact subsets of $\mathbb{N}^{\mathbb{Z}^d}$ are easy to describe. We leave to the reader to check that a subset K of $\mathbb{N}^{\mathbb{Z}^d}$ is compact if and only if K is closed and there exists a collection of positive numbers $\{n_x, \, x \in \mathbb{Z}^d\}$ such that $\eta(x) \le n_x$ for all η in K.

Denote by $C(\mathbb{N}^{\mathbb{Z}^d})$ (resp. $C_b(\mathbb{N}^{\mathbb{Z}^d})$) the space of continuous (resp. bounded continuous) functions on $\mathbb{N}^{\mathbb{Z}^d}$ and by $\mathcal{C}$ (resp. $\mathcal{C}_b$) the space of cylinder (resp. bounded cylinder) functions, i.e., functions that depend on the configurations only through a finite set of coordinates. To clarify ideas and illustrate that continuous functions can not be uniformly approximated by cylinder functions, we present an example of a continuous function that is not uniformly continuous.

In dimension 1 and for $k \geq 1$, denote by Λ_k the set $\{-k, \dots, k\}$ and by A_k the set of configurations with k particles on each site of Λ_k: $A_k = \{\eta;\ \eta(x) = k\, ,\ x \in \Lambda_k\}$. Define $f \colon \mathbb{N}^{\mathbb{Z}} \to \mathbb{R}$ by $f(\eta) = 1$ if η belongs to $\cup_{k \geq 0} A_k$ and $f(\eta) = 0$ otherwise. We leave to the reader to check that f is continuous but not uniformly continuous. In particular, f can not be approximated by cylinder or uniformly continuous functions in the uniform topology.

Denote by $\mathcal{M}_1(\mathbb{N}^{\mathbb{Z}^d})$ the space of probability measures on $\mathbb{N}^{\mathbb{Z}^d}$ endowed with the weak topology defined as follows:

Definition 1.1 A sequence of probability measures μ^k on $\mathbb{N}^{\mathbb{Z}^d}$ converges weakly to a probability measure μ if $E_{\mu^k}[f]$ converges to $E_\mu[f]$ for every bounded continuous function f.

To prove the weak convergence of probability measures, we need only to investigate the limit behavior of expectations of cylinder functions:

Lemma 1.2 *A sequence of probability measures μ^k on $\mathbb{N}^{\mathbb{Z}^d}$ converges weakly to a probability measure μ if and only if $E_{\mu^k}[\Psi]$ converges to $E_\mu[\Psi]$ for every bounded cylinder function Ψ.*

Proof. Consider a sequence of probability measures μ^k such that $E_{\mu^k}[\Psi]$ converges to $E_\mu[\Psi]$ for every bounded cylinder function Ψ and fix $\varepsilon > 0$ and a bounded continuous function f. It follows from the convergence of expectations of bounded cylinder functions that for every positive integer ℓ, there exists $A_\ell = A_\ell(\varepsilon) > 0$ such that

$$\mu^k\Big\{\eta,\ \sum_{x \in \Lambda_\ell} \eta(x) > A_\ell\Big\} \;\leq\; \frac{\varepsilon}{2^\ell}$$

for every $1 \leq k \leq \infty$, where to keep notation simple we denoted μ by μ^∞. In this formula Λ_ℓ stands for $\{-\ell, \dots, \ell\}^d$. In particular,

$$\mu^k\Big\{\eta,\ \sum_{x \in \Lambda_\ell} \eta(x) \leq A_\ell \text{ for all } \ell \geq 1\Big\} \;\geq\; 1 - \varepsilon$$

for every $1 \leq k \leq \infty$. The set K_ε of configurations η such that $\sum_{x \in \Lambda_\ell} \eta(x) \leq A_\ell$ for all $\ell \geq 1$ is compact. On a compact set, a continuous function is uniformly continuous and may therefore be uniformly approximated by a cylinder function: there exists a bounded cylinder function Ψ_ε such that $\|\Psi_\varepsilon\|_\infty \leq \|f\|_\infty$ and

$$\sup_{\eta \in K_\varepsilon} |\Psi_\varepsilon(\eta) - f(\eta)| \;\leq\; \varepsilon\, .$$

In particular, for each fixed $1 \leq k \leq \infty$, $|E_{\mu^k}[f] - E_{\mu^k}[\Psi_\varepsilon]|$ is bounded above by

$$E_{\mu^k}[f\mathbf{1}\{K_\varepsilon^c\}] + E_{\mu^k}[\Psi_\varepsilon \mathbf{1}\{K_\varepsilon^c\}] + E_{\mu^k}\big[|f-\Psi_\varepsilon|\mathbf{1}\{K_\varepsilon\}\big] \le 2\|f\|_\infty \varepsilon + \varepsilon .$$

The convergence of $E_{\mu^k}[f]$ to $E_\mu[f]$ follows therefore from the convergence of expectations of bounded cylinder functions. □

Lemma 1.3 *A sequence of probability measures μ^k on $\mathbb{N}^{\mathbb{Z}^d}$ converges weakly to a probability measure μ if and only if for every finite subset Λ of $\mathbb{Z}^d$ and every sequence $\{a_x,\, x\in\Lambda\}$, $\mu^k\{\eta,\, \eta(x)=a_x, x\in\Lambda\}$ converges to $\mu\{\eta,\, \eta(x)=a_x, x\in\Lambda\}$.*

Proof. By Lemma 1.2, we just need to show that $E_{\mu^k}[\Psi]$ converges to $E_\mu[\Psi]$ for every bounded cylinder function Ψ. So fix a bounded cylinder function Ψ and $\varepsilon>0$. Denote by Λ_Ψ the support of the cylinder function Ψ. There exists $B>0$ such that

$$\mu\Big\{\eta,\ \sum_{x\in\Lambda_\Psi}\eta(x)>B\Big\} \le \varepsilon .$$

Denote by ξ the configurations of $\mathbb{N}^{\Lambda_\Psi}$. With this notation we may rewrite $E_{\mu^k}[\Psi]$ as

$$\sum_{\xi,\, \sum\xi(x)\le B} \Psi(\xi)\mu^k\{\eta,\, \eta(y)=\xi(y),\, y\in\Lambda_\Psi\} + E_{\mu^k}\Big[\Psi(\eta)\mathbf{1}\{\sum_{y\in\Lambda_\Psi}\eta(y)>B\}\Big] .$$

By assumption, as $k\uparrow\infty$, the first term converges to

$$\sum_{\xi,\, \sum\xi(x)\le B} \Psi(\xi)\mu\{\eta,\, \eta(y)=\xi(y),\, y\in\Lambda_\Psi\} , \tag{1.1}$$

while the absolute value of the second is bounded above by

$$\|\Psi\|_\infty\mu^k\{\eta,\ \sum_{y\in\Lambda_\Psi}\eta(y)>B\} = \|\Psi\|_\infty\Big[1-\mu^k\{\eta,\ \sum_{y\in\Lambda_\Psi}\eta(y)\le B\}\Big]$$

that converges, as $k\uparrow\infty$, to

$$\|\Psi\|_\infty\Big[1-\mu\{\eta,\ \sum_{y\in\Lambda_\Psi}\eta(y)\le B\}\Big] \le \varepsilon\|\Psi\|_\infty$$

by definition of B. Since the absolute value of the difference between (1.1) and $E_\mu[\Psi]$ is also bounded by $\varepsilon\|\Psi\|_\infty$, we proved that $E_{\mu^k}[\Psi]$ converges to $E_\mu[\Psi]$. □

Introduce on $\mathbb{N}^{\mathbb{Z}^d}$ the natural partial order: $\eta\le\xi$ if and only if $\eta(x)\le\xi(x)$ for every x in $\mathbb{Z}^d$. Denote by $\mathcal{C}_m$ (resp. $\mathcal{C}_{m,b}$) the space of monotone (resp. bounded monotone) cylinder functions in the sense that $f(\eta)\le f(\xi)$ for all $\eta\le\xi$. The partial order extends to the space of probability measures over $\mathbb{N}^{\mathbb{Z}^d}$ in a natural way:

$$\mu_1 \le \mu_2 \quad \text{provided} \quad \int f\, d\mu_1 \le \int f\, d\mu_2 \tag{1.2}$$

for all functions f in $C_{b,m}$.

Theorem 1.4 *Let μ_1 and μ_2 be two probability measures on $\mathbb{N}^{\mathbb{Z}^d}$. The following two statements are equivalent:*

(a) $\mu_1 \le \mu_2$.

(b) There exists a probability measure $\bar\mu$ on $\mathbb{N}^{\mathbb{Z}^d} \times \mathbb{N}^{\mathbb{Z}^d}$ such that the first (resp. second) marginal is equal to μ_1 (resp. μ_2) and $\bar\mu$ is concentrated on configurations "above the diagonal":

$$\bar\mu\{(\eta,\xi);\ \eta \le \xi\} = 1 .$$

The reader will have no difficulty in adapting the proof of Theorem II.2.4 in Liggett (1985) to the present setting. One first prove the result on $\mathbb{N}^S \times \mathbb{N}^S$, where S is a finite set, and then use Kolmogorov theorem and a sequence of finite set S_k increasing to $\mathbb{Z}^d$ to conclude the proof. $\bar\mu$ is called the coupling measure of μ_1 and μ_2.

Lemma 1.5 *A sequence of probability measures μ^k on $\mathbb{N}^{\mathbb{Z}^d}$ converges weakly to a probability measure μ if and only if $E_{\mu^k}[\Psi]$ converges to $E_\mu[\Psi]$ for every bounded monotone cylinder function Ψ.*

Proof. By Lemma 1.3, we just need to show the convergence of the expected value of cylinder functions of type $\prod_{x\in\Lambda} \mathbf{1}\{\eta(x) = a_x\}$ for finite sets Λ. The result follows therefore from the representation of $\mathbf{1}\{\eta(x) = a_x\}$ as the difference of the two bounded monotone cylinder functions $\mathbf{1}\{\eta(x) \ge a_x\}$ and $\mathbf{1}\{\eta(x) \ge a_x + 1\}$. □

Remark 1.6 In Chapter 1, 3 and 9 we face the problem of proving the convergence of probability measures defined in different spaces. More precisely, for large positive integers N, we consider probability measures μ^N defined on $\mathbb{N}^{\mathbb{T}^d_N}$ and wish to prove that this sequence converges weakly to some probability measure ν defined on $\mathbb{N}^{\mathbb{Z}^d}$. It is easy to build a mathematical framework to render the argument rigorous. For each N, extend the measure μ^N to the space $\mathbb{N}^{\mathbb{Z}^d}$ in the natural way: denote by $\tilde\mu^N$ the periodic measure on $\mathbb{N}^{\mathbb{Z}^d}$ with period $\mathbb{T}^d_N$ defined by the following two properties. We first require the projection of $\tilde\mu^N$ on $\mathbb{T}^d_N$ to be equal to μ^N:

$$\begin{aligned}\tilde\mu^N\Big\{\eta,\ \eta(x) = a_x, x \in \{-[(N-1)/2], \dots, [N/2]\}^d\Big\} \\ = \mu^N\{\eta,\ \eta(x) = a_x, x \in \mathbb{T}^d_N\Big\}\end{aligned}$$

and then impose the measure to be periodic:

$$\tilde{\mu}^N \Big\{ \eta,\ \eta(x) = \eta(x + Ny), y \in \mathbb{Z}^d \Big\} \ = \ 1 \ .$$

In the first formula, $[r]$ stands for the integer part of r. We have now a sequence of probability measures $\tilde{\mu}^N$ defined on $\mathbb{N}^{\mathbb{Z}^d}$ and we may investigate the weak convergence of $\tilde{\mu}^N$ to ν. It corresponds to the convergence of $E_{\tilde{\mu}^N}[\Psi]$ (which is equal to $E_{\mu^N}[\Psi]$ provided N is large enough for $\{-[N/2], \dots, [N/2]\}^d$ to contain the support of Ψ) to $E_\nu[\Psi]$, for every bounded cylinder function Ψ.

Remark 1.7 All analysis developed above extends in a straightforward manner to the configuration space $\{0, \dots, \kappa\}^{\mathbb{Z}^d}$ and to probability measures defined on this space. The situation is in fact even simpler because the space $\{0, \dots, \kappa\}^{\mathbb{Z}^d}$ endowed with the product topology (and therefore the measure space $\mathcal{M}_1(\{0, \dots, \kappa\}^{\mathbb{Z}^d})$ endowed with the weak topology) is compact.

2. Simple Exclusion Processes (Hard Core Interaction)

Among the simplest and most widely studied interacting particle systems is the exclusion process. In contrast with superpositions of random walks presented in Chapter 1, the exclusion process allows at most one particle per site. The state space is therefore $\{0, 1\}^{\mathbb{T}_N^d}$.

To prevent the occurrence of more than one particle per site we introduce an exclusion rule that suppresses each jump to an already occupied site. In fact, we shall focus only on the simplest class of exclusion processes: systems where particles jump, whenever the jump is allowed, independently of the others and according to the same translation invariant elementary transition probability.

Definition 2.1 Let p be a finite range, translation invariant, irreducible transition probability on $\mathbb{Z}^d$:

$$p(x, y) \ = \ p(0, y - x) \ =: \ p(y - x)$$

for all pair (x, y) of d-dimensional integers and for some finite range probability measure $p(\cdot)$ on $\mathbb{Z}^d$:

$$\sum_{z \in \mathbb{Z}^d} p(z) \ = \ 1 \quad \text{and } p(x) = 0 \text{ for } |x| \text{ large enough .}$$

We shall refer to $p(\cdot)$ as the elementary jump probability. The generator

$$(L_N f)(\eta) \ := \ \sum_{x \in \mathbb{T}_N^d} \sum_{z \in \mathbb{T}_N^d} \eta(x)[1 - \eta(x + z)]\, p^N(z) \,[f(\eta^{x, x+z}) - f(\eta)] \ ,$$

where $\eta^{x,y}$ is the configuration obtained from η letting a particle jump from x to y:

$$\eta^{x,y}(z) = \begin{cases} \eta(z) & \text{if } z \neq x, y\,, \\ \eta(x) - 1 & \text{if } z = x\,, \\ \eta(y) + 1 & \text{if } z = y \end{cases} \qquad \text{and} \quad p^N(z) := \sum_{y \in \mathbb{Z}^d} p(z + yN) \qquad (2.1)$$

defines a Markov process called simple exclusion process with elementary jump probability $p(\cdot)$. In the particular case where $p(z) = p(-z)$ we say that it is a symmetric simple exclusion process.

The interpretation is clear. Between 0 and dt each particle tries, independently from the others, to jump from x to $x + z$ with rate $p^N(z)$. The jump is suppressed if it leads to an already occupied site.

Note. Irreducibility of the transition probability $p(\cdot)$ means that the set $\{x\,, p(x) > 0\}$ generates $\mathbb{Z}^d$, i.e., that for any pair of sites x, y in $\mathbb{Z}^d$, there exists $M \geq 1$ and a sequence $x = x_0, \ldots, x_M = y$ such that $p(x_i, x_{i+1}) > 0$ for $0 \leq i \leq M - 1$. We usually renormalize the transition probability in a slightly different way (we assume $\sum_x p(x) = d$ instead of 1) in order to get the heat equation as hydrodynamic equation for the symmetric simple exclusion process. Finally, since the transition probability is assumed to be of finite range, there exists A_0 in $\mathbb{N}$ such that $p(z) = 0$ for all sites z outside the cube $[-A_0, A_0]^d$. In particular, $p^N(\cdot)$ and $p(\cdot)$ coincide provided $N \geq A_0$. For this reason, from now on we omit the superscript N in the elementary jump probability.

We denote by $\{S^N(t),\ t \geq 0\}$ the semigroup of the Markov process with generator L_N. We use the same notation for semigroups acting on continuous functions or on the space $\mathcal{M}_1(\{0,1\}^{\mathbb{T}_N^d})$ of probability measures on $\{0,1\}^{\mathbb{T}_N^d}$.

It might be worthwhile to justify the terminology. The rule that forbids jumps to occupied sites explains the term exclusion. Notice, on the other hand, that the rate at which a particle jumps from x to y depends on the configuration η only through the occupation variables $\eta(x)$ and $\eta(y)$. This last dependence on $\eta(x)$ and $\eta(y)$ reflects the exclusion rule. To distinguish this class from processes where the jump rate depends in a more complicated way on the configuration, we call the first family simple exclusion processes. Finally, notice that the total number of particles is conserved by the dynamics.

For $0 \leq \alpha \leq 1$, we denote by $\nu_\alpha = \nu_\alpha^N$ the Bernoulli product measure of parameter α, that is, the product and translation invariant measure on $\{0,1\}^{\mathbb{T}_N^d}$ with density α. In particular, under ν_α, the variables $\{\eta(x),\ x \in \mathbb{T}_N^d\}$ are independent with marginals given by

$$\nu_\alpha\{\eta(x) = 1\} \;=\; \alpha \;=\; 1 - \nu_\alpha\{\eta(x) = 0\}\,.$$

In Chapter 1, we have used the same notation ν_α to designate the family of invariant measures for superpositions of independent random walks. In fact, throughout this book we will represent the equilibrium states of various processes by the symbol ν_α. The context will clarify to which measure we are referring to.

Proposition 2.2 *The Bernoulli measures $\{\nu_\alpha\,, 0 \leq \alpha \leq 1\}$ are invariant for simple exclusion processes. In addition, with respect to each ν_α, exclusion processes with* *elementary jump probability* $\check{p}(z) := p(-z)$ *are adjoint to processes with elementary jump probability* $p(z)$. *In particular symmetric simple exclusion processes are self-adjoint with respect to each* ν_α.

Proof. Notice that by a simple change of variables

$$\int f(\eta^{0,z})\, g(\eta)\, \eta(0)[1 - \eta(z)]\, \nu_\alpha(d\eta) \;=\; \int f(\eta)\, g(\eta^{z,0})\, \eta(z)[1 - \eta(0)]\, \nu_\alpha(d\eta)\,.$$

This identity, the fact that $1 = \sum_{z \in \mathbb{Z}^d} p(z) = \sum_{z \in \mathbb{Z}^d} p(-z)$ and a change in the order of summation prove the proposition. □

The family of invariant measures ν_α is parametrized by the density, for

$$E_{\nu_\alpha}[\eta(0)] \;=\; \nu_\alpha\{\eta(0) = 1\} \;=\; \alpha\,.$$

Terminology 2.3 We shall say that an interacting particle system is *symmetric* if the transition probability $p(\cdot)$ is symmetric ($p(x) = p(-x)$ for x in $\mathbb{Z}^d$), that it is *mean-zero asymmetric* if the transition probability is not symmetric but $\sum_x x p(x) = 0$ and that it is *asymmetric* if $\sum_x x p(x) \neq 0$.

Remark 2.4 Since the total number of particles is conserved by the dynamics the measures

$$\nu_{N,K}(\,\cdot\,) \;:=\; \nu_\alpha\left(\,\cdot \,\Big|\, \sum_{x \in \mathbb{T}_N^d} \eta(x) = K\right)$$

are invariant and it could have seemed more natural to consider them instead of the Bernoulli product measures ν_α. Nevertheless, a simple computation on binomials shows that for all finite subsets E of $\mathbb{Z}^d$, for all sequences $\{\varepsilon_x;\ x \in E\}$ with values in $\{0,1\}$ and for all $0 \le \alpha \le 1$,

$$\lim_{N \to \infty} \nu_\alpha\Big\{\eta(x) = \varepsilon_x\,,\ x \in E \,\Big|\, \sum_{y \in \mathbb{T}_N^d} \eta(y) = [\alpha_0 N^d]\Big\}$$
$$= \nu_{\alpha_0}\left\{\eta(x) = \varepsilon_x\,,\ x \in E\right\}$$

uniformly in α_0. Therefore, the Bernoulli product measures are obtained as limits of the invariant measures $\nu_{N,K}$, as the total number of sites increases to infinity.

For each $0 \le K \le N^d$, denote by $\Sigma_{N,K}^1$ the "hyperplanes" of all configurations with K particles:

$$\Sigma_{N,K}^1 \;=\; \Big\{\eta \in \{0,1\}^{\mathbb{T}_N^d};\ \sum_{x \in \mathbb{T}_N^d} \eta(x) = K\Big\}\,.$$

Invariant measures of density preserving particle systems that are concentrated on hyperplanes with a fixed total number of particles are called canonical measures (the family $\{\nu_{N,K},\ 0 \le K \le N^d\}$ in the context of simple exclusion processes, for instance). In contrast, the measures obtained as weak limits of the canonical measures, as the number of sites increases to infinity, are called the grand canonical measures (here the Bernoulli measures as we have just seen in Remark 2.4). We return to this point in Appendix 2, where we investigate in a general context the asymptotic behavior of canonical measures and deduce some estimates on the total variation distance between the canonical measures and the grand canonical ones.

3. Zero Range Processes

This time we consider evolutions without restrictions on the total number of particles per site. The state space will therefore be $\mathbb{N}^{\mathbb{T}_N^d}$. The process is defined through a function $g: \mathbb{N} \to \mathbb{R}_+$ vanishing at 0, which represents the rate at which one particle leaves a site, and a translation invariant transition probability $p(\cdot,\cdot)$ on $\mathbb{Z}^d$. It can be described as follows. If there are k particles at a site x, independently of the number of particles on other sites, at rate $g(k)p(x,y)$ one of the particles at x jumps to y. In this way particles only interact with particles sitting on the same site. For this reason these processes are called zero range processes.

Definition 3.1 Let $g: \mathbb{N} \to \mathbb{R}_+$ be a function vanishing at 0 and $p(\cdot,\cdot)$ be a finite range, irreducible, translation invariant transition probability. We assume that g is strictly positive on the set of positive integers and that it has bounded variation in the following sense:

$$g^* := \sup_{k \ge 0} |g(k+1) - g(k)| < \infty . \tag{3.1}$$

Denote by φ^* the radius of convergence of the partition function $Z: \mathbb{R}_+ \to \mathbb{R}_+$ defined by

$$Z(\varphi) = \sum_{k \ge 0} \frac{\varphi^k}{g(k)!} \cdot$$

In this last formula $g(k)!$ stands for $\prod_{1 \le j \le k} g(j)$ and by convention $g(0)! = 1$. Notice that Z is analytic and strictly increasing on $[0, \varphi^*)$. Assume that $Z(\cdot)$ increases to ∞ as φ converges to φ^*:

$$\lim_{\varphi \uparrow \varphi^*} Z(\varphi) = \infty . \tag{3.2}$$

The generator

$$(L_N f)(\eta) = \sum_{x \in \mathbb{T}_N^d} \sum_{z \in \mathbb{T}_N^d} p^N(z)\, g(\eta(x))\, [f(\eta^{x,x+z}) - f(\eta)]$$

defines a Markov process on $\mathbb{N}^{\mathbb{T}_N^d}$ called zero range process with parameters (g,p). Here, as in the previous section, $\eta^{x,y}$ represents the configuration η where one particle jumped from x to y and $p^N(\cdot)$ represents the transition probability translated to the origin and restricted to the torus:

$$p^N(z) := p^N(0,z) = \sum_{y\in\mathbb{Z}^d} p(0, z+yN)$$

for every d-dimensional integer z.

The assumption (3.2) is not necessary to define the process but will always be required to prove the hydrodynamic behavior of zero range processes. We therefore preferred to include it in the definition.

In these processes each particle jumps, independently of particles sitting at other sites, from x to y at a rate $p^N(y-x)g(\eta(x))(\eta(x))^{-1}$. In particular, if $g(k) = k$ for every $k \ge 0$, we obtain the superposition of independent random walks studied in Chapter 1. On the other hand, the case $g(k) = \mathbf{1}\{k \ge 1\}$ models a system of queues with mean-one exponential random times of service.

We now describe some invariant measures of the process. For each $0 \le \varphi < \varphi^*$, let $\bar\nu_{\varphi,g} = \bar\nu^N_{\varphi,g}$ denote the product measure on $\mathbb{N}^{\mathbb{T}^d_N}$ with marginals given by

$$\bar\nu_{\varphi,g}\{\eta;\ \eta(x) = k\} = \frac{1}{Z(\varphi)}\frac{\varphi^k}{g(k)!} \tag{3.3}$$

for each $k \ge 0$ and x in $\mathbb{T}^d_N$.

Proposition 3.2 *For each $0 \le \varphi < \varphi^*$ the product measure $\bar\nu_{\varphi,g}$ is invariant for the zero range process with parameters (g, p). Moreover, the adjoint process with respect to any of the measures $\bar\nu_{\varphi,g}$ is the zero range process with parameters $(g, \check{p})$. In particular, if p is symmetric the process is self-adjoint.*

Proof. The proof relies on the same computations that were made for the simple exclusion process and on the following identity

$$g(k)\frac{\varphi^k}{g(k)!}\frac{\varphi^j}{g(j)!} = g(j+1)\frac{\varphi^{k-1}}{g(k-1)!}\frac{\varphi^{j+1}}{g(j+1)!}\ . \qquad \square$$

Since the function $g(\cdot)$ will always be fixed, to keep notation as simple as possible, we hide the dependence on g of the measure $\bar\nu_{\varphi,g}$ and denote it simply by $\bar\nu_\varphi$. Let $R(\varphi)$ denote the expected value of the occupation variable under $\bar\nu_\varphi$:

$$R(\varphi) = E_{\bar\nu_\varphi}\big[\eta(0)\big] = \frac{1}{Z(\varphi)}\sum_{k\ge0} k\frac{\varphi^k}{g(k)!}\ . \tag{3.4}$$

From the last equality we obtain a relation that will be often used in the sequel:

$$R(\varphi) = \frac{Z'(\varphi)\varphi}{Z(\varphi)} = \varphi\partial_\varphi \log Z(\varphi)\ . \tag{3.5}$$

Computing the first derivative of $R(\cdot)$ we show that it is strictly increasing. In fact, it is easy to show that $\log Z(e^\lambda)$ is strictly convex in λ.

Since we want to parametrize the invariant measures by the conserved quantity, which is here the density of particles, we change variables in the definition of the invariant measures $\bar\nu_\varphi$ as follows. For $\alpha \ge 0$, define the product measure ν_α by

$$\nu_\alpha(\cdot) \;=\; \bar\nu_{\Phi(\alpha)}(\cdot)\,. \tag{3.6}$$

In this formula $\Phi(\cdot)$ stands for the inverse function of $R(\cdot)$ defined in (3.4). In the next lemma we show that assumption (3.2) guarantees that the range of the function $R(\cdot)$ is all $\mathbb{R}_+$. We obtained in this way a family $\{\nu_\alpha,\ \alpha \ge 0\}$ of invariant measures parametrized by the density since the expected value of the occupation variables $\eta(x)$ under ν_α is equal to α:

$$E_{\nu_\alpha}[\eta(x)] \;=\; \alpha \tag{3.7}$$

for every $\alpha \ge 0$. Moreover, a simple computation shows that the function $\Phi(\alpha)$ is the expected value of the jump rate $g(\eta(0))$ under the measure ν_α:

$$\Phi(\alpha) \;=\; E_{\nu_\alpha}[g(\eta(0))]\,. \tag{3.8}$$

Lemma 3.3 *Recall that we denoted by φ^* the radius of convergence of the partition function Z.*

$$\lim_{\varphi\uparrow\varphi^*} R(\varphi) \;=\; \infty\,.$$

Furthermore, for each $0 \le \varphi < \varphi^*$ the measure $\bar\nu_\varphi$ has a finite exponential moment: there exists $\theta(\varphi) > 0$ such that

$$E_{\bar\nu_\varphi}\left[e^{\theta\eta(0)}\right] \;<\; \infty\,. \tag{3.9}$$

Proof. We consider separately two different cases. Assume first that Z is defined for all positive reals or, equivalently, that the radius of convergence φ^* is infinite. Suppose, by contradiction, that the function R is bounded by some constant C_0. From identity (3.5) we obtain that

$$\partial_\varphi \log Z(\varphi) \;\le\; C_0\varphi^{-1}\,.$$

Hence, integrating over φ we get that for every $\varphi > 1$,

$$Z(\varphi) \;\le\; Z(1)\varphi^{C_0}\,.$$

But this is in contradiction with the fact that $Z(\varphi) \ge \varphi^k[g(k)!]^{-1}$ for every integer k by the definition of Z.

Assume now that the radius of convergence is finite. Fix some positive $\varphi_0 <$ φ^*. Since $Z(\cdot)$ is a smooth increasing function, for $\varphi \ge \varphi_0$,

$$\log Z(\varphi) \;\le\; \log Z(\varphi_0) + \frac{1}{\varphi_0}\int_{\varphi_0}^{\varphi} \psi\,\partial_\psi \log Z(\psi)\,d\psi\,.$$

Since, on the other hand, by relation (3.5)

$$R(\varphi) \;=\; \varphi \partial_\varphi \log Z(\varphi) \,,$$

we obtain that

$$\varphi_0 \log \left(\frac{Z(\varphi)}{Z(\varphi_0)} \right) \;\le\; \int_{\varphi_0}^{\varphi} R(\psi)\, d\psi \,.$$

Since the left hand side of this inequality, by assumption (3.2), increases to ∞ as $\varphi \uparrow \varphi^*$, it follows that

$$\lim_{\varphi \uparrow \varphi^*} \int_{\varphi_0}^{\varphi} R(\psi) d\psi \;=\; \infty \,.$$

Since the function R is increasing the first statement of the lemma is proved.

Notice that $E_{\bar\nu_\varphi}[\exp\{\theta\eta(0)\}]$ is equal to $Z(\varphi e^\theta)/Z(\varphi)$. Thus (3.9) follows from assumption (3.2). $\Box$

Before proceeding, we present an example of zero range dynamics that does not possess an invariant product measure for each density $\alpha \ge 0$. In virtue of Lemma 3.3, the partition function $Z(\cdot)$ can not satisfy assumption (3.2).

Example 3.4 Consider a one-dimensional, nearest neighbor, symmetric zero range process with jump rate $g(k) = (1+k^{-1})^3$ for $k \ge 1$. Then, $\varphi^* = 1$ and the partition function is

$$Z(\varphi) \;=\; 1 \;+\; \sum_{k \ge 1} \frac{\varphi^k}{(k+1)^3}$$

so that

$$\lim_{\varphi \to 1} Z(\varphi) \;=\; 1 \;+\; \sum_{k \ge 1} \frac{1}{(k+1)^3} \;<\; \infty \,.$$

Consider a product invariant measure ν. Since ν is invariant, we have that $\int L_N \eta(x) d\nu = 0$ for every x. Denote by φ_x the expectation of $g(\eta(x))$ under ν: $\varphi_x = E_\nu[g(\eta(x))]$. Since $L_N \eta(x) = (1/2)\{g(\eta(x+1)) + g(\eta(x-1)) - 2g(\eta(x))\}$, the previous identity gives that $(\Delta_N \varphi)_x = 0$ if Δ_N stands for the discrete Laplacian. This identity forces φ_x to be constant, equal, say, to φ.

On the other hand, for every x in $\mathbb{T}_N^d$ and $a > 0$, $\int L_N \mathbf{1}\{\eta(x) = a\} \nu(d\eta) = 0$. Since

$$\begin{aligned} L_N \mathbf{1}\{\eta(x) = a\} \;&=\; -g(a) \mathbf{1}\{\eta(x) = a\} \\ &+\; (1/2) \mathbf{1}\{\eta(x) = a - 1\} \big\{ g(\eta(x+1)) + g(\eta(x-1)) \big\} \,, \end{aligned}$$

since the measure ν is assumed to be product and since $E_\nu[g(\eta(x))] = \varphi$ is constant, we have that

$$g(a)\, \nu\{\eta, \eta(x) = a\} \;=\; \varphi\, \nu\{\eta, \eta(x) = a - 1\} \,.$$

In particular, an invariant product measure must be of the form (3.3). In this example, since $g(k) = (1 + k^{-1})^3$,

$$R(\varphi) = \sum_{k\geq 1} k \frac{\varphi^k}{(k+1)^3}$$

so that

$$\lim_{\varphi \to 1} R(\varphi) = \sum_{k\geq 1} \frac{k}{(k+1)^3} = R_* < \infty .$$

Thus for $\alpha > R_*$ there is no invariant product measure with density α.

We now present a few results concerning the family of invariant measures ν_α that will be needed later. These technical lemmas might be skipped in a first reading. We start by showing that the family of invariant measures $\{\nu_\alpha;\ \alpha \geq 0\}$ is an increasing continuous sequence of measures for the order defined in section 1 of this chapter. Continuous family means that ν_{α_k} converges to ν_α for every $\alpha \geq 0$ and every α_k that approaches α as $k \uparrow \infty$: $\lim_{\alpha' \to \alpha} E_{\nu_{\alpha'}}[\Psi] = E_{\nu_\alpha}[\Psi]$ for every bounded continuous function Ψ.

Lemma 3.5 *Suppose that* $\alpha_1 \leq \alpha_2$. *Then* $\nu_{\alpha_1} \leq \nu_{\alpha_2}$.

Proof. Since these measures are product measures, it is enough to prove that the marginals are ordered. Define therefore, for each $0 \leq \varphi < \varphi^*$, a measure m_φ on $\mathbb{N}$ by

$$m_\varphi(k) = \frac{1}{Z(\varphi)} \frac{\varphi^k}{g(k)!} .$$

We need to show that the family $\{m_\varphi;\ 0 \leq \varphi < \varphi^*\}$ is an increasing set of measures. In order to do it, we have to prove that for every $A \geq 1$, the function $F_A\colon [0, \varphi^*) \to [0,1]$ defined by

$$F_A(\varphi) := m_\varphi\{k;\ k \geq A\}$$

is increasing. A simple computation shows that the derivative of F_A is equal to

$$\frac{1}{\varphi Z(\varphi)} \left\{ \sum_{k\geq A} k \frac{\varphi^k}{g(k)!} - R(\varphi) \sum_{k \geq A} \frac{\varphi^k}{g(k)!} \right\} .$$

We denote by $R_A(\varphi)$ the expression inside parentheses. To conclude the proof of the lemma it is enough to show that $R_A(\varphi)$ is positive. We prove it by induction on A. Fix $0 < \varphi < \varphi^*$. Since $R_1(\varphi)$ is equal to $R(\varphi)$ it is positive. On the other hand,

$$R_{A+1}(\varphi) - R_A(\varphi) = \frac{\varphi^A}{g(A)!} \big[R(\varphi) - A \big] .$$

Therefore, for each φ, $R_\bullet(\varphi)$ is increasing in the set $\{1, \ldots, [R(\varphi)] + 1\}$ and decreasing in the complementary set. In particular,

$$R_A(\varphi) \geq \min \left\{ R_1(\varphi), \lim_{A\to\infty} R_A(\varphi) \right\} .$$

Since

$$\lim_{A\to\infty} R_A(\varphi) = 0 ,$$

we proved that $R_A(\varphi)$ is nonnegative for every A and φ. $\square$

Corollary 3.6 *Recall the definition of g^* given in (3.1). The function $\Phi: \mathbb{R}_+ \to [0, \varphi^*)$ appearing in (3.8) is uniformly Lipschitz on $\mathbb{R}_+$ with constant g^*:*

$$\left|\Phi(\alpha_2) - \Phi(\alpha_1)\right| = \left| E_{\nu_{\alpha_2}}[g(\eta(0))] - E_{\nu_{\alpha_1}}[g(\eta(0))]\right| \le g^* |\alpha_2 - \alpha_1|$$

for all non negative reals α_1 and α_2.

Proof. Fix $\alpha_1 \le \alpha_2$. Recall from (3.8) that $\Phi(\alpha)$ is the expected value of $g(\eta(0))$ under the measure ν_α:

$$\Phi(\alpha) = E_{\nu_\alpha}[g(\eta(0))] .$$

Denote by ν_{α_1,α_2} a measure on $\mathbb{N}^{\mathbb{T}_N^d} \times \mathbb{N}^{\mathbb{T}_N^d}$ whose first marginal is equal to ν_{α_1}, whose second marginal is equal to ν_{α_2} and which is concentrated on configurations (ξ_1, ξ_2) such that $\xi_1 \le \xi_2$. The existence of this measure is guaranteed by Theorem 1.4 because $\nu_{\alpha_1} \le \nu_{\alpha_2}$. Hence, we have

$$\begin{aligned}\left|\Phi(\alpha_2) - \Phi(\alpha_1)\right| &\le E_{\nu_{\alpha_1,\alpha_2}}\left[\left|g(\xi_2(0)) - g(\xi_1(0))\right|\right] \\ &\le g^* E_{\nu_{\alpha_1,\alpha_2}}\left[\left|\xi_2(0) - \xi_1(0)\right|\right] .\end{aligned}$$

Since the measure ν_{α_1,α_2} is concentrated on configurations $\xi_1 \le \xi_2$ we can remove the absolute value in the last expression and obtain that it is equal to $g^*(\alpha_2 - \alpha_1)$. $\square$

The cylinder function $g(\eta(0))$ does not play any particular role in Corollary 3.6. The statement applies to a broad class of cylinder functions that we now introduce. A cylinder function Ψ is called Lipschitz if there exists a constant C_0 and finite subset Λ of $\mathbb{Z}^d$ such that

$$\left|\Psi(\eta) - \Psi(\xi)\right| \le C_0 \sum_{x\in\Lambda} \left|\eta(x) - \xi(x)\right| \tag{3.10}$$

for every configurations η and ξ. If Ψ is Lipschitz, then, taking in the last formula ξ to be the configuration with no particles, we obtain that there are finite constants C_0 and C_1 and a finite subset Λ of $\mathbb{Z}^d$ such that

$$\left|\Psi(\eta)\right| \le C_1 + C_0 \sum_{x\in\Lambda} \eta(x) \tag{3.11}$$

for every configuration η.

A cylinder function Ψ is said to have sublinear growth if there exists a finite subset Λ of $\mathbb{Z}^d$ such that for each $\delta > 0$ there exists a finite constant $C(\delta)$ such that

$$\left|\Psi(\eta)\right| \;\le\; C(\delta) \,+\, \delta \sum_{x\in\Lambda} \eta(x)$$

for every configuration η. Every bounded cylinder function has a sublinear growth.

For a cylinder function Ψ satisfying (3.11) we denote by $\tilde{\Psi}: \mathbb{R}_+ \to \mathbb{R}$ the function whose value at some density $\alpha \ge 0$ is equal to the expectation of Ψ with respect to the invariant measure ν_α:

$$\tilde{\Psi}(\alpha) := \; E_{\nu_\alpha}\left[\Psi(\eta)\right] . \tag{3.12}$$

With the notation just introduced, we have the identity $\tilde{g} = \Phi$.

The proof of Corollary 3.6 shows that $\tilde{\Psi}$ is a uniform Lipschitz function provided Ψ is a Lipschitz cylinder function:

Corollary 3.7 *Let Ψ be a Lipschitz cylinder function. The function $\tilde{\Psi}: \mathbb{R}_+ \to \mathbb{R}$ defined by (3.12) is uniformly Lipschitz on $\mathbb{R}_+$: there exists a finite constant $C_0 = C_0(\Psi)$ such that*

$$\left|\tilde{\Psi}(\alpha_2) - \tilde{\Psi}(\alpha_1)\right| \;\le\; C_0\left|\alpha_2 - \alpha_1\right|$$

for all α_1, α_2 in $\mathbb{R}_+$.

Lemma 3.8 *$\{\nu_\alpha,\ \alpha \ge 0\}$ is a continuous family: $\tilde{\Psi}(\alpha) = E_{\nu_\alpha}[\Psi]$ is a continuous bounded real function for every bounded cylinder function Ψ.*

Proof. Fix α_0 in $\mathbb{R}_+$ and a sequence $\{\alpha_k,\ k \ge 1\}$ converging to α_0. By Lemma 1.3, we just need to show that for every finite subset Λ of $\mathbb{Z}^d$ and every sequence $\{a_x,\ x \in A\}$, $\nu_{\alpha_k}\{\eta,\ \eta(x) = a_x, x \in \Lambda\}$ converges to $\nu_{\alpha_0}\{\eta,\ \eta(x) = a_x, x \in \Lambda\}$. This is obvious because the measures ν_α are product and Φ is Lipschitz. □

We conclude this section showing that the set of probability measures on $\mathbb{N}^{\mathbb{T}_N^d}$ bounded by some product measure ν_α is weakly relatively compact.

Lemma 3.9 *For every $\alpha \ge 0$, the set of probability measures $\mathcal{A}_\alpha = \{\mu,\ \mu \le \nu_\alpha\}$ is weakly relatively compact.*

Proof. By Prohorov theorem, we just need to show that the collection of probability measures $\{\mu,\ \mu \le \nu_\alpha\}$ is tight. Fix a positive $\varepsilon > 0$. For each positive integer ℓ, there exists A_ℓ such that

$$\sup_{\mu\in\mathcal{A}_\alpha} \mu\Big\{\eta,\ \sum_{x\in\Lambda_\ell} \eta(x) > A_\ell\Big\} \;\le\; \nu_\alpha\Big\{\eta,\ \sum_{x\in\Lambda_\ell} \eta(x) > A_\ell\Big\} \;\le\; \frac{\varepsilon}{2^\ell}\,\cdot$$

Therefore,

$$\inf_{\mu\in\mathcal{A}_\alpha} \mu\Big\{\eta,\ \sum_{x\in\Lambda_\ell} \eta(x) \le A_\ell \text{ for every } \ell \ge 1\Big\} \;\ge\; \varepsilon\,,$$

what concludes the proof of the lemma because $\cap_{\ell \ge 1}\{\eta,\ \sum_{x\in\Lambda_\ell} \eta(x) \le A_\ell\}$ is a compact set (cf. characterization of compact sets given at the beginning of section 1). □

4. Generalized Exclusion Processes

In this section we consider a third type of interacting particle systems that will appear in the next chapters: a mixture of zero range and simple exclusion processes. We admit this time no more than κ particles per site for some positive integer κ. Like in the simple exclusion process the jump rate of a particle from x to y depends exclusively on the occupancy at x and y and on a translation invariant transition probability. More precisely, we are given a function $r : \{0,\dots,\kappa\}^2 \to \mathbb{R}_+$ so that $r(a,b)$ represents the rate at which a particle jumps from a site occupied by a particles to a site occupied by b particles. Since no particles may jump from a site x if there are no particles at x, $r(0,\cdot) \equiv 0$. Our exclusion rule also imposes that $r(\cdot,\kappa) \equiv 0$ to avoid sites with more than κ particles. In conclusion, a particle jumps from site x to site y at rate $r(\eta(x),\eta(y))p(x,y)$ independently of the number of particles at other sites.

Definition 4.1 Let p be a finite range, translation invariant, irreducible transition probability on $\mathbb{Z}^d$. The generator

$$(L_N f)(\eta) = \sum_{x,y\in\mathbb{T}_N^d} p^N(y) r(\eta(x),\eta(x+y))[f(\eta^{x,x+y}) - f(\eta)] ,$$

where $\eta^{x,x+y}$ is the configuration obtained from η by letting a particle jump from x to $x+y$, defines a Markov process on $\{0,\dots,\kappa\}^{\mathbb{T}_N^d}$ called the generalized exclusion process with elementary jump probability $p(\cdot)$. In the particular case where $p(z) = p(-z)$ we say that it is a symmetric generalized exclusion process.

As the reader will notice in next sections *all* proofs presented in this book of hydrodynamic behavior of interacting particle systems rely on the *explicit* knowledge of the invariant measures. Moreover, we shall only consider processes with *product* invariant measures. In order to guarantee the existence of product invariant measures for generalized exclusion processes we need to impose some restrictions on the jump rate $r(\cdot,\cdot)$. We shall restrict our analysis to the special case where particles jump whenever a jump is allowed and the transition probability is symmetric: $r(a,b) = \mathbf{1}\{a > 0, b < \kappa\}$, $p(x) = p(-x)$. In this case the generator is given by

$$(L_N f)(\eta) = \sum_{x,y\in\mathbb{T}_N^d} p(y)\mathbf{1}\{\eta(x) > 0, \eta(x+y) < \kappa\}[f(\eta^{x,x+y}) - f(\eta)] . \tag{4.1}$$

As in the previous section, for $\varphi \geq 0$, consider the product measure $\bar\nu_\varphi = \bar\nu_\varphi^N$ on $\{0, \dots, \kappa\}^{\mathbb{T}_N^d}$ with marginals given by

$$\bar\nu_\varphi\{\eta;\ \eta(x) = k\} = \frac{\varphi^k}{Z(\varphi)}$$

for $0 \leq k \leq \kappa$, where $Z(\varphi) = \sum_{0 \leq j \leq \kappa} \varphi^j$ is the normalizing constant.

Proposition 4.2 *Assume the elementary jump probability $p(\cdot)$ to be symmetric. Then the Markov process with generator given in (4.1) is self–adjoint with respect to each product measure $\bar\nu_\varphi$.*

The proof is identical to the proof of Propositions 2.2 or 3.2 and is therefore omitted. To parametrize the invariant measures by the density observe that the density of particles $R(\varphi)$ for the measure $\bar\nu_\varphi$, which is equal to

$$R(\varphi) = E_{\bar\nu_\varphi}[\eta(0)] = \frac{1}{Z(\varphi)} \sum_{k=0}^{\kappa} k\varphi^k , \tag{4.2}$$

is strictly increasing. Denote by $\Phi : [0, \infty] \to [0, \kappa]$ the inverse function of $R(\cdot)$ and, for α in $[0, \kappa]$, define ν_α by

$$\nu_\alpha := \bar\nu_{\Phi(\alpha)} .$$

We have now a one-parameter family of invariant measures indexed by the density since, by definition, $E_{\nu_\alpha}[\eta(0)] = \alpha$.

5. Attractive Systems

In sake of completeness, in this section we briefly define and state the main properties of attractive interacting particle systems. To fix ideas, we shall consider zero range processes but every definition or statement can be translated to generalized exclusion processes. On the other hand, since we have nothing to add to section II.2 of Liggett (1985), we refer the reader to this book for a complete discussion on attractiveness and the corresponding coupling features of some particle systems. Recall the partial order on $\mathcal{M}_1(\mathbb{N}^{\mathbb{Z}^d})$ introduced in (1.2). A similar order can be defined on $\mathcal{M}_1(\mathbb{N}^{\mathbb{T}_N^d})$.

Definition 5.1 An interacting particle system is said to be attractive if its semigroup preserves the partial order:

$$\mu_1 \leq \mu_2 \Rightarrow S^N(t)\mu_1 \leq S^N(t)\mu_2$$

for all $t \geq 0$.

Theorem 5.2 below is the main result of this section. It gives a sufficient (and in fact necessary) condition for zero range processes to be attractive. It also shows how closely related are the partial order and the coupling of two copies of the process. This coupling technique is the main tool to prove the hydrodynamic behavior of asymmetric interacting particle systems beyond the appearance of the first shock.

Theorem 5.2 *A zero range process with parameters (g, p) is attractive provided the jump rate $g(\cdot)$ is a non decreasing function.*

Proof. Fix two probability measures μ_1 and μ_2 on $\mathbb{N}^{\mathbb{T}_N^d}$ such that $\mu_1 \le \mu_2$. The idea is to construct two copies of zero range processes on the same probability space whose evolution preserves the order. Denote therefore by $\bar{\mu}$ the coupling measure given by Theorem 1.4 and consider the Markov process (η_t, ξ_t) on $\mathbb{N}^{\mathbb{T}_N^d} \times \mathbb{N}^{\mathbb{T}_N^d}$ starting from $\bar{\mu}$ and with generator $\bar{L}_N$ given by

$$
\begin{aligned}
(\bar{L}_N f)(\eta, \xi) &= \\
&= \sum_{x,y \in \mathbb{T}_N^d} p(y) \min\Big\{g(\eta(x)), g(\xi(x))\Big\} \left[f(\eta^{x,x+y}, \xi^{x,x+y}) - f(\eta, \xi)\right] \\
&+ \sum_{x,y \in \mathbb{T}_N^d} p(y) \Big\{g(\eta(x)) - g(\xi(x))\Big\}^+ \left[f(\eta^{x,x+y}, \xi) - f(\eta, \xi)\right] \\
&+ \sum_{x,y \in \mathbb{T}_N^d} p(y) \Big\{g(\xi(x)) - g(\eta(x))\Big\}^+ \left[f(\eta, \xi^{x,x+y}) - f(\eta, \xi)\right] .
\end{aligned}
\tag{5.1}
$$

Notice that for the dynamics generated by $\bar{L}_N$, the η-particles and the ξ-particles try to jump as much as possible together. On the other hand, both coordinates evolve as zero range processes with parameters (g, p). Indeed, consider a function $f : \mathbb{N}^{\mathbb{T}_N^d} \times \mathbb{N}^{\mathbb{T}_N^d}$ depending, say, on the first coordinate only. It is simple to check that $\bar{L}_N f$ is also a function that depends only on the first coordinate. Moreover, $(\bar{L}_N f)(\eta, \xi) = (L_N f_0)(\eta)$ if f_0 denotes the restriction of f to the first coordinate and L_N the generator of a zero range process with parameters (g, p).

For the coupling measure $\bar{\mu}$ on $\mathbb{N}^{\mathbb{T}_N^d} \times \mathbb{N}^{\mathbb{T}_N^d}$, denote by $\bar{\mathbb{P}}_{\bar{\mu}} = \bar{\mathbb{P}}_{N,\bar{\mu}}$ the measure on the path space $D([0, \infty), \mathbb{N}^{\mathbb{T}_N^d} \times \mathbb{N}^{\mathbb{T}_N^d})$ corresponding to the Markov process of generator $\bar{L}_N$ starting from $\bar{\mu}$. Expectation with respect to $\bar{\mathbb{P}}_{\bar{\mu}}$ is denoted by $\bar{\mathbb{E}}_{\bar{\mu}}$.

We prove now the main feature of the coupled process defined above: starting from two configurations $\eta \le \xi$ the dynamics keeps them ordered at later times. Consider two configurations η and ξ with $\eta \le \xi$ and denote by $\delta_{(\eta,\xi)}$ the Dirac probability on $\mathbb{N}^{\mathbb{T}_N^d} \times \mathbb{N}^{\mathbb{T}_N^d}$ concentrated on (η, ξ). We claim that

$$
\bar{\mathbb{P}}_{\delta_{(\eta,\xi)}} \Big[\eta_t \le \xi_t \text{ for all } t \ge 0 \Big] = 1 .
$$

Since (η_t, ξ_t) is right continuous (cf. section 2 of Appendix 1) and $F_0 = \{(\eta, \xi);\ \eta \le \xi\}$ is closed for the product topology, to prove the last statement it is enough to show that $\bar{\mathbb{P}}_{\delta_{(\eta,\xi)}}[\eta_t \le \xi_t] = 1$ for all $t \ge 0$.

Notice that $\bar{L}_N$ does not admit a jump out of F_0 because the jump rate $g(\cdot)$ is non decreasing. In other words, F_0 is an absorbing set: $\bar{L}_N \mathbf{1}\{F_0\} \geq 0$. This inequality can be checked by a direct computation on the generator $\bar{L}_N$. It follows that

$$\partial_t \bar{\mathbb{E}}_{\delta_{(\eta,\xi)}} \Big[\mathbf{1}\{F_0\}(\eta_t, \xi_t) \Big] = \bar{\mathbb{E}}_{\delta_{(\eta,\xi)}} \Big[\bar{L}_N \mathbf{1}\{F_0\}(\eta_t, \xi_t) \Big] \geq 0 .$$

Therefore,

$$1 \geq \bar{\mathbb{E}}_{\delta_{(\eta,\xi)}} \Big[\mathbf{1}\{F_0\}(\eta_t, \xi_t) \Big] \geq \bar{\mathbb{E}}_{\delta_{(\eta,\xi)}} \Big[\mathbf{1}\{F_0\}(\eta_0, \xi_0) \Big] = 1$$

because $\eta \leq \xi$. This proves the claim.

In conclusion, we proved that the Markov process with generator $\bar{L}_N$ preserves the partial order in the following sense: $\eta_t \leq \xi_t$ for all $t \geq 0$ a.s. as soon as $\eta_0 \leq \xi_0$. To conclude the proof of the theorem it remains to show that this property implies that the semigroup preserves the order. Denote by $\{\bar{S}^N(t),\ t \geq 0\}$ the semigroup of the Markov process with generator $\bar{L}_N$.

We showed above that F_0 is an absorbing set. Since the coupled measure $\bar{\mu}$ is concentrated on F_0 ($\bar{\mu}\{F_0\} = 1$), the measure $\bar{S}^N(t)\bar{\mu}$ shares the same property. On the other hand, since both coordinates evolve as zero range processes with parameters (g, p), the first (resp. second) marginal of $\bar{S}^N(t)\bar{\mu}$ is equal to $S^N(t)\mu_1$ (resp. $S^N(t)\mu_2$) if $S^N(t)$ represents the semigroup of a zero range process with parameters (g, p). In particular, $\bar{S}^N(t)\bar{\mu}$ is a coupling measure for $\mu_1 S^N(t)$ and $\mu_2 S^N(t)$. Therefore, by Theorem 1.4, $\mu_1 S^N(t) \leq \mu_2 S^N(t)$. □

Remark 5.3 Theorem 1.4 extends to the compact space $\{0, \dots, \kappa\}^{\mathbb{T}_N^d}$ and the proof of Theorem 5.2 applies to generalized exclusion processes provided the jump rate $r(\cdot,\cdot)$ is non decreasing in the first variable and non increasing in the second. In particular, the simple exclusion process is attractive.

6. Zero Range Processes in Infinite Volume

In Chapter 9, while proving the conservation of local equilibrium for attractive zero range processes, we shall need some results on the invariant measures for the process on $\mathbb{N}^{\mathbb{Z}^d}$. We briefly summarize these results in this section and refer to Liggett (1973), (1985) or Andjel (1982) for proofs and further details.

Fix a jump rate $g(\cdot)$ satisfying assumption (3.1) and a finite range, translation invariant, irreducible transition probability $p(\cdot,\cdot)$:

$$p(x, y) = p(0, y - x) =: p(y - x) \geq 0$$

for all pair (x, y) of d-dimensional integers,

$$\sum_{z \in \mathbb{Z}^d} p(z) = 1 \quad \text{and} \quad p(z) = 0 \text{ for all } z \geq A_0 .$$

We desire to prove the existence of a Markov process on $\mathbb{N}^{\mathbb{Z}^d}$ whose generator acts on cylinder functions as

$$(Lf)(\eta) = \sum_{x\in\mathbb{Z}^d}\sum_{z\in\mathbb{Z}^d} p(z)\, g(\eta(x))\, [f(\eta^{x,x+z}) - f(\eta)] \,. \tag{6.1}$$

The process is well defined for any initial configuration with a finite number of particles since in this case it is a Markov process on a countable state space with a bounded transition rate. Thus, for a configuration η with a finite number of particles, denote by $\mathbb{P}_\eta$ the probability on the path space $D(\mathbb{R}_+, \mathbb{N}^{\mathbb{Z}^d})$ corresponding to the Markov process with generator (6.1) starting from η and by $\mathbb{E}_\eta$ expectation with respect to $\mathbb{P}_\eta$.

To extend the definition to configurations with infinitely many particles, we need to impose some conditions to avoid interference coming from infinity. Consider a function $l\colon \mathbb{Z}^d \to \mathbb{R}_+$ for which there exists a finite constant M such that

$$\sum_{y,\, |y-x|\le A_0} l(y) \;\le\; M l(x)$$

for every x in $\mathbb{Z}^d$. $l(x) = e^{-|x|}$ is an example of function satisfying the above growth restriction. For such fixed function l, on the configuration space $\mathbb{N}^{\mathbb{Z}^d}$, define the norm $\|\cdot\|_l$ by

$$\|\eta - \xi\|_l = \sum_{x\in\mathbb{Z}^d} |\eta(x) - \xi(x)| l(x) \,.$$

Denote by $\mathcal{E}_l$ the space of configurations with finite norm: $\mathcal{E}_l = \{\eta,\ \|\eta\|_l < \infty\}$ and by $\mathcal{L}_l$ the space of Lipschitz functions for this norm, i.e., the space of functions $f\colon \mathcal{E}_l \to \mathbb{R}$ for which there exists c_f such that $|f(\eta) - f(\xi)| \le c_f \|\eta - \xi\|_l$ for every η, ξ in $\mathcal{E}_l$. Denote by ℓ_f the smallest constant satisfying this inequality. Andjel (1982) proved the following existence result:

Theorem 6.1 *There exists a semigroup $\{S(t),\, t\ge 0\}$ defined on $\mathcal{L}_l$ such that*

$$S(t)f(\eta) = \mathbb{E}_\eta[f(\eta_t)]$$

for every f in $\mathcal{L}_l$ and finite η. Moreover, $\mathcal{L}_l$ is closed under $S(t)$:

$$\Big| S(t)f(\eta) - S(t)f(\xi) \Big| \;\le\; \ell_f e^{g^*(M+2)t} \|\eta - \xi\|_l$$

and

$$(Lf)(\eta) = \lim_{t\to 0} \frac{S(t)f(\eta) - f(\eta)}{t}$$

for all f in $\mathcal{L}_l$ and η in $\mathcal{E}_l$.

The existence of a probability $\mathbb{P}_\eta$ on $D(\mathbb{R}_+, \mathbb{N}^{\mathbb{Z}^d})$, for every η in $\mathcal{E}_l$, corresponding to a Markov process whose generator acts on functions in $\mathcal{L}_l$ as (6.1) follows from this theorem and general Markov process arguments that can be found in Chapter IV of Ethier and Kurtz (1986) or in Chapter III of Revuz and Yor (1991), for instance.

From this point on we assume the process to be attractive, i.e., the jump rate to be non decreasing:

$$0 = g(0) < g(k) \le g(k+1)$$

for all $k \ge 1$. Note that in this case coupling two copies of the process permits to prove the existence of a Markov process on $\mathbb{N}^{\mathbb{Z}^d}$ with generator given by (6.1) and starting from any initial state η in $\mathcal{E}_l$. One just needs to consider a sequence of states η_k, $k \ge 1$, with finite total number of particles and increasing to η.

Denote by $\mathcal{I}$ the set of invariant probability measures for the zero range process with generator given by (6.1) and by $\mathcal{S}$ the space of translation invariant probability measures. The next result states that all invariant and translation invariant measures are convex combinations of the product measures $\{\nu_\alpha, \alpha \ge 0\}$. Its proof can be found in Theorem VIII.3.9 of Liggett (1985) for exclusion processes and in Theorem 1.9 of Andjel (1982) for zero range processes.

Theorem 6.2

$$\mathcal{I} \cap \mathcal{S} = \left\{ \int \nu_\alpha m(d\alpha), \text{ where } m \text{ is a probability measure on } \mathbb{R}_+ \right\} .$$

In particular, if $(\mathcal{I} \cap \mathcal{S})_e$ stands for the extremal points of the set $\mathcal{I} \cap \mathcal{S}$,

$$(\mathcal{I} \cap \mathcal{S})_e = \{\nu_\alpha, \alpha \ge 0\} .$$

We conclude this section with a characterization of translation invariant stationary measures.

Theorem 6.3 *A translation invariant probability measure μ^* bounded above by some product measure ν_{α_0} is invariant if and only if for each density $\alpha \ge 0$ there exists a measure $\bar{\mu}_\alpha$ on the product space $\mathbb{N}^{\mathbb{Z}^d} \times \mathbb{N}^{\mathbb{Z}^d}$ with first marginal equal to μ^*, second marginal equal to ν_α and concentrated on ordered configurations:*

(i) the first marginal of $\bar{\mu}_\alpha$ is μ^*,

(ii) the second marginal of $\bar{\mu}_\alpha$ is ν_α and

(iii) $\bar{\mu}_\alpha\{(\eta, \xi);\ \eta \le \xi \text{ or } \xi \le \eta\} = 1$.

Proof. One direction follows straightforwardly from last theorem and Theorem 1.4. To prove the oder direction, first observe that the proof of Lemma VIII.3.6 in Liggett (1985) applies to the zero range processes considered here because we assumed μ^* to be bounded above by some measure ν_{α_0}.The result now follows from the proof of Theorem VIII.3.9 of Liggett (1985). □

3. Weak Formulations of Local Equilibrium

In Chapter 1 we presented a mathematical model to describe the evolution of a gas. In this respect we introduced the notions of local equilibrium and conservation of local equilibrium. Unfortunately, at the present state of knowledge, there is no satisfactory general proof of conservation of local equilibrium for large classes of interacting particle systems. All existing approaches to this problem rely too much on particular features of the model, such as the existence of dual processes (for symmetric simple exclusion processes by De Masi, Ianiro, Pellegrinotti and Presutti (1984), superposed with a Glauber dynamics by De Masi, Ferrari and Lebowitz (1986) or for the voter model Presutti and Spohn (1983)); explicit computations (for superpositions of independent random walks Dobrushin and Siegmund–Schultze (1982)); resolvent equation methods (for Ginzburg–Landau processes Fritz (1987a,b) and (1989a)); or on the attractiveness (Rost (1981), Andjel and Kipnis (1984) Benassi and Fouque (1987, 1988), Andjel and Vares (1987)). Furthermore, we would like to prove the hydrodynamic behavior of systems starting from initial states more general than local equilibrium measures. This leads us to weaken the concept of local equilibrium.

The purpose of this chapter is to introduce two alternative versions of local equilibrium, to briefly study some relations among these versions and to present an idea of how one may deduce conservation of local equilibrium from a weaker version.

Throughout this chapter we use the notation introduced in Chapter 1. Thus, for instance, for $\alpha \ge 0$, $\nu_\alpha = \nu_\alpha^N$ denotes the translation invariant product measure on $\mathbb{N}^{\mathbb{T}_N^d}$ whose marginals are Poisson measures with parameter α. Of course, all concepts extend naturally to generalized exclusion processes and zero range processes.

In Chapter 1 we presented a quite restrictive definition of local equilibrium associated to a density profile. We required the marginal probabilities of the state of the system at a macroscopic time t, that is, of $S^N(t\theta_N)\mu^N$, to converge to one of the extremal invariant and translation invariant measures of the infinite system at each continuity point of the profile. Even if this definition is "physically" natural, it is difficult to prove it. With this in mind, we introduce two weaker notions of local equilibrium. We first recall the notion of product measures with slowly varying parameter associated to a profile.

Definition 0.1 (Product measures with slowly varying parameter associated to a profile). Given a continuous profile $\rho_0\colon \mathbb{T}^d \to \mathbb{R}_+$, we denote by $\nu^N_{\rho_0(\cdot)}$ the product measure on $\mathbb{N}^{\mathbb{T}^d_N}$ with marginals given by

$$\nu^N_{\rho_0(\cdot)}\{\eta,\ \eta(x) = k\} \;=\; \nu^N_{\rho_0(x/N)}\{\eta,\ \eta(0) = k\}$$

for all x in $\mathbb{T}^d_N$, $k \geq 0$. This measure is called the product measure with slowly varying parameter associated to $\rho_0(\cdot)$.

Measures on $\mathbb{N}^{\mathbb{Z}^d}$ are characterized by the way they integrate bounded cylinder functions. In order to keep notation simple, to each bounded cylinder function Ψ in $\mathbb{N}^{\mathbb{Z}^d}$ we associate the real bounded function $\tilde{\Psi} : \mathbb{R}_+ \to \mathbb{R}$ that at α is equal to the expected value of Ψ under ν_α:

$$\tilde{\Psi}(\alpha) := E_{\nu_\alpha}[\Psi] \;=\; \int \Psi(\eta)\,\nu_\alpha(d\eta)\,. \tag{0.1}$$

By Lemma 2.3.8 $\tilde{\Psi}$ is a bounded continuous function.

In the same way that we have defined in Chapter 1 translations of configurations, for all continuous functions Ψ, we denote by $\tau_x\Psi$ the translation of Ψ by x units:

$$(\tau_x\Psi)(\eta) \;=\; \Psi(\tau_x\eta)$$

for every configuration η.

By Chebychev inequality and the dominated convergence theorem, for every smooth profile $\rho_0\colon \mathbb{T}^d \to \mathbb{R}_+$ and every sequence $\nu^N_{\rho_0(\cdot)}$ of Poisson measures with slowly varying parameter associated to this profile,

$$\lim_{N\to\infty} \nu^N_{\rho_0(\cdot)}\left[\left|\frac{1}{N^d}\sum_{x\in\mathbb{T}^d_N} G(x/N)\,(\tau_x\Psi)(\eta) \;-\; \int_{\mathbb{T}^d} G(u)\;\tilde{\Psi}(\rho_0(u))\,du\right| > \delta\right] \;=\; 0$$

for all continuous functions $G\colon \mathbb{T}^d \to \mathbb{R}$, all bounded cylinder functions Ψ and all strictly positive δ. We leave to the reader to check this statement. Details can be found in the proof of a slightly stronger result stated as Proposition 0.4 below.

This last statement asserts that the sequence of measures $\nu^N_{\rho_0(\cdot)}$ integrates the cylinder function Ψ around the macroscopic point u in $\mathbb{T}^d$ in the same way as an equilibrium measure of density $\rho_0(u)$ would do. We do not require anymore the marginals of the sequence of measures to converge to an extremal equilibrium measure at every continuity point of the profile ρ_0. We only impose that its spatial mean converges to the corresponding spatial mean. This notion is therefore weaker than the one of local equilibrium introduced in Chapter 1. The main difference is the spatial average over small macroscopic boxes of volume of order N^d that is implicit in this new concept and absent in the definition of local equilibrium.

Definition 0.2 . (Weak local equilibrium) A sequence $(\mu^N)_{N\geq 1}$ of probability measures on $\mathbb{N}^{\mathbb{T}_N^d}$ is a weak local equilibrium of profile $\rho_0: \mathbb{T}^d \to \mathbb{R}_+$ if for every continuous function $G: \mathbb{T}^d \to \mathbb{R}$, every bounded cylinder function Ψ and every $\delta > 0$ we have

$$\lim_{N\to\infty} \mu^N \left[\left| \frac{1}{N^d} \sum_{x\in\mathbb{T}_N^d} G(x/N)\tau_x\Psi(\eta) - \int_{\mathbb{T}^d} G(u)\,\tilde{\Psi}(\rho_0(u))\,du \right| > \delta \right] = 0\,.$$

The function $\Psi(\eta) = \eta(0)$ plays a special role in the whole study of hydrodynamics being closely related to the conserved quantity. Though it is generally not bounded (for zero range processes, for instance), the definition that follows is a particularly weakened notion since it demands only convergence for this cylinder function. Since it will appear repeatedly throughout the book, we introduce a special terminology:

Definition 0.3 (Probability measures associated to a density profile) A sequence $(\mu^N)_{N\geq 1}$ of probability measures on $\mathbb{N}^{\mathbb{T}_N^d}$ is associated to a profile $\rho_0: \mathbb{T}^d \to \mathbb{R}_+$ if for every continuous function $G: \mathbb{T}^d \to \mathbb{R}$, and for every $\delta > 0$ we have

$$\lim_{N\to\infty} \mu^N \left[\left| \frac{1}{N^d} \sum_{x\in\mathbb{T}_N^d} G(x/N)\eta(x) - \int_{\mathbb{T}^d} G(u)\,\rho_0(u)\,du \right| > \delta \right] = 0\,.$$

The quantity just introduced in the definition above can be reformulated in terms of empirical measures. Let π^N be the positive measure on the torus $\mathbb{T}^d$ obtained by assigning to each particle a mass N^{-d}:

$$\pi^N(\eta, du) := \frac{1}{N^d} \sum_{x\in\mathbb{T}_N^d} \eta(x)\,\delta_{x/N}(du). \tag{0.2}$$

In this formula, for a d-dimensional vector u, δ_u represents the Dirac measure concentrated on u. The measure $\pi^N(\eta, du)$ is called the empirical measure associated to the configuration η. The dependence in η will frequently be omitted to keep notation as simple as possible. With this notation $N^{-d}\sum_x G(x/N)\eta(x)$ is the integral of G with respect to the empirical measure π^N, denoted by $<\pi^N, G>$.

Let $\mathcal{M}_+(\mathbb{T}^d)$ be the space of finite positive measures on the torus $\mathbb{T}^d$ endowed with the weak topology. Notice that for each N, π^N is a continuous function from $\mathbb{N}^{\mathbb{T}_N^d}$ to $\mathcal{M}_+$. For a probability measure μ on $\mathbb{N}^{\mathbb{T}_N^d}$, throughout this book, we shall also denote by μ the measure $\mu(\pi^N)^{-1}$ on $\mathcal{M}_+$ induced by μ and π^N ($\mu(\pi^N)^{-1}[A] = \mu[\pi^N \in A]$). With this convention, a sequence of probability measures $(\mu^N)_{N\geq 1}$ in $\mathbb{N}^{\mathbb{T}_N^d}$ is associated to a density profile ρ_0 if the sequence of random measures $\pi^N(du)$ converges in probability to the deterministic measure $\rho_0(u)du$.

In the study of the hydrodynamic behavior of interacting particle systems we will sometimes be forced to reduce our goals and to content ourselves in proving that starting from a sequence of measures associated to a density profile ρ_0 then, at a later suitably renormalized time, we obtain a state $(S^N(t\theta_N)\mu^N$ in the notation of Chapter 1) associated to a new density profile $\rho(t,\cdot)$ which is the solution of some partial differential equation. More precisely, we want to prove that for all sequences of probability measures μ^N associated to a profile $\rho_0 : \mathbb{T}^d \to \mathbb{R}_+$ and not too far, in a sense to be specified later, from a global equilibrium ν_α^N,

$$\lim_{N\to\infty} \mu^N \left[\left| \frac{1}{N^d} \sum_{x\in\mathbb{T}_N^d} G(x/N)\eta_{t\theta_N}(x) \; - \; \int_{\mathbb{T}^d} G(u)\,\rho(t,u)\,du \right| > \delta \right] = 0$$

for a suitable renormalization θ_N, for every continuous function $G\colon \mathbb{T}^d \to \mathbb{R}$ and every δ strictly positive. As in the case of local equilibrium, in this formula $\rho(\cdot,\cdot)$ will be the solution of a Cauchy problem with initial condition $\rho_0(\cdot)$. This constitutes the program of the next chapters and deserves therefore a special terminology. We shall say that we proved the hydrodynamic behavior of an interacting particle system whenever we are able to achieve the goal just described.

We will now compare the strong notion of local equilibrium introduced in Chapter 1 with the weaker versions just presented. We start showing that it is really a weaker notion. In this direction we first present two results that are not stated in the greatest possible generality.

Proposition 0.4 *Let $(\mu^N)_{N\ge 1}$ be a local equilibrium of profile ρ_0 almost surely continuous. Then the sequence (μ^N) is a weak local equilibrium of profile ρ_0.*

Proof. Let Ψ be a bounded cylinder function. For a positive integer ℓ, we introduce the real positive function $h_{N,\ell}$ defined on the torus $\mathbb{T}^d$ by

$$\begin{aligned} h_{N,\ell}(u) \;=\; & \sum_{x\in\mathbb{T}_N^d} \mathbf{1}\{x \le Nu < x + \underline{1}\} \times \\ & \times\; E_{\mu^N}\left[\left| \frac{1}{(2\ell+1)^d} \sum_{|y-x|\le\ell} \tau_y\Psi(\eta) - \tilde{\Psi}(\rho_0(x/N)) \right| \right] . \end{aligned}$$

Here $\underline{1}$ represents the vector of $\mathbb{R}^d$ with all coordinates equal to 1 and "$\le$" is the natural partial order of $\mathbb{R}^d$: $a \le b$ provided $a_i \le b_i$ for $1 \le i \le d$.

$h_{N,\ell}$ is bounded everywhere because Ψ is bounded. Moreover, when N increases to infinity, $h_{N,\ell}(u)$ converges to

$$E_{\nu_{\rho_0(u)}}\left[\left| \frac{1}{(2\ell+1)^d} \sum_{|y|\le\ell} \tau_y\Psi(\eta) - \tilde{\Psi}(\rho_0(u)) \right| \right] .$$

at all continuity points u of ρ_0. By the law of large numbers, when ℓ tends to infinity, this expression converges to 0. On the other hand, the integral of $h_{N,\ell}$ with respect to the Lebesgue measure on the torus is

$$E_{\mu^N}\left[\frac{1}{N^d}\sum_{x\in\mathbb{T}_N^d}\left|\frac{1}{(2\ell+1)^d}\sum_{|y-x|\le\ell}\tau_y\Psi(\eta)-\tilde{\Psi}\left(\rho_0(x/N)\right)\right|\right].$$

The dominated convergence theorem permits to conclude the proof of the proposition. □

In the same way a weak assumption of equiintegrability suffices to prove an analogous result for probability measures associated to density profiles.

Proposition 0.5 *Let $(\mu^N)_{N\ge 1}$ be a local equilibrium of profile almost surely continuous and bounded. Then, the sequence (μ^N) is associated to the density profile ρ_0 if*

$$\lim_{M\to\infty}\lim_{N\to\infty}\frac{1}{N^d}\sum_{x\in\mathbb{T}_N^d}E_{\mu^N}\left[\eta(x)\mathbf{1}\{\eta(x)>M\}\right] = 0\,.$$

The weak notion of local equilibrium is not however too far from the strong form as it is shown by the following result.

Proposition 0.6 *Let $(\mu^N)_{N\ge 1}$ be a weak local equilibrium of profile ρ_0. Assume that there exists a sequence $(\epsilon_k)_{k\ge 1}$ decreasing to 0 such that for each integer k there exists two sequences of measures $(\mu^N_{\epsilon_k,+})_{N\ge 1}$ and $(\mu^N_{\epsilon_k,-})_{N\ge 1}$ having the following properties:*

(a) $(\mu^N_{\epsilon_k,\pm})_{N\ge 1}$ is a weak local equilibrium of profile $\rho_{\epsilon_k,\pm}$;

(b) $\tau_x\mu^N_{\epsilon_k,+}\ge\mu^N\ge\tau_x\mu^N_{\epsilon_k,-}$ for every d-dimensional integer x of absolute value bounded by $\epsilon_k N$;

(c) For every continuity points u of ρ_0,

$$\limsup_{k\to\infty}\sup_{|v-u|\le\epsilon_k}|\rho_{\epsilon_k,\pm}(v)-\rho_0(u)| = 0\,.$$

Then the sequence $(\mu^N)_{N\ge 1}$ is a strong local equilibrium of profile ρ_0.

Proof. Let Ψ be a bounded monotone cylinder function. From hypotheses (a), (b) and (c) the expectation of Ψ under the measure μ^N translated by $[uN]$ can be bounded below and above by expressions that converge to $\tilde{\Psi}(\rho_0(u))$. Indeed, as Ψ is monotone, by property (b), for every integer k

$$E_{\tau_{[uN]}\mu^N}[\Psi]\le\frac{1}{(2[\epsilon_k N]+1)^d}\sum_{|y-[uN]|\le[\epsilon_k N]}E_{\tau_y\mu^N_{\epsilon_k,+}}[\Psi]\,.$$

By hypothesis (a), as N increases to infinity, the right hand side converges to

$$\frac{1}{(2\epsilon_k)^d} \int_{[u-\epsilon_k, u+\epsilon_k]^d} \tilde{\Psi}\left(\rho_{\epsilon_k,+}(v)\right) dv .$$

Notice that in the definition of weak local equilibrium, we assume convergence only for continuous functions G. It is not difficult, however, to prove convergence for an indicator function $\mathbf{1}\{[u-\epsilon, u+\epsilon]\}$, by approximating the indicator by continuous functions, keeping in mind that Ψ is bounded.

Finally, since by Lemma 2.3.8 the function $\tilde{\Psi}$ is continuous and bounded, by hypothesis (c) when k converges to infinity this expression converges to $\tilde{\Psi}(\rho_0(u))$ for every continuity points u of ρ_0.

In this way we have proved that

$$\limsup_{N\to\infty} \tau_{[uN]} E_{\mu^N}[\Psi] \;\leq\; \tilde{\Psi}(\rho_0(u)) \;=\; E_{\nu_{\rho_0(u)}}[\Psi] .$$

The same arguments replacing the sequence $\mu^N_{\epsilon_k,+}$ by the sequence $\mu^N_{\epsilon_k,-}$ prove the reverse inequality. It follows from Lemma 2.1.5 that the sequence $\tau_{[uN]}\mu^N$ converges weakly to $\nu_{\rho_0(u)}$ for every continuity point u of ρ_0. □

A similar argument to this one will permit to prove conservation of local equilibrium for attractive processes from a weak local equilibrium statement. Details are given in Chapter 9.

Remark 0.7 The definitions of product measure with slowly varying parameter associated to a continuous profile, of local equilibrium in strong and weak sense and the one of density profile depend uniquely on the existence of a parametrized family of invariant measures. These definitions and the propositions stated above extend therefore in a natural way to the generalized exclusion processes and to the zero range processes defined in the previous chapter.

4. Hydrodynamic Equation of Symmetric Simple Exclusion Processes

In this chapter we prove the hydrodynamic behavior of nearest neighbor symmetric simple exclusion processes and show that the hydrodynamic equation is the heat equation:

$$\partial_t \rho \;=\; (1/2)\,\Delta\rho\;.$$

In this formula, $\Delta\rho$ stands for the Laplacian of ρ: $\Delta\rho = \sum_{1\le i\le d} \partial^2_{u_i}\rho$.

We briefly present the strategy of the proof. We shall first prove that the empirical measure solves the heat equation in a weak sense in an integral form. More precisely, for a positive measure π on $\mathbb{T}^d$ of finite total mass and for a continuous function $G\colon \mathbb{T}^d \to \mathbb{R}$, denote by $<\pi, G>$ the integral of G with respect to π:

$$<\pi, G> = \int_{\mathbb{T}^d} G(u)\,\pi(du)\;.$$

We shall prove that the empirical measure π^N_t associated to the symmetric simple exclusion process converges, in a way to be specified later, to a measure π_t absolutely continuous with respect to the Lebesgue measure and satisfying:

$$<\pi_t, G> = <\pi_0, G> \; + (1/2)\int_0^t <\pi_s, \Delta G> \,ds \tag{0.1}$$

for a sufficiently large class of functions $G: \mathbb{T}^d \to \mathbb{R}$ and for every t in an interval $[0,T]$ fixed in advance.

Recall from Chapter 3 that we denoted by $\mathcal{M}_+ = \mathcal{M}_+(\mathbb{T}^d)$ the space of finite positive measures on $\mathbb{T}^d$ endowed with the weak topology. In order to work in a fixed space as N increases, we consider the time evolution of the empirical measure π^N_t associated to the particle system defined by:

$$\pi^N_t(du) \;=\; \pi^N(\eta_t, du) \;:=\; \frac{1}{N^d}\sum_{x\in\mathbb{T}^d_N} \eta_t(x)\,\delta_{x/N}(du)\;. \tag{0.2}$$

Notice that there is a one to one correspondence between configurations η and empirical measures $\pi^N(\eta, du)$. In particular, π^N_t inherits the Markov property from η_t.

We consider the distribution of the empirical measure as a sequence of probability measures on a fixed space. Since there are jumps this space must be

$D([0,T],\mathcal{M}_+)$, the space of right continuous functions with left limits taking values in $\mathcal{M}_+$.

Fix a profile $\rho_0 : \mathbb{T}^d \to [0,1]$ and denote by μ^N a sequence of probability measures associated to ρ_0. For each $N \geq 1$, let Q^N be the probability measure on $D([0,T],\mathcal{M}_+)$ corresponding to the Markov process π_t^N speeded up by N^2 and starting from μ^N. We speeded up the process by N^2 because we have seen in Chapter 1 that to obtain a non trivial hydrodynamic evolution for mean-zero processes we need to consider time scales of order N^2.

Our goal is to prove that, for each fixed time t, the empirical measure π_t^N converges in probability to $\rho(t,u)du$ where $\rho(t,u)$ is the solution of the heat equation with initial condition ρ_0. We shall proceed in two steps. We first prove that the process π_t^N converges in distribution to the probability measure concentrated on the deterministic path $\{\rho(t,u)du,\ 0 \leq t \leq T\}$ and then argue that convergence in distribution to a deterministic weakly continuous trajectory implies convergence in probability at any fixed time $0 \leq t \leq T$.

A deterministic trajectory can be interpreted as the support of a Dirac probability measure on $D([0,T],\mathcal{M}_+)$ concentrated on this trajectory. The proof of the hydrodynamic behavior of symmetric simple exclusion processes is therefore reduced to show the convergence of the sequence of probability measures Q^N to the Dirac measure concentrated on the solution of the heat equation.

An indirect standard method to prove the convergence of a sequence is to show that this sequence is relatively compact and then to show that all converging subsequences converge to the same limit. To show the relative compactness we will use Prohorov's criterion. At this point it will remain the identification of all limit points of subsequences.

Prohorov's criterion states that a sequence of probability measures $\{Q^N, N \geq 1\}$ in a reasonable topological space (e.g. a polish space cf. Billingsley (1968)) is weakly relatively compact if and only if for every $\varepsilon > 0$ there is a compact set K_ε such that for every N,

$$Q^N(K_\varepsilon) \geq 1 - \varepsilon .$$

We must therefore examine the compact subsets of $D([0,T],\mathcal{M}_+)$. In this respect, notice that we have the choice of the topology to be attributed to $D([0,T],\mathcal{M}_+)$. We just need to be able to consider Dirac measures concentrated on a trajectory. Our choice under this restriction is guided by the estimates we have and by the simplicity of the corresponding compact sets. The simplest and strongest topology is of course the topology of uniform convergence. However, the identification of compact sets forces us to use the Skorohod topology.

To characterize all limit points of the sequence Q^N, we have to investigate how we may use the random evolution to make an equation of type (0.1) appear. Notice first that, under Q^N, for every function $G: \mathbb{T}^d \to \mathbb{R}$, the quantities

$$< \pi_t^N, G > = \frac{1}{N^d} \sum_{x \in \mathbb{T}_N^d} G(x/N)\, \eta_t(x) \tag{0.3}$$

verify the identity

$$< \pi_t^N, G > = < \pi_0^N, G > + \int_0^t N^2 L_N < \pi_s^N, G > ds + M_t^{G,N}$$

where $M_t^{G,N}$ are martingales with respect to the natural filtration $\mathcal{F}_t = \sigma(\eta_s,\ s \leq t)$. The factor N^2 in front of the generator L_N appears because we speeded up the process by N^2. In the particular case of nearest neighbor symmetric simple exclusion processes, the second term on the right hand side may be rewritten as a function of the empirical measure. Indeed, applying the generator to the function $\eta \to \eta(x)$ we have:

$$L_N\, \eta(x) = (1/2) \sum_{j=1}^{d} \Big[\eta(x + e_j) + \eta(x - e_j) - 2\eta(x) \Big] .$$

After two summations by parts we obtain that under Q^N

$$< \pi_t^N, G > = < \pi_0^N, G > + (1/2) \int_0^t < \pi_s^N, \Delta_N G > ds + M_t^{G,N}$$

where Δ_N is the discrete Laplacian:

$$(\Delta_N G)(x/N) = N^2 \sum_{j=1}^{d} \Big[G\big((x + e_j)/N\big) + G\big((x - e_j)/N\big) - 2G(x/N) \Big] .$$

To conclude the proof of the hydrodynamic behavior of symmetric simple exclusion processes, it remains to show, on the one hand, an uniqueness theorem for solutions of equations (0.1); uniqueness theorem that will require to prove identity (0.1) for a certain class of functions G; and, on the other hand, to prove that the martingales $M_t^{G,N}$ vanish in the limit as $N \uparrow \infty$ for this family of functions G. From these two results it follows that the sequence Q^N has a unique limit point Q^* which is the probability measure concentrated on the unique solution of (0.1).

1. Topology and Compactness

We work on $\mathcal{M}_+(\mathbb{T}^d)$, the space of finite positive measures on $\mathbb{T}^d$ endowed with the weak topology. Denote by $C(\mathbb{T}^d)$ the space of continuous functions on $\mathbb{T}^d$. Recall that we may define a metric on $\mathcal{M}_+(\mathbb{T}^d)$ by introducing a dense countable family $\{f_k;\ k \geq 1\}$ of continuous functions on $\mathbb{T}^d$ and by defining the distance $\delta(\cdot, \cdot)$ by

$$\delta(\mu, \nu) = \sum_{k=1}^{\infty} \frac{1}{2^k} \frac{| < \mu, f_k > - < \nu, f_k > |}{1 + | < \mu, f_k > - < \nu, f_k > |} . \tag{1.1}$$

We assume hereafter that $f_1 = 1$. $\mathcal{M}_+$ endowed with the weak topology is complete and a set $A \subset \mathcal{M}_+$ is relatively compact if and only if

$$\sup_{\mu \in A} < \mu, 1 > < +\infty , \tag{1.2}$$

that is, if the total masses are uniformly bounded.

On the other hand, from the existence of jumps, the natural space to consider for the evolution of the empirical measure π_t is $D([0,T], \mathcal{M}_+)$, the set of right continuous functions with left limits taking values in $\mathcal{M}_+$. To endow this space with a reasonable topology notice that we cannot demand the uniform convergence because a small time change in a jump should not modify too much the proximity of two paths.

The results presented below are a short recapitulation of results from Chapter 3 of Billingsley (1968) or Chapter 3 of Ethier and Kurtz (1986). Since in next chapters we will sometimes be interested in other jump processes than the empirical measure and since all results do not depend on the general structure of $\mathcal{M}_+$, we state them in a general setting.

Consider, therefore, a complete separable metric space $\mathcal{E}$ with metric $\delta(\cdot, \cdot)$ and a sequence P^N of probability measures on $D([0,T], \mathcal{E})$. Elements of $\mathcal{E}$ are denoted by the Greek letters μ and ν. Let Λ be the set of strictly increasing continuous functions λ of $[0,T]$ into itself. We define then

$$\|\lambda\| = \sup_{s \neq t} \left| \log \frac{\lambda(t) - \lambda(s)}{t - s} \right|$$

and

$$d(\mu, \nu) = \inf_{\lambda \in \Lambda} \max \left\{ \|\lambda\| , \sup_{0 \le t \le T} \delta(\mu_t, \nu_{\lambda(t)}) \right\} .$$

Proposition 1.1 *$D([0,T], \mathcal{E})$ endowed with the metric d is a complete separable metric space.*

Even though this definition of distance is important because it allows to use the property of completeness to identify compact sets, it is not very useful in practice because it takes into account all functions λ of Λ. (Remember that a complete set A is compact if and only if it is precompact, that is, if for every $\varepsilon > 0$ A can be covered by a finite number of balls of radius less than or equal to ε).

The main tool will be a modified uniform modulus of continuity that allows to extend the Ascoli Theorem to the set $D([0,T], \mathcal{E})$. For this reason we introduce

$$w'_\mu(\gamma) := \inf_{\{t_i\}_{0 \le i \le r}} \max_{0 \le i < r} \sup_{t_i \le s < t < t_{i+1}} \delta(\mu_s, \mu_t) ,$$

where the first infimum is taken over all partitions $\{t_i, \, 0 \le i \le r\}$ of the interval $[0,T]$ such that

$$\begin{cases} 0 = t_0 < t_1 < \cdots < t_r = T \\ t_i - t_{i-1} > \gamma \qquad i = 1, \dots, r . \end{cases}$$

We can easily verify that the trajectory $\{\mu_t \, , 0 \le t \le T\}$ belongs to $D([0,T], \mathcal{E})$ if and only if the modified uniform modulus of continuity $w'_\mu(\gamma)$ vanishes as $\gamma \downarrow 0$.

Moreover, we can characterize the compact sets of $D([0,T],\mathcal{E})$ in terms of the modified modulus of continuity.

Proposition 1.2 *A set A of $D([0,T],\mathcal{E})$ is relatively compact if and only if*

(1) $\{\mu_t\,;\ \mu\in A\,,\ t\in[0,T]\}$ *is relatively compact on* $\mathcal{E}$,

(2) $\lim_{\gamma\to 0}\ \sup_{\mu\in A}\ w'_\mu(\gamma)\ =\ 0\,.$

With this result we obtain a statement of Prohorov's theorem.

Theorem 1.3 *Let P^N be a sequence of probability measures on $D([0,T],\mathcal{E})$. The sequence is relatively compact if and only if*

(1) For every t in $[0,T]$ and every ε strictly positive, there is a compact $K(t,\varepsilon)\subset\mathcal{E}$ such that $\sup_N\ P^N[\mu_t\notin K(t,\varepsilon)]\ \le\ \varepsilon\,.$

(2) For every ε strictly positive, $\lim_{\gamma\to 0}\ \limsup_{N\to\infty}\ P^N\left[\mu;\ w'_\mu(\gamma)>\varepsilon\right]\ =\ 0\,.$

Remark 1.4 The second condition that appears in the previous theorem, that depends on all path $\{\mu_t,\ 0\le t\le T\}$ and not only on the behavior at a fixed time t, is the most difficult to verify. Nevertheless, for any positive γ, the modified uniform modulus of continuity $w'_\mu(\gamma)$ is naturally smaller than the usual uniform modulus of continuity w_μ calculated at 2γ: $w'_\mu(\gamma)\le w_\mu(2\gamma)$, where

$$w_\mu(\gamma)\ =\ \sup_{|t-s|\le\gamma}\ \delta(\mu_s,\mu_t)\,.$$

In all cases we investigate in the following chapters, instead of condition (2) of the previous theorem, we prove:

(2') For every ε strictly positive, $\lim_{\gamma\to 0}\ \limsup_{N\to\infty}\ P^N\left[\mu;\ w_\mu(\gamma)>\varepsilon\right]\ =\ 0\,.$

Remark 1.5 It is easy to see that all limit points of a sequence P^N satisfying (2') are concentrated on continuous paths.

We also have a very useful sufficient condition due to Aldous (1978). To state this result denote by $\mathfrak{T}_T$ the family of all stopping times bounded by T.

Proposition 1.6 *A sequence of probability measures P^N on $D([0,T],\mathcal{E})$ satisfies condition (2) of Theorem 1.3 provided*

$$\lim_{\gamma\to 0}\ \limsup_{N\to\infty}\ \sup_{\substack{\tau\in\mathfrak{T}_T\\ \theta\le\gamma}}\ P^N\Big[\ \delta(\mu_\tau,\mu_{\tau+\theta})\ >\ \varepsilon\ \Big]\ =\ 0 \tag{1.3}$$

for every $\varepsilon>0$.

Here and in the proof of this result, by convention, all times are assumed bounded by T so, for instance, $\tau+\theta$ should be read as $(\tau+\theta)\wedge T$.

Proof of Proposition 1.6. Fix $\varepsilon > 0$. By assumption, there exists $\gamma_0 > 0$ and N_0 such that for $N > N_0$

$$P^N\Big[\delta(\mu_{\tau+\gamma}, \mu_\tau) > \varepsilon\Big] \;\le\; \varepsilon \tag{1.4}$$

for each $\gamma \le 2\gamma_0$ and each stopping time τ. Let $M > 2T/\gamma_0$ be a fixed integer. By the same reasons, there exists $\sigma_0 > 0$ and $N_1 \ge N_0$ such that for $N > N_1$,

$$P^N\Big[\delta(\mu_{\tau+\sigma}, \mu_\tau) > \varepsilon\Big] \;\le\; \frac{\varepsilon}{M} \tag{1.5}$$

for each $\sigma \le 2\sigma_0$ and each stopping time τ.

For $0 \le i \le M$, define recursively the stopping times τ_i by $\tau_0 = 0$ and

$$\tau_{i+1} \;=\; \min\Big\{t > \tau_i\, ;\, \delta(\mu_{\tau_i}, \mu_t) \ge 2\varepsilon\Big\}\, .$$

By convention if the set on the right hand side is empty, we fix τ_{i+1} to be T.

To prove the proposition, it is enough to show that for $N \ge N_1$,

$$\begin{aligned} &P^N[\tau_M < T] \;\le\; 16\varepsilon \quad \text{and} \\ &P^N\Big[\tau_{i+1} < T \wedge (\tau_i + \sigma) \text{ for some } 0 \le i \le M-1\Big] \;\le 8\varepsilon\, . \end{aligned} \tag{1.6}$$

Indeed, it follows from the definition of the stopping time $\{\tau_i,\, 0 \le i \le M\}$ and (1.6) that

$$P\Big[w'_\mu(\sigma) > 4\varepsilon\Big] \;\le\; 24\varepsilon$$

for all $N \ge N_1$, which is exactly condition (2) of Theorem 1.3.

It remains to prove (1.6). Denote by $\mathcal{F}$ the σ-algebra generated by $\{\mu_t,\, 0 \le t \le T\}$. Consider a random variable U distributed uniformly on $[0, 2\gamma_0]$ and independent of $\mathcal{F}$. We abuse of notation and denote by P^N the probability measure on $D([0,T], \mathcal{E}) \times [0, 2\gamma_0]$ corresponding to the variable (μ, U). For a fixed trajectory μ, and $0 \le t_1 \le t_2 \le T$, assume that

$$t_2 - t_1 < \gamma_0 \quad \text{and} \quad P^N\Big[\delta(\mu_{t_j+U}, \mu_{t_j}) < \varepsilon \,\Big|\, \mathcal{F}\Big] \;>\; 3/4$$

for $j = 1, 2$. In this formula the integration is made with respect to the variable U. Then, there exists t^* in $[t_1 + \gamma_0, t_1 + 2\gamma_0]$ such that

$$\delta(\mu_{t^*}, \mu_{t_j}) < \varepsilon$$

for $j = 1, 2$. In particular, $\delta(\mu_{t_1}, \mu_{t_2}) < 2\varepsilon$ and

$$\begin{aligned} &\Big\{(\mu, t_1, t_2),\; \delta(\mu_{t_1}, \mu_{t_2}) \ge 2\varepsilon,\; t_1 \le t_2 < t_1 + \gamma_0\Big\} \\ &\quad\subset \bigcup_{j=i}^{i+1} \Big\{(\mu, t_1, t_2),\; P^N\Big[\delta(\mu_{t_j+U}, \mu_{t_j}) \ge \varepsilon \,\Big|\, \mathcal{F}\Big] \;\ge\; 1/4\Big\}\, . \end{aligned}$$

Thus for $0 \le i \le M - 1$,

$$
\begin{aligned}
P^N & \Big[\delta(\mu_{\tau_i}, \mu_{\tau_{i+1}}) \ge 2\varepsilon \,,\; \tau_{i+1} < \tau_i + \gamma_0\Big] \\
& \le \sum_{j=i}^{i+1} P^N \Big\{\mu\,;\; P^N \Big[\delta(\mu_{\tau_j+U}, \mu_{\tau_j}) \ge \varepsilon \big| \mathcal{F}\Big] \ge 1/4\Big\}\,.
\end{aligned}
$$

By Chebychev inequality and (1.4), last expression is bounded above by

$$
4 \sum_{j=i}^{i+1} P^N \Big[\delta(\mu_{\tau_j+U}, \mu_{\tau_j}) \ge \varepsilon\Big] \;\le\; 8\varepsilon\,.
$$

Since, for $0 \le i \le M-1$ the set $\{\tau_{i+1} < T\}$ is contained in $\{\delta(\mu_{\tau_i}, \mu_{\tau_{i+1}}) \ge 2\varepsilon\}$ we obtain from the two previous estimates that

$$
P^N \Big[\tau_{i+1} < \tau_i + \gamma_0 \,,\; \tau_{i+1} < T\Big] \;\le\; 8\varepsilon \tag{1.7}
$$

for $0 \le i \le M-1$. A similar argument taking advantage of (1.5) instead of (1.4) shows that

$$
P^N \Big[\tau_{i+1} < \tau_i + \sigma_0 \,,\; \tau_{i+1} < T\Big] \;\le\; \frac{8\varepsilon}{M}
$$

for $0 \le i \le M-1$, what proves the second statement in (1.6).

To prove the first claim in (1.6), notice that $\{\tau_M < T\} \subset \{\tau_{i+1} < T\}$ for $0 \le i \le M-1$. In particular, by Chebychev inequality and estimate (1.7) we have that

$$
\begin{aligned}
E^N \Big[\tau_{i+1} - \tau_i \,\big|\, \tau_M < T\Big] \;&\ge\; \gamma_0 P\Big[\tau_{i+1} - \tau_i \ge \gamma_0 \,\big|\, \tau_M < T\Big] \\
= \gamma_0 \Big\{1 - \frac{P^N[\tau_{i+1} - \tau_i < \gamma_0 \,,\; \tau_M < T]}{P^N[\tau_M < T]}\Big\} \;&\ge\; \gamma_0 \Big\{1 - \frac{8\varepsilon}{P^N[\tau_M < T]}\Big\}\,.
\end{aligned}
$$

Thus,

$$
\begin{aligned}
T \;\ge\; E^N \Big[\tau_M \,\big|\, \tau_M < T\Big] \;&=\; \sum_{i=0}^{M-1} E^N \Big[\tau_{i+1} - \tau_i \,\big|\, \tau_M < T\Big] \\
&\ge\; M\gamma_0 \Big\{1 - \frac{8\varepsilon}{P^N[\tau_M < T]}\Big\}\,.
\end{aligned}
$$

This proves the first claim of (1.6) because we chose $M > 2T/\gamma_0$. □

In the context of this book, i.e., in the case of $\mathcal{M}_+$ endowed with the weak topology, to prove the relative compactness of a sequence of measures Q^N defined on $D([0,t], \mathcal{M}_+)$ it is enough to check conditions of Theorem 1.3 or the one of Remark 1.4 for each real process obtained by "projecting" the empirical measure π_t^N with functions of a dense countable set of $C(\mathbb{T}^d)$. More precisely, the following result asserts that to establish the relative compactness of a family Q^N of probability measures on $D([0,T], \mathcal{M}_+)$ it is enough to study the same problem for the processes $< \pi_t^N, g_k >$ for a dense countable family $\{g_k,\; k \ge 1\}$ of functions of $C(\mathbb{T}^d)$.

Proposition 1.7 *Let $\{g_k; k \ge 1\}$ be a dense subfamily of $C(\mathbb{T}^d)$ with $g_1 = 1$. A family of probability measures $(Q^N)_{N\ge 1}$ on $D([0,T], \mathcal{M}_+)$ is relatively compact if for every positive integer k the family $Q^N g_k^{-1}$ of probabilities on $D([0,T], \mathbb{R})$ has this property. Here, $Q^N g_k^{-1}$ is the sequence of measures obtained by "projecting" Q^N with a function g_k and is defined by*

$$Q^N g_k^{-1}[A] = Q^N \left[\pi^N; < \pi^N, g_k > \in A\right] .$$

Proof. We have already observed in (1.2) that a set $A \subset \mathcal{M}_+$ is weakly relatively compact if $\sup_{\mu \in A} < \mu, 1 >< \infty$. Fix $\varepsilon > 0$ and $0 \le t \le T$. By assumption, there exists $A > 0$ such that

$$P^N\Big[| < \mu_t, g_1 > | > A\Big] \le \varepsilon$$

for every $N \ge 1$. In particular, since $g_1 = 1$ and since the set $\{\mu, | < \mu, g_1 > | \le A\}$ is weakly relatively compact, the first condition of Theorem 1.3 is proved.

To prove the second condition, fix $\varepsilon > 0$ and $\beta > 0$. Let k_ε such that $2^{1-k_\varepsilon} \le \varepsilon$. It follows from the definition of the distance δ introduced in (1.1) and of k_ε that for each $\gamma > 0$,

$$w'_\mu(\gamma) \le \sum_{k=1}^{k_\varepsilon} \frac{1}{2^k} w'_{<\mu, g_k>}(\gamma) + \frac{\varepsilon}{2} \cdot \tag{1.8}$$

By assumption, there exists γ_0 such that

$$P^N\Big[w'_{<\mu, g_k>}(\gamma) > \varepsilon/2\Big] \le \frac{\beta}{2^k} \cdot$$

for each $k \le k_\varepsilon$, $\gamma \le \gamma_0$ and $N \ge 1$. Therefore,

$$P^N\Big[\sum_{k=1}^{k_\varepsilon} \frac{1}{2^k} w'_{<\mu, g_k>}(\gamma) > \varepsilon/2\Big] \le \beta$$

for each $\gamma \le \gamma_0$ and $N \ge 1$. This together with (1.8) shows that

$$P^N\Big[w'_\mu(\gamma) \ge \varepsilon\Big] \le \beta$$

for each $\gamma \le \gamma_0$ and $N \ge 1$. □

2. The Hydrodynamic Equation

We are now in possession of all elements to examine the hydrodynamic behavior of the symmetric simple exclusion process.

Theorem 2.1 *Let $\rho_0 : \mathbb{T}^d \to [0,1]$ be an initial density profile and let μ^N be the sequence of Bernoulli product measures of slowly varying parameter associated to the profile ρ:*

$$\mu^N\{\eta;\, \eta(x) = 1\} \;=\; \rho_0(x/N), \quad x \in \mathbb{T}_N^d .$$

Then, for every $t > 0$, the sequence of random measures

$$\pi_t^N(du) \;=\; \frac{1}{N^d} \sum_{x \in \mathbb{T}_N^d} \eta_t(x)\, \delta_{x/N}(du)$$

converges in probability to the absolutely continuous measure $\pi_t(du) = \rho(t,u)\, du$ whose density is the solution of the heat equation:

$$\begin{cases} \partial_t \rho \;=\; (1/2)\, \Delta \rho \\ \rho(0,\cdot) \;=\; \rho_0(\cdot)\,. \end{cases} \tag{2.1}$$

Proof. We start fixing a time $T > 0$ and considering a sequence of probability measures Q^N on $D([0,T], \mathcal{M}_+)$ corresponding to the Markov process π_t^N, defined by (0.2), speeded up by N^2 and starting from μ^N.

First step (Relative compactness). We have seen in section 1 that the first step in the proof of the hydrodynamic behavior consists in showing that the sequence Q^N is relatively compact. Denote by $C^2(\mathbb{T}^d)$ the space of twice continuously differentiable functions $G : \mathbb{T}^d \to \mathbb{R}$. Of course $C^2(\mathbb{T}^d)$ is dense in $C(\mathbb{T}^d)$ for the uniform topology. By Proposition 1.7 it suffices to check that the sequence of measures corresponding to the real processes $< \pi_t^N, G >$ is relatively compact for all G in $C^2(\mathbb{T}^d)$.

Fix therefore a function G in $C^2(\mathbb{T}^d)$ and denote by $Q^{N,G}$ the probability measure $Q^N G^{-1}$ on $D([0,T], \mathbb{R})$. Since $< \pi_t^N, G >$ is a real process, we shall apply Theorem 1.3 and Proposition 1.6 with $\mathcal{E} = \mathbb{R}$ and δ the usual distance in $\mathbb{R}$. Since the total mass of the empirical measure π_t^N is bounded by 1, condition (1) of Theorem 1.3 is trivially verified. It remains to prove the second condition or, as we shall do, the one of Proposition 1.6. Recall from (0.3) that under Q^N we have the identity:

$$< \pi_t^N, G > = < \pi_0^N, G > + (1/2) \int_0^t < \pi_s^N, \Delta_N G > ds + M_t^{G,N} \tag{2.2}$$

where $M_t^G = M_t^{G,N}$ is a martingale. In particular, in order to prove (1.3) for $< \pi_t^N, G >$, we just need to prove the same relation for the integral term $\int_0^t < \pi_s^N, \Delta_N G > ds$ and for the martingale term M_t^G. We start with the former.

Of course, since G is of class C^2 and since the total mass of π_s^N is bounded by 1, the integral

$$\int_\tau^{\tau+\theta} < \pi_s^N , \Delta_N G > ds$$

is bounded above by $C(G)\theta$, whether τ is a stopping time or not. (From now on, $C(G)$ represents a finite constant depending only on G and that may change from line to line). In particular, the integral term on the right hand side of (2.2) satisfies the condition of Proposition 1.6.

To check condition (1.3) for the martingale $M_t^{G,N}$ we show that the expected value of its square converges to 0. Denote by $B_t^G = B_t^{G,N}$ the process given by

$$B_t^G \;=\; N^2 L_N < \pi_t^N, G >^2 \; - 2N^2 < \pi_t^N, G > L_N < \pi_t^N, G > \; .$$

By Lemma A1.5.1, the process $N_t^G = N_t^{G,N}$ defined by:

$$N_t^G \;=\; \left(M_t^G\right)^2 - \int_0^t B_s^G \, ds \; .$$

is a new martingale. Straightforward computations show that

$$B_s^G \;=\; \frac{1}{N^{2(d-1)}} \sum_{|x-y|=1} \left[G(y/N) - G(x/N)\right]^2 \eta_s(x)(1 - \eta_s(y)) \; .$$

Since τ is a stopping time we have that

$$E_{Q^N}\left[(M_{\tau+\theta} - M_\tau)^2\right] \;=\; E_{Q^N}\left[\int_\tau^{\tau+\theta} B_s^G \, ds\right] \;\le\; \frac{C(G)\theta}{N^d} \; .$$

Condition (1.3) follows from this estimate and Chebychev inequality. This concludes the proof that the sequence Q^N is relatively compact.

Second step (Uniqueness of limit points). Now that we know that the sequence Q^N is weakly relatively compact, it remains to characterize all limit points of Q^N. Let Q^* be a limit point and let Q^{N_k} be a sub-sequence converging to Q^*.

First of all, notice that the application from $D([0,T],\mathcal{M}_+)$ to $\mathbb{R}$ that associates to a trajectory $\{\pi_t, 0 \le t \le T\}$ the number

$$\sup_{t\le T} \left| < \pi_t, G > - < \pi_0, G > - (1/2)\int_0^t < \pi_s, \Delta G > ds \right|$$

is continuous as long as G is of class C^2. We therefore have that for every $\varepsilon > 0$

$$\begin{aligned}
&\liminf_{k\to\infty} Q^{N_k}\left(\sup_{t\le T}\left| < \pi_t, G > - < \pi_0, G > - (1/2)\int_0^t < \pi_s, \Delta G > \right| > \varepsilon\right) \\
&\quad \ge Q^*\left(\sup_{t\le T}\left| < \pi_t, G > - < \pi_0, G > - (1/2)\int_0^t < \pi_s, \Delta G > \right| > \varepsilon\right) \; .
\end{aligned}$$

since Q^{N_k} converges weakly to Q^* and since the above set is open.

At this point the same estimate on the martingale $(M_t^G)^2 - \int_0^t B_s^G \, ds$ obtained in the first part of the proof shows that every limit point Q^* is concentrated on trajectories such that

$$< \pi_t, G > = < \pi_0, G > + (1/2) \int_0^t < \pi_s, \Delta G > ds \tag{2.3}$$

for all $0 \le t \le T$. Indeed, by Chebychev and Doob inequality,

$$\begin{aligned} Q^N \left[\sup_{0 \le t \le T} \left| M_t^G \right| \ge \varepsilon \right] &\le 4\varepsilon^{-2} E_{Q^N} \left[(M_T^G)^2 \right] \\ &= 4\varepsilon^{-2} E_{Q^N} \left[\int_0^T B_s^G \, ds \right] \le \frac{C(G)T}{\varepsilon^2 N^d} \, . \end{aligned}$$

To conclude that all limit points are concentrated on trajectories π_t satisfying relation (2.3), we just need to point out that the difference between $< \pi_t^N, \Delta_N G >$ and $< \pi_t^N, \Delta G >$ is of order $o_N(1)$ because the total mass of π_t^N is bounded by 1 and G is of class C^2.

We now prove that all limit points Q^* of the sequence Q^N are concentrated on absolutely continuous measures with respect to the Lebesgue measure. Notice that

$$\sup_{0 \le t \le T} \left| < \pi_t^N, G > \right| \le \frac{1}{N^d} \sum_{x \in \mathbb{T}_N^d} |G(x/N)|$$

because there is at most one particle per site. Since, on the other hand, for fixed continuous functions G, the application that associates to a trajectory π the value $\sup_{0 \le t \le T} | < \pi_t, G > |$ is continuous, by weak convergence, all limit points are concentrated on trajectories π_t such that

$$\left| < \pi_t, G > \right| \le \int |G(u)| \, du$$

for all continuous function G and for all $0 \le t \le T$. Thus all limit points are concentrated on absolutely continuous trajectories with respect to the Lebesgue measure:

$$Q^* \left[\pi ; \ \pi_t(du) = \pi_t(u) du \right] = 1 \, .$$

Moreover, all limit points of the sequence Q^N are concentrated on trajectories that at time 0 are equal to $\rho_0(u)du$. Indeed, by weak convergence, for every $\varepsilon > 0$

$$Q^* \left[\left| \frac{1}{N^d} \sum_{x \in \mathbb{T}_N^d} G(x/N) \eta_0(x) \; - \; \int G(u)\rho_0(u)\, du \right| > \varepsilon \right]$$

$$\le \liminf_{k\to\infty} Q^{N_k} \left[\left| \frac{1}{N^d} \sum_{x \in \mathbb{T}_N^d} G(x/N) \eta_0(x) \; - \; \int G(u)\rho_0(u)\, du \right| > \varepsilon \right]$$

$$= \lim_{k\to\infty} \mu^{N_k} \left[\left| \frac{1}{N^d} \sum_{x \in \mathbb{T}_N^d} G(x/N) \eta(x) \; - \; \int G(u)\rho_0(u)\, du \right| > \varepsilon \right] = 0\, .$$

The two previous results show that every limit point is concentrated on absolutely continuous trajectories $\pi_t(du) = \pi_t(u)du$ whose density is a weak solution, in the sense (2.3), of the heat equation. To prove, however, an uniqueness result of weak solutions for the heat equation, we need to prove relation (2.3) for time dependent functions G. For positive integers m and n, denote by $C^{m,n}([0,T] \times \mathbb{T}^d)$ the space of continuous functions with m continuous derivatives in time and n continuous derivatives in space. For $G : [0,T] \times \mathbb{T}^d \to \mathbb{R}$ of class $C^{1,2}$, consider the martingale $M_t^G = M_t^{G,N}$ given by

$$M_t^G \;= < \pi_t^N, G_t > \; - \; < \pi_0^N, G_0 > \; - \; \int_0^t (\partial_s + N^2 L_N) < \pi_s^N, G_s > ds\, .$$

By Lemma A1.5.1, the process $N_t^G = N_t^{G,N}$ defined by

$$N_t^G \;=\; \left(M_t^G\right)^2 - N^2 \int_0^t A_G(s)\, ds$$

$$\text{with} \quad A_G(s) \;=\; L_N < \pi_s^N, G_s >^2 \; - 2 < \pi_s^N, G_s > L_N < \pi_s^N, G_s >$$

is a martingale with respect to the natural filtration. In particular, as before,

$$\limsup_{N\to\infty} Q^N \left[\left| M_t^G \right| \; > \; \varepsilon \right] \;=\; 0$$

for every $\varepsilon > 0$. Since, on the other hand,

$$M_t^G \;= < \pi_t^N, G_t > \; - \; < \pi_0^N, G_0 > \; - \; \int_0^t < \pi_s^N, \partial_s G_s - (1/2)\Delta_N G_s > ds\, ,$$

all limit points are concentrated on paths $\{\pi_t,\ 0 \le t \le T\}$ such that

$$< \pi_t, G_t > = < \pi_0, G_0 > \; + \; \int_0^t < \pi_s, \partial_s G_s - (1/2)\Delta G_s > ds\, , \qquad (2.4)$$

In conclusion, all limit points are concentrated on absolutely continuous trajectories $\pi_t(du) = \pi_t(u)du$ that are weak solutions of the heat equation in the sense of (2.4) and whose density at time 0 is $\rho_0(\cdot)$.

Third step (Uniqueness of weak solutions of the heat equation). We turn now to the question of uniqueness of weak solutions. We first fix the terminology on weak solutions of partial differential equations. Let $\mathcal{L}$ be a second order differential operator acting only on space variables:

$$\mathcal{L}\rho \;=\; \sum_{1\le i,j\le d} \partial^2_{u_i,u_j} A_{i,j}(\rho) \;+\; \sum_{1\le i\le d} \partial_{u_i} B_i(\rho) \;+\; C(\rho)$$

for smooth functions $A_{i,j}$, B_i and C. In this chapter, for example, $\mathcal{L}$ represents one half of the Laplacian.

Definition 2.2 For a bounded initial profile $\rho_0 \colon \mathbb{T}^d \to \mathbb{R}$, a bounded function $\rho \colon [0,T] \times \mathbb{T}^d \to \mathbb{R}$ is a weak solution of the Cauchy problem

$$\begin{cases} \partial_t \rho \;=\; \mathcal{L}\rho \\ \rho(0,\cdot) = \rho_0(\cdot) \end{cases}$$

if for every function $G \colon [0,T] \times \mathbb{T}^d \to \mathbb{R}$ of class $C^{1,2}$

$$\begin{aligned} &\int_{\mathbb{T}^d} G(T,u)\rho(T,u)\, du \;-\; \int_{\mathbb{T}^d} G(0,u)\rho_0(u)\, du \\ &\quad = \int_0^T dt \int_{\mathbb{T}^d} du \Big\{ \rho\partial_t G \;+\; \sum_{1\le i,j\le d} A_{i,j}(\rho)\partial^2_{u_i,u_j} G \Big\} \\ &\quad\quad - \int_0^T dt \int_{\mathbb{T}^d} du \sum_{1\le i\le d} \Big\{ B_i(\rho)\partial_{u_i} G \;+\; C(\rho) G \Big\} \,. \end{aligned}$$

It was proved in the previous section that every limit point of the sequence Q^N is concentrated on weak solutions of the heat equation with initial profile ρ_0. Therefore, to conclude the proof of the uniqueness of limit points, it remains to show that there exists only one weak solution of this equation.

There exists many methods. Brezis and Crandall (1979) proved such a result for a class of quasi-linear second order equations. Their theorem gives us immediately the result since Definition 2.2 of weak solutions implies their condition (1.23) and since the limit condition on the boundedness in L^1 is automatically satisfied because the total mass in simple exclusion processes is conserved and is at most 1.

We present in Appendix A2.4 a uniqueness result based on the investigation of the time evolution of the H_{-1} norm. This method requires, however, supplementary properties of weak solutions that are not difficult to check in the case of symmetric simple exclusion processes.

Finally, since the hydrodynamic equation of symmetric simple exclusion is linear, the methods developed by Oelschläger (1985) give a third possible approach.

Note that for any bounded profile ρ_0 the heat equation admits a strong solution given by

$$\rho(t,u) = \int \bar{\rho}_0(v)\, G_t(u-v)\, dv$$

if $\bar{\rho}_0 : \mathbb{R}^d \to \mathbb{R}$ represents the $\mathbb{T}^d$-periodic function identically equal to ρ_0 on $\mathbb{T}^d$ and if $G_t(w)$ is the usual d-dimensional Gaussian kernel: $G_t(w) = (2\pi t)^{-d/2} \exp\{-(1/2t)|w|^2\}$. In particular the weak solution is in fact a strong solution.

In conclusion, with any of these uniqueness results, we proved that the sequence Q^N converges to the Dirac measure concentrated on this strong solution.

Fourth step (Convergence in probability at fixed time). Even if in general it is false that the application from $D([0,T], \mathcal{M}_+)$ to $\mathcal{M}_+$ obtained by taking the value at time $0 < t < T$ of the process is continuous, this statement is true if the process is almost surely continuous at time t for the limiting probability measure. In the present context, the limiting probability measure is concentrated on weakly continuous trajectories. Thus π_t^N converges in distribution to the deterministic measure $\pi_t(u)du$. Since convergence in distribution to a deterministic variable implies convergence in probability, the theorem is proved. □

In the previous proof, the initial state μ^N appeared only in the second step. It was necessary to show that the limit points Q^* were concentrated on trajectories $\pi_t(du)$ that at time 0 were given by

$$\pi_0(du) = \rho_0(u)\, du\ .$$

Therefore, the special structure of the measure μ^N did not play any particular role in the proof and the hypothesis of Theorem 2.1 concerning the initial state can be considerably relaxed:

Theorem 2.3 *For $\rho_0 : \mathbb{T}^d \to [0,1]$, consider a sequence of measures $(\mu^N)_{N\ge 1}$ on $\{0,1\}^{\mathbb{T}_N^d}$ associated to the profile ρ_0:*

$$\limsup_{N\to\infty} \mu^N \left[\left| \frac{1}{N^d} \sum_x G(x/N)\eta(x) - \int G(u)\rho_0(u)\, du \right| > \delta \right] = 0\ .$$

for every $\delta > 0$ and every continuous function $G : \mathbb{T}^d \to \mathbb{R}$. The conclusions of Theorem 2.1 remain in force.

Thus, under the hypothesis of a weak law of large numbers at time 0 for the empirical measure π^N, we have proved a law of large numbers for any later time t.

Remark 2.4 In the proof of the hydrodynamic behavior, we took advantage of many of the special features of symmetric simple exclusion processes. We will try to point out here the elements that were used.

(a) From the conservation of the total mass, condition (1) of the compactness criterion presented in Theorem 1.3 refers only to the total initial mass. This condition was therefore easy to check.

(b) By a summation by parts we have been able to rewrite the integral term $\int_0^T N^2 L_N < \pi_t^N, G > dt$ as a sum involving the discrete Laplacian of G. This property is shared by many processes. Indeed, for a nearest neighbor system conserving the total number of particles we often have (in dimension one) that

$$L\eta(x) = W_{x-1,x}(\eta) - W_{x,x+1}(\eta)$$

where $W_{x,x+1}$ stands for the instantaneous current of particles from x to $x+1$. In few particular cases, like in symmetric simple exclusion processes where $W_{x,x+1}(\eta) = \eta(x) - \eta(x+1)$, the current is itself a discrete gradient of another cylinder function. This condition known as the "gradient condition" is the one that permitted to perform a second discrete integration by parts to cancel the factor N^2 that appeared from time renormalization.

(c) After these two summations by parts we were able to rewrite the integral term $\int_0^T N^2 L_N < \pi_t^N, G > dt$ as a function of the empirical measure. We obtained in this way a closed equation for the empirical measure. This is a very special feature of symmetric simple exclusion processes. In general, we obtain a correlation field and the main difficulty in the proof is to close the equation, that is, to replace the correlation field by a function of the empirical measure.

In the next two chapters we will examine the hydrodynamic behavior of a gradient system whose correlation field is not a function of the density field and in Chapter 7 we investigate a nongradient system.

Since gradient systems appear several times in the next chapters we introduce the following terminology. For two sites x and y we denote by $W_{x,y}(\eta)$ the instantaneous algebraic current between sites x and y, that is, the rate at which a particle jumps from x to y minus the rate at which a particle jumps from y to x. In the nearest neighbor symmetric simple exclusion process, for example, the current $W_{0,e_1}(\eta)$ is equal to $p(e_1)[\eta(0) - \eta(e_1)]$.

Definition 2.5 Let E denote a subset of $\mathbb{N}$. A translation invariant nearest neighbor particle system (η_t) on $E^{\mathbb{T}_N^d}$ with generator L_N is said to be gradient if there exists a positive integer n_0, cylinder functions $\{h_{i,n},\ 1 \le i \le d,\ 1 \le n \le n_0\}$ and finite range functions $\{p_{i,n},\ 1 \le i \le d,\ 1 \le n \le n_0\}$ such that

$$W_{0,e_i}(\eta) = \sum_{n=1}^{n_0} \sum_{x \in \mathbb{T}_N^d} p_{i,n}(x) \tau_x h_{i,n}(\eta) \quad \text{and}$$

$$\sum_{x \in \mathbb{T}_N^d} p_{i,n}(x) = 0 \qquad \text{for every } 1 \le n \le n_0$$

for every $1 \le i \le d$.

This definition can of course be extended to processes that have finite range transition probabilities. Notice that for a function $G\colon \mathbb{T}^d \to \mathbb{R}$ of class C^2 the gradient condition allows two discrete integrations by part in the integral term of the martingale

$$M_t^{G,N} = < \pi_t^N, G > - < \pi_0^N, G > - \int_0^t N^2 L_N < \pi_s^N, G > ds$$

defined in the beginning of this section.

3. Comments and References

Tracer particles and self diffusion. The diffusion of a tracer particle is a standard problem in non equilibrium statistical mechanics. Consider a large system of interacting particles and add one, the tracer particle. Denote by X_t its position at time t. One expects X_t to behave as a Brownian motion with some diffusion coefficient on a large space and time scale.

To fix ideas consider a mean-zero, asymmetric simple exclusion process $\{\eta_t, t \geq 0\}$ on $\mathbb{Z}^d$ with transition probability $p(\cdot)$ ($\sum_x p(x) = 1$, $\sum_x x p(x) = 0$). The generator L of this process acts on cylinder functions f as

$$(Lf)(\eta) = \sum_{x,y \in \mathbb{Z}^d} p(y)\eta(x)[1 - \eta(x+y)][f(\eta^{x,x+y}) - f(\eta)] \, .$$

We refer to Liggett (1985) for a proof of the existence of this process. We start the evolution with one tagged particle at the origin and a Bernoulli product measure with density α on $\mathbb{Z}^d - \{0\}$. Denote by X_t the position of the tracer particle at time t. One would like to prove that X_{tN^2}/N converges, as $N \uparrow \infty$, to a Brownian motion with diffusion matrix $D = D(\alpha)$ (the so called self diffusion matrix).

To investigate the behavior of the tracer particle it is more convenient to describe the system in terms of the position of the tagged particle and the configuration ξ seen from the tagged particle. It is easy to check that the process as seen from the tagged particle, i. e., in which the tagged particle is fixed to be at the origin, is itself Markovian. Its generator has two pieces, the first one corresponds to the jumps of the tracer particle and the second one corresponds to all other jumps. The generator of the process becomes

$$\begin{aligned}(\mathcal{L}f)(\xi) = & \sum_{x \in \mathbb{Z}^d} p(x)[1 - \xi(x)][f(T_x\xi) - f(\xi)] \\ & + \sum_{\substack{x \neq 0 \\ y \in \mathbb{Z}^d}} p(y)\xi(x)[1 - \xi(x+y)][f(\xi^{x,x+y}) - f(\xi)] \, ,\end{aligned}$$

where $T_x\xi$ is the configuration obtained from ξ letting one particle jump from 0 to x and then translating the new configuration by x: $T_x\xi = \tau_x\xi^{0,x}$. If the transition probability is symmetric and of finite range ($p(x) = p(-x)$, $p(x) = 0$ for x large enough), the Bernoulli product measure with density α conditioned to have a particle at the origin is reversible and ergodic for this process. (For zero range processes with the same assumptions on the transition probability, the probability measure $\hat{\nu}_\alpha(d\xi) = \{\xi(0)/\alpha\}\nu_\alpha(d\xi)$ is ergodic and reversible for the process as seen from the tagged particle.)

Recall that X_t stands for the position of the tracer particle at time t. For a fixed θ in $\mathbb{R}^d$, a simple computation shows that $\mathcal{L}(X_t \cdot \theta) = \sum_z p(z)(z \cdot \theta)[1 - \xi(z)] = \Psi_\theta(\xi)$. In this formula $(u \cdot v)$ stands for the inner product of u and v in $\mathbb{R}^d$. It follows from Lemma A1.5.1 that $X_t \cdot \theta = \int_0^t \Psi_\theta(\xi_s)\, ds + N_t^\theta$, where N^θ is a martingale. Since there exists a wide class of central limit theorems for martingales (cf. Helland (1982) or Durrett (1991)), the problem is reduced to the analysis of the asymptotic behavior of $t^{-1/2} \int_0^t \Psi_\theta(\xi_s)\, ds$. To this end Kipnis and Varadhan (1986) proved a central limit theorem for additive functionals of reversible Markov processes:

Theorem 3.1 (Kipnis and Varadhan). *Let Y_t be a Markov process on a state space E with generator L. Let ν be an invariant measure and assume that ν is reversible and ergodic. Let $V: E \to \mathbb{R}$ be a mean-zero function in $L^2(\nu)$ $(E_\nu[V] = 0)$ for which there exists a finite constant $C = C(V)$ such that for every f in $L^2(\nu)$,*

$$\big| < V, f >_\nu \big| \;\le\; C < -Lf, f >_\nu \;. \tag{3.1}$$

Here $< \cdot, \cdot >_\nu$ stands for the inner product on $L^2(\nu)$. Then, there exists a $\mathbb{P}_\nu$-square integrable martingale M_t with stationary increments, measurable with respect to the natural σ-algebra and such that $M_0 = 0$ and

$$\lim_{t\to\infty} \frac{1}{\sqrt{t}} \sup_{0\le s\le t} \Big| \int_0^s V(Y_r)\, dr - M_s \Big| \;=\; 0$$

in $\mathbb{P}_\nu$ probability.

Since in the symmetric case the Bernoulli measures on $\{0,1\}^{\mathbb{Z}^d - \{0\}}$ are ergodic and reversible, this theorem reduces the proof of the central limit theorem for the the tagged particle to the verification of property (3.1) for the function $\Psi_\theta(\xi)$. This follows from an elementary computation involving the generator $\mathcal{L}$.

The proof of Theorem 3.1 relies on spectral analysis. For $\lambda > 0$, consider the resolvent equation

$$\lambda u_\lambda - L u_\lambda = V \;.$$

With this notation we may rewrite the integral $\int_0^t V(Y_s)\, ds$ as

$$\int_0^t V(Y_s)\, ds \;=\; M_\lambda(t) \;+\; \big\{u_\lambda(Y_0) - u_\lambda(Y_t)\big\} \;+\; \int_0^t \lambda u_\lambda(Y_s)\, ds \;,$$

where $M_\lambda(t)$ is a martingale with stationary increments. It follows from assumption (3.1) that $M_\lambda(t)$ (resp. $\int_0^t \lambda u_\lambda(Y_s)\, ds$ and $u_\lambda(Y_0) - u_\lambda(Y_t)$) converges in $L^2(\mathbb{P}_\nu)$, as $\lambda \downarrow 0$, to some martingale $M(t)$ (resp. 0 and some process ζ_t). To conclude the proof of the theorem it remains to show that $t^{-1/2} \sup_{0\le s\le t} |\zeta_s|$ converges to 0 in probability as $t \uparrow \infty$. This is more demanding. The argument can be found in Kipnis and Varadhan (1986).

This is in substance the method that allowed Kipnis and Varadhan to prove a central limit theorem for the position of a tagged particle for symmetric simple exclusion processes. It follows from their proof that the self diffusion matrix D is non degenerated in dimension $d \geq 2$ and in dimension $d = 1$ if $p(1) + p(-1) < 1$. We discuss below the behavior of the tracer particle in the one-dimensional nearest neighbor exclusion process.

De Masi, Ferrari, Goldstein and Wick (1985, 1989) extended Kipnis and Varadhan central limit theorem for additive functionals to antisymmetric random variables of reversible Markov processes. Goldstein (1995) presents a simpler proof of this latter result. Varadhan (1995) extended Kipnis and Varadhan's method to a non reversible context, the case of mean-zero asymmetric simple exclusion processes.

Spohn (1990) obtains a variational formula for the diffusion matrix $D(\alpha)$. (cf. also Quastel (1992)). This variational formula permits to show that for exclusion processes where the jump rates depend on the configuration, the so called Kawasaki dynamics, the diffusion matrix is still non degenerated for all reversible dynamics that are not nearest neighbor one-dimensional. Varadhan (1994b) proves that the self diffusion coefficient $D(\alpha)$ is Lipschitz continuous in dimension $d \geq 3$ and that there exist constants $0 < C_0 < C_1 < \infty$, such that $C_0(1-\alpha) \leq D(\alpha) \leq C_1(1-\alpha)$ in the matrix sense in all dimensions. Asselah, Brito and Lebowitz (1997) prove bounds on $D(\alpha)$ in terms of the range of the jumps.

Grigorescu (1997a,b) investigates the evolution of one-dimensional independent Brownian motions. Particles evolve independently until they touch. At this point, with a small probability ε particles collide and with probability $1-\varepsilon$ particles do not collide. Grigorescu (1997a,b) proves a central limit theorem for a tagged particle starting from a nonequilibrium state. In this context the evolution of the tagged particle depends on the evolution of the all macroscopic profile, described by the hydrodynamic equation.

Siri (1996) considers space inhomogeneous symmetric zero range processes where a particle at site x jumps at rate $g(x/N, \eta(x))p(y)$ to site $x + y$. Here $p(\cdot)$ is a finite range symmetric transition probability and $g: \mathbb{R}^d \times \mathbb{N} \to \mathbb{R}_+$ a jump rate smooth in the first coordinate. She proves a central limit theorem for the tracer particle in this context and obtains a space dependent diffusion matrix. Shiga (1988) investigates a class of particle systems on $\mathbb{Z}$ in which the tagged particle converges to a Gaussian random variable with variance of order $t^{1/a}$ for $1 < a \leq 2$. In these models more than one particle may jump simultaneously. Carlson, Grannan and Swindle (1993) prove a central limit theorem for a tagged particle in a one-dimensional symmetric exclusion process in which a particle at x jumps to $x + y$ at rate $c(|y|)$ provided all sites between x and y are occupied.

For the one-dimensional, symmetric, nearest neighbor simple exclusion process Arratia (1983) proved that the right space rescaling is $t^{1/4}$ instead of $\sqrt{t}$. Harris (1965) already observed that this subdiffusive behavior is peculiar to one-dimensional, nearest neighbor processes. Arratia (1983) proved that starting from a Bernoulli product measure with density α and conditioned to have a particle at the origin, $X_t/t^{1/4}$ converges to a mean-zero Gaussian distribution with variance equal to

$$\sqrt{\frac{2}{\pi}}\frac{1-\alpha}{\alpha} \,. \tag{3.2}$$

Rost and Vares (1985), relating this process to a zero range process, prove that $X_{tN^2}/\sqrt{N}$ converges in the Skorohod space to a fractional Brownian motion. Landim, Olla and Volchan (1997), (1998), considered the evolution of an asymmetric tagged particle, jumping with probability p to the right and $q = 1 - p$ to the left, in an infinite, one-dimensional system of particles evolving according to nearest neighbor, symmetric simple exclusion process. They proved that the diffusively rescaled position of the asymmetric particle, $(X_{tN^2} - X_0)/N$, converges in probability to a deterministic function v_t that depends only on the initial profile. In the particular case where the initial profile is constant equal to α, they proved that $v(t) = v\sqrt{t}$, where

$$\lim_{p-q\to 0} \frac{v}{p-q} = \sqrt{\frac{2}{\pi}}\frac{1-\alpha}{\alpha} \,.$$

This result, which relates the mobility $v/(p-q)$ of an asymmetric particle in a symmetric environment to the diffusivity (3.2) of a tagged symmetric particle, can be interpreted as an Einstein relation.

Einstein relations. Consider a tagged particle, whose position at time t is denoted by X_t, in a symmetric environment. Denote by D its self diffusion coefficient. When the tracer particle is submitted to an external field E, that does not alter the evolution of the environment, it starts to move with a drift. Denote by $v(E)$ the mean drift when the environment is in equilibrium: $v(E) =< X_t > /t$ and set $\sigma = \lim_{E\to 0} v(E)/E$. Einstein relation states that $\sigma = \beta D$, where β is the inverse temperature.

Ferrari, Goldstein and Lebowitz (1985) proved Einstein relation for particle systems for which there exists a unique invariant measure for the process as seen from the tagged particle. They considered, for instance, exclusion processes on cubes with periodic boundary conditions. Buttà (1993) proved the Einstein relation in the case of an interface dynamics. Lebowitz and Rost (1994) proposed a general approach to derive the Einstein relation for interacting particle systems and applied it to several distinct class of models.

We refer to Lebowitz and Spohn (1982a), Spohn (1991) for the physical aspects of the evolution of tracer particles. Ferrari (1996) presents a review on limit theorems for tagged particles. We discuss in the last section of Chapter 8 the behavior of tagged particles in asymmetric interacting particle systems.

Propagation of chaos. Rezakhanlou (1994b) proved the propagation of chaos for symmetric simple exclusion processes and Quastel, Rezakhanlou and Varadhan (1997) proved the corresponding large deviations principle in dimension $d \geq 3$. Consider a symmetric simple exclusion process with transition probability $p(\cdot)$

evolving on $\mathbb{T}_N^d$. Fix a profile $\rho_0 : \mathbb{T}^d \to [0,1]$ and denote by μ^N a sequence of probability measures associated to ρ_0 in the sense that

$$\lim_{N\to\infty} E_{\mu^N}\Big[\Big| \frac{1}{N^d} \sum_{x\in\mathbb{T}_N^d} H(x/N)\eta(x) - \int H(u)\rho_0(u)du \Big| \Big] = 0$$

for every continuous function $H : \mathbb{T}^d \to \mathbb{R}_+$.

Fix a time $T > 0$ and let L stand for the (random) total number of particles at time 0 ($L/N^d \to \int_{\mathbb{T}^d} \rho_0(u)du$ as $N \uparrow \infty$). Tag all particles at time 0. Denote by $x_i(t)$, $1 \le i \le L$, the position of the i-th particle at time t and by $y_i^N(t)$ the rescaled position at time t of the i-th particle: $y_i^N(t) = N^{-1}x_i(tN^2)$. Let R_N be the empirical measure associated to these trajectories up to time T:

$$R_N = \frac{1}{N^d}\Big\{ \delta_{y_1^N(\cdot)} + \cdots + \delta_{y_L^N(\cdot)} \Big\} .$$

Notice that for each $N \ge 1$, R_N is a measure on $D([0,T], \mathbb{T}^d)$. We may thus interpret R_N as a random variable taking values in $\mathcal{M}(D([0,T], \mathbb{T}^d))$. Denote by Q_N the probability measure on $\mathcal{M}(D([0,T], \mathbb{T}^d))$ induced by R_N and μ^N.

Denote by R_{ρ_0} the distribution on $C(\mathbb{R}_+, \mathbb{T}^d)$ corresponding to a diffusion process on $\mathbb{T}^d$ with generator $\mathcal{L}$ given by

$$\mathcal{L} = \frac{1}{2}\nabla \cdot S(\rho(t,u))\nabla + \frac{1}{2}[S(\rho(t,u)) - D]\frac{\nabla \rho(t,u)}{\rho(t,u)} \cdot \nabla$$

and initial distribution $\rho_0(u)du$. Here $\rho(t,u)$ is the solution of the hydrodynamic equation $\partial_t \rho = (1/2)\nabla \cdot D\nabla\rho$, D is the covariance matrix of the transition probability $p(\cdot)$ and $S(\alpha)$ is the self–diffusion coefficient of the symmetric simple exclusion process with density α.

Rezakhanlou (1994b) proved that R_N converges in distribution (and thus in probability) to the (constant) random variable $\delta_{R_{\rho_0}}$: For every continuous function $J : D(\mathbb{R}_+, \mathbb{T}^d) \to \mathbb{R}$,

$$\lim_{N\to\infty} E_{Q_N}\Big[\Big| \frac{1}{L}\sum_{j=1}^{L} J(y_j^N(\cdot)) - \int J dR_{\rho_0} \Big| \Big] = 0 .$$

Quastel, Rezakhanlou and Varadhan (1997) proved a large deviations principle for R_N in dimension $d \ge 3$. More precisely, they proved the existence of a lower semi–continuous rate function $\mathcal{I} : \mathcal{M}(D([0,T], \mathbb{T}^d)) \to \mathbb{R}_+$ which has compact level sets such that

$$\limsup_{N\to\infty} Q_N[R_N \in F] \le - \inf_{m\in F} \mathcal{I}(m)$$
$$\liminf_{N\to\infty} Q_N[R_N \in G] \ge - \inf_{m\in G} \mathcal{I}(m)$$

for all closed subsets F and open subsets G of $\mathcal{M}(D([0,T], \mathbb{T}^d))$.

5. An Example of Reversible Gradient System: Symmetric Zero Range Processes

In this chapter we investigate the hydrodynamic behavior of reversible gradient interacting particle systems. To keep notation simple and to avoid minor technical difficulties, we consider the simplest prototype of reversible gradient system: the nearest neighbor symmetric zero range process. The generator of this Markov process is given by

$$(L_N f)(\eta) = (1/2) \sum_{x\in\mathbb{T}_N^d} \sum_{|z|=1} g(\eta(x))\,[f(\eta^{x,x+z}) - f(\eta)] \tag{0.1}$$

for cylinder functions $f : \mathbb{N}^{\mathbb{T}_N^d} \to \mathbb{R}$.

Recall from the last chapter the strategy adopted to prove the hydrodynamic behavior of symmetric simple exclusion processes. The argument relied on the analysis of some martingales associated to the empirical measure. More precisely, to each smooth function $G : \mathbb{T}^d \to \mathbb{R}$ of class $C^2(\mathbb{T}^d)$, we considered the martingales $M_t^{G,N} = M_t^G$ and its quadratic variation $N_t^{G,N} = N_t^G$ defined by

$$\begin{aligned} M_t^G &= <\pi_t^N, G> \;-\; <\pi_0^N, G> \;-\; \int_0^t N^2 L_N <\pi_s^N, G>\, ds \\ N_t^G &= (M_t^G)^2 \\ &\quad - \int_0^t \Big\{ N^2 L_N <\pi_s^N, G>^2 \;-\; 2 <\pi_s^N, G> N^2 L_N <\pi_s^N, G> \Big\}\, ds\,. \end{aligned}$$

In the context of nearest neighbor symmetric zero range processes, these martingales can be rewritten as

$$\begin{aligned} M_t^G &= \frac{1}{N^d} \sum_{x\in\mathbb{T}_N^d} G(x/N)\,\eta_t(x) - \frac{1}{N^d} \sum_{x\in\mathbb{T}_N^d} G(x/N)\,\eta_0(x) \\ &\quad - \int_0^t \frac{1}{2N^{d-2}} \sum_{x\in\mathbb{T}_N^d} G(x/N) \sum_{|y-x|=1} \Big[g(\eta_s(y)) - g(\eta_s(x)) \Big]\, ds \end{aligned}$$

and

$$N_t^G = (M_t^G)^2 - \int_0^t \frac{1}{2N^{2d-2}} \sum_{|y-x|=1} \big[G(y/N) - G(x/N) \big]^2\, g(\eta_s(x))\, ds\,.$$

In the same way as we did for symmetric simple exclusion processes, we perform discrete integrations by parts to rewrite the integral term of the martingale M_t^G as:

$$\int_0^t \frac{1}{2N^d} \sum_{x \in \mathbb{T}_N^d} \Delta_N G(x/N)\, g(\eta_s(x))\, ds\,,$$

where Δ_N represents the discrete Laplacian defined in the previous chapter. Notice that it is the gradient condition that allowed us to perform two integrations by parts. This is easy to see in dimension 1 where the instantaneous current over the bond $\{0,1\}$, i.e., the rate at which a particle jumps from 0 to 1 minus the rate at which a particle jumps from 1 to 0, is equal to

$$\frac{1}{2}\Big\{ g(\eta(0)) - g(\eta(1)) \Big\}\,.$$

This time, however, it is not the density field itself that appears in the integral part of the martingale M_t^G but another local function of the configuration: $g(\eta(0))$. The main problem consists therefore in using the particular characteristics of our model to replace $g(\eta(x))$ by an ad hoc function of the density field in order to "close" the equation.

Recall that we have parametrized the invariant and translation invariant measures ν_α of the zero range processes by the density:

$$E_{\nu_\alpha}\big[\eta(0)\big] \;=\; \alpha\,.$$

Moreover, we denoted by $\Phi(\alpha)$ the expected value of the jump rate $g(\eta(0))$:

$$E_{\nu_\alpha}\big[g(\eta(0))\big] \;=\; \Phi(\alpha)\,.$$

The function Φ plays a crucial role in the hydrodynamic limit. We shall present a rigorous proof of the following intuitive argument. Consider a box of length εN around a macroscopic point $u \in \mathbb{T}^d$. Since the variation of the total number of particles in this box is due to surface effects, the system quickly reaches equilibrium in $[(u-\varepsilon)N, (u-\varepsilon)N]^d$ before the total number of particles has had time to change in a discernible way. We would like therefore to replace, in the average,

$$\frac{1}{(2N\varepsilon+1)^d} \sum_{|y-x|\le \varepsilon N} g(\eta_s(y))$$

by the expectation of the function g under the invariant measure of density

$$\frac{1}{(2\varepsilon N+1)^d} \sum_{|y-x|\le \varepsilon N} \eta_s(y)$$

that is, from the definition of Φ, by

$$\Phi\left(\frac{1}{(2N\varepsilon+1)^d} \sum_{|y-x|\le \varepsilon N} \eta_s(y) \right)\,.$$

Such a replacement allows to close the equation since the density of particles in a box of macroscopic length (εN above) is a function of the empirical measure. In fact, if $\iota_\varepsilon(\cdot)$ stands for the approximation of the identity defined by $\iota_\varepsilon(\cdot) = (2\varepsilon)^{-d}\mathbf{1}\{[-\varepsilon,\varepsilon]^d\}(\cdot)$, $(2\varepsilon N+1)^{-d}\sum_{|y|\le\varepsilon N}\eta(y)$ is equal to the integral of ι_ε with respect to the empirical measure:

$$\frac{1}{(2\varepsilon N+1)^d}\sum_{|y|\le\varepsilon N}\eta(y) = \frac{(2N\varepsilon)^d}{(2\varepsilon N+1)^d} < \pi^N, \iota_\varepsilon > .$$

From this heuristic argument we may guess the hydrodynamic equation of nearest neighbor symmetric zero range processes. The macroscopic behavior of the system should be described by the non linear heat equation:

$$\begin{cases} \partial_t\rho = (1/2)\,\Delta(\Phi(\rho)) \\ \rho(0,\cdot) = \rho_0(\cdot)\,. \end{cases}$$

Once we performed the replacement of the local field $N^{-d}\sum_x \Delta_N G(x/N)\, g(\eta(x))$ by the density field, to conclude the proof of the hydrodynamic behavior, we just need to repeat the argument presented in last chapter. There is just an additional difficulty: we have to prove that the empirical measure converges to an absolutely continuous measure with respect to the Lebesgue measure since nothing a priori prevents particles to accumulate at one site creating a Dirac measure. This phenomenon could be discarded for exclusion processes because we allowed at most one particle per site.

Throughout this chapter we repeatedly apply results on the relative entropy and the Dirichlet form presented in Appendix 1.

1. The Law of Large Numbers

To state the main theorem of this chapter, we lack just a few notations and certain hypotheses.

At Lemma 1.3, where we prove that all limit probability measures are concentrated on the subset of absolutely continuous measures with respect to the Lebesgue measure, and at Lemma 3.2, where we replace cylinder fields by functions of the empirical measure, we need the family of invariant measures $\{\nu_\alpha,\ \alpha\ge 0\}$ to have all exponential moments or, in an equivalent way, the partition function $Z(\cdot)$ to be finite on $\mathbb{R}_+$. We therefore suppose throughout this section that

(**FEM**) The partition function $Z(\cdot)$ introduced in Definition 2.3.1 is finite on $\mathbb{R}_+$.

Throughout this chapter, for a fixed time T and for a probability μ^N in $\mathbb{N}^{\mathbb{T}_N^d}$, we represent by $\mathbb{P}^N_{\mu^N} = \mathbb{P}_{\mu^N}$ the probability on the path space $D([0,T],\mathbb{N}^{\mathbb{T}_N^d})$ corresponding to the nearest neighbor symmetric zero range process with generator

L_N, defined at (0.1) above, accelerated by N^2 and starting from the initial measure μ^N.

Theorem 1.1 *Assume the jump rate $g(\cdot)$ to increase at least linearly: there exists a positive constant a_0 such that $g(k) \geq a_0 k$ for all $k \geq 0$. Let $\rho_0: \mathbb{T}^d \to \mathbb{R}_+$ be an integrable function with respect to the Lebesgue measure. Let μ^N be a sequence of probability measures on $\mathbb{N}^{\mathbb{T}_N^d}$ associated to the profile ρ_0 and for which there exist positive constants K_0, K_1 and α^* such that the relative entropy of μ^N with respect to ν_{α^*} is bounded by $K_0 N^d$:*

$$H(\mu^N|\nu_{\alpha^*}) \leq K_0 N^d \tag{1.1}$$

and

$$\limsup_{N\to\infty} E_{\mu^N}\Big[N^{-d} \sum_{x\in\mathbb{T}_N^d} \eta(x)^2\Big] \leq K_1 . \tag{1.2}$$

Then, for every $t \leq T$, for every continuous function $G: \mathbb{T}^d \to \mathbb{R}$ and for every δ strictly positive,

$$\lim_{N\to\infty} \mathbb{P}_{\mu^N}\left[\left|\frac{1}{N^d}\sum_{x\in\mathbb{T}_N^d} G(x/N)\eta_t(x) - \int_{\mathbb{T}^d} G(u)\rho(t,u)\,du\right| > \delta\right] = 0 ,$$

where $\rho(t,u)$ is the unique weak solution of the non linear heat equation

$$\begin{cases} \partial_t \rho = (1/2)\,\Delta(\Phi(\rho)) , \\ \rho(0,\cdot) = \rho_0(\cdot) . \end{cases} \tag{1.3}$$

Remark 1.2 In the assumption regarding the relative entropy, the parameter α^* does not play any special role in the sense that if this condition is verified for α^*, we can easily show that it is also satisfied for every other choice of the parameter. Indeed, for each density $\alpha > 0$, the entropy inequality gives that

$$H(\mu^N|\nu_\alpha) \leq (1+\gamma^{-1})H(\mu^N|\nu_{\alpha^*}) + \frac{1}{\gamma}\log\int\Big(\frac{d\nu_{\alpha^*}}{d\nu_\alpha}\Big)^\gamma d\nu_{\alpha^*}$$

for every $\gamma > 0$. Since the measures ν_{α^*} and ν_α are product, the second term of the right hand side can be explicitly computed. It is not difficult to show that this term is finite and of order N^d for $\gamma = \gamma(\alpha, \alpha^*)$ small enough because by (2.3.9) each measure ν_α has finite exponential moments.

On the other hand, this assumption on the relative entropy is naturally satisfied by every product measure with slowly varying parameter associated to a bounded profile.

Remark 1.3 We have seen in Chapter 4 that the proof of the hydrodynamic behavior relies on an uniqueness theorem for weak solutions of the partial differential equation that describes the macroscopic evolution of the system, equation (1.3) in the present context. We present such a result in Appendix A2.4 for weak solutions belonging to $L^2([0,T] \times \mathbb{T}^d)$. It is only in the proof that all limit points of the sequence Q^N (defined right below) are concentrated on trajectories $\pi(t, du) = \rho(t, u)du$ whose density is in $L^2([0,T] \times \mathbb{T}^d)$ that the assumption on the jump rate ($g(k) \geq a_0 k$) and assumption (1.2) are required.

There are, however, stronger uniqueness results. We shall prove in section 7, for instance, that in dimension 1 there exists a unique solution ρ satisfying the energy estimate:$\Phi(\rho(t,u))^{-1/2}\nabla\Phi(\rho(t,u))$ belongs to $L^2([0,T]\times\mathbb{T}^d)$. This energy estimate is proved in section 7 assuming only (**FEM**) and the bound (1.1). Brezis and Crandall (1979) present an alternative uniqueness result of weak solutions of (1.3).

Remark 1.4 Assumption (1.2) is weakened in Remark 6.4.

Proof of Theorem 1.1. We fix once for all a time $T > 0$. As in the previous chapter we denote by $\pi_t^N(du)$ the empirical measure defined in (4.0.2) and by $Q^N_{\mu^N}$ or simply Q^N the measure on the trajectories space $D([0,T], \mathcal{M}_+)$ associated to the process π_t^N starting from μ^N.

To enable the reader to discern each step of the proof, we will state them as separate lemmas. The step that consists in proving the relative compactness of the sequence of probabilities Q^N being almost always technical, even if it is instructive since it gives information about the limit trajectories, will systematically be postponed to the end of the proof in the following chapters. In this chapter however, we will start by this point because it provides a first opportunity to show the power of the entropy inequality.

Lemma 1.5 *The sequence of probabilities Q^N is relatively compact.*

Proof. As we have already seen in Chapter 4, we have to prove the compactness of the marginals at every fixed time t and a condition regarding the oscillations.

In the case of positive measures on a compact set, the condition on the compactness of marginals at fixed times t is reduced to estimates on the total mass. Since the total mass is conserved by the evolution, it is enough to check it at time zero. We have therefore to prove that

$$\lim_{A\to\infty} \limsup_{N\to\infty} \mu^N\Big\{\eta; \frac{1}{N^d} \sum_{x\in\mathbb{T}^d_N} \eta(x) \geq A\Big\} = 0 . \tag{1.4}$$

This equality holds because the sequence μ^N is associated to the profile ρ_0 that we assumed integrable.

In our context the Aldous condition regarding the oscillations leads us to prove that

$$\lim_{\gamma\to 0} \limsup_{N\to\infty} \sup_{\substack{\tau\in\mathfrak{T}_T \\ \theta\leq\gamma}}$$
$$\mathbb{P}_{\mu^N}\left[\left|\frac{1}{N^d}\sum_{x\in\mathbb{T}^d_N} G(x/N)\,\eta_{\tau+\theta}(x) - \frac{1}{N^d}\sum_{x\in\mathbb{T}^d_N} G(x/N)\,\eta_\tau(x)\right| > \delta\right] = 0$$

for every function G of class C^2 and every $\delta > 0$. Recall the definition of the martingale M_t^G. To prove last equality it is enough to show that

$$\lim_{\gamma\to 0}\limsup_{N\to\infty}\sup_{\substack{\tau\in\mathfrak{T}_T\\ \theta\le\gamma}} \mathbb{P}_{\mu^N}\left[\left|\int_\tau^{\tau+\theta}\frac{1}{2N^d}\sum_{x\in\mathbb{T}_N^d}\Delta_N G(x/N) g(\eta_s(x))\,ds\right| > \delta\right] = 0$$

$$\text{and}\quad \lim_{\gamma\to 0}\limsup_{N\to\infty}\sup_{\substack{\tau\in\mathfrak{T}_T\\ \theta\le\gamma}} \mathbb{P}_{\mu^N}\left[\left|M_{\tau+\theta}^G - M_\tau^G\right| > \delta\right] = 0\,.$$

Since G is of class C^2 and the function g increases at most linearly (cf. Assumption (2.3.1)), the absolute value of the integral term is bounded by

$$C_1(g^*, G)\int_\tau^{\tau+\theta} N^{-d}\sum_{x\in\mathbb{T}_N^d}\eta_s(x)\,ds$$

that, by conservation of the total number of particles, is equal to

$$\theta C_1(g^*, G) N^{-d}\sum_{x\in\mathbb{T}_N^d}\eta_0(x)\,.$$

The stopping time τ having disappeared and since θ converges to 0 we are brought back to the very same estimate (1.4) necessary to prove the relative compactness of marginals at fixed times.

For the martingale term we use Chebychev's inequality and the explicit formula for the quadratic variation N_t^G to bound it above by

$$\begin{aligned}
&\delta^{-2}\mathbb{E}_{\mu^N}\left[\left(M_{\tau+\theta}^G - M_\tau^G\right)^2\right]\\
&\quad = \delta^{-2}\mathbb{E}_{\mu^N}\left[\int_\tau^{\tau+\theta}\frac{1}{2N^{2d-2}}\sum_{|y-x|=1}[G(y/N)-G(x/N)]^2 g(\eta_s(x))\,ds\right]\\
&\quad \le \frac{C_2(g^*, G)\delta^{-2}\theta}{2N^d} E_{\mu^N}\left[\frac{1}{N^d}\sum_{x\in\mathbb{T}_N^d}\eta(x)\right]
\end{aligned}$$

because the total number of particles is conserved by the evolution. From the presence of an additional factor N^{-d}, we have only to show that the expectation of the total density is bounded by a constant independent of N. The fact that the sequence μ^N is associated to an integrable or bounded profile cannot guaranty it. It is in fact the assumption on the entropy that permits to prove that the expectation of the total mass is bounded. Indeed, by the entropy inequality,

$$E_{\mu^N}\left[\frac{1}{N^d}\sum_{x\in\mathbb{T}_N^d}\eta(x)\right] \le \frac{1}{\gamma N^d}\log E_{\nu_{\alpha^*}}\left[e^{\gamma\sum_x\eta(x)}\right] + \frac{H(\mu^N|\nu_{\alpha^*})}{\gamma N^d} \qquad (1.5)$$

for every $\gamma > 0$. Since the measure ν_{α^*} is product and the entropy of μ^N with respect to ν_{α^*} was supposed bounded above by $K_0 N^d$, the right hand side of the last expression is bounded by

$$\gamma^{-1}\left(\log E_{\nu_{\alpha^*}}\left[e^{\gamma\eta(0)}\right] + K_0\right) .$$

By assumption (2.3.2), this expression is finite for all γ sufficiently small. The proof of the relative compactness of the sequence Q^N is thus concluded. □

Notice that assumption (**FEM**) is not necessary in Lemma 1.5. We just need the existence of some exponential moments and this is guaranteed by hypothesis (2.3.2).

After having established the relative compactness of the sequence Q^N it remains to show the existence of at most one limit point. We shall do it by proving some regularity properties of all possible limit points. We start, for instance, showing that all limit points are concentrated on absolutely continuous measures. This result is a second application of the entropy inequality.

Recall from Appendix 1 that the entropy with respect to some invariant reference state decreases in time. In particular, $H(\mu^N S_t^N \mid \nu_{\alpha^*}) \le K_0 N^d$ for each $0 \le t \le T$ if S_t^N stands for the semigroup associated to the generator L_N defined in (0.1) accelerated by N^2. Next lemma states that any sequence of probability measures μ^N on $\mathcal{M}_+$ with entropy with respect to ν_{α^*} bounded by $K_0 N^d$ and that converges must converge to a probability measure concentrated on absolutely continuous measures.

Lemma 1.6 *Under the hypothesis (***FEM***) stated in the beginning of this chapter, let μ^N be a sequence of probability measures on $\mathbb{N}^{\mathbb{T}_N^d}$ with entropy with respect to ν_{α^*} bounded by $K_0 N^d$:*

$$H(\mu^N | \nu_{\alpha^*}) \le K_0 N^d .$$

Recall from (3.0.1) that we denote by $\pi^N : \mathbb{N}^{\mathbb{T}_N^d} \to \mathcal{M}_+$ the function that associates to each configuration η the positive measure obtained assigning mass N^{-d} to each particle. Let R_{μ^N} be the probability measure $\mu^N(\pi^N)^{-1}$ on $\mathcal{M}_+$ defined by

$$R_{\mu^N}[\mathcal{A}] \;=\; \mu^N\{\,\eta;\ \pi^N(\eta) \in \mathcal{A}\,\}$$

for every Borel subset $\mathcal{A}$ of $\mathcal{M}_+$. Then, all limit points R^* of the sequence R_{μ^N} are concentrated on absolutely continuous measures with respect to the Lebesgue measure:

$$R^*\left\{\pi;\ \pi(du) = \pi(u)du\right\} \;=\; 1 .$$

Proof. The strategy consists in obtaining a positive lower semi–continuous functional $I\colon \mathcal{M}_+ \to \mathbb{R}_+$ such that

(a) $\limsup_{N\to\infty} E_{R_{\mu^N}}[I(\pi)] < \infty$.

(b) $I(\pi) = \infty$ if $\pi(du)$ is not absolutely continuous with respect to the Lebesgue measure.

Indeed, we would then have that

$$E_{R^*}[I(\pi)] \ < \ \infty$$

for every limit point R^* because I was supposed lower semi–continuous. In particular by property (b), R^* is concentrated on absolutely continuous measures.

All the problem is therefore to find such an appropriate functional I. Consider for a while a continuous and bounded function $J: \mathcal{M}_+ \to \mathbb{R}_+$. By the entropy inequality,

$$E_{R_{\mu^N}}[J(\pi)] \ \le \ N^{-d} H(\mu^N | \nu_{\alpha^*}) \ + \ \frac{1}{N^d} \log E_{R_{\nu_{\alpha^*}}} \left[e^{N^d J(\pi)} \right] \ .$$

By hypothesis, the first expression on the right hand side of this inequality is bounded above by K_0. The second one, since J is bounded, by the Laplace–Varadhan theorem (cf. Theorem A2.3.1), converges to

$$\sup_{\pi \in \mathcal{M}_+} \ [J(\pi) - I_0(\pi)]$$

if I_0 represents the large deviations rate function for the random measure π under $R_{\nu_{\alpha^*}}$. This rate function can be easily computed. In order to define it, let $M_{\alpha^*}: \mathbb{R} \to [0, \infty]$ be the Laplace transform of $\eta(0)$ under ν_{α^*}:

$$M_{\alpha^*}(\theta) \ = \ E_{\nu_{\alpha^*}}[e^{\theta \eta(0)}] \ = \ \frac{Z(e^\theta \Phi(\alpha^*))}{Z(\Phi(\alpha^*))} \ . \tag{1.6}$$

The function $\log M_{\alpha^*}$ is convex as well as its Legendre transform h defined by

$$\begin{aligned} h(\alpha) \ &= \ \sup_\theta \{ \theta \alpha - \log M_{\alpha^*}(\theta) \} \\ &= \ \alpha \log \frac{\Phi(\alpha)}{\Phi(\alpha^*)} \ - \ \log \frac{Z(\Phi(\alpha))}{Z(\Phi(\alpha^*))} \ . \end{aligned}$$

The large deviations rate function $I_0: \mathcal{M}_+ \to [0, \infty]$ is equal to

$$I_0(\pi) \ = \ \sup_{f \in C(\mathbb{T}^d)} \left\{ \int_{\mathbb{T}^d} f(u) \pi(du) - \int_{\mathbb{T}^d} \log M_{\alpha^*}(f(u)) \, du \right\} \ .$$

This supremum can be computed. Under the assumption (**FEM**) stated just before Theorem 1.1, we have that

$$I_0(\pi) \ = \ \begin{cases} \displaystyle\int_{\mathbb{T}^d} h(\pi(u)) \, du & \text{if} \quad \pi(du) = \pi(u) du \ , \\ \infty & \text{otherwise.} \end{cases}$$

In particular property (b) is satisfied by the rate function I_0. Moreover, it follows from the variational formula for I_0 that this functional is lower semi–continuous. We have now to check that our candidate has finite expectation with respect to all limit points of R_{μ^N}.

The variational representation of I_0 shows that I_0 is the increasing limit of the bounded and continuous functionals J_k defined by

$$J_k(\pi) = \max_{1 \le j \le k} \left\{ < \pi, f_j > - \int_{\mathbb{T}^d} \log M_{\alpha^*}(f_j(u))\, du \right\} \wedge k$$

if $\{f_k;\ k \ge 1\}$ stands for a dense sequence in $C(\mathbb{T}^d)$ with f_1 being the function identically equal to 0 to ensure the positiveness of J_k.

Using the entropy inequality for each functional J_k, we obtain by the Laplace–Varadhan theorem that

$$\limsup_{N \to \infty} E_{R_{\mu^N}}[J_k(\pi)] \le K_0$$

because $J_k \le I_0$ for each $k \ge 1$. In particular for all limit point R^* of the sequence R_{μ^N},

$$E_{R^*}[J_k(\pi)] \le K_0 .$$

By the monotone convergence theorem and by the variational representation of the rate function I_0,

$$E_{R^*}[I_0(\pi)] = \lim_{k \to \infty} E_{R^*}[J_k(\pi)] \le K_0 . \quad \square$$

Remark 1.7 We have in fact established a slightly stronger result. Under the hypotheses of the previous lemma, if I_0 represents the large deviations rate function, all limit points R^* of the sequence R_{μ^N} are such that

$$E_{R^*}[I_0(\pi)] = E_{R^*}\left[\int_{\mathbb{T}^d} h(\pi(u))\, du \right] \le K_0 .$$

We shall take advantage of this property later to prove that all limit points of the sequence Q^N are concentrated on weak solutions of the non–linear heat equation (1.3).

Remark 1.8 The previous result applied to the marginal at time t of the measure Q^N shows that all limit points Q^* of the sequence (Q^N) are such that

$$E_{Q^*}[I_0(\pi_t)] \le K_0$$

for all $0 \le t \le T$. Since the rate function I_0 is positive, by Fubini's lemma,

$$E_{Q^*}\left[\int_0^T dt\, I_0(\pi_t) \right] = \int_0^T dt\, E_{Q^*}\left[I_0(\pi_t) \right] \le K_0 T .$$

In particular, changing if necessary $\pi_t(du)$ in a time set of measure 0, all limit points Q^* are concentrated on absolutely continuous trajectories:

$$Q^* \{ \pi;\ \pi_t(du) = \pi_t(u) du\ ,\ 0 \le t \le T \} = 1 .$$

Remark 1.9 Notice that the symmetry assumption on the transition probabilities does not play any role in the proof of Lemma 1.6. In particular, under assumption (**FEM**), in the asymmetric case, limit points are concentrated on absolutely continuous trajectories provided the entropy at time 0 is bounded by $K_0 N^d$.

We have just proved that all limit points are concentrated on absolutely continuous trajectories with integrable densities. By Theorem A2.4.4, there exists a unique weak solution of the Cauchy's problem (1.3) in $L^2([0,T] \times \mathbb{T}^d)$. To conclude the proof of Theorem 1.1, it remains therefore to show that all limit points are concentrated on weak solutions of (1.3) in $L^2([0,T] \times \mathbb{T}^d)$.

We prove in this section that all limit points of the sequence $\{Q^N, N \ge 1\}$ are concentrated on weak solutions of (1.3) and in section 6 that all limit points are concentrated on trajectories $\pi(t,du) = \rho(t,u)du$ whose density belongs to $L^2([0,T] \times \mathbb{T}^d)$.

We now return to the strategy adopted to prove the hydrodynamic behavior of symmetric simple exclusion processes. For each fixed smooth function $G\colon [0,T] \times \mathbb{T}^d \to \mathbb{R}$ of class $C^{1,2}$, $M_t^{G,N} = M_t^G$ defined by

$$
\begin{aligned}
M_t^G \;&=\; \frac{1}{N^d} \sum_{x \in \mathbb{T}_N^d} G(t,x/N)\, \eta_t(x) - \frac{1}{N^d} \sum_{x \in \mathbb{T}_N^d} G(0,x/N)\, \eta_0(x) \\
&\quad - \int_0^t \frac{1}{N^d} \sum_{x \in \mathbb{T}_N^d} \Big\{ \partial_s G(s,x/N) \eta_s(x) \;-\; (1/2) \Delta_N G(s,x/N) g(\eta_s(x)) \Big\}\, ds
\end{aligned}
$$

is a martingale with quadratic variation $< M^G >_t$ equal to

$$
< M^G >_t = \frac{1}{2N^{2d-2}} \sum_{|x-y|=1} \int_0^t \Big[G(s,y/N) - G(s,x/N) \Big]^2 g(\eta_s(x))\, ds\,.
$$

Since the jump rate $g(\cdot)$ is at most linear ($g(k) \le g^* k$) and the total number of particles is conserved, by estimate (1.5), the expected value of the quadratic variation $< M^G >_t$ vanishes as $N \uparrow \infty$. In particular, by Doob's inequality, for each $\delta > 0$,

$$
\begin{aligned}
\lim_{N\to\infty} \mathbb{P}_{\mu^N} \Big[\sup_{0 \le t \le T} \Big| &< \pi_t^N, G_t > - < \pi_0^N, G_0 > - \int_0^t ds \; < \pi_s^N, \partial_s G_s > \\
&- \int_0^t ds\, (1/2) \frac{1}{N^d} \sum_{x \in \mathbb{T}_N^d} \Delta_N G(s,x/N) g(\eta_s(x)) \Big| > \delta \Big] \;=\; 0\,.
\end{aligned}
\tag{1.7}
$$

In contrast with symmetric simple exclusion processes, the integral term

$$
\int_0^t ds\, (1/2) N^{-d} \sum_{x \in \mathbb{T}_N^d} \Delta_N G(s,x/N) g(\eta_s(x))
$$

is not a function of the empirical measure. This time the equation is not closed anymore and a new argument is required.

The next lemma allows the replacement of the local function $g(\eta(x))$ by a function of the empirical density of particles in a small macroscopic box. More precisely, it states that the difference

$$\int_0^t ds \frac{1}{N^d} \sum_{x \in \mathbb{T}_N^d} H(s, x/N) \Big\{ g(\eta_s(x)) - \Phi\Big(\frac{1}{(2\varepsilon N + 1)^d} \sum_{|y-x| \le \varepsilon N} \eta_s(y) \Big) \Big\}$$

vanishes in probability as $N \uparrow \infty$ and then $\varepsilon \downarrow 0$ for every continuous function H. Notice that the argument of $\Phi(\cdot)$ in the previous integral is a function of the empirical measure. Indeed, for each $\varepsilon > 0$, denote by ι_ε the approximation of the identity

$$\iota_\varepsilon(\cdot) = (2\varepsilon)^{-d} \mathbf{1}\{[-\varepsilon, \varepsilon]^d\}(\cdot) . \tag{1.8}$$

We have that

$$\begin{aligned} \frac{1}{(2[\varepsilon N] + 1)^d} \sum_{|y-x| \le \varepsilon N} \eta(y) &= \frac{(2\varepsilon N)^d}{(2[\varepsilon N] + 1)^d} < \pi^N, \iota_\varepsilon(\cdot - x/N) > \\ &=: C_{N,\varepsilon} (\pi^N * \iota_\varepsilon)(x/N) . \end{aligned} \tag{1.9}$$

To keep notation simple, for each positive integer ℓ and d-dimensional integer x, denote by $\eta^\ell(x)$ the empirical density of particles in a box of length $2\ell + 1$ centered at x:

$$\eta^\ell(x) = \frac{1}{(2\ell + 1)^d} \sum_{|y-x| \le \ell} \eta(y) . \tag{1.10}$$

Lemma 1.10 **(Replacement lemma)**. *For every* $\delta > 0$,

$$\limsup_{\varepsilon \to 0} \limsup_{N \to \infty} \mathbb{P}_{\mu^N} \Big[\int_0^T \frac{1}{N^d} \sum_{x \in \mathbb{T}_N^d} \tau_x V_{\varepsilon N}(\eta_s) \, ds \ge \delta \Big] = 0 .$$

where,

$$V_\ell(\eta) = \Big| \frac{1}{(2\ell + 1)^d} \sum_{|y| \le \ell} g(\eta(y)) - \Phi\left(\eta^\ell(0)\right) \Big| .$$

To keep notation simple, in the previous statement and hereafter we are writing εN for $[\varepsilon N]$, the integer part of εN. Lemma 1.10 is the main step in the proof of Theorem 1.1. Its proof is postponed to section 3.

Notice that the constant introduced in (1.9) is equal to $1 + O(N^{-1})$. Since by Corollary 2.3.6 Φ is Lipschitz continuous and since by (1.5) the total density of particles has bounded expectation, in last formula we may replace $\eta_s^{\varepsilon N}(x)$ by $(\pi_s^N * \iota_\varepsilon)(x/N)$. Therefore, from Lemma 1.10 and equation (1.7), we have that

$$\limsup_{\varepsilon\to 0}\limsup_{N\to\infty} Q^N \Big[\Big| < \pi_T, G_T > - < \pi_0, G_0 > - \int_0^T < \pi_s, \partial_s G_s > ds$$
$$- \int_0^T \frac{1}{N^d} \sum_{x\in\mathbb{T}_N^d} (1/2)\Delta G(s,x/N)\Phi\left((\pi_s * \iota_\varepsilon)(x/N)\right) ds \Big| \ge \delta \Big] = 0 .$$

We have also replaced the discrete Laplacian Δ_N by the continuous one. This replacement is allowed because Φ increases at most linearly and the expected value of the total density of particles is bounded in virtue of (1.5).

By the same reasons and because G is of class $C^{1,2}$, we may replace the sum

$$N^{-d}\sum_x \Delta G(s,x/N)\Phi((\pi_s^N * \iota_\varepsilon)(x/N))$$

by the integral

$$\int_{\mathbb{T}^d} du\, \Delta G(s,u)\Phi((\pi_s^N * \iota_\varepsilon)(u)) .$$

By the dominated convergence theorem, for each $\varepsilon > 0$, the function that associates to a trajectory π the expression

$$< \pi_T, G_T > - < \pi_0, G_0 > - \int_0^T < \pi_s, \partial_s G_s > ds$$
$$- \int_0^T ds \int_{\mathbb{T}^d} du\,(1/2)\Delta G(s,u)\Phi\left((\pi_s * \iota_\varepsilon)(u)\right)$$

is continuous. In particular, all limit point Q^* of the sequence Q_{μ^N} are such that

$$\limsup_{\varepsilon\to 0} Q^* \Big[\Big| < \pi_T, G_T > - < \pi_0, G_0 > - \int_0^T < \pi_s, \partial_s G_s > ds$$
$$- \int_0^T ds \int_{\mathbb{T}^d} du\,(1/2)\Delta G(s,u)\Phi\left((\pi_s * \iota_\varepsilon)(u)\right) \Big| \ge \delta \Big] = 0 .$$

To show that all limit points Q^* are concentrated on weak solutions of equation (1.3), it remains to prove that we may replace the convolution $\pi_s * \iota_\varepsilon$ by π_s as $\varepsilon \downarrow 0$:

$$\limsup_{\varepsilon\to 0}$$
$$Q^* \Big[\Big| \int_0^T ds \int_{\mathbb{T}^d} du\, \Delta G(s,u) \Big\{ \Phi\left((\pi_s * \iota_\varepsilon)(u)\right) - \Phi\left(\pi(s,u)\right) \Big\} \Big| \ge \delta \Big] = 0$$

for all $\delta > 0$.

By Remark 1.8, π_s has a density with respect to the Lebesgue measure. Therefore,

$$(\pi_s * \iota_\varepsilon)(u) = (2\varepsilon)^{-d} \int_{[u-\varepsilon, u+\varepsilon]^d} \pi(s,v)\, dv$$

for all $0 \le s \le T$. Q^* is concentrated on integrable trajectories π. On the other hand, for each integrable function $f \colon \mathbb{T}^d \to \mathbb{R}$, $f * \iota_\varepsilon$ converges in $L^1(\mathbb{T}^d)$ to f as $\varepsilon \downarrow 0$. In particular, since Φ is Lipschitz, the random variable

$$\int_0^T ds \int_{\mathbb{T}^d} du\, \Delta G(s,u) \Phi\left((\pi_s * \iota_\varepsilon)(u)\right)$$

converges Q^* almost surely to

$$\int_0^T ds \int_{\mathbb{T}^d} du\, \Delta G(s,u) \Phi\left(\pi(s,u)\right) .$$

Finally, letting $\delta \downarrow 0$, we obtain that Q^* is concentrated on weak solutions of the non linear parabolic equation (1.3):

$$Q^* \left[\Big| < \pi_T, G_T > - < \pi_0, G_0 > - \int_0^T < \pi_s, \partial_s G_s > ds - \int_0^T ds \int_{\mathbb{T}^d} du\, (1/2) \Delta G(s,u) \Phi\left(\pi(s,u)\right) \Big| = 0 \right] = 1 . \quad \square$$

2. Entropy Production

Recall from Appendix 1 the definition of the relative entropy and the Dirichlet form of the state of the process with respect to some reference invariant measure. Hereafter, to keep terminology simple, we refer to these two concepts as the entropy and the Dirichlet form.

This section is devoted to the examination of the entropy time evolution. We prove not only that the entropy decreases in time but that its evolution is closely connected to the time evolution of the Dirichlet form. Since the entropy decrease is not a particularity of reversible processes, in this section, we place ourselves in a more general context to include asymmetric evolutions.

Denote by L_N the generator of a zero range process as defined in section 2.3. Consider a sequence $(\mu^N)_{N \ge 1}$ of probability measures on the configuration space $\mathbb{N}^{\mathbb{T}_N^d}$ and denote by S_t^N the semigroup associated to the generator L_N accelerated by $\theta(N)$. Here $\theta(N) = N^2$ in case the drift of each elementary particle vanishes ($\sum_x x p(x) = 0$) and N otherwise.

Let f_t^N be the density of $\mu^N S_t^N$ with respect to a reference invariant measure ν_{α^*}. f_t^N is the solution of the forward Kolmogorov equation

$$\begin{cases} \partial_t f_t^N &= \theta(N) L_N^* f_t^N \\ f_0^N &= d\mu^N / d\nu_{\alpha^*} \, , \end{cases}$$

where L_N^* represents the adjoint operator of L_N in $L^2(\nu_{\alpha^*})$.

We have seen in Appendix 1 that the entropy of $\mu^N S_t^N$ with respect to ν_{α^*} is given by

$$H(\mu^N S_t^N \,|\, \nu_{\alpha^*}) \;=\; \int f_t^N \log f_t^N \, d\nu_{\alpha^*} \, .$$

To keep notation simple, we shall abbreviate it by $H_N(f_t^N)$.

The time derivative of the entropy, known as the entropy production, is therefore equal to

$$\partial_t \int f_t^N \log f_t^N \, d\nu_{\alpha^*} \;=\; \theta(N) \int \log f_t^N L_N^* f_t^N \, d\nu_{\alpha^*} \;+\; \theta(N) \int L_N^* f_t^N \, d\nu_{\alpha^*} \, . \tag{2.1}$$

Since ν_{α^*} is invariant, the second term on the right hand side vanishes. Since L_N^* is the adjoint of L_N in $L^2(\nu_{\alpha^*})$, the first term can be rewritten as

$$\theta(N) \int f_t^N L_N \log f_t^N \, d\nu_{\alpha^*} \, .$$

From the elementary inequality

$$a \log (b/a) \;\le\; 2\sqrt{a}\,(\sqrt{b} - \sqrt{a})$$

that holds for all positives a, b, we can bound above the last integral by

$$2\theta(N) \int \sqrt{f_t^N} L_N \sqrt{f_t^N} \, d\nu_{\alpha^*} \;=\; 2\theta(N) \int \sqrt{f_t^N} L_N^{sym} \sqrt{f_t^N} \, d\nu_{\alpha^*} \, .$$

In this formula, L_N^{sym} stands for the symmetric part of the generator L_N:

$$L_N^{sym} \;=\; 2^{-1}(L_N + L_N^*) \, .$$

Recall the definition of the Dirichlet form introduced in section A1.10. We have bounded above the entropy production by the Dirichlet form associated to the symmetric part of generator:

$$\partial_t H_N(f_t^N) \;\le\; 2\theta(N) \int \sqrt{f_t^N} L_N^{sym} \sqrt{f_t^N} \, d\nu_{\alpha^*} \;=\; -2\,\theta(N)\, D_N(f_t^N) \, .$$

Integrating in time we get

$$H_N(f_t^N) \;+\; 2\theta(N) \int_0^t D_N(f_r^N) \, dr \;\le\; H_N(f_0^N) \, .$$

In particular, if the entropy at time 0 is bounded by $K_0 N^d$, by convexity of the entropy and of the Dirichlet form,

$$H_N\left(\frac{1}{t}\int_0^t f_r^N\,dr\right) \le K_0N^d\,, \qquad D_N\left(\frac{1}{t}\int_0^t f_r^N\,dr\right) \le \frac{K_0N^d}{2t\,\theta(N)}\,. \tag{2.2}$$

To keep terminology simple, up to the end of this section we shall refer to interacting particle systems whose elementary particles evolve, when interaction is suppressed, according to mean-zero random walks ($\sum_x xp(x) = 0$) as mean-zero processes. All other systems are called asymmetric processes. In Chapter 1 we have seen that in order to observe interesting hydrodynamic behavior of mean-zero processes we have to rescale time by N^2 and by N otherwise. In particular, in the mean-zero case we proved a bound on the Dirichlet form of order CN^{d-2} and of order CN^{d-1} in the asymmetric case.

The proof of the replacement lemma relies exclusively on the above two estimates. In particular, the symmetry of the transition probabilities, assumed to clarify the exposition, is not required. All proofs extend therefore to mean-zero asymmetric zero range processes.

The proof of the replacement lemma is divided in two steps. We first prove that we are allowed to replace large *microscopic* spatial averages of $g(\cdot)$, that is, averages over cubes of length ℓ, a parameter independent of N and that increases to infinity after N, by $\Phi(\eta^\ell(\cdot))$. This statement is just the one of Lemma 1.10 with εN replaced by ℓ and letting $\ell \uparrow \infty$ instead of $\varepsilon \downarrow 0$. This result stated in Lemma 3.1, called the one block estimate, requires a bound on the Dirichlet form of order K_0N^{d-1}. The proof presented in next section applies therefore to asymmetric processes.

In contrast, the second step which consists in showing that the particles density over large microscopic boxes is close to the particles density over small macroscopic boxes strongly relies on an estimate of order CN^{d-2} of the Dirichlet form. It applies therefore only to mean-zero processes rescaled by N^2.

3. Proof of the Replacement Lemma

The proof is divided in several steps. We first reduce the dynamical problem involving the stochastic evolution to a static one by means of the estimates obtained in last section. Let $\mu^N(T)$ be the Cesaro mean of $\mu^N S_t^N$:

$$\mu^N(T) := \frac{1}{T}\int_0^T \mu^N S_t^N\,dt$$

and $\bar f_T^N$ the Radon–Nikodym density of $\mu^N(T)$ with respect to ν_{α^*}. In last section we showed that the entropy of $\mu^N(T)$ with respect to ν_{α^*} is bounded by K_0N^d and that its Dirichlet form is bounded by $K_0N^{d-2}(2T)^{-1}$. We claim that to prove the replacement lemma it is enough to show that

$$\limsup_{\varepsilon\to 0}\limsup_{N\to\infty}\sup_{\substack{D_N(f)\le C_0N^{d-2}\\ H_N(f)\le C_0N^d}}\int \frac{1}{N^d}\sum_{x\in\mathbb{T}_N^d}\tau_x V_{\varepsilon N}(\eta)f(\eta)\,\nu_{\alpha^*}(d\eta) \le 0 \tag{3.1}$$

for every positive constant C_0. In this formula, for a positive integer ℓ, V_ℓ represents the cylinder function

$$V_\ell(\eta) := \left| \frac{1}{(2\ell+1)^d} \sum_{|y| \le \ell} g(\eta(y)) \; - \; \Phi\left(\eta^\ell(0)\right) \right|$$

and τ_x translation by x.

Indeed, by Markov inequality,

$$\mathbb{P}_{\mu^N} \left[\frac{1}{N^d} \int_0^T \sum_{x \in \mathbb{T}_N^d} \left| \frac{1}{(2N\varepsilon+1)^d} \sum_{|y-x| \le \varepsilon N} g(\eta_s(y)) \; - \; \Phi\left(\eta_s^{\varepsilon N}(x)\right) \right| ds \ge \delta \right]$$

is less than or equal to

$$\delta^{-1} \mathbb{E}_{\mu^N} \left[\frac{1}{N^d} \int_0^T \sum_{x \in \mathbb{T}_N^d} \left| \frac{1}{(2N\varepsilon+1)^d} \sum_{|y-x| \le \varepsilon N} g(\eta_s(y)) \; - \; \Phi\left(\eta_s^{\varepsilon N}(x)\right) \right| ds \right] .$$

With the notation just introduced, we may rewrite this expectation as

$$\delta^{-1} T \int \frac{1}{N^d} \sum_{x \in \mathbb{T}_N^d} \tau_x V_{\varepsilon N}(\eta) \bar{f}_T^N(\eta) \, \nu_{\alpha^*}(d\eta)$$

and (3.1) is proved.

The proof of inequality (3.1) is divided in two steps. We first show that we may replace the spatial average of g over large *microscopic* boxes, that is of length ℓ independent of N and that increases to infinity after N, by $\Phi(\eta^\ell(\cdot))$. It consists therefore in proving inequality (3.1) with ℓ in place of εN. This is the content of Lemma 3.1.

In the proof of this result we need weaker assumptions than the one formulated in the beginning of this chapter. Since this replacement of spatial averages of cylinder functions over large microscopic hypercubes will be required several times in next chapters, we prove it here in the greatest possible generality. For this reason, consider the hypothesis of sub–linear growth of the jump rate $g(\cdot)$:

(SLG) A zero range process is said to have a sub–linear jump rate if

$$\limsup_{k \to \infty} \frac{g(k)}{k} \; = \; 0 \, .$$

Lemma 3.1 **(One block estimate)**. *Under hypothesis* (**SLG**) *or* (**FEM**), *for every finite constant* C_0,

$$\limsup_{\ell \to \infty} \limsup_{N \to \infty} \sup_{\substack{D_N(f) \le C_0 N^{d-2} \\ H_N(f) \le C_0 N^d}} \int \frac{1}{N^d} \sum_{x \in \mathbb{T}_N^d} (\tau_x V_\ell)(\eta) f(\eta) \, \nu_{\alpha^*}(d\eta) \; = \; 0 \, .$$

The second step in the proof of inequality (3.1) consists in showing that the particles density over large microscopic boxes are close to the particles density over small macroscopic boxes:

Lemma 3.2 **(Two blocks estimate)**. *Under hypothesis* **(FEM)**, *for every finite constant* C_0,

$$\limsup_{\ell\to\infty}\limsup_{\varepsilon\to 0}\limsup_{N\to\infty}\sup_{\substack{D_N(f)\le C_0N^{d-2}\\ H_N(f)\le C_0N^d}}\sup_{|y|\le\varepsilon N}$$
$$\int N^{-d}\sum_{x\in\mathbb{T}_N^d}\left|\eta^\ell(x+y)-\eta^{N\varepsilon}(x)\right| f(\eta)\nu_{\alpha^*}(d\eta) = 0\,.$$

Before turning to the proof of Lemmas 3.1 and 3.2, we check that inequality (3.1) follows from these two statements. Add and subtract the expression

$$\frac{1}{(2\varepsilon N+1)^d}\sum_{|y|\le N\varepsilon}\left\{\frac{1}{(2\ell+1)^d}\sum_{|z-y|\le\ell} g(\eta(z)) - \Phi(\eta^\ell(y))\right\}$$

inside the absolute value that appears in the definition of $V_{\varepsilon N}$. We shall estimate three terms separately.

The first and also the simplest one is equal to

$$\int\frac{1}{N^d}\sum_{x\in\mathbb{T}_N^d}\tau_x V_{\varepsilon N,\ell}(\eta)f(\eta)\,\nu_{\alpha^*}(d\eta)\,,$$

where

$$V_{\varepsilon N,\ell}(\eta) = \left|\frac{1}{(2\varepsilon N+1)^d}\sum_{|y|\le N\varepsilon}\left\{g(\eta(y)) - \frac{1}{(2\ell+1)^d}\sum_{|z-y|\le\ell} g(\eta(z))\right\}\right|\,.$$

This expression is bounded above by

$$\frac{C_1(d)\ell}{N\varepsilon}\int\frac{1}{N^d}\sum_{x\in\mathbb{T}_N^d} g(\eta(x))f(\eta)\,\nu_{\alpha^*}(d\eta)\,,$$

where $C_1(d)$ is a constant that depends only on dimension. Since the jump rate g increases at most linearly and $H_N(f)\le C_0N^d$, by the entropy inequality, the last integral is bounded above by

$$\frac{g^*C_1(d)}{\gamma}\left\{C_0 + \log E_{\nu_{\alpha^*}}\left[\exp\left\{(\gamma\ell/N\varepsilon)\eta(0)\right\}\right]\right\} \tag{3.2}$$

for every $\gamma>0$ because ν_{α^*} is a product measure. For N sufficiently large the expectation that appears in this expression is finite in virtue of (2.1.2). Moreover, this sum vanishes as $N\uparrow\infty$ and then $\gamma\uparrow\infty$.

The second term is bounded by

$$\int \frac{1}{N^d} \sum_{x \in \mathbb{T}_N^d} \tau_x \left| \frac{1}{(2\ell+1)^d} \sum_{|y| \le \ell} g(\eta(y)) - \Phi(\eta^\ell(0)) \right| f(\eta)\, \nu_{\alpha^*}(d\eta) .$$

According to Lemma 3.1, it vanishes as $N \uparrow \infty$ and $\ell \uparrow \infty$.

Finally, the third term is bounded above by

$$\sup_{|y| \le \varepsilon N} \int N^{-d} \sum_{x \in \mathbb{T}_N^d} \left| \Phi(\eta^\ell(x+y)) - \Phi(\eta^{N\varepsilon}(x)) \right| f(\eta) \nu_{\alpha^*}(d\eta)$$

$$\le g^* \sup_{|y| \le \varepsilon N} \int N^{-d} \sum_{x \in \mathbb{T}_N^d} \left| \eta^\ell(x+y) - \eta^{N\varepsilon}(x) \right| f(\eta) \nu_{\alpha^*}(d\eta)$$

because, by Corollary 2.3.6, Φ is Lipschitz. This expression is estimated in Lemma 3.2.

4. The One Block Estimate

We prove in this section Lemma 3.1. To detach the main ideas, we divide the proof in several steps. We first assume hypothesis (**SLG**), the adjustments needed for the case (**FEM**) will be presented at the end of the proof.

Step 1: Cut off of large densities. We start with a technical lemma. It allows to introduce an indicator function that later will guarantee that some set of probability measures is compact for the weak topology.

Lemma 4.1 *For every positive constant C_0, there exists a finite constant $C_3 = C_3(C_0, \alpha^*)$ such that for every integer N,*

$$\sup_{H_N(f) \le C_0 N^d} \int \frac{1}{N^d} \sum_{x \in \mathbb{T}_N^d} \eta(x) f(\eta) \nu_{\alpha^*}(d\eta) \le C_3 .$$

Proof. By the entropy inequality, since ν_{α^*} is a translation invariant product measure, for every positive γ, the supremum is bounded above by

$$\gamma^{-1} C_0 + \gamma^{-1} \log E_{\nu_{\alpha^*}} \left[e^{\gamma \eta(0)} \right] .$$

By (2.1.2), the Laplace transform of $\eta(0)$ under ν_{α^*} is finite for γ sufficiently small. □

This result permits to restrict the integral that appears in the statement of Lemma 3.1 to configurations with bounded particles density over large microscopic boxes. Indeed, from the previous result, to prove Lemma 3.1, it is enough to show that

$$\limsup_{\ell\to\infty}\limsup_{N\to\infty}\sup_{D_N(f)\le C_0N^{d-2}} \frac{1}{N^d}\sum_{x\in\mathbb{T}_N^d}\int\{(\tau_xV_\ell)(\eta)-a\eta^\ell(x)\}\,f(\eta)\nu_{\alpha^*}(d\eta)\ \le\ 0$$

for every $a>0$. For each $b>0$, hypothesis **(SLG)** guarantees the existence of a constant $C_4(b)$ such that

$$g(k)\ \le\ C_4(b)+bk\ .$$

Replacing k by $\eta(0)$ and taking expectations with respect to ν_α, we get a bound for $\Phi(\cdot)$:

$$\Phi(\alpha)\ \le C_4(b)+b\alpha\ .$$

In particular, the cylinder function V_ℓ is bounded above by

$$C_4(a/2)\ +\ (a/2)\eta^\ell(0)\ .$$

Therefore, $V_\ell(\eta)-a\eta^\ell(0)$ is negative as soon as $\eta^\ell(0)\ge 2C_4(a/2)a^{-1}=:C_5(a)=C_5$. We can thus restrict last integral to configurations η satisfying the reversed inequality. In conclusion, to prove the one block estimate, it is enough to show that

$$\limsup_{\ell\to\infty}\limsup_{N\to\infty}\sup_{D_N(f)\le C_0N^{d-2}} \frac{1}{N^d}\sum_{x\in\mathbb{T}_N^d}\int(\tau_xV_\ell)(\eta)\mathbf{1}\{\eta^\ell(x)\le C_5\}\,f(\eta)\nu_{\alpha^*}(d\eta)\ \le\ 0 \tag{4.1}$$

for every positive constants C_0 and C_5.

Step2: Reduction to microscopic cubes. Notice that $V_\ell(\eta)\mathbf{1}\{\eta^\ell(0)\le C_5\}$ depends on the configuration η only through $\{\eta(x),\ |x|\le\ell\}$. The second step in the proof consists in taking advantage of this fact to project the density f over a configuration space that does not depend on the scale parameter N.

Since the measure ν_{α^*} is translation invariant, we can rewrite last sum as

$$\int V_\ell(\eta)\mathbf{1}\{\eta^\ell(0))\le C_5\}\Big(\frac{1}{N^d}\sum_{x\in\mathbb{T}_N^d}\tau_xf\Big)(\eta)\nu_{\alpha^*}(d\eta)$$
$$=\ \int V_\ell(\eta)\mathbf{1}\{\eta^\ell(0)\le C_5\}\,\bar f(\eta)\nu_{\alpha^*}(d\eta)\ ,$$

where $\bar f$ stands for the space average of all translations of f:

$$\bar f(\eta)\ =\ \frac{1}{N^d}\sum_{x\in\mathbb{T}_N^d}\tau_xf(\eta)\ .$$

Before proceeding, we introduce some notation. For a fixed positive integer ℓ we represent by Λ_ℓ a cube of length $2\ell+1$ centered at the origin:

$$\Lambda_\ell\ =\ \{-\ell,\dots,\ell\}^d\ , \tag{4.2}$$

by X^ℓ the configuration space $\mathbb{N}^{\Lambda_\ell}$, by the Greek letter ξ the configurations of X^ℓ and by $\nu^\ell_{\alpha^*}$ the product measure ν_{α^*} restricted to X^ℓ. For a density $f: \mathbb{N}^{\mathbb{T}^d_N} \to \mathbb{R}_+$, f_ℓ stands for the conditional expectation of f with respect to the σ-algebra generated by $\{\eta(z);\ z \in \Lambda_\ell\}$, that is, obtained by integrating all coordinates outside this hypercube:

$$f_\ell(\xi) = \frac{1}{\nu^\ell_{\alpha^*}(\xi)} \int \mathbf{1}\{\eta; \eta(z) = \xi(z),\ z \in \Lambda_\ell\} f(\eta) \nu_{\alpha^*}(d\eta) \quad \text{for} \quad \xi \in X^\ell\,. \tag{4.3}$$

Since $V_\ell(\eta)\mathbf{1}\{\eta^\ell(0) \le C_5\}$ depends on the configuration η only through the occupation variables $\{\eta(x),\ x \in \Lambda_\ell\}$, in last integral we can replace $\bar f$ by $\bar f_\ell$. In particular, we may rewrite inequality (4.1) as

$$\limsup_{\ell\to\infty} \limsup_{N\to\infty} \sup_{D_N(f) \le C_0 N^{d-2}} \int V_\ell(\xi)\mathbf{1}\{\xi^\ell(0) \le C_5\}\, \bar f_\ell(\xi) \nu^\ell_{\alpha^*}(d\xi) \;\le\; 0\,.$$

Step 3: Estimates on the Dirichlet form of $\bar f_\ell$. The third step consists in obtaining information concerning the density $\bar f_\ell$ from the estimate on the Dirichlet form of f. For a fixed pair of neighbor sites x, y, denote by $L_{x,y}$ the piece of the generator corresponding to jumps across the bond with ends x and y:

$$(L_{x,y} f)(\eta) \;=\; (1/2) g(\eta(x))\{f(\eta^{x,y}) - f(\eta)\} \;+\; (1/2) g(\eta(y))\{f(\eta^{y,x}) - f(\eta)\}\,.$$

Denote furthermore by $I_{x,y}(f)$ the piece of the Dirichlet form corresponding to jumps over the bond $\{x, y\}$:

$$\begin{aligned} I_{x,y}(f) \;&=\; -< L_{x,y}\sqrt{f}, \sqrt{f} >_{\nu_{\alpha^*}} \\ &=\; (1/2) \int g(\eta(x))\{\sqrt{f(\eta^{x,y})} - \sqrt{f(\eta)}\}^2 \nu_{\alpha^*}(d\eta)\,. \end{aligned}$$

A simple change of variables ($\zeta = \eta^{x,y}$) shows that we may interchange x and y in the previous integral. With this notation, the Dirichlet form $D_N(f)$ may be written as

$$D_N(f) \;=\; \sum_{|x-y|=1} I_{x,y}(f)\,,$$

where summation is carried over all non oriented pairs of neighbors. By non oriented we understand that each bond $\{x, y\}$ appears only once in the previous sum.

Keep in mind that the Dirichlet form is translation invariant:

$$D_N(\tau_x f) = D_N(f)$$

for every x in $\mathbb{Z}^d$ because it is defined through a sum over all bonds of $\mathbb{T}^d_N$. In particular, since the Dirichlet form is also convex, we have that

$$D_N(\bar f) \;=\; D_N\Big(N^{-d} \sum_{x\in\mathbb{T}^d_N} \tau_x f\Big) \;\le\; N^{-d} \sum_{x\in\mathbb{T}^d_N} D_N(\tau_x f) \;=\; D_N(f)\,.$$

Taking advantage again of the convexity of the Dirichlet form, we prove now that the Dirichlet form restricted to bonds in Λ_ℓ of $\bar{f}_\ell$ is bounded by $D_N(f)$. For a positive integer ℓ, denote by D^ℓ the Dirichlet form defined on all densities $h: X^\ell \to \mathbb{R}_+$ by

$$D^\ell(h) = \sum_{\substack{|y-z|=1 \\ y,z\in\Lambda_\ell}} I^\ell_{y,z}(h),$$

where,

$$I^\ell_{y,z}(h) = (1/2) \int g(\xi(y)) \left[\sqrt{h(\xi^{y,z})} - \sqrt{h(\xi)}\right]^2 \nu^\ell_{\alpha^*}(d\xi) \,. \tag{4.4}$$

Like before summation is carried here over all pairs of non oriented bonds of Λ_ℓ.

With this notation, since the Dirichlet form is convex and since conditional expectation is an average,

$$I^\ell_{y,z}(\bar{f}_\ell) \;\le\; I_{y,z}(\bar{f}) \tag{4.5}$$

for every pair of neighbors y, z in Λ_ℓ. This inequality can also be deduced applying Schwarz inequality to the explicit expression for $I^\ell_{y,z}(\bar{f}_\ell)$ presented in (4.4) and keeping in mind the definition of $\bar{f}_\ell$ given in (4.3)

By inequality (4.5), we have that

$$D^\ell(\bar{f}_\ell) \;\le\; \sum_{\substack{|z-y|=1 \\ z,y\in\Lambda_\ell}} I_{y,z}(\bar{f}).$$

On the other hand, by translation invariance of $\bar{f}$, $I_{z,y}(\bar{f}) = I_{z+x,y+x}(\bar{f})$ for every d-dimensional integer x. Thus,

$$\sum_{\substack{|z-y|=1 \\ z,y\in\Lambda_\ell}} I_{y,z}(\bar{f}) \;=\; (2\ell+1)^{d-1}(2\ell)\sum_{j=1}^{d} I_{0,e_j}(\bar{f}) \;=\; (2\ell+1)^{d-1}(2\ell)N^{-d}D_N(\bar{f})\,.$$

Therefore, for every density f with Dirichlet form bounded by C_0N^{d-2} we have that

$$D^\ell(\bar{f}_\ell) \;\le\; C_0(2\ell+1)^d N^{-2} \;=: C_6(C_0,\ell)N^{-2}\,. \tag{4.6}$$

Notice that we bounded the Dirichlet form of $\bar{f}_\ell$ by a constant vanishing as $N \uparrow \infty$. In conclusion, to prove Lemma 3.1, it remains to show that

$$\limsup_{\ell\to\infty}\; \limsup_{N\to\infty}\; \sup_{D^\ell(f)\le C_6(C_0,\ell)N^{-2}} \int V_\ell(\xi)\mathbf{1}\{\xi^\ell(0)\le C_5\}\, f(\xi)\nu^\ell_{\alpha^*}(d\xi) \;\le\; 0\,, \tag{4.7}$$

where this time the supremum is carried over all densities with respect to $\nu^\ell_{\alpha^*}$. Notice that the scaling parameter N appears now only on the bound of the Dirichlet form.

Step 4: The Limit as $N \uparrow \infty$. The fourth step consists to examine the behavior of last expression as $N \uparrow \infty$. Relying on the lower semicontinuity of the Dirichlet form and on the relative compactness provided by the indicator function, we bound last expression by another expression in which the supremum over all densities is replaced by one over all densities with Dirichlet form equal to 0.

From the presence of the indicator function and since V_ℓ is positive, we can restrict last supremum to densities concentrated on the set $\{\xi;\ \xi^\ell(0) \le C_5\}$ or on the set of all densities such that

$$\int_{X^\ell} \xi^\ell(0)\, f(\xi)\nu^\ell_{\alpha^*}(d\xi) \ \le\ C_5\ .$$

This subset of $\mathcal{M}_{1,+}(X^\ell)$ is compact for the weak topology. This explains the reason we introduced the indicator function $\mathbf{1}\{\eta^\ell(0) \le C_5\}$ in the beginning of the proof.

Since this set is compact, for each fixed N, there exists a density f^N with Dirichlet form bounded by $C_6 N^{-2}$ that reaches the supremum. Consider now a subsequence N_k such that

$$\begin{aligned}\lim_{k\to\infty} &\int V_\ell(\xi)\mathbf{1}\{\xi^\ell(0) \le C_5\}\, f^{N_k}(\xi)\nu^\ell_{\alpha^*}(d\xi)\\ &= \limsup_{N\to\infty} \int V_\ell(\xi)\mathbf{1}\{\xi^\ell(0) \le C_5\}\, f^{N}(\xi)\nu^\ell_{\alpha^*}(d\xi)\ .\end{aligned}$$

To keep notation simple, assume, without loss of generality, that the sequences N_k and N coincide. By compactness, we can find a convergent subsequence f^{N_k}. Denote by f^∞ the weak limit. Since the Dirichlet form is lower semicontinuous,

$$D^\ell(f^\infty) = 0\ .$$

Moreover, by weak continuity,

$$\int \xi^\ell(0) f^\infty(\xi)\nu^\ell_{\alpha^*}(d\xi) \ \le\ C_5 \quad \text{and}$$

$$\begin{aligned}\lim_{k\to\infty} &\int V_\ell(\xi)\mathbf{1}\{\xi^\ell(0) \le C_5\}\, f^{N_k}(\xi)\nu^\ell_{\alpha^*}(d\xi)\\ &= \int V_\ell(\xi)\mathbf{1}\{\xi^\ell(0) \le C_5\}\, f^{\infty}(\xi)\nu^\ell_{\alpha^*}(d\xi)\ .\end{aligned}$$

In conclusion, expression (4.7) is bounded above by

$$\limsup_{\ell\to\infty}\ \sup_{D^\ell(f)=0} \int V_\ell(\xi)\mathbf{1}\{\xi^\ell(0) \le C_5\} f(\xi)\nu^\ell_{\alpha^*}(d\xi)\ .$$

Step 5: Decomposition along hyperplanes with a fixed number of particles. A probability density with Dirichlet form equal to 0 is constant on each hyperplane with a fixed total number of particles. It is convenient therefore to decompose each density f along these hyperplanes with particles density bounded by C_5.

For each integer $j \ge 0$, denote by $\nu^{\ell,j}$ the measure $\nu^\ell_{\alpha^*}$ conditioned to the hyperplane $\{\xi;\ \sum_{z\in\Lambda_\ell} \xi(z) = j\}$:

$$\nu^{\ell,j}(\cdot) = \nu^{\ell}_{\alpha^*}\Big(\cdot \mid \sum_{z\in\Lambda_\ell} \xi(z) = j \Big) = \nu^{\ell}_{\alpha^*}(\cdot) \Big[\nu^{\ell}_{\alpha^*} \Big\{ \xi ;\ \sum_{z\in\Lambda_\ell} \xi(z) = j \Big\} \Big]^{-1} .$$

Notice that this measure does not depend on the parameter α^*.

With this notation we have that

$$\int V_\ell(\xi) \mathbf{1}\{\xi^\ell(0) \le C_5\} f(\xi) \nu^{\ell}_{\alpha^*}(d\xi) = \sum_{j=0}^{[(2\ell+1)^d C_5]} C_j(f) \int V_\ell(\xi) \nu^{\ell,j}(d\xi) ,$$

where

$$C_j(f) = \int \mathbf{1}\{\sum_{z\in\Lambda_\ell} \xi(z) = j\} f(\xi) \nu^{\ell}_{\alpha^*}(d\xi) .$$

Since $\sum_{j\ge 0} C_j(f) = 1$, to conclude the proof of the one block estimate we have to show that

$$\limsup_{\ell\to\infty} \sup_{j\le[(2\ell+1)^d C_5]} \int V_\ell(\xi)\, \nu^{\ell,j}(d\xi) \le 0 .$$

Step 6: An application of the local central limit theorem. The previous inequality follows from the equivalence of ensembles presented in Appendix 2. Since the measure $\nu^{\ell,j}$ is concentrated on configurations with j particles, the last integral is equal to

$$\int \Big| \frac{1}{(2\ell+1)^d} \sum_{|x|\le\ell} g(\xi(x)) - E_{\nu_{j/(2\ell+1)^d}} \big[g(\xi(0)) \big] \Big| \nu^{\ell,j}(d\xi) .$$

Fix a positive integer k, that shall increase to infinity after ℓ. Decompose the set Λ_ℓ in cubes of length $2k+1$: Consider the set $A = \{(2k+1)x,\ x \in \mathbb{Z}^d\} \cap \Lambda_{\ell-k}$ and enumerate its elements: $A = \{x_1, \dots, x_q\}$ in such a way that $|x_i| \le |x_j|$ for $i \le j$. For $1 \le i \le q$, let $B_i = x_i + \Lambda_k$. Notice that $B_i \cap B_j = \phi$ for $i \ne j$ and that $\cup_{1\le i\le q} B_i \subset \Lambda_\ell$. Let $B_0 = \Lambda_\ell - \cup_{1\le i\le q} B_i$. By construction, $|B_0| \le Ck\ell^{d-1}$ for some universal constant C. The previous integral is bounded above by

$$\sum_{i=0}^{q} \frac{|B_i|}{|\Lambda_\ell|} \int \Big| \frac{1}{|B_i|} \sum_{x\in B_i} g(\xi(x)) - E_{\nu_{j/(2\ell+1)^d}} \big[g(\xi(0)) \big] \Big| \nu^{\ell,j}(d\xi) .$$

Since $|B_0| \le Ck\ell^{d-1}$, since $g(m) \le g^* m$ and since under $\nu^{\ell,j}$ the variables $\xi(x)$ have mean $j/(2\ell+1)^d$, this sum is equal to

$$\frac{|\Lambda_k|}{|\Lambda_\ell|} \sum_{i=1}^{q} \int \Big| \frac{1}{|\Lambda_k|} \sum_{x\in B_i} g(\xi(x)) - E_{\nu_{j/(2\ell+1)^d}} \big[g(\xi(0)) \big] \Big| \nu^{\ell,j}(d\xi) + O(k/\ell) .$$

Since the distribution of the vectors $\{\xi(z),\ z \in B_i\}$ does not depend on i, the previous sum is equal to

$$\int \left| \frac{1}{(2k+1)^d} \sum_{|x| \le k} g(\xi(x)) \;-\; E_{\nu_{j/(2\ell+1)^d}}\Big[g(\xi(0))\Big] \right| \nu^{\ell,j}(d\xi) \;+\; O\Big(k/\ell\Big)\,.$$

By the equivalence of ensembles (cf. Corollary A2.1.7), as $\ell \uparrow \infty$ and $j/(2\ell+1)^d \to \alpha$, last integral converges to

$$\int \left| \frac{1}{(2k+1)^d} \sum_{|x| \le k} g(\xi(x)) \;-\; E_{\nu_\alpha}\Big[g(\xi(0))\Big] \right| \nu_\alpha(d\xi)\,.$$

This convergence is uniform in α on every compact subset of $\mathbb{R}_+$. On the other hand, as $k \uparrow \infty$, by the law of large numbers, this integral converges to 0 uniformly on every compact subset of $\mathbb{R}_+$. The proof of Lemma 3.1 when the jump rate satisfies assumption (**SLG**) is thus concluded.

It remains to consider the case (**FEM**). Notice in this respect that we used assumption (**SLG**) only to justify the introduction of the indicator function in the beginning of the proof. Next lemma states that this indicator function can also be introduced under assumption (**FEM**). □

Lemma 4.2 *Under hypothesis* (**FEM**),

$$\limsup_{A\to\infty} \limsup_{\ell\to\infty} \limsup_{N\to\infty} \sup_{H_N(f)\le C_0 N^d} \int \frac{1}{N^d} \sum_{x\in\mathbb{T}_N^d} \eta^\ell(x) \mathbf{1}\{\eta^\ell(x) \ge A\}\, f(\eta)\, \nu_{\alpha^*}(d\eta) \;=\; 0\,.$$

Proof. By the entropy inequality, the integral in the statement of the lemma is bounded above by

$$C_0 \gamma^{-1} \;+\; \frac{1}{\gamma N^d} \log \int \exp\left\{ \gamma \sum_{x\in\mathbb{T}_N^d} \eta^\ell(x) \mathbf{1}\{\eta^\ell(x) \ge A\} \right\} \nu_{\alpha^*}(d\eta)$$

for each $\gamma > 0$, because $H_N(f) \le C_0 N^d$ by assumption.

With respect to the product measure ν_{α^*}, the random variables $\eta^\ell(x)$ and $\eta^\ell(y)$ are independent if $|x-y| > 2\ell$. Recall from (4.2) the definition of the hypercube Λ_ℓ. For each x in Λ_ℓ, denote by Ω_x the set of sites z in $\mathbb{T}_N^d$ equal to x modulo $2\ell+1$: $\Omega_x = \{z \in \mathbb{T}_N^d;\; z - x \in (2\ell+1)\mathbb{Z}^d\}$. With this notation,

$$\sum_{x\in\mathbb{T}_N^d} \Psi_A(\eta^\ell(x)) \;=\; \sum_{x\in\Lambda_\ell} \sum_{y\in\Omega_x} \Psi_A(\eta^\ell(y))\,.$$

In this formula, to keep notation simple, we abbreviated $\eta^\ell(x)\mathbf{1}\{\eta^\ell(x) \ge A\}$ by $\Psi_A(\eta^\ell(x))$.

Notice that for each fixed x, $\{\Psi_A(\eta^\ell(y));\ y \in \Omega_x\}$ are independent random variables with respect to ν_{α^*}. Therefore, by Hölder inequality and the translation invariance of ν_{α^*}, last integral is bounded above by

$$C_0\gamma^{-1} + \frac{1}{\gamma(2\ell+1)^d} \log \int \exp\Big\{\gamma(2\ell+1)^d \eta^\ell(0) \mathbf{1}\{\eta^\ell(0) \ge A\}\Big\} \nu_{\alpha^*}(d\eta) .$$

By Cauchy–Schwarz inequality and Chebychev exponential inequality, the last integral is bounded by

$$\begin{aligned} 1 \,+\, & E_{\nu_{\alpha^*}}\Big[\mathbf{1}\{\eta^\ell(0) \ge A\}\, e^{\gamma \sum_{|x|\le \ell} \eta(x)}\Big] \\ & \le\, 1 + \Big\{\nu_{\alpha^*}\{\eta,\, \eta^\ell(0) \ge A\}\, E_{\nu_{\alpha^*}}\Big[e^{2\gamma \sum_{|x|\le \ell} \eta(x)}\Big]\Big\}^{1/2} \\ & \le\, 1 + \exp\Big\{-(1/2)(2\ell+1)^d\big[A \,-\, \log M_{\alpha^*}(1) \,-\, \log M_{\alpha^*}(2\gamma)\big]\Big\}, \end{aligned}$$

where M_{α^*} is defined in (1.6). Since $\log(1+u) \le u$ for all $u > 0$, up to this point we proved that the supremum appearing in the statement of the lemma is bounded above by

$$C_0\gamma^{-1} + \frac{1}{\gamma(2\ell+1)^d} \exp\Big\{-(1/2)(2\ell+1)^d\big[A \,-\, \log M_{\alpha^*}(1) \,-\, \log M_{\alpha^*}(2\gamma)\big]\Big\},$$

for every $\gamma > 0$. This expression is finite because by assumption **(FEM)**, $M_{\alpha^*}(\theta) < \infty$ for every θ in $\mathbb{R}$. Moreover, the right hand term vanishes in the limit as $\ell \uparrow \infty$ for all A large enough. In particular, we may choose $\gamma = \gamma(A)$ so that $\lim_{A\to\infty} \gamma(A) = \infty$ and the right hand term vanishes as $\ell \uparrow \infty$ for every A (take for instance $A/2 = \log M_{\alpha^*}(2\gamma)$). □

Going back to the proof of the one block estimate, the reader should notice that the bound on the Dirichlet form was used only in (4.6) to estimate $D^\ell(\bar{f}^\ell)$ by a constant $C(\ell, N)$ that vanishes as $N \uparrow \infty$. In particular the estimate $D_N(f) \le C_0 N^{d-2}$ is not crucial. In fact any bound on the Dirichlet form of order $o(N^d)$ would be enough to prove the lemma. Such a bound on the Dirichlet form, of order $O(N^{d-1})$, was obtained in section 2 for asymmetric zero range processes. In particular, we have the following result.

Lemma 4.3 *Let L_N be the generator of an asymmetric zero range process satisfying hypothesis* **(SLG)** *or* **(FEM)***, let μ^N be a sequence if initial measures with entropy bounded by $C_0 N^d$:*

$$H(\mu^N | \nu_{\alpha^*}) \,\le\, C_0 N^d$$

and let $\mathbb{P}_{\mu^N}$ be the probability on the paths space $D([0,T], \mathbb{N}^{\mathbb{T}_N^d})$ corresponding to the Markov process with generator L_N accelerated by N and starting from μ^N. For every $\delta > 0$ and $0 < t \le T$,

$$\limsup_{\ell\to\infty}\limsup_{N\to\infty}$$
$$\mathbb{P}_{\mu^N}\left[\frac{1}{N^d}\int_0^t\sum_{x\in\mathbb{T}_N^d}\left|\frac{1}{(2\ell+1)^d}\sum_{|y-x|\le\ell}g(\eta_s(y))\;-\;\Phi\left(\eta_s^\ell(x)\right)\right|ds\ge\delta\right]\;=\;0\,.$$

5. The Two Blocks Estimate

The proof of the two blocks estimate follows closely the proof of the one block estimate. We thus keep all notation introduced in the previous section and leave somc details to the reader. The unique novelty appears in the proof of an estimate for the restricted Dirichlet form of the conditional expectation of f.

The first step in the proof of the two blocks estimate consists in replacing the density average over a small macroscopic box by an average of densities average over large microscopic boxes. More precisely, for every N sufficiently large, the integral appearing in the statement of Lemma 3.2 is bounded above by

$$\int\frac{1}{N^d}\sum_{x\in\mathbb{T}_N^d}\frac{1}{(2N\varepsilon+1)^d}\sum_{\substack{|z|\le N\varepsilon\\ 2\ell<|y-z|}}\left|\eta^\ell(x+y)-\eta^\ell(x+z)\right|f(\eta)\,\nu_{\alpha^*}(d\eta)$$
$$+\;C_1(d)\frac{\ell}{N\varepsilon}\int\frac{1}{N^d}\sum_{x\in\mathbb{T}_N^d}\eta(x)f(\eta)\,\nu_{\alpha^*}(d\eta)\,,$$

where $C_1(d)$ is a constant that depends only on the dimension. Notice that summation over z in the first line is carried over all d-dimensional integers at a distance at least 2ℓ from y. All other terms are included in the second line. In this way the averages $\eta^\ell(x+y)$ and $\eta^\ell(x+z)$ are performed over disjoint sets of sites. The entropy inequality shows that the limit, as $N\uparrow\infty$, of the second line vanishes (cf. (3.2)). It remains therefore to estimate the first line.

The second step of the proof consists in introducing an indicator function to avoid possible large values of particles density. In fact, by Lemma 4.2, in order to prove Lemma 3.2, it is enough to show that

$$\limsup_{\ell\to\infty}\limsup_{\varepsilon\to0}\limsup_{N\to\infty}\sup_{D_N(f)\le C_0N^{d-2}}\sup_{2\ell<|y|\le2N\varepsilon}N^{-d}$$
$$\int\sum_{x\in\mathbb{T}_N^d}\left|\eta^\ell(x)-\eta^\ell(x+y)\right|\mathbf{1}\{\eta^\ell(x)\vee\eta^\ell(x+y)\le A\}\,f(\eta)\,\nu_{\alpha^*}(d\eta)\;\le\;0 \tag{5.1}$$

for every $A>0$.

With the notation introduced in the proof of Lemma 3.1, this integral can be rewritten as

$$\int\left|\eta^\ell(0)-\eta^\ell(y)\right|\mathbf{1}\{\eta^\ell(0)\vee\eta^\ell(y)\le A\}\,\bar f(\eta)\,\nu_{\alpha^*}(d\eta)\,,$$

where $\bar{f}$ stands for the average of all space translations of f. $\eta^\ell(0)$ and $\eta^\ell(y)$ depend on the configuration η only through the occupation variables $\eta(x)$ in the set

$$\Lambda_{y,\ell} := \{-\ell,\dots,\ell\}^d \cup \left[y+\{-\ell,\dots,\ell\}^d\right].$$

We can thus replace in the last integral $\bar{f}$ by its conditional expectation with respect to the σ-algebra generated by $\{\eta(z);\ z\in\Lambda_{y,\ell}\}$.

Before proceeding we introduce some notation. For a positive integer ℓ, we represent by $X^{2,\ell}$ the configuration space $\mathbb{N}^{\Lambda_\ell}\times\mathbb{N}^{\Lambda_\ell}$, by the couple $\xi=(\xi_1,\xi_2)$ configurations of $X^{2,\ell}$ and by $\nu_{\alpha^*}^{2,\ell}$ the product measure ν_{α^*} restricted to $X^{2,\ell}$. For a density $f:\mathbb{N}^{\mathbb{T}_N^d}\to\mathbb{R}_+$, $f_{y,\ell}$ stands for the conditional expectation of f with respect to the σ-algebra generated by $\{\eta(z);\ z\in\Lambda_{y,\ell}\}$, that is, obtained integrating all coordinates outside $\Lambda_{y,\ell}$:

$$f_{y,\ell}(\xi_1,\xi_2) = \frac{1}{\nu_{\alpha^*}^{2,\ell}(\xi_1,\xi_2)}\int \mathbf{1}\{\eta;\eta(z)=\xi(z),\ z\in\Lambda_{y,\ell}\}\, f(\eta)\nu_{\alpha^*}(d\eta)$$

for ξ in $X^{2,\ell}$. $f_{y,\ell}$ is really a density on $X^{2,\ell}$ because y has absolute value larger than 2ℓ.

With these notations we can rewrite (5.1) as

$$\limsup_{\ell\to\infty}\limsup_{\varepsilon\to0}\limsup_{N\to\infty}\sup_{D_N(f)\le C_0N^{d-2}}\sup_{2\ell<|y|\le2\varepsilon N}$$
$$\int\left|\xi_1^\ell(0)-\xi_2^\ell(0)\right|\mathbf{1}\{\xi_1^\ell(0)\vee\xi_2^\ell(0)\le A\}\,\bar{f}_{y,\ell}(\xi)\nu_{\alpha^*}^{2,\ell}(d\xi) \le 0\,.$$

Next step consists in obtaining information concerning the density $\bar{f}_{y,\ell}$ from the bound on the Dirichlet form of f. It is here that the proofs of Lemma 3.1 and 3.2 differ sensibly. First of all, as we have seen in the proof of the one block estimate, the Dirichlet form of $\bar{f}$ is bounded above by the one of f:

$$D_N(\bar{f}) \le D_N(f)\,.$$

Let $D^{2,\ell}$ be the Dirichlet form defined on positive densities $h: X^{2,\ell}\to\mathbb{R}_+$ by

$$D^{2,\ell}(h) = I_{0,0}^\ell(h) + \sum_{\substack{|x-z|=1\\ x,z\in\Lambda_\ell}} I_{x,z}^{1,\ell}(h) + \sum_{\substack{|x-z|=1\\ x,z\in\Lambda_\ell}} I_{x,z}^{2,\ell}(h)\,,$$

where

$$I_{x,z}^{1,\ell}(h) = (1/2)\int g(\xi(x))\left[\sqrt{h(\xi_1^{x,z},\xi_2)}-\sqrt{h(\xi)}\right]^2\nu_{\alpha^*}^{2,\ell}(d\xi)\,,$$

$$I_{x,z}^{2,\ell}(h) = (1/2)\int g(\xi(x))\left[\sqrt{h(\xi_1,\xi_2^{x,z})}-\sqrt{h(\xi)}\right]^2\nu_{\alpha^*}^{2,\ell}(d\xi)$$

$$\text{and}\quad I_{0,0}^\ell(h)=(1/2)\int g(\xi_1(0))\left[\sqrt{h(\xi_1^{0,-},\xi_2^{0,+})}-\sqrt{h(\xi)}\right]^2\nu_{\alpha^*}^{2,\ell}(d\xi)\,.$$

In last formula, for a configuration $\xi = (\xi_1, \xi_2)$ of $X^{2,\ell}$ and for $i = 1, 2$, $\xi_i^{0,\pm}$ is defined as

$$\xi_i^{0,\pm}(x) = \begin{cases} \xi_i(0) \pm 1 & \text{if } x = 0 \\ \xi_i(x) & \text{otherwise .} \end{cases}$$

This Dirichlet form corresponds to an interacting particle system on $\Lambda_\ell \times \Lambda_\ell$ where particles evolve according to nearest neighbor symmetric zero range processes on each coordinate and where particles can jump from the origin of one of the coordinates to the origin of the other.

Recall that for a density f on $\mathbb{N}^{\mathbb{T}_N^d}$, we denote by $f_{y,\ell}$ the conditional expectation of f with respect to the σ-algebra generated by $\{\eta(z),\ z \in \Lambda_{y,\ell}\}$. In the same way as we did in (4.5), we can show that for each pair of neighbor sites x, x' in Λ_ℓ (resp. z, z' in in Λ_ℓ) the Dirichlet form $I^{1,\ell}_{x,x'}(f_{y,\ell})$ (resp. $I^{2,\ell}_{z,z'}(f_{y,\ell})$) is bounded above by the Dirichlet form of f:

$$I^{1,\ell}_{x,x'}(f_{y,\ell}) \le I_{x,x'}(f) \quad \text{and} \quad I^{2,\ell}_{z,z'}(f_{y,\ell}) \le I_{y+z,y+z'}(f)$$

for each neighbor sites x, x' in Λ_ℓ and each neighbor sites z, z' in Λ_ℓ. In particular,

$$\sum_{\substack{|x-z|=1 \\ x,z\in\Lambda_\ell}} I^{1,\ell}_{x,z}(\bar f_{y,\ell}) + \sum_{\substack{|x-z|=1 \\ x,z\in\Lambda_\ell}} I^{2,\ell}_{x,z}(\bar f_{y,\ell}) \le 2C_0(2\ell+1)^d N^{-2}$$

for every density with Dirichlet form $D_N(f)$ bounded by $C_0 N^{d-2}$.

It remains to show that we can also estimate the Dirichlet form $I^\ell_{0,0}(\bar f_{y,\ell})$ by the Dirichlet form of f. In order to do it we first derive an equivalent expression for the Dirichlet forms $I^\ell_{0,0}$ and $I_{x,z}$. A change of variable $\zeta = \eta - \mathfrak{d}_0$ and $\zeta = \eta - \mathfrak{d}_x$ permits to rewrite these Dirichlet forms as

$$I^\ell_{0,0}(h) = (1/2)\Phi(\alpha^*) \int \left(\sqrt{h(\xi_1^{0,+}, \xi_2)} - \sqrt{h(\xi_1, \xi_2^{0,+})} \right)^2 \nu^{2,\ell}_{\alpha^*}(d\xi_1, d\xi_2) ,$$

$$I_{x,z}(f) = (1/2)\Phi(\alpha^*) \int \left(\sqrt{f(\eta^{x,+})} - \sqrt{f(\eta^{z,+})} \right)^2 \nu_{\alpha^*}(d\eta)$$

where, as before, for a configuration η and an integer x, $\eta^{x,+}$ represents the configuration obtained from η adding a particle to site x:

$$\eta^{x,+}(z) = \begin{cases} \eta(x) + 1 & \text{if } z = x \\ \eta(z) & \text{otherwise .} \end{cases}$$

In the same way as we proved that $I^\ell_{x,z}(f_\ell) \le I_{x,z}(f)$, for each density f with respect to ν_{α^*}, $I^\ell_{0,0}(\bar f_{y,\ell})$ is bounded above by

$$(1/2)\Phi(\alpha^*) \int \left(\sqrt{\bar f(\eta^{0,+})} - \sqrt{\bar f(\eta^{y,+})} \right)^2 \nu_{\alpha^*}(d\eta) . \tag{5.2}$$

Let $(x_k)_{0\le k\le \|y\|}$ be a path from the origin to y, that is, a sequence of sites such that the first one is the origin, the last one is y and the distance between two consecutive sites is equal to 1:

$$x_0 = 0\,, \quad x_{\|y\|} = y \quad \text{and} \quad |x_{k+1} - x_k| = 1 \quad \text{for every } 0 \le k \le \|y\| - 1\,.$$

In these formulas, $\|\cdot\|$ represents the sum norm:

$$\|(y_1,\dots,y_d)\| = \sum_{1\le i\le d} |y_i|\,.$$

We use the telescopic identity

$$\sqrt{f(\eta^{0,+})} - \sqrt{f(\eta^{y,+})} = \sum_{k=0}^{\|y\|-1} \sqrt{f(\eta^{x_k,+})} - \sqrt{f(\eta^{x_{k+1},+})}$$

and the Cauchy–Schwarz inequality

$$\left(\sum_{k=0}^{\|y\|-1} a_k\right)^2 \le \|y\| \sum_{k=0}^{\|y\|-1} a_k^2$$

to estimate the integral (5.2) by

$$\begin{aligned}(1/2)\Phi(\alpha^*)\|y\| &\sum_{k=0}^{\|y\|-1} \int \left(\sqrt{\bar f(\eta^{x_k,+})} - \sqrt{\bar f(\eta^{x_{k+1},+})}\right)^2 \nu_{\alpha^*}(d\eta) \\ &= \|y\| \sum_{k=0}^{\|y\|-1} I_{x_k,x_{k+1}}(\bar f)\,.\end{aligned}$$

Since $\bar f$ is translation invariant, for each k, $I_{x_k,x_{k+1}}(\bar f) = I_{x_k+z,x_{k+1}+z}(\bar f)$ for all z in $\mathbb{Z}^d$ so that $I_{x_k,x_{k+1}}(\bar f) \le N^{-d} D_N(f)$. In particular,

$$I^{\ell}_{0,0}(\bar f_{y,\ell}) \le \|y\|^2 N^{-d} D_N(f)\,.$$

Recall that $|y| \le 2N\varepsilon$ so that $\|y\| \le d|y| \le 2dN\varepsilon$. Since the Dirichlet form is assumed bounded by $C_0 N^{d-2}$, we have proved that

$$I^{\ell}_{0,0}(\bar f_{y,\ell}) \le 4C_0 d^2 \varepsilon^2\,.$$

In conclusion, for every density f with Dirichlet form bounded by C_0N^{d-2} and for every d-dimensional integer y with max norm between 2ℓ and $2N\varepsilon$,

$$D^{2,\ell}(\bar f_{y,\ell}) \le C_7(C_0, d, \ell)\varepsilon^2\,.$$

Notice that a factor ε^2 that vanishes in the limit as $\varepsilon \downarrow 0$ appeared. In particular to conclude the proof it is enough to show that

$$\limsup_{\ell\to\infty} \limsup_{\varepsilon\to 0} \sup_{D^{2,\ell}(f)\le C_7(C_0,\ell)\varepsilon^2} \int \left|\xi_1^\ell(0) - \xi_2^\ell(0)\right| \mathbf{1}\{\xi_1^\ell(0) \vee \xi_2^\ell(0) \le A\}\, f(\xi)\nu_{\alpha^*}^{2,\ell}(d\xi) \;\le\; 0\,,$$

for every $A > 0$. This time the supremum is taken over all densities with respect to $\nu_{\alpha^*}^{2,\ell}$.

We may now follow the arguments presented in the proof of Lemma 3.1 to conclude. The unique worthwhile mentioning slight difference is that every density f with Dirichlet form equal to 0 ($D^{2,\ell}(f) = 0$) is constant on hyperplanes having a fixed total number of particles on $\Lambda_\ell \cup \{y + \Lambda_\ell\}$ because particles can jump from the origin of one of the coordinates to the origin of the other. □

Remark 5.1 Notice that in the proof of Lemmas 3.1 and 3.2 only the bounds on the Dirichlet form of order N^{d-2} and on the entropy of order N^d were used. In particular, by section 2, Lemmas 3.1 and 3.2 apply to mean-zero asymmetric zero range processes satisfying assumption (**FEM**) or (**SLG**).

Remark 5.2 The time integral in the statement of Lemmas 3.1 and 3.2 is crucial. Indeed, we have no a priori information on the order of magnitude of the Dirichlet form at a fixed time. We are only able to prove in the mean-zero case a bound of order $O(N^{d-2})$ for the time integral of the Dirichlet form and this gives no information on its value at a fixed time.

The cylinder function $g(\eta(0))$ does not play any particular role in Lemmas 3.1 and 3.2. The statement applies to a broad class of cylinder functions. Recall from (2.3.10) the definition of Lipschitz cylinder functions and Lipschitz cylinder functions with sublinear growth and recall from (2.3.12) the definition of the real function $\tilde{\Psi}$. We proved in Corollary 2.3.7 that $\tilde{\Psi}$ is uniformly Lipschitz for every Lipschitz cylinder function Ψ. The next two results follow from the proof of Lemma 3.1.

Lemma 5.3 *Under assumptions of Lemma 3.1, for every cylinder function Ψ with sublinear growth and for every C_0,*

$$\limsup_{\ell\to\infty} \limsup_{N\to\infty} \sup_{\substack{D_N(f)\le C_0 N^{d-2} \\ H_N(f)\le C_0 N^d}} \int \frac{1}{N^d} \sum_{x\in\mathbb{T}_N^d} \left| \frac{1}{(2\ell+1)^d} \sum_{|y-x|\le\ell} \tau_y\Psi(\eta) - \tilde{\Psi}(\eta^\ell(x)) \right| f(\eta)\,\nu_{\alpha^*}(d\eta) \;=\; 0\,.$$

Lemma 5.4 *Under hypothesis* (**FEM**), *the statement of the previous lemma holds for cylinder Lipschitz functions Ψ.*

We mentioned above that $\tilde{\Psi}$ is a Lipschitz function for all cylinder Lipschitz functions Ψ. In particular, we have the following result.

Lemma 5.5 *Under hypothesis* (**FEM**)*, the statement of Lemma 1.10 remains in force if g is replaced by a cylinder Lipschitz function Ψ and Φ by its homologue $\tilde{\Psi}$.*

We conclude this section with a remark concerning the hypotheses (**FEM**) and (**SLG**) and few words about the replacement lemma for generalized simple exclusion processes.

Remark 5.6 Attractive zero range processes, i.e., systems with non decreasing jump rate g, satisfy either hypothesis (**FEM**) or (**SLG**) because if the jump rate $g(\cdot)$ is unbounded it fulfills hypothesis (**FEM**) and if it is bounded it satisfies assumption (**SLG**). In particular, the one block estimates can be proved for all attractive zero range processes.

At last, notice that in the case of generalized exclusion processes, the entropy of any probability measure μ^N on $\{0,\dots,\kappa\}^{\mathbb{T}_N^d}$ with respect to an invariant product state ν_α^N is bounded by $C(\alpha)N^d$. Indeed, by convexity of the entropy,

$$H\Big(\mu^N \,\big|\, \nu_\alpha^N\Big) \;\le\; \max_\eta H\Big(\delta_\eta \,\big|\, \nu_\alpha^N\Big)\,.$$

Here δ_η stands for the Dirac measure concentrated on the configuration η. Since

$$H\Big(\delta_\eta \,\big|\, \nu_\alpha^N\Big) \;=\; -\log \nu_\alpha^N(\eta) \;=\; Z(\Phi(\alpha))N^d - \sum_{x\in\mathbb{T}_N^d} \log \Phi(\alpha)\eta(x)$$

and since the total number of particles per site is at most κ, the previous expression is bounded above by $C(\alpha)N^d$ for some finite constant $C(\alpha)$. Thus, for generalized exclusion processes the bound on the entropy and on the time integral of the Dirichlet form proved in section 2 apply to any sequence of initial measures μ^N. Moreover, in the statement of the replacement lemma the bound on the entropy is unnecessary:

Lemma 5.7 *For generalized exclusion processes, for every cylinder function Ψ and for every C_0,*

$$\limsup_{\varepsilon\to 0}\limsup_{N\to\infty} \sup_{D_N(f)\le C_0 N^{d-2}}$$
$$\int \frac{1}{N^d}\sum_{x\in\mathbb{T}_N^d}\left|\frac{1}{(2\varepsilon N+1)^d}\sum_{|y-x|\le\varepsilon N} \tau_y\Psi(\eta) - \tilde{\Psi}(\eta^{\varepsilon N}(x))\right| f(\eta)\,\nu_{\alpha^*}(d\eta) \;=\; 0\,.$$

6. A L^2 Estimate

We prove in this section that all limit points Q^* of the sequence $\{Q^N, N \geq 1\}$ are concentrated on absolutely continuous measures whose density is in $L^2([0,T] \times \mathbb{T}^d)$. We start introducing some notation related to Fourier transforms. Fix a positive integer N and consider the space $L^2(\mathbb{T}_N^d)$ of complex functions on $\mathbb{T}_N^d$ endowed with the inner product $< f, g >= N^{-d} \sum_{x \in \mathbb{T}_N^d} f(x) g(x)^*$, where a^* stands for the conjugate of a. For each z in $\mathbb{T}_N^d$, denote by $\psi_z = \psi_{N,z}$ the $L^2(\mathbb{T}_N^d)$ function defined by

$$\psi_z(x) \;=\; \exp\left\{ \frac{2\pi i}{N} (z \cdot x) \right\} ,$$

where $z \cdot x$ stands for the usual inner product in $\mathbb{R}^d$. It is easy to check that $\{\psi_z, \, z \in \mathbb{T}_N^d\}$ forms an orthonormal basis of $L^2(\mathbb{T}_N^d)$. In particular, each function f in $L^2(\mathbb{T}_N^d)$ can be written as

$$f \;=\; \sum_{z \in \mathbb{T}_N^d} < f, \psi_z > \psi_z \, .$$

We shall repeatedly use the following three properties of the Fourier transform. Since $\{\psi_z, \, z \in \mathbb{T}_N^d\}$ is an orthonormal basis, for f, g in $L^2(\mathbb{T}_N^d)$,

$$< f, g > = \sum_{z \in \mathbb{T}_N^d} < f, \psi_z > < g, \psi_z >^* \, . \tag{6.1}$$

Denote by Δ_N the discrete Laplacian: $(\Delta_N f)(x) = \sum_{1 \leq j \leq d} \{ f(x + e_j) + f(x - e_j) - 2 f(x) \}$. A double summation by parts gives that

$$< N^2 \Delta_N f, \psi_z > = \; -2N^2 \sum_{j=1}^{d} \left\{ 1 - \cos\left(\frac{2\pi z_j}{N} \right) \right\} < f, \psi_z > \tag{6.2}$$

because $\Delta_N \psi_z = -2 \sum_{1 \leq j \leq d} \{1 - cos(2\pi z_j / N)\} \psi_z$. For two functions f, g in $L^2(\mathbb{T}_N^d)$, denote by $f * g$ the convolution of f and g:

$$(f * g)(z) \;=\; \sum_{x \in \mathbb{T}_N^d} f(x) g(z - x) \, .$$

From this definition it is easy to deduce that

$$< (f * g), \psi_z > = \; N^d < f, \psi_z > < g, \psi_z > \, . \tag{6.3}$$

We now introduce two additional norms in $L^2(\mathbb{T}_N^d)$ and investigate the relations between them. For a function f in $L^2(\mathbb{T}_N^d)$, define the $\mathcal{H}_1$ norm $\|f\|_1$ of f by

$$\|f\|_1^2 \; = < f, (I - N^2 \Delta_N) f > \, .$$

By properties (6.1) and (6.2), the $\mathcal{H}_1$ norm of f is equal to

$$\sum_{z\in\mathbb{T}_N^d} < f,\psi_z >< (I - N^2\Delta_N)f,\psi_z >^* = \sum_{z\in\mathbb{T}_N^d} \big| < f,\psi_z > \big|^2 a_N(z)\,, \quad (6.4)$$

where $a_N\colon \mathbb{T}_N^d \to \mathbb{R}_+$ is the positive function given by

$$a_N(z) \;=\; 1 + 2N^2\sum_{j=1}^d \{1 - \cos(2\pi z_j N^{-1})\}\,.$$

Denote by $A_N\colon \mathbb{T}_N^d \to \mathbb{R}$, the inverse Fourier transform of the function a_N: $A_N(z) =< a_N,\psi_z^* >$. Since A_N is the inverse Fourier transform of a_N,

$$A_N \;=\; \frac{1}{N^d}\sum_{z\in\mathbb{T}_N^d} a_N(z)\,\psi_z\,,$$

what can be confirmed by an elementary computation. Notice that A_N is an even real function because a_N is even. We claim that the $\mathcal{H}_1$ norm can be rewritten as

$$\|f\|_1^2 \;=< f, A_N * f >= \frac{1}{N^d}\sum_{z,y} f(z)A_N(z-y)f(y)^*\,. \quad (6.5)$$

Indeed, by property (6.1) and (6.3),

$$< f, A_N * f >= N^d \sum_{z\in\mathbb{T}_N^d} \big| < f,\psi_z > \big|^2 < A_N,\psi_z >^*\,.$$

Since $A_N = N^{-d}\sum_{y\in\mathbb{T}_N^d} a_N(y)\psi_y$ and $< \psi_y,\psi_z >= \delta_{y,z}$, we have that $< A_N,\psi_z >= N^{-d}a_N(z)$, what proves (6.5) in view of (6.4) because a_N is a real function.

We now introduce the dual norm of $\mathcal{H}_1$ with respect to $L^2(\mathbb{T}_N^d)$. For each function f in $L^2(\mathbb{T}_N^d)$, define

$$\|f\|_{-1}^2 \;=\; \sup_g \Big\{2\big| < f,g > \big| - \|g\|_1^2\Big\}\,,$$

where the supremum is taken over all functions g in $L^2(\mathbb{T}_N^d)$. We claim that the $\mathcal{H}_{-1}$ norm is

$$\|f\|_{-1}^2 \;=\; \sum_{z\in\mathbb{T}_N^d} \big| < f,\psi_z > \big|^2 \frac{1}{a_N(z)}\,. \quad (6.6)$$

Indeed, for fixed real functions f, g, by properties (6.1), (6.4),

$$\begin{aligned} &2\big| < f,g > \big| - \|g\|_1^2 \\ &\quad = 2\Big|\sum_{z\in\mathbb{T}_N^d} < f,\psi_z >< g,\psi_z >^*\Big| - \sum_{z\in\mathbb{T}_N^d} \big| < g,\psi_z > \big|^2 a_N(z)\,. \end{aligned} \quad (6.7)$$

Take $g = \sum_z a_N(z)^{-1} < f, \psi_z > \psi_z$ (so that $< g, \psi_z >= a_N(z)^{-1} < f, \psi_z >$) to obtain that the left hand side of (6.6) is bounded below by the right hand side. To prove the reverse inequality, notice that (6.7) is bounded above by

$$\sum_{z\in\mathbb{T}_N^d} \Big\{2\big| < f, \psi_z > \big|\big| < g, \psi_z > \big| - a_N(z)\big| < g, \psi_z > \big|^2\Big\} .$$

Since $2ab - Rb^2 \le a^2R^{-1}$, this expression is less than or equal to $\sum_z \big| < f, \psi_z > \big|^2 a_N(z)^{-1}$, what proves (6.6).

Denote by $K_N: \mathbb{T}_N^d \to \mathbb{R}$ the inverse Fourier transform of a_N^{-1}: $K_N(z) =< a_N^{-1}, \psi_z^* >$ or

$$K_N \;=\; \frac{1}{N^d} \sum_{z\in\mathbb{T}_N^d} a_N^{-1}(z)\, \psi_z \, . \tag{6.8}$$

Since $a_N(\cdot)$ is an even function, $K_N(\cdot)$ is an even real function. Repeating the arguments that lead to (6.5), we deduce that the $\mathcal{H}_{-1}$ norm can be expressed as

$$\|f\|_{-1}^2 \;=< f, K_N * f > = \frac{1}{N^d} \sum_{x,y} f(x) K_N(x-y) f(y)^* \tag{6.9}$$

because K_N is a real function.

Two properties of the kernel $K_N(\cdot)$ will be used in the proof of the L^2 estimate for the density of the empirical measure. On the one hand, it follows from formula (6.8) and the identity $(I - N^2\Delta_N)\psi_z = a_N(z)\psi_z$ that

$$(I - N^2\Delta_N) K_N(y) \;=\; \delta_{0,y} \, , \tag{6.10}$$

where $\delta_{x,y}$ stands for the delta of Kronecker. On the other hand, the same formula (6.8) for the kernel $K_N(\cdot)$ gives that

$$N^2 \sum_{j=1}^{d} \{K_N(0) - K_N(e_j)\} \;\le\; \frac{1}{2} \, . \tag{6.11}$$

We conclude this preamble with two observations. It follows from formula (6.6) for the $\mathcal{H}_{-1}$ norm and from (6.2) that $\|f\|_{-1}^2 \le \|f\|_0^2$ because $a_N(\cdot) \ge 1$. Furthermore, by Schwarz inequality, for any real functions f, g in $L^2(\mathbb{T}_N^d)$,

$$2\big| < f, K_N * g > \big| \;\le\; \varepsilon\|f\|_{-1}^2 \;+\; \varepsilon^{-1}\|g\|_{-1}^2 \tag{6.12}$$

for every $\varepsilon > 0$. Indeed, by (6.1) and (6.3), the left hand side is bounded above by

$$2N^d \sum_{z\in\mathbb{T}_N^d} \big| < f, \psi_z > \big|\big| < g, \psi_z > \big|\big| < K_N, \psi_z > \big| \, .$$

Since $< K_N, \psi_z >= N^{-d} a_N(z)^{-1}$ and $2ab \le a^2 + b^2$, the previous expression is bounded above by

$$\varepsilon \sum_{z \in \mathbb{T}_N^d} | < f, \psi_z > |^2 \frac{1}{a_N(z)} + \varepsilon^{-1} \sum_{z \in \mathbb{T}_N^d} | < g, \psi_z > |^2 \frac{1}{a_N(z)}$$

for every $\varepsilon > 0$. This concludes the proof of (6.12) in view of (6.6).

We are now ready to state the main result of this section.

Proposition 6.1 *Fix $T > 0$ and a sequence of probability measures $\{\mu^N, N \geq 1\}$ such that*

$$\limsup_{N \to \infty} E_{\mu^N} \Big[\|\eta\|_{-1}^2 \Big] = K_2 < \infty . \qquad (6.13)$$

There exists a finite constant C depending only on K_2, the jump rate g and T such that

$$\mathbb{E}_{\mu^N} \Big[\|\eta(t)\|_{-1}^2 \Big] + \mathbb{E}_{\mu^N} \Big[\int_0^t ds \frac{1}{2N^d} \sum_{x \in \mathbb{T}_N^d} g(\eta_s(x)) \{\eta_s(x) - 1\} \Big] \leq C$$

for all $t \leq T$.

Proof. Consider the martingale $M(t)$ defined by

$$M(t) = \|\eta(t)\|_{-1}^2 - \|\eta(0)\|_{-1}^2 - \int_0^t ds \, N^2 L_N \|\eta(s)\|_{-1}^2 .$$

A simple computation relying on the explicit formula (6.9) for the $\mathcal{H}_{-1}$ norm permits to compute $N^2 L_N \|\eta\|_{-1}^2$. Since the kernel K_N is an even function, $N^2 L_N \|\eta\|_{-1}^2$ is equal to

$$\frac{1}{N^d} \sum_{x,y \in \mathbb{T}_N^d} \eta(x) (N^2 \Delta_N K_N)(x - y) g(\eta(y))$$
$$+ 2N^2 \sum_{j=1}^d \{K_N(0) - K_N(e_j)\} \frac{1}{N^d} \sum_{x \in \mathbb{T}_N^d} g(\eta(x)) .$$

By property (6.11) of the kernel, the second term of the previous expression is bounded above by $N^{-d} \sum_{x \in \mathbb{T}_N^d} g(\eta(x))$. On the other hand, the first term can be rewritten as

$$- \frac{1}{N^d} \sum_{x,y \in \mathbb{T}_N^d} \eta(x) ([I - N^2 \Delta_N] K_N)(x - y) g(\eta(y))$$
$$+ \frac{1}{N^d} \sum_{x,y \in \mathbb{T}_N^d} \eta(x) K_N(x - y) g(\eta(y)) .$$

By (6.10), (6.9) and (6.12), this expression is bounded above by

$$- \frac{1}{N^d} \sum_{x \in \mathbb{T}_N^d} \eta(x) g(\eta(x)) + \frac{A}{2} \|\eta\|_{-1}^2 + \frac{1}{2A} \|g(\eta)\|_{-1}^2$$

for every $A > 0$. Since the $\mathcal{H}_{-1}$ norm is bounded above by the L^2 norm the third term of this sum is bounded above by $(2A)^{-1}\|g(\eta)\|_0^2$. Recollect all the previous estimates. Up to this point we proved that $N^2 L_N \|\eta\|_{-1}^2$ is bounded above by

$$-\frac{1}{N^d}\sum_{x\in\mathbb{T}_N^d} g(\eta(x))\{\eta(x)-1\} \;+\; \frac{A}{2}\|\eta\|_{-1}^2 \;+\; \frac{1}{2A}\|g(\eta)\|_0^2$$

for every $A > 0$. Since $g(k) \le g^* k$, $g(k) \le 2g^*(k-1)$ for $k \ge 2$.. In particular, $g(k)^2$ is bounded above by $g(1)^2 + 2g^* g(k)(k-1)$. Choosing therefore $A = 2g^*$, we obtain that $N^2 L_N \|\eta\|_{-1}^2$ is less than or equal to

$$-\frac{1}{2N^d}\sum_{x\in\mathbb{T}_N^d} g(\eta(x))\{\eta(x)-1\} \;+\; g^*\|\eta\|_{-1}^2 \;+\; C(g)$$

for some finite constant $C(g)$ that depends only on the jump rate.

Let $R_N(t) = \mathbb{E}_{\mu^N}[\|\eta(t)\|_{-1}^2]$. Since $M(t)$ is a mean-zero martingale, we just proved that

$$R_N(t) \;+\; \mathbb{E}_{\mu^N}\Big[\int_0^t ds\, \frac{1}{2N^d}\sum_{x\in\mathbb{T}_N^d} g(\eta_s(x))\{\eta_s(x)-1\}\Big]$$
$$\le\; R_N(0) \;+\; g^*\int_0^t ds\, R_N(s) \;+\; C(g)T$$

for all $t \le T$. We conclude the proof of the proposition applying Gronwall inequality. □

Corollary 6.2 *In the case where the jump rate g is such that $g(k) \ge a_0 k$ for some positive constant a_0, under the assumptions of the previous proposition, for each $T > 0$, there exists a finite constant C depending only on g, K_2 and T such that*

$$\mathbb{E}_{\mu^N}\Big[\|\eta(t)\|_{-1}^2\Big] \;+\; \mathbb{E}_{\mu^N}\Big[\int_0^t ds\, \frac{1}{N^d}\sum_{x\in\mathbb{T}_N^d} \eta_s(x)^2\Big] \;\le\; C$$

for all $t \le T$.

Corollary 6.3 *Under the assumptions of Theorem 1.1, all limit points Q^* of the sequence $\{Q^N, N \ge 1\}$ are concentrated on paths $\pi(t, du) = \rho(t, u)du$ such that*

$$\int_0^T dt \int_{\mathbb{T}^d} du\, \rho(t,u)^2 \;<\; \infty$$

almost surely.

Proof. It follows from the previous corollary and Schwarz inequality that

$$\limsup_{\varepsilon\to 0}\limsup_{N\to\infty} \mathbb{E}_{\mu^N}\Big[\int_0^T ds\, \frac{1}{N^d}\sum_{x\in\mathbb{T}_N^d} \big(\eta_s^{\varepsilon N}(x)\big)^2\Big] \;\le\; C$$

for some finite constant C. It is now easy to conclude the proof of the corollary. □

Remark 6.4 In view of Proposition 6.1, we may replace assumption (1.2) on the sequence of initial measures $\{\mu^N,\, N\ge 1\}$ by the weaker one (6.13)

7. An Energy Estimate

We prove in this section an energy estimate for the trajectories $\rho(t,u)$. At the end of the section we present a simple proof of uniqueness of weak solutions of equation (1.3) in dimension 1 in the class of paths satisfying an energy estimate.

Fix a limit point Q^* of the sequence Q^N and assume, without loss of generality, that the sequence Q_N converges to Q^*. The main theorem of this section can be stated as follows.

Theorem 7.1 *The probability measure Q^* is concentrated on paths $\rho(t,u)du$ with the property that there exist $L^1([0,T]\times\mathbb{T}^d)$ functions denoted by $\{\partial_{u_j}\Phi(\rho(s,u)),\ 1\le j\le d\}$ such that*

$$\int_0^T ds\int_{\mathbb{T}^d} du\,(\partial_{u_j}G)(s,u)\Phi(\rho(s,u)) \;=\; -\int_0^T ds\int_{\mathbb{T}^d} du\, G(s,u)\partial_{u_j}\Phi(\rho(s,u))$$

for all smooth functions G and all $1\le j\le d$. Moreover,

$$\int_0^T ds\int_{\mathbb{T}^d} du\,\frac{\|\nabla\Phi(\rho(s,u))\|^2}{\Phi(\rho(s,u))} \;<\;\infty\,. \tag{7.1}$$

The proof of this theorem relies on the following estimate.

Lemma 7.2 *Recall the definition of the constant K_0 introduced in Theorem 1.1. For $1\le j\le d$,*

$$E_{Q^*}\Big[\sup_H\Big\{\int_0^T ds\int_{\mathbb{T}^d} du\,(\partial_{u_j}H)(s,u)\Phi(\rho(s,u)) \\ -2\int_0^T ds\int_{\mathbb{T}^d} du\, H(s,u)^2\Phi(\rho(s,u))\Big\}\Big] \;\le\; K_0\,.$$

In this formula the supremum is taken over all functions H in $C^{0,1}([0,T]\times\mathbb{T}^d)$.

Before proving this lemma, we show how to deduce Theorem 7.1 from this statement.

Proof of Theorem 7.1. From Lemma 7.2, for Q^* almost every path ρ, there exists a finite constant $B = B(\rho)$ such that

$$\int_0^T ds \int_{\mathbb{T}^d} du\, (\partial_{u_j} H)(s,u)\Phi(\rho(s,u)) - 2\int_0^T ds \int_{\mathbb{T}^d} du\, H(s,u)^2\Phi(\rho(s,u)) \;\le B \tag{7.2}$$

for every $1 \le j \le d$ and every $C^{0,1}([0,T]\times\mathbb{T}^d)$ function H. Fix such a path ρ and consider on $C([0,T]\times\mathbb{T}^d)$ the inner product $<\cdot,\cdot>_\rho$ defined by

$$< F, G >_\rho = \int_0^T ds \int_{\mathbb{T}^d} du\, H(s,u)G(s,u)\Phi(\rho(s,u))\,.$$

Denote by L^2_ρ the Hilbert space induced by $C([0,T]\times\mathbb{T}^d)$ and this inner product.

For $1 \le j \le d$, let $\ell_j(H)$ be the linear functional on $C^{0,1}([0,T]\times\mathbb{T}^d)$ defined by

$$\ell_j(H) \;=\; \int_0^T ds \int_{\mathbb{T}^d} du\, (\partial_{u_j} H)(s,u)\Phi(\rho(s,u))\,.$$

It follows from estimate (7.2) that

$$a\ell_j(H) - 2a^2\int_0^T ds \int_{\mathbb{T}^d} du\, H(s,u)^2\Phi(\rho(s,u)) \;\le B$$

for every a in $\mathbb{R}$. Maximizing over a we show that the linear operator ℓ_j is bounded in L^2_ρ. In particular, it can be extended to a bounded linear functional in L^2_ρ. By Riesz representation theorem there exists a L^2_ρ function, denoted by $\partial_{u_j}\log\Phi(\rho(s,u))$, such that

$$\begin{aligned}\int_0^T ds \int_{\mathbb{T}^d} du\, (\partial_{u_j} G)(s,u)\Phi(\rho(s,u))\\ = -\int_0^T ds \int_{\mathbb{T}^d} du\, G(s,u)\partial_{u_j}\log\Phi(\rho(s,u))\Phi(\rho(s,u))\end{aligned}$$

for every smooth function $G\colon [0,T]\times\mathbb{T}^d \to \mathbb{R}$. Moreover,

$$\sum_{j=1}^d \int_0^T ds \int_{\mathbb{T}^d} du \left(\partial_{u_j}\log\Phi(\rho(s,u))\right)^2 \Phi(\rho(s,u)) \;<\; \infty\,.$$

To conclude the proof of the theorem, define $\partial_{u_j}\Phi(\rho(s,u))$ as $\Phi(\rho(s,u))\partial_{u_j}$ $\log\Phi(\rho(s,u))$. It is straightforward to check the properties of $\partial_{u_j}\Phi(\rho(s,u))$. □

The proof of Lemma 7.2 relies on the following estimate. Fix $1 \le j \le d$ and recall that $\{e_j,\ 1 \le j \le d\}$ stands for the canonical basis of $\mathbb{R}^d$. For a smooth function $H\colon \mathbb{T}^d \to \mathbb{R}$, $\delta > 0$, $\varepsilon > 0$ and a positive integer N, define $W_N(\varepsilon,\delta,H,\eta)$ by

$$\begin{aligned} &W_N(\varepsilon,\delta,H,\eta) \\ &\quad = N^{1-d}\sum_{x\in\mathbb{T}_N^d} H(x/N)(\varepsilon N)^{-1}\Big\{\Phi(\eta^{\delta N}(x)) - \Phi(\eta^{\delta N}(x+\varepsilon N e_j))\Big\} \\ &\qquad - 2N^{-d}\sum_{x\in\mathbb{T}_N^d} H(x/N)^2(\varepsilon N)^{-1}\sum_{k=0}^{\varepsilon N}\Phi(\eta^{\delta N}(x+ke_j))\,. \end{aligned}$$

Lemma 7.3 *Consider a sequence $\{H_\ell,\ \ell\ge 1\}$ dense in $C^{0,1}([0,T]\times\mathbb{T}^d)$. For every $k\ge 1$, and every $\varepsilon>0$,*

$$\limsup_{\delta\to 0}\limsup_{N\to\infty}\mathbb{E}_{\mu^N}\left[\max_{1\le i\le k}\left\{\int_0^T ds\, W_N(\varepsilon,\delta,H_i(s,\cdot),\eta_s)\right\}\right]\le K_0\,.$$

Proof. It follows from the replacement lemma that in order to prove the lemma we just need to show that

$$\limsup_{N\to\infty}\mathbb{E}_{\mu^N}\left[\max_{1\le i\le k}\left\{\int_0^T ds\, W_N(\varepsilon,H_i(s,\cdot),\eta_s)\right\}\right]\le K_0\,,$$

where

$$\begin{aligned} W_N(\varepsilon,H,\eta) \;=\; & N^{1-d}\sum_{x\in\mathbb{T}_N^d} H(x/N)(\varepsilon N)^{-1}\{g(\eta(x)) - g(\eta(x+\varepsilon N e_j))\} \\ & - 2N^{-d}\sum_{x\in\mathbb{T}_N^d} H(x/N)^2(\varepsilon N)^{-1}\sum_{k=0}^{\varepsilon N} g(\eta(x+ke_j))\,. \end{aligned}$$

By the entropy inequality and the Jensen inequality, for each fixed N, the previous expectation is bounded above by

$$\frac{H(\mu^N|\nu^N_{\alpha^*})}{N^d} + \frac{1}{N^d}\log\mathbb{E}_{\nu^N_{\alpha^*}}\left[\exp\left\{\max_{1\le i\le k}\left\{N^d\int_0^T ds\, W_N(\varepsilon,H_i(s,\cdot),\eta_s)\right\}\right\}\right].$$

Since $\exp\{\max_{1\le j\le k} a_j\}\le\sum_{1\le j\le k} e^{a_j}$ and since $\limsup_N N^{-d}\log\{a_N+b_N\}$ is bounded above by $\max\{\limsup_N N^{-d}\log a_N, \limsup_N N^{-d}\log b_N\}$, the limit, as $N\uparrow\infty$, of the second term of the previous expression is less than or equal to

$$K_0 + \max_{1\le i\le k}\limsup_{N\to\infty}\frac{1}{N^d}\log\mathbb{E}_{\nu^N_{\alpha^*}}\left[\exp\left\{N^d\int_0^T ds\, W_N(\varepsilon,H_i(s,\cdot),\eta_s)\right\}\right].$$

We now prove that for each fixed i the limit of the second term is nonpositive.

Fix $1\le i\le k$. By Feynman–Kac formula and the variational formula for the largest eigenvalue of a symmetric operator, for each fixed N, the second term of the previous expression is bounded above by

$$\int_0^T ds \sup_f \Big\{ \int W_N(\varepsilon, H_i(s,\cdot), \eta) f(\eta) \nu_{\alpha^*}^N(d\eta) - N^{2-d} D_N(f) \Big\} . \tag{7.3}$$

In this formula the supremum is taken over all densities f with respect to $\nu_{\alpha^*}^N$.

Recall the formula for $W_N(\varepsilon, H_i(s,\cdot), \eta)$ and that for an integer x, $\mathfrak{d}_x$ stands for the configuration with no particles but one at x. The change of variables $\xi = \eta - \mathfrak{d}_x$ shows that

$$\int g(\eta(x)) f(\eta) \nu_{\alpha^*}^N(d\eta) = \Phi(\alpha^*) \int f(\eta + \mathfrak{d}_x) \nu_{\alpha^*}^N(d\eta) .$$

In particular, we have that

$$\begin{aligned} &H(s, x/N) \int \{g(\eta(x)) - g(\eta(x + \varepsilon N e_j))\} f(\eta) \nu_{\alpha^*}^N(d\eta) \\ &\quad = \Phi(\alpha^*) H(s, x/N) \int \{f(\eta + \mathfrak{d}_x) - f(\eta + \mathfrak{d}_{x+\varepsilon N e_j})\} \nu_{\alpha^*}^N(d\eta) \\ &\quad = \Phi(\alpha^*) H(s, x/N) \sum_{k=0}^{\varepsilon N - 1} \int \{f(\eta + \mathfrak{d}_{x+ke_j}) - f(\eta + \mathfrak{d}_{x+(k+1)e_j})\} \nu_{\alpha^*}^N(d\eta) \end{aligned}$$

Writing $f(\eta + \mathfrak{d}_y) - f(\eta + \mathfrak{d}_z)$ as $(\sqrt{f(\eta + \mathfrak{d}_y)} - \sqrt{f(\eta + \mathfrak{d}_z)})(\sqrt{f(\eta + \mathfrak{d}_y)} + \sqrt{f(\eta + \mathfrak{d}_z)})$ and applying the elementary inequality $2ab \le \beta a^2 + \beta^{-1} b^2$ that holds for every a, b in $\mathbb{R}$ and $\beta > 0$, we obtain that the previous expression is bounded above by $\Phi(\alpha^*)$ times

$$\begin{aligned} &\frac{H(s, x/N)^2}{2\beta} \sum_{k=0}^{\varepsilon N - 1} \int \Big\{ \sqrt{f(\eta + \mathfrak{d}_{x+(k+1)e_j})} + \sqrt{f(\eta + \mathfrak{d}_{x+ke_j})} \Big\}^2 \nu_{\alpha^*}^N(d\eta) \\ &\quad + \frac{\beta}{2} \sum_{y=x}^{x+\varepsilon N - 1} \int \Big\{ \sqrt{f(\eta + \mathfrak{d}_{x+(k+1)e_j})} - \sqrt{f(\eta + \mathfrak{d}_{x+ke_j})} \Big\}^2 \nu_{\alpha^*}^N(d\eta) \end{aligned}$$

for every $\beta > 0$. The inequality $(a + b)^2 \le 2a^2 + 2b^2$ and the change of variables $\zeta = \eta + \mathfrak{d}_z$ for $z = x + ke_j$, $x + (k + 1)e_j$ permit to show that the first term is bounded above by

$$\frac{2H(s, x/N)^2}{\beta} \sum_{k=0}^{\varepsilon N} \int g(\eta(x + ke_j)) \nu_{\alpha^*}^N(d\eta) .$$

Therefore, setting $\beta = N$, recalling the definition of the Dirichlet form $D_N(f)$ and summing over x we obtain that

$$\begin{aligned} &N^{1-d} \sum_{x \in \mathbb{T}_N^d} H(s, x/N)(\varepsilon N)^{-1} \int \{g(\eta(x)) - g(\eta(x + \varepsilon N e_j))\} f(\eta) \nu_{\alpha^*}^N(d\eta) \\ &\quad \le \frac{2}{N^d} \sum_{x \in \mathbb{T}_N^d} H(s, x/N)^2 \frac{1}{\varepsilon N} \sum_{k=0}^{\varepsilon N} \int g(\eta(x + ke_j)) f(\eta) \nu_{\alpha^*}^N(d\eta) \\ &\quad + N^{2-d} D_N(f) . \end{aligned}$$

This proves that (7.3) is nonpositive because the Dirichlet forms cancel and the second term of $W_N(\varepsilon, H, \eta)$ is just the first term on the right hand side of the previous inequality. □

Proof of Lemma 7.2. Fix $1 \le j \le d$. Since Q_N converges weakly to Q^*, it follows from Lemma 7.3 that for every $k \ge 1$

$$\limsup_{\delta\to 0} E_{Q^*}\Big[\max_{1\le i\le k} \Big\{ \int_0^T ds \int_{\mathbb{T}^d} du \\ \Big\{ H_i(s,u)\varepsilon^{-1}\Big(\Phi(\rho_s * \iota_\delta(u)) - \Phi(\rho_s * \iota_\delta(u+\varepsilon e_j))\Big) \\ - 2\, H_i(s,u)^2\varepsilon^{-1} \int_{[u,u+\varepsilon e_j]} dv\, \Phi(\rho_s * \iota_\delta(v))\Big\}\Big\}\Big] \le K_0\,,$$

where ι_δ is the approximation of the identity $\iota_\delta(\cdot) = (2\delta)^{-1}\mathbf{1}\{[-\delta,\delta]\}(\cdot)$.

Letting $\delta \downarrow 0$, changing variables and then letting $\varepsilon \downarrow 0$, we obtain that

$$E_{Q^*}\Big[\max_{1\le i\le k} \Big\{ \int_0^T ds \int_{\mathbb{T}^d} du \\ \Big\{(\partial_{u_j} H_i)(s,u)\Phi(\rho(s,u)) - 2\, H_i(s,u)^2\Phi(\rho(s,u))\Big\}\Big\}\Big] \le K_0\,.$$

To conclude the proof it remains to apply the monotone convergence theorem and recall that $\{H_\ell,\, \ell \ge 1\}$ is a dense sequence in $C^{0,1}([0,T]\times\mathbb{T}^d)$ for the norm $\|H\|_\infty + \|(\partial_u H)\|_\infty$. □

We conclude this section with a proof in dimension 1 of the uniqueness of weak solutions in $\mathcal{H}_1$ of the Cauchy problem (1.3). We start introducing some terminology.

Definition 7.4 A measurable function $\rho\colon [0,T]\times\mathbb{T} \to \mathbb{R}_+$ is said to be a weak solution in $\mathcal{H}_1$ of (1.3) provided

(a) $\rho(t,\cdot)$ belongs to $L^1(\mathbb{T}^d)$ for every $0 \le t \le T$ and $\sup_{0\le t\le T} \|\rho_t\|_{L^1} < \infty$.

(b) There exists a function in $L^2([0,T]\times\mathbb{T})$, denoted by $\partial_u\Phi(\rho(s,u))$, such that for every smooth function $G\colon [0,T]\times\mathbb{T}\to\mathbb{R}$

$$\int_0^T ds \int_{\mathbb{T}} du\, (\partial_u G)(s,u)\Phi(\rho(s,u)) \\ = -\int_0^T ds \int_{\mathbb{T}} du\, G(s,u)\partial_u\Phi(\rho(s,u))\,.$$

(c) For every smooth function $G\colon \mathbb{R}\to\mathbb{R}$ and for every $0 < t \le T$,

$$\int_{\mathbb{T}} du\, \rho(t,u)G(u) \;-\; \int_{\mathbb{T}} du\, \rho_0(u)G(u)$$
$$= -\int_0^t ds \int_{\mathbb{T}} du\, G'(u)\partial_u \Phi(\rho(s,u))\,.$$

Notice that we require here $\partial_u \Phi(\rho(s,u))$ to belong to $L^2([0,T]\times\mathbb{T})$, while we proved only Q^* to be concentrated on paths satisfying (7.1). We need therefore further estimates on the trajectories in order to show that for zero range processes all limit points are concentrated on weak solutions in the $\mathcal{H}_1$ sense. However, for other models like generalized exclusion processes or Ginzburg–Landau processes, the same proof of Lemma 7.2 provides stronger estimates from which it is easy to deduce that all limit points are concentrated on weak solutions in the $\mathcal{H}_1$ sense.

To prove the uniqueness of weak solutions we need to introduce some notation. On the torus $\mathbb{T}$ the kernel of Δ^{-1}, the inverse of the Laplacian, is given by:

$$K(u,v) \;=\; \begin{cases} u + v(u-1) & \text{provided} \quad 0 \le v \le u \le 1\,, \\ uv & \text{provided} \quad 0 \le u \le v \le 1\,. \end{cases}$$

For a smooth function $H\colon\mathbb{T}\to\mathbb{R}$, denote by $K * H\colon\mathbb{T}\to\mathbb{R}$ the convolution of K with H:

$$(K * H)(u) \;=\; \int_{\mathbb{T}} K(u,v)H(v)\, dv\,.$$

A simple computation shows that $\Delta(K * H) = H$, what confirms that K is the kernel of Δ^{-1}.

On $C^2(\mathbb{T})$ consider the inner product $< H, G >_1$ defined by

$$< H, G >_1 = -\int_{\mathbb{T}} H(u)(\Delta G)(u)\, du\,.$$

Let $\mathcal{H}_1$ be the Hilbert space induced by $C^2(\mathbb{T})$ and the inner product $< \cdot, \cdot >_1$. The $\mathcal{H}_1$ norm is denoted by $\| \cdot \|_1$.

Denote by $\mathcal{H}_{-1}$ the dual Hilbert space of $\mathcal{H}_1$ with respect to $L^2(\mathbb{T})$. Since K is the kernel of the inverse of the Laplacian, the $\mathcal{H}_{-1}$ norm is given by

$$\|H\|_{-1}^2 \;=\; -\int H(u)(K * H)(u)\, du\,.$$

We are now in a position to investigate the uniqueness of weak solutions of (1.3).

Theorem 7.5 *There is at most one weak solution in $\mathcal{H}_1$ of equation (1.3).*

Proof. Consider two weak solutions ρ_1 and ρ_2 and denote by $\bar\rho$ their difference: $\bar\rho = \rho_2 - \rho_1$. It follows from property (c) of weak solutions that

$$\|\bar\rho(t,\cdot)\|_{-1}^2 \;=\; \|\bar\rho(0,\cdot)\|_{-1}^2 \;+\; 2\int_0^t ds \int_{\mathbb{T}} du\, \bar\rho(s,u) \int_{\mathbb{T}} dv\, (\partial_v K)(u,v)(\partial_v \bar\Phi_s)(v)$$

for every $0 \le t \le T$. In this formula, $\bar\Phi_s(v)$ stands for $\Phi(\rho_2(s,v)) - \Phi(\rho_1(s,v))$. The explicit formula for the kernel K permits to rewrite the right hand side as

$$\|\bar\rho(0,\cdot)\|_{-1}^2 \;-\; 2\int_0^t ds \int_{\mathbb{T}} du\, \bar\rho(s,u)\{\bar\Phi_s(u) - \bar\Phi_s(0)\}\,.$$

Since the total mass is conserved ($\partial_t \int_{\mathbb{T}} \rho(t,u)\, du = 0$) and ρ_1, ρ_2 are weak solutions, the second term of the previous formula is equal to

$$-\,2\int_0^t ds \int_{\mathbb{T}} du\, \bar\rho(s,u)\bar\Phi(s,u)\,,$$

which is negative because $\Phi(\cdot)$ is an increasing function and

$$\bar\Phi_s(v) = \Phi(\rho_2(s,v)) - \Phi(\rho_1(s,v))\,.$$

This computation proves that the $\mathcal{H}_{-1}$ norm of the difference of two weak solutions with the same total mass does not increase in time. Uniqueness follows. □

8. Comments and References

The entropy method presented in this chapter is due to Guo, Papanicolaou and Varadhan (1988). It permitted to prove the hydrodynamic behavior of a large class of interacting particle systems and interacting diffusion processes through the investigation of the time evolution of the entropy.

Fritz (1990) extended the entropy estimate to infinite volume for Ginzburg–Landau processes. Yau (1994) proposed an alternative method to estimate the infinite volume entropy. Landim and Mourragui (1997) adapted Yau's approach to zero range processes.

Lu (1995) proved that starting from a class of deterministic configurations, there is a *microscopic* time t_0 and a finite constant C_0 with the property that at time t_0 the entropy of the process with respect to a reference invariant measure is bounded by $C_0 N$ in dimension 1. This estimate permits to extend the entropy argument to processes starting from deterministic configurations.

Yau (1994) introduced a method, based on the time evolution of the $\mathcal{H}_{-1}$ norm, to prove a law of large numbers for the density field at a finer scale than the hydrodynamic scale for dissipative systems. It permits to show the existence of $0 < a < 1$ such that

$$N^{-d} \sum_{x \in \mathbb{Z}^d} H(x/N, \eta_t^M(x))$$

converges to $\int H(u, \rho(t,u))\, du$, with $M = N^a$. Landim and Vares (1996) applied this method to the superposition of a speeded up Kawasaki dynamics and a Glauber dynamics.

Applying the entropy method Funaki, Handa and Uchiyama (1991) prove the hydrodynamic behavior of a one-dimensional, reversible, gradient, symmetric exclusion process with speed change. Suzuki and Uchiyama (1993) show that the macroscopic evolution of a gradient reversible and conservative $[0,\infty)$-valued spin process is described by a nonlinear parabolic equation. In these models the usual entropy bound can be relaxed and a moment estimate is required. Ekhaus and Seppäläinen (1996) and Feng, Iscoe and Seppäläinen (1997) consider similar models giving rise to the porous medium equations $\partial_t \rho = \kappa \Delta \rho^\beta$, for $\kappa > 0$, $\beta > 1$.

Fritz (1989b) consider the hydrodynamic behavior of a one-dimensional Ginzburg–Landau process in random environment. Quastel (1995b) examines the same question for exclusion processes. In this case the process turns out to be nongradient. Koukkous (1997) investigates the hydrodynamic behavior of mean-zero asymmetric zero range processes in random environment. The model is the following. Consider a sequence of independent random variables $\{a_x,\, x \in \mathbb{Z}^d\}$ taking values in some interval $[a, b]$ with $a > 0$. Denote by m the distribution of a_0. For a fixed realization of the environment, consider the zero range process η_t in which a particle at x jumps to $x + y$ at rate $p(y)g(\eta(x))a_x$. In this model, for each x in $\mathbb{Z}^d$, the jump rate is speeded up or slowed down by the random factor a_x. The hydrodynamic equation of this model is shown to be

$$\partial_t \rho = \Delta_\sigma \hat{\Phi}(\rho) ,$$

where $\hat{\Phi}$ is defined as follows. Recall the definition of the function $R(\cdot)$. Let $\hat{R}(\varphi) = E_m[R(\varphi a_0^{-1})]$. $\hat{\Phi}$ is the inverse of $\hat{R}$ and Δ_σ is the second order differential operator defined just before (1.1).

Systems in contact with stochastic reservoirs. The question is to characterize the density profile in a pipe connecting two infinite reservoirs containing a fluid with two different densities in a stationary regime.

To fix ideas, consider a simple exclusion process on $\mathbb{Z}_N = \{0, 1, \dots, N - 1\}$ with symmetric jump rates in the interior of $\mathbb{Z}_N$ and with jump rates at the boundary chosen in order to obtain there a priori fixed densities. The generator of this process is:

$$(L_N f)(\eta) \;=\; \sum_{\substack{x,y \in \mathbb{Z}_N \\ |x-y|=1}} \eta(x)[1 - \eta(y)][f(\eta^{x,y}) - f(\eta)] \;+\; (L_- f)(\eta) \;+\; (L_+ f)(\eta) ,$$

where L_-, L_+ are the boundary generators given by

$$\begin{aligned}
(L_+ f)(\eta) \;&=\; \eta(N-1)[f(\eta - \mathfrak{d}_{N-1}) - f(\eta)] \\
&\quad +\; \alpha_+[1 - \eta(N-1)][f(\eta + \mathfrak{d}_{N-1}) - f(\eta)] , \\
(L_- f)(\eta) \;&=\; \eta(0)[f(\eta - \mathfrak{d}_0) - f(\eta)] \;+\; \alpha_-[1 - \eta(0)][f(\eta + \mathfrak{d}_0) - f(\eta)] .
\end{aligned}$$

Here α_- and α_+ are two positive constants that stand for the rate at which particles are created at the boundary. A simple computation shows that the Bernoulli product measure $\nu_{\alpha_+/(1+\alpha_+)}$ (resp. $\nu_{\alpha_-/(1+\alpha_-)}$) is reversible for the process with generator L_+ (resp. L_-).

Since the process is indecomposable, there exists a unique stationary measure, denoted by μ^N. Only in very special cases is this measure explicitly computable. The problem is to investigate the density profile associated to this stationary state. More precisely, to prove the existence of a profile ρ_0: $[0,1] \to \mathbb{R}_+$ such that

$$\limsup_{N\to\infty} \mu^N\Big\{\eta, \Big| < \pi^N, G > - \int_0^1 G(u)\rho_0(u)\, du\Big| > \delta\Big\} = 0$$

for every continuous function $G : [0,1] \to \mathbb{R}$ and every $\delta > 0$. One expects ρ_0 to be the solution of an elliptic equation with boundary conditions:

$$\begin{cases} \partial_u(D(\rho(u))\partial_u \rho) = 0 \,, \\ \rho(0) = \alpha_-/(1+\alpha_-)\,, \quad \rho(1) = \alpha_+/(1+\alpha_+)\,, \end{cases}$$

where $D(\cdot)$ is the diffusion coefficient defined by the Green–Kubo formula.

Fick's law of transport for the expected value of the current in the stationary regime can also be examined. For $0 \le x \le N-2$, denote by $W_{x,x+1}$ the current over the bond $\{x, x+1\}$, i.e., the rate at which a particle jumps from x to $x+1$ minus the rate at which a particle jumps from $x+1$ to x. In the example we introduced above, the current is equal to $\eta(x) - \eta(x+1)$. One would like to prove that for every u in $[0,1]$,

$$\lim_{N\to\infty} E_{\mu^N}\Big[N W_{[uN],[uN]+1}\Big] = -D(\rho_0(u))\partial_u \rho_0(u)\,.$$

Finally, we may also investigate the relaxation to equilibrium starting from a state associated to some profile. To illustrate this question, in the example introduced above, fix a profile $\gamma : [0,1] \to [0,1]$, consider a sequence of probability measures $\{\mu^N_{\gamma(\cdot)},\, N \ge 1\}$ on $\{0,1\}^{\mathbb{Z}_N}$ associated to the profile γ and denote by $\mathbb{P}^N_\gamma$ the probability on the path space corresponding to the Markov process with generator L_N speeded up by N^2 and starting from $\mu^N_{\gamma(\cdot)}$. It is natural to prove a law of large numbers for the empirical measure under $\mathbb{P}^N_\gamma$. More precisely, to show that for every $t > 0$

$$\limsup_{N\to\infty} \mathbb{P}^N_\gamma\Big\{\Big| < \pi^N_t, G > - \int_0^1 G(u)\rho(t,u)\, du\Big| > \delta\Big\} = 0$$

for every continuous function $G : [0,1] \to \mathbb{R}$ and every $\delta > 0$, provided $\rho(t,u)$ stands for the solution of the nonlinear parabolic equation

$$\begin{cases} \partial_t \rho = \partial_u(D(\rho(u))\partial_u \rho)\,, \\ \rho(0,\cdot) = \gamma(\cdot)\,, \\ \rho(\cdot,0) = \alpha_-/(1+\alpha_-)\,, \quad \rho(\cdot,1) = \alpha_+/(1+\alpha_+)\,. \end{cases}$$

Goldstein, Lebowitz and Presutti (1981) and Goldstein, Kipnis and Ianiro (1985) investigate the stationary state of N particles moving on a bounded region Λ of $\mathbb{R}^3$ according to a deterministic Hamiltonian equation in which particles are thermalized at the boundary. They prove the existence of a stationary state, which is equivalent to the Lebesgue measure and show convergence in variation norm of any probability measure under the time evolution. Goldstein, Lebowitz and Ravishankar (1982) and Farmer, Goldstein, Speer (1984) prove the existence of a nonequilibrium steady state for a one-dimensional infinite system of molecules confined in a region Λ in interaction with atoms which flow to Λ from two semi-infinite reservoirs separated by Λ.

Kipnis, Marchioro and Presutti (1982) consider a one-dimensional system of harmonic oscillators in contact with reservoirs at different temperature. They obtain the stationary measure, the temperature profile and prove the local convergence to the Gibbs measure. De Masi, Ferrari, Ianiro and Presutti (1982) proved the same statement in the context of symmetric simple exclusion processes in contact with stochastic reservoirs at different temperature.

Ferrari and Goldstein (1988) consider a symmetric simple exclusion process on $\mathbb{Z}^3$ with creation and destruction of particles at the origin. They deduce the density profile of the nonequilibrium stationary measure, that turns out to be non product, and compute the decay of the two point correlation function. Lebowitz, Neuhauser and Ravishankar (1996) deduce asymptotic occupation properties of the stationary measure of a semi–infinite asymmetric one-dimensional particle system with a source at the origin, coupled jumps and annihilation. This is a first approximation of the so-called Toom cellular automaton.

For zero range processes, as noticed by De Masi and Ferrari (1984), the stationary measure of a system in contact with an infinite reservoir is a product measure with slowly varying parameter. All computations are thus explicit.

Based on the entropy method introduced by Guo, Papanicolaou and Varadhan (1988), Eyink, Lebowitz and Spohn (1990, 1991) obtained the macroscopic profile of the stationary measure and proved the hydrodynamic behavior of the system for a gradient exclusion process where the jump rates depend locally on the configuration. Kipnis, Landim, Olla (1995) extended this result to a nongradient generalized exclusion process relying on Varadhan's nongradient method (Varadhan (1994a), Quastel (1992)). Systems in contact with stochastic reservoirs have never been considered in higher dimensions.

Onsager's reciprocity relations. Consider a zero range process with two types of particles. For $a = 1, 2$, fix jump rates $g_a\colon \mathbb{N} \times \mathbb{N} \to \mathbb{R}_+$ and mean-zero transition probabilities $p_a\colon \mathbb{Z}^d \to \mathbb{R}_+$ ($\sum_y p_a(y) = 1$, $\sum_y y p_a(y) = 0$). Define the generator of the Markov process (η_t, ξ_t) on $\mathbb{N}^{\mathbb{T}_N^d} \times \mathbb{N}^{\mathbb{T}_N^d}$ by

$$(L_N f)(\eta, \xi) = \sum_{a=1}^{2} \sum_{x,y \in \mathbb{T}_N^d} p_a(y) g_a(\eta(x), \xi(x)) (\sigma_a^{x,x+y} f)(\eta, \xi) ,$$

where

$$(\sigma_1^{x,x+y} f)(\eta,\xi) = [f(\eta^{x,x+y},\xi) - f(\eta,\xi)]$$
$$\text{and}\quad (\sigma_2^{x,x+y} f)(\eta,\xi) = [f(\eta,\xi^{x,x+y}) - f(\eta,\xi)]\ .$$

If the jump rates are not degenerated, this process has only two conserved quantities, the total number of η and ξ particles. Moreover, for each $\alpha = (\alpha_1, \alpha_2)$ in $\mathbb{R}_+ \times \mathbb{R}_+$, there exists an invariant measure, denoted by $\nu^N_{\alpha_1,\alpha_2}$, with global density of η-particles (resp. ξ-particles) equal to α_1 (resp. α_2): $E_{\nu^N_{\alpha_1,\alpha_2}}[N^{-d}\sum_x \eta(x)] = \alpha_1$, $E_{\nu^N_{\alpha_1,\alpha_2}}[N^{-d}\sum_x \xi(x)] = \alpha_2$.

Assume that this family of invariant measures has good regularity properties in order to be able to define local equilibrium states. For each profile $\rho = (\rho_1,\rho_2)\colon \mathbb{T}^d \to (\mathbb{R}_+)^2$ of density $\alpha = (\alpha_1,\alpha_2)$ ($\int_{\mathbb{T}^d} \rho_a(u)du = \alpha_a$, $a = 1, 2$), denote by $\mu^N_{\rho_1,\rho_2}$ a local equilibrium of profile ρ. Assume that the specific entropy of $\mu^N_{\rho_1,\rho_2}$ with respect to $\nu^N_{\alpha_1,\alpha_2}$ converges, as $N \uparrow \infty$, and denote by $S(\rho_1,\rho_2)$ its limit: $S(\rho_1,\rho_2) = \lim_{N\to\infty} N^{-d} H(\mu^N_{\rho_1,\rho_2} | \nu^N_{\alpha_1,\alpha_2})$. Suppose that the entropy is written as the integral of a density $s(\rho)$: $S(\rho_1,\rho_2) = \int_{\mathbb{T}^d} s(\rho_1(u),\rho_2(u))du$.

The macroscopic behavior of a system starting from $\mu^N_{\rho_1,\rho_2}$ is expected to be described by a system of diffusion equations: denote by $\rho_a(t,u)$ the density of (η,ξ)–particles at the macroscopic point u at time t, $\rho(t,u)$ should evolve according to the equation

$$\begin{cases} \partial_t \rho = \sum_{i,j=1}^{d} \partial_{u_i}\{D_{i,j}(\rho)\partial_{u_j}\rho\}\ , \\ \rho(0,\cdot) = \rho(\cdot)\ . \end{cases}$$

Here ρ stands for the vector (ρ_1,ρ_2) and $D_{i,j}(\rho)$ is a two by two matrix for each $1 \le i,j \le d$.

The Onsager coefficients are defined in this context by

$$L_{i,j}(\rho) = D_{i,j}(\rho)\cdot R(\rho)\ ,$$

where the matrix R is determined by the entropy density $s(\rho(u))$ in the following way

$$(R^{-1})_{a,b} = \frac{\partial^2}{\partial\rho_a(u)\partial\rho_b(u)} s(\rho(u))\ ,$$

which is by definition a symmetric matrix. Onsager's reciprocity relations (cf. Onsager (1931a,b) mean that the matrices $L_{i,j} = \{L^{a,b}_{i,j},\ 1 \le a,b \le 2\}$ are such that $L^{a,b}_{i,j} = L^{b,a}_{j,i}$ for $1 \le i,j \le d$, $1 \le a,b \le 2$. The exact microscopic conditions required to prove these relations are far to be understood.

Gabrielli, Jona–Lasinio and Landim (1996) proved Onsager's reciprocity relations for mean-zero, non reversible zero range processes in the case where the jump rates are $g_1(\eta(x),\xi(x)) = g(\eta(x)+\xi(x))\{\eta(x)/(\eta(x)+\xi(x))\}$, $g_2(\eta(x),\xi(x)) = g(\eta(x)+\xi(x))\{\xi(x)/(\eta(x)+\xi(x))\}$ for some jump rate $g\colon \mathbb{N} \to \mathbb{R}_+$ and $p_1 = p_2$. In this case the invariant measures are product and all computations presented in this chapter can be done explicitly. Moreover, the Onsager matrices L are diagonal.

Lebowitz and Spohn (1997) observed that the previous models belong in fact to a larger class that has a mirror type symmetry: the dynamics is invariant by the exchange of η and ξ particles. Gabrielli, Jona–Lasinio and Landim (1998) present sufficient conditions to guarantee the validity of Onsager's reciprocity relations for a general class of interacting particle systems. This question is discussed in great generality in Eyink, Lebowitz and Spohn (1996)

6. The Relative Entropy Method

In Chapter 1 we introduced in the context of interacting particle systems the physical concepts of local equilibrium and conservation of local equilibrium and we proved the persistence of local equilibrium in a model where particles evolve independently. Consider a particle system η_t evolving on the torus $\mathbb{T}_N^d$ and possessing a family $\{\nu_\alpha^N, \alpha \ge 0\}$ of product invariant measures indexed by the density. Fix a profile $\rho_0: \mathbb{T}^d \to \mathbb{R}_+$ and assume that the process η_t has a hydrodynamic behavior described by the solution $\rho(t,u)$ of some partial differential equation with initial condition ρ_0. Denote by μ^N a sequence of initial states associated to the profile ρ_0 and by μ_t^N the state at the macroscopic time t of the process that started from μ^N. The conservation of local equilibrium states that μ_t^N should be close to the product measure $\nu_{\rho(t,\cdot)}^N$ with slowly varying parameter associated to $\rho(t,\cdot)$.

In contrast with the entropy method, where the hydrodynamic behavior is deduced from the investigation of the time evolution of the entropy $H(\mu_t^N|\nu_\alpha^N)$ of the state of the process with respect to a fixed invariant measure, the relative entropy method examine the time evolution of the entropy $H(\mu_t^N|\nu_{\rho(t,\cdot)}^N)$ of the state of the process with respect to the product measure $\nu_{\rho(t,\cdot)}^N$ with slowly varying parameter associated to the solution of the hydrodynamic equation.

The relative entropy method requires some regularity of the solution of the hydrodynamic equation. Of course, this is not a restriction for systems where the average displacement of each elementary particle has mean zero since the macroscopic behavior of these processes are described by second order quasi–linear equations, whose weak solutions are smooth. This is not the case, however, for asymmetric processes described by first order hyperbolic equations whose solutions develop shocks. In this latter case, the relative entropy method allows to deduce the hydrodynamic limit of the system up to the appearance of the first shock. In fact, the relative entropy approach is the unique method that derives the hydrodynamic behavior of non attractive asymmetric processes.

A last remark concerns the assumptions on the solutions of the hydrodynamic equation. While the entropy method to be implemented requires a theorem asserting the uniqueness of weak solutions and proves the existence of weak solutions, the relative entropy method requires the existence of a smooth solution and proves the uniqueness of such smooth solutions.

1. Weak Conservation of Local Equilibrium

To illustrate the relative entropy method, we consider a mean-zero asymmetric zero range process on the torus $\mathbb{T}_N^d$. This is the Markov process introduced in Chapter 2 whose generator is

$$(L_N f)(\eta) = \sum_{x,y\in\mathbb{T}_N^d} p(y)g(\eta(x))\{f(\eta^{x,x+y}) - f(\eta)\} .$$

To avoid minor technical difficulties, we assume the transition probability $p(\cdot)$ to be associated to a finite range mean-zero random walk:

(i)

$$\sum_{x\in\mathbb{Z}^d} x_j p(x) = 0 \qquad \text{for} \quad 1 \le j \le d .$$

(ii) There exists an integer A_0 such that $p(x) = 0$ if $|x| \ge A_0$.

Notice that we did not assume the matrix $p(\cdot)$ to be symmetric and thus the process to be reversible.

In the proof of the hydrodynamic behavior by means of the relative entropy method, we shall need the one block estimate. This result (Lemma 5.3.1) was proved in the previous chapter under hypotheses (**FEM**) or (**SLG**) on the jump rate $g(\cdot)$. We shall therefore assume either one of these hypotheses throughout this chapter without mentioning it again.

We now describe the hydrodynamic equation. Denote by $\sigma = (\sigma_{i,j})_{1\le i,j\le d}$ the matrix of correlations of the displacement of an elementary particle:

$$\sigma_{i,j} = \sum_{x\in\mathbb{Z}^d} x_i x_j \, p(x)$$

and by Δ_σ the second order differential operator

$$\Delta_\sigma = \sum_{1\le i,j\le d} \sigma_{i,j} \, \partial_{u_i} \partial_{u_j} .$$

Let $0 < e < 1$ and $\rho_0 \colon \mathbb{T}^d \to \mathbb{R}_+$ be a profile of class $C^{2+e}(\mathbb{T}^d)$. The Cauchy problem

$$\begin{cases} \partial_t \rho = \Delta_\sigma \Phi(\rho) \\ \rho(0,\cdot) = \rho_0(\cdot) \end{cases} \tag{1.1}$$

admits a classic solution, that we denote by $\rho(t,u)$, twice continuously differentiable in space and once continuously differentiable in time (cf. Oleinik et Kružkov (1961)). Moreover, for each $t \ge 0$, the profile $\rho(t,\cdot)$ is of class $C^{2+e}(\mathbb{T}^d)$.

To avoid uninteresting technical difficulties, we assume that the initial profile $\rho_0(\cdot)$ is bounded below by a strictly positive constant:

$$K_1 \;:=\; \inf_{u\in\mathbb{T}^d} \rho_0(u) \;>\; 0\,.$$

By the maximum principle, for every $t > 0$, $\rho(t,\cdot)$ is bounded below by K_1:

$$\inf_{t\ge 0}\inf_{u\in\mathbb{T}^d} \rho(t,u) \;=\; K_1$$

and above by the L^∞ norm of $\rho_0(\cdot)$, denoted by K_2:

$$\sup_{t\ge 0}\sup_{u\in\mathbb{T}^d} \rho(t,u) \;=\; \sup_{u\in\mathbb{T}^d} \rho_0(u) \;=:\; K_2 < \infty\,.$$

Hereafter, for $t \ge 0$, we denote by $\nu^N_{\rho(t,\cdot)}$ the product measure with slowly varying parameter associated to the profile $\rho(t,\cdot)$ (cf. Definition 3.0.1):

$$\nu^N_{\rho(t,\cdot)}\{\eta; \eta(x) = n\} \;=\; \nu_{\rho(t,x/N)}\{\eta, \eta(0) = n\}\,, \quad \text{for} \quad x\in\mathbb{T}^d_N \quad \text{and} \quad n\in\mathbb{N}\,.$$

We may now state the main result of this chapter.

Theorem 1.1 *Under the assumption* **(FEM)** *or* **(SLG)**, *let* $(\mu^N)_{N\ge1}$ *be a sequence of probability measures on* $\mathbb{N}^{\mathbb{T}^d_N}$ *whose entropy with respect to* $\nu^N_{\rho(0,\cdot)}$ *is of order* $o(N^d)$:

$$H(\mu^N|\nu^N_{\rho(0,\cdot)}) \;=\; o(N^d)\,.$$

Then, the relative entropy of the state of the process at the macroscopic time t with respect to $\nu^N_{\rho(t,\cdot)}$ *is also of order* $o(N^d)$:

$$H\big(\mu^N S^N_t|\nu^N_{\rho(t,\cdot)}\big) \;=\; o(N^d) \qquad \textit{for every} \quad t\ge 0\,.$$

In this formula, S^N_t *stands for the semi-group associated to the generator* L_N *speeded up by* N^2.

Remark 1.2 Fix a bounded profile $\rho_0\colon \mathbb{T}^d \to \mathbb{R}_+$. The computation performed in Remark 5.1.2 shows that every sequence of probability measures μ^N with entropy $H(\mu^N|\nu^N_{\rho_0(\cdot)})$ of order $o(N^d)$ is such that

$$H(\mu^N|\nu^N_\alpha) \;=\; O(N^d)$$

for every $\alpha > 0$.

Before proving Theorem 1.1, we deduce the conservation of local equilibrium in the weak sense, as defined in Chapter 3.

Corollary 1.3 *Under the assumptions of the theorem, for every continuous function* $H\colon \mathbb{T}^d \to \mathbb{R}$ *and every bounded cylinder function* Ψ,

$$\lim_{N\to\infty} E_{\mu^N S^N_t}\Big[\Big|N^{-d}\sum_{x\in\mathbb{T}^d_N} H(x/N)\tau_x\Psi(\eta) \;-\; \int_{\mathbb{T}^d} H(u)E_{\nu_{\rho(t,u)}}[\Psi]\,du\Big|\Big] \;=\; 0\,.$$

Proof. To concentrate exclusively on the essential problems, we assume that the cylinder function Ψ depends on the configuration η only through $\eta(0)$:

$$\Psi(\eta) = \Psi(\eta(0)) .$$

Since both functions $H(\cdot)$ and $\rho(t, \cdot)$ are continuous and since the cylinder function Ψ is bounded, a summation by parts shows that in order to prove the corollary we only need to check that

$$\limsup_{\ell\to\infty} \limsup_{N\to\infty} E_{\mu^N S_t^N} \Big[N^{-d} \sum_{x\in\mathbb{T}_N^d} \Big| (2\ell+1)^{-d} \sum_{|y-x|\le\ell} \Psi(\eta(y)) - E_{\nu_{\rho(t,x/N)}}[\Psi] \Big| \Big] \le 0 .$$

By the entropy inequality, for every $\gamma > 0$, the expectation in the previous formula is bounded above by

$$\frac{1}{\gamma N^d} H(\mu^N S_t^N | \nu_{\rho(t,\cdot)}^N) + \frac{1}{\gamma N^d} \log E_{\nu_{\rho(t,\cdot)}^N} \Big[\exp\Big\{ \gamma \sum_{x\in\mathbb{T}_N^d} \Big| (2\ell+1)^{-d} \sum_{|y-x|\le\ell} \Psi(\eta(y)) - E_{\nu_{\rho(t,x/N)}}[\Psi] \Big| \Big\} \Big] .$$

At the end of the proof, we shall choose γ as a function of ℓ. By Theorem 1.1, the first term converges to 0 as $N \uparrow \infty$. On the other hand, since the measure $\nu_{\rho(t,\cdot)}^N$ is product, the random variables $(2\ell+1)^{-d} \sum_{|y-x_1|\le\ell} \Psi(\eta(y))$ and $(2\ell+1)^{-d} \sum_{|y-x_2|\le\ell} \Psi(\eta(y))$ are independent as soon as $|x_1 - x_2| > 2\ell$. In particular, by Hölder inequality, the second term is bounded above by

$$\frac{1}{\gamma N^d} \sum_{x\in\mathbb{T}_N^d} \frac{1}{(2\ell+1)^d} \log E_{\nu_{\rho(t,\cdot)}^N} \Big[\exp \gamma \Big| \sum_{|y-x|\le\ell} \Big\{ \Psi(\eta(y)) - E_{\nu_{\rho(t,x/N)}}[\Psi] \Big\} \Big| \Big] .$$

This step is explained in more details in the proof of Lemma 1.8. Since the profile $\rho(t, \cdot)$ is continuous, as $N \uparrow \infty$, this sum converges to

$$\int_{\mathbb{T}^d} du \, \frac{1}{\gamma(2\ell+1)^d} \log E_{\nu_{\rho(t,u)}} \Big[\exp \gamma \Big| \sum_{|y|\le\ell} \Big\{ \Psi(\eta(y)) - E_{\nu_{\rho(t,u)}}[\Psi] \Big\} \Big| \Big] .$$

Since the cylinder function Ψ is bounded, it follows from the elementary identities $e^x \le 1 + x + 2^{-1} x^2 e^{|x|}$, $\log(1+x) \le x$, that this integral is bounded above by

$$\int_{\mathbb{T}^d} du \, \frac{1}{\gamma(2\ell+1)^d} \Big\{ \gamma E_{\nu_{\rho(t,u)}} \Big[\Big| \sum_{|y|\le\ell} \Big\{ \Psi(\eta(y)) - E_{\nu_{\rho(t,u)}}[\Psi] \Big\} \Big| \Big] + 2\gamma^2 (2\ell+1)^{2d} \|\Psi\|_\infty^2 \exp\big\{ 2\gamma(2\ell+1)^d \|\Psi\|_\infty \big\} \Big\} .$$

To conclude the proof, it remains to choose $\gamma = (2\ell+1)^{-d}\epsilon$. In this case, by the law of large numbers, this expression converges to 0 as $\ell \uparrow \infty$ and than $\epsilon \downarrow 0$. □

The proof of Theorem 1.1 is divided in several lemmas. We start introducing some notation used throughout this chapter. For $\alpha > 0$, ν_α^N stands for a reference invariant measure and $\psi_N(t)$ is the Radon–Nikodym derivative of $\nu_{\rho(t,\cdot)}^N$ with respect to ν_α^N:

$$\psi_N(t) \;:=\; \frac{d\nu_{\rho(t,\cdot)}^N}{d\nu_\alpha^N}\,.$$

A simple computation allows to obtain an explicit formula for $\psi_N(t)$ because the measures $\nu_{\rho(t,\cdot)}^N$, ν_α^N are product and the profile $\rho(t,\cdot)$ is bounded below by a strictly positive constant uniformly in time:

$$\psi_N(t) \;=\; \exp\left\{\sum_x \Big[\eta(x)\log\Phi_\alpha(\rho(t,x/N)) - \log Z_\alpha(\rho(t,x/N))\Big]\right\}.$$

In this formula Φ_α and Z_α are given by

$$\Phi_\alpha(\beta) = \frac{\Phi(\beta)}{\Phi(\alpha)} \qquad \text{and} \qquad Z_\alpha(\beta) = \frac{Z(\Phi(\beta))}{Z(\Phi(\alpha))}\,,$$

where $\Phi(\cdot)$ and the partition function $Z(\cdot)$ have been defined in section 2 of Chapter 3. To keep notation as simple as possible, hereafter we denote by μ_t^N the measure μ^N at macroscopic time t:

$$\mu_t^N \;:=\; \mu^N S_t^N\,,$$

by $f_N(t) = f_t^N$ the Radon–Nikodym derivative of μ_t^N with respect to the reference measure ν_α^N:

$$f_N(t) \;:=\; \frac{d\mu_t^N}{d\nu_\alpha^N} \;=\; \frac{d\mu^N S_t^N}{d\nu_\alpha^N}$$

and by $H_N(t)$ the relative entropy of μ_t^N with respect to $\nu_{\rho(t,\cdot)}^N$:

$$H_N(t) \;=\; H\big(\mu_t^N \,|\, \nu_{\rho(t,\cdot)}^N\big)\,.$$

Since μ_t^N is absolutely continuous with respect to $\nu_{\rho(t,\cdot)}^N$, the explicit formula for the relative entropy presented in Theorem A1.8.3 gives that

$$\begin{aligned} H_N(t) \;&=\; \int \frac{d\mu_t^N}{d\nu_{\rho(t,\cdot)}^N}\log\frac{d\mu_t^N}{d\nu_{\rho(t,\cdot)}^N}\,d\nu_{\rho(t,\cdot)}^N \\ &=\; \int f_t^N(\eta)\log\left[\frac{f_t^N(\eta)}{\psi_t^N(\eta)}\right]\,\nu_\alpha^N(d\eta)\,. \end{aligned}$$

We turn now to the proof of Theorem 1.1. The strategy consists in estimating the relative entropy $H_N(t)$ by a term of order $o(N^d)$ and the time integral of the entropy multiplied by a constant:

$$H_N(t) \;\le\; o(N^d) \;+\; \gamma^{-1}\int_0^t H_N(s)\,ds$$

and apply Gronwall lemma to conclude. The first step stated in Lemma 1.4 below gives an upper bound for the entropy production.

Lemma 1.4 *For every $t \ge 0$,*

$$\partial_t H_N(t) \;\le\; \int \frac{1}{\psi_t^N} \left\{ N^2 L_N^* \psi_t^N - \partial_t \psi_t^N \right\} f_t^N \, d\nu_\alpha^N \,,$$

where L_N^* is the adjoint of L_N in $L^2(\nu_\alpha^N)$.

Proof. We have seen in Chapter 5 that $f_N(t)$ is the solution of the Kolmogorov forward equation

$$\partial_t f_N(t) \;=\; N^2 L_N^* f_N(t) \,.$$

Since the profile $\rho(t,\cdot)$ is smooth, a simple computation shows that

$$\begin{aligned} \partial_t H_N(t) \;&=\; \int N^2 L_N^* f_t^N \cdot \log\left[\frac{f_t^N}{\psi_t^N}\right] d\nu_\alpha^N \\ &+\; \int \left\{ N^2 L_N^* f_t^N - f_t^N \, \frac{\partial_t \psi_t^N}{\psi_t^N} \right\} d\nu_\alpha^N \,. \end{aligned}$$

Since L_N^* is the adjoint of L_N in $L^2(\nu_\alpha)$,

$$\int L_N^* f_t^N \, d\nu_\alpha^N \;=\; 0 \,.$$

By the same reason, the first expression on the right hand side may be rewritten as

$$N^2 \int \psi_t^N \, \frac{f_t^N}{\psi_t^N} \, L_N \left(\log \frac{f_t^N}{\psi_t^N} \right) d\nu_\alpha \,.$$

The elementary inequality

$$a\,[\log b - \log a] \;\le\; (b-a)$$

that holds for positive reals a, b, shows that for every positive function h and for every generator L of a jump process,

$$h\, L(\log h) \;\le\; L h \,.$$

In particular, the last integral is bounded above by

$$N^2 \int \psi_t^N \, L_N \left(\frac{f_t^N}{\psi_t^N} \right) d\nu_\alpha^N \;=\; N^2 \int \frac{f_t^N}{\psi_t^N} \, L_N^* \psi_t^N \, d\nu_\alpha^N \,. \qquad \square$$

We now estimate the upper bound for the entropy production obtained in the previous lemma using the explicit formula for ψ_t^N. First of all, a simple computation shows that $(\psi_t^N)^{-1} N^2 L_N^* \psi_t^N$ is given by

$$N^2 \sum_{x,y\in\mathbb{T}_N^d} g(\eta(x))\, p(x-y) \left[\frac{\Phi(\rho(t,y/N))}{\Phi(\rho(t,x/N))} - 1\right],$$

that is well defined because $\rho(t,\cdot)$ is strictly positive. On the other hand, the sum

$$N^2 \sum_{x,y\in\mathbb{T}_N^d} \Phi(\rho(t,x/N))\, p(x-y) \left[\frac{\Phi(\rho(t,y/N))}{\Phi(\rho(t,x/N))} - 1\right] \tag{1.2}$$

clearly vanishes. Therefore, if Δ_σ stands for the second order differential operator defined in the beginning of this chapter, since the solution of (1.1) is of class $C^{2+e}(\mathbb{T}^d)$, Taylor expansion gives that

$$\begin{aligned}&\left(\psi_t^N(\eta)\right)^{-1} N^2 L_N^* \psi_t^N(\eta)\\ &\quad= \sum_{x\in\mathbb{T}_N^d} \frac{(\Delta_\sigma\Phi)(\rho(t,x/N))}{\Phi(\rho(t,x/N))} \left[g(\eta(x)) - \Phi(\rho(t,x/N))\right] + o(N^d)\,.\end{aligned}$$

We shall see in a while the reason for adding a vanishing term to $(\psi_t^N)^{-1}$ $N^2 L_N^* \psi_t^N$.

The identity

$$\frac{Z'(\varphi)}{Z(\varphi)} = \frac{R(\varphi)}{\varphi} \tag{1.3}$$

proved in section 2.3 and the fact that ρ is the solution of equation (1.1) gives that

$$\begin{aligned}&(\psi_t^N(\eta))^{-1}\, \partial_t \psi_t^N(\eta) = \partial_t \left(\log \psi_t^N\right)(\eta)\\ &\quad= \sum_{x\in\mathbb{T}_N^d} \frac{\Delta_\sigma(\Phi(\rho(t,x/N)))}{\Phi(\rho(t,x/N))}\, \Phi'(\rho(t,x/N))\,[\eta(x) - \rho(t,x/N)]\,.\end{aligned}$$

To keep notation as simple as possible, we denote by $F(t,\cdot)$ the function of class $C^e(\mathbb{T}^d)$, $\Phi(\rho(t,\cdot))^{-1}\Delta_\sigma(\Phi(\rho(t,\cdot)))$. Up to this point we proved that

$$\begin{aligned}&\left(\psi_t^N(\eta)\right)^{-1} [N^2 L_N^* - \partial_t]\,(\psi_t^N(\eta))\\ &\quad= \sum_{x\in\mathbb{T}_N^d} F(t,x/N) \Big\{g(\eta(x)) - \Phi(\rho(t,x/N)) \\ &\qquad - \Phi'(\rho(t,x/N))\,[\eta(x) - \rho(t,x/N)]\Big\} + o(N^d)\,.\end{aligned} \tag{1.4}$$

It is important to stress that a microscopic Taylor expansion up to the second order appeared in this formula since, by local equilibrium, the mean value of $g(\eta(\cdot))$ at the microscopic point x is given by $E_{\nu_{\rho(t,x/N)}}[g(\eta(0))] = \Phi(\rho(t,x/N))$. This explains why we introduced above the term (1.2).

To fully take advantage of the Taylor expansion that appeared, the next step consists in applying the one block estimate to replace the cylinder function $g(\eta(x))$ by $\Phi(\eta_t^\ell(x))$, the expected value of $g(\eta(x))$ under the invariant measure with density equal to the empirical density of particles in a microscopic box centered at x. We shall obtain in this way a second order Taylor expansion of $\Phi(\eta_t^\ell(x))$ around $\Phi(\rho(t,x/N))$ in formula (1.4).

Lemma 1.5 *Under the assumptions* (**FEM**) *or* (**SLG**)*, for every* $t > 0$

$$\lim_{\ell\to\infty}\lim_{N\to\infty} \mathbb{E}_{\mu^N}\left[\int_0^t \frac{1}{N^d}\sum_{x\in\mathbb{T}_N^d} F(s,x/N)\left\{g(\eta_s(x)) - \Phi(\eta_s^\ell(x))\right\} ds\right] = 0 .$$

Here $\mathbb{E}_{\mu^N}$ stands for the expectation with respect to $\mathbb{P}_{\mu^N}$, the probability measure on the path space $D([0,T],\mathbb{N}^{\mathbb{T}_N^d})$ induced by the Markov process with generator L_N speeded up by N^2 starting from μ^N and, for a positive integer ℓ, $\eta^\ell(x)$ stands for the empirical density of particles in a cube of length ℓ centered at x:

$$\eta^\ell(x) = \frac{1}{(2\ell+1)^d}\sum_{|y-x|\le\ell}\eta(y) .$$

Lemma 1.5 is proved in section 5.4. It permits to replace in formula (1.4) the cylinder function $g(\eta(x))$ by its mean value $\Phi(\eta^\ell(x))$. On the other hand, since $F(t,\cdot)\Phi'(\rho(t,\cdot))$ is a continuous function, a summation by parts permits to replace $\eta(x)$ in the same formula by $\eta^\ell(x)$. We may thus rewrite $(\psi_t^N)^{-1}\{N^2L_N^* - \partial_t\}\psi_t^N$ as

$$\sum_{x\in\mathbb{T}_N^d} F(t,x/N)\Big\{\Phi(\eta^\ell(x)) - \Phi(\rho(t,x/N))$$
$$- \Phi'(\rho(t,x/N))\,[\eta^\ell(x) - \rho(t,x/N)] + o(N^d) .$$

Here $o(N^d)$ stands for an expression whose expectation is of order $o(N^d)$ as $N\uparrow\infty$ and $\ell\uparrow\infty$. In conclusion, it follows from Lemma 1.4, Lemma 1.5 and the computations just performed that for every $t > 0$ the entropy $H_N(t)$ is bounded above by

$$H_N(0) + \mathbb{E}_{\mu^N}\Big[\int_0^t \sum_{x\in\mathbb{T}_N^d} F(s,x/N)M(\eta_s^\ell(x),\rho(s,x/N))\,ds\Big] + o(N^d) ,$$

where

$$M(a,b) = \Phi(a) - \Phi(b) - \Phi'(b)\,(a-b) .$$

Besides the expectation, all terms in this expression are of order $o(N^d)$ since we assumed the initial entropy to be of this order. To conclude the proof of the theorem it remains to show that the expectation is bounded by the sum of a term of order $o(N^d)$ and the time integral of the entropy multiplied by a constant. This estimate is obtained through the entropy inequality. We start rewriting the last expectation as

$$\int_0^t ds\, E_{\mu_s^N}\Big[\sum_{x\in\mathbb{T}_N^d} F(s,x/N)\,M(\eta^\ell(x),\rho(s,x/N))\Big] .$$

By the entropy inequality, for every $\gamma > 0$, this integral is bounded above by

$$\gamma^{-1} \int_0^t ds\, H_N(s)$$
$$+ \gamma^{-1} \int_0^t ds \log E_{\nu^N_{\rho(s,\cdot)}} \Big[\exp\Big\{ \gamma \sum_{x \in \mathbb{T}^d_N} F(s, x/N) M(\eta^\ell(x), \rho(s, x/N)) \Big\} \Big] .$$

The next result concludes the proof of Theorem 1.1.

Proposition 1.6 *There exists $\gamma_0 > 0$ such that for all $0 \le s \le t$*

$$\limsup_{\ell \to \infty} \limsup_{N \to \infty} \frac{1}{N^d} \log E_{\nu^N_{\rho(s,\cdot)}} \Big[\exp \Big\{ \gamma_0 \sum_x F(s, x/N) M(\eta^\ell(x), \rho(s, x/N)) \Big\} \Big] \le 0 . \tag{1.5}$$

A rigorous and complete proof of this result is a bit long. The idea is however simple, and relies on large deviations arguments. Since the measure $\nu^N_{\rho(s,\cdot)}$ are product, the random variables $\eta^\ell(x_1)$, $\eta^\ell(x_2)$ are independent as soon as $|x_1 - x_2| > 2\ell$. In particular, the Laplace–Varadhan theorem and a large deviations principle for i.i.d. random variables give an upper bound for the left hand side of (1.5) of the form

$$\int_{\mathbb{T}^d} du \sup_\lambda \Big\{ \gamma_0 F(s, u) M(\lambda, \rho(s, u)) - J_{\rho(s,u)}(\lambda) \Big\} .$$

where $J_\beta(\cdot)$ is a rate function strictly convex vanishing at β. Since $M(\cdot, \beta)$ also vanishes at β, is quadratic close to β and linear at infinity, it will not be difficult to show that the supremum vanishes for γ_0 small enough.

We conclude this section with a rigorous proof of Proposition 1.6. Fix a sequence of i.i.d. random variables with distribution

$$P^0_\beta[X_1 = k] = \frac{\Phi(\beta)^k}{g(1) \cdots g(k)} \frac{1}{Z(\Phi(\beta))} , \qquad k \in \mathbb{N} \tag{1.6}$$

and recall the large deviations principle:

Lemma 1.7 *The sequence $\big(N^{-1} \sum_{k=1}^N X_k\big)$ satisfies a large deviations principle with rate function given by*

$$J^1_\beta(\lambda) = \begin{cases} \lambda \log \left(\dfrac{\Phi(\lambda)}{\Phi(\beta)} \right) - \log \left(\dfrac{Z(\Phi(\lambda))}{Z(\Phi(\beta))} \right) & \textit{for} \ \ \lambda \ge 0 \\ \infty & \textit{otherwise} . \end{cases} \tag{1.7}$$

We refer to Deuschel and Stroock (1989) for a proof of this large deviations principle. This lemma provides an upper bound for the left hand side of (1.5).

In order to deduce this bound, we need to introduce some notation. Recall that φ^* (which might be infinite) stands for the radius of convergence of the partition function $Z(\cdot)$ defined in section 2.3. Let $\rho: \mathbb{T}^d \to \mathbb{R}_+$ be a continuous function bounded by K_2. We shall denote by $\nu^N_{\rho(\cdot)}$ the product measure with slowly varying parameter associated to ρ.

Lemma 1.8 *Let $G : \mathbb{T}^d \times \mathbb{R}_+ \to \mathbb{R}$ be a continuous function such that*

$$\sup_{u\in\mathbb{T}^d} |G(u,\lambda)| \le C_0 + C_1 \lambda \quad \textit{for all} \quad \lambda \in \mathbb{R}_+ \tag{1.8}$$

where C_0 is a finite constant and C_1 a constant bounded by $\log[\varphi^*/\Phi(K_2)]$*:*

$$C_1 < \log \frac{\varphi^*}{\Phi(K_2)} .$$

Then,

$$\limsup_{\ell\to\infty} \limsup_{N\to\infty} \frac{1}{N^d} \log E_{\nu^N_{\rho(\cdot)}} \Big[\exp \sum_{x\in\mathbb{T}^d_N} G(x/N, \eta^\ell(x)) \Big]$$
$$\le \int_{\mathbb{T}^d} du \sup_{\lambda\ge 0} \big\{ G(u,\lambda) - J^1_{\rho(u)}(\lambda) \big\} ,$$

where the rate function $J^1_\beta(\cdot)$ is defined in (1.7).

If the partition function $Z(\cdot)$ is finite on $\mathbb{R}_+$, the assumption in the previous lemma on C_1 requires only C_1 to be a finite constant.

Proof. Since $\eta^\ell(0)$ depends on the variables $\eta(x)$ only for $|x| \le \ell$, $\eta^\ell(x)$ and $\eta^\ell(y)$ are independent under $\nu^N_{\rho(\cdot)}$ for $|y-x| \ge 2\ell+1$. We shall take advantage of this property to decompose the expectation in a product of simpler terms. Assume, without loss of generality, that $2\ell+1$ divides N.

The sum $\sum_x G(x/N, \eta^\ell(x))$ can be rewritten as

$$\sum_{x\in\Lambda_\ell} \sum_{y; x+(2\ell+1)y\in\mathbb{T}^d_N} G\Big(\frac{x+(2\ell+1)y}{N}, \eta^\ell(x+(2\ell+1)y)\Big) ,$$

where Λ_ℓ is a cube of length $2\ell+1$ centered at the origin:

$$\Lambda_\ell = \{-\ell, \dots, \ell\}^d .$$

It is important to remark that the variables $\{\eta^\ell(x+(2\ell+1)y),\ y\}$ are independent under $\nu^N_{\rho(\cdot)}$ for each x fixed. Therefore, by Hölder inequality and by independence,

$$\log E_{\nu^N_{\rho(\cdot)}} \Big[\exp \sum_{x\in\mathbb{T}^d_N} G(x/N, \eta^\ell(x)) \Big]$$

is bounded by

$$\frac{1}{(2\ell+1)^d} \sum_{x \in \Lambda_\ell} \log E_{\nu^N_{\rho(\cdot)}} \Big[$$

$$\exp (2\ell+1)^d \sum_y G\Big(\frac{x+(2\ell+1)y}{N}, \eta^\ell(x+(2\ell+1)y)\Big)\Big]$$

$$= \frac{1}{(2\ell+1)^d} \sum_{x \in \mathbb{T}^d_N} \log E_{\nu^N_{\rho(\cdot)}} \Big[\exp (2\ell+1)^d G(x/N, \eta^\ell(x))\Big] .$$

For a positive β, let P_β be the probability in $(\mathbb{R}_+)^{\mathbb{N}}$ corresponding to a sequence of i.i.d. random variables with distribution given by (1.6) and denote by E_β expectation with respect to P_β. Since, by assumption, the function $G(\cdot,\cdot)$ and the profile $\rho(\cdot)$ are continuous and the family $\{\nu^N_a, a \ge 0\}$ of product measures defined in (2.3.6) is weakly continuous in virtue of Lemma 2.3.8; the last line divided by N^d converges, as $N \uparrow \infty$, to

$$\frac{1}{(2\ell+1)^d} \int_{\mathbb{T}^d} du \, \log E_{\rho(u)} \left[\exp (2\ell+1)^d G(u, \bar{X}_{(2\ell+1)^d})\right] , \tag{1.9}$$

where, for a positive integer k, $\bar{X}_k$ stands for the average of the first k elements:

$$\bar{X}_k = \frac{1}{k} \sum_{j=1}^k X_j .$$

If G was bounded, it would follow from the large deviations principle for the sequence $(X_j)_{j \ge 1}$ stated in Lemma 1.7 and the Laplace–Varadhan theorem (cf. Theorem A2.3.1), that the limit, as $\ell \uparrow \infty$, of the last line is equal to

$$\int_{\mathbb{T}^d} du \sup_{\lambda > 0} \left\{G(u,\lambda) - J^1_{\rho(u)}(\lambda)\right\} .$$

A cut off argument permits to reduce the general case to the case of bounded functions. We first compute, for each fixed u, the limit of the expression in (1.9). We shall than argue to exchange the limit with the integral.

We start with the upper bound. Fix $\beta = \rho(u)$. For $A > 0$, denote by $G_A(\cdot,\cdot)$ the function G truncated at the level A:

$$G_A(u,\lambda) = G(u,\lambda)\mathbf{1}\{|\lambda| \le A\} + G(u,A)\mathbf{1}\{|\lambda| > A\} .$$

By Hölder inequality and by the assumption (1.8),

$$E_\beta \Big[\exp\{\ell G(u, \bar{X}_\ell)\}\Big]$$
$$\le E_\beta \left[\exp\{\ell G_A(u,\bar{X}_\ell)\}\right] + E_\beta \left[\exp\{\ell[C_0 + C_1 \bar{X}_\ell]\} \mathbf{1}\{|\bar{X}_\ell| \ge A\}\right]$$
$$\le E_\beta \Big[\exp\Big\{\ell G_A(u,\bar{X}_\ell)\Big\}\Big]$$
$$+ \, e^{C_0 \ell} E_\beta^{1/q} \left[\mathbf{1}\{|\bar{X}_\ell| \ge A\}\right] E_\beta^{1/p} \Big[\exp\Big\{pC_1 \sum_{k=1}^\ell X_k\Big\}\Big] .$$

In this last formula, p and q are conjugates: $p > 1$ and $p^{-1} + q^{-1} = 1$ and p is chosen such that $E_{\nu_\beta}[e^{pC_1\eta(0)}] = Z(e^{pC_1}\Phi(\beta))/Z(\Phi(\beta)) < \infty$. This choice is possible because $C_1 < \log[\varphi^*/\Phi(K_2)]$ and the function ρ is bounded by K_2. Therefore, since the variables X_k are independent, we have that

$$\frac{1}{\ell} \log E_\beta \left[\exp\left\{pC_1 \sum_{k=1}^{\ell} X_k\right\}\right] = \log E_\beta \left[e^{pC_1 X_1}\right]$$

is finite. On the other hand, by the large deviations principle for the sequence $(X_k)_{k\geq 1}$,

$$\lim_{A\to\infty} \limsup_{\ell\to\infty} \frac{1}{\ell} \log P_\beta [|\bar{X}_\ell| \geq A] \leq \lim_{A\to\infty} -J^1_\beta(A) = -\infty .$$

The last equality follows from the explicit formula for the rate function J^1_β.

Finally, G_A being bounded, by the large deviations principle for the sequence $(X_k)_{k\geq 1}$ and by the Laplace–Varadhan theorem, for every A,

$$\lim_{\ell\to\infty} \frac{1}{\ell} \log E_\beta \left[\exp\{\ell G_A(u, \bar{X}_\ell)\}\right] = \sup_{\lambda>0} \left\{G_A(u,\lambda) - J^1_\beta(\lambda)\right\} .$$

We have thus proved that

$$\limsup_{\ell\to\infty} \frac{1}{\ell} \log E_\beta \left[\exp\{\ell G(u, \bar{X}_\ell)\}\right] \leq \lim_{A\to\infty} \sup_{\lambda>0} \left\{G_A(u,\lambda) - J^1_\beta(\lambda)\right\} .$$

To derive the upper bound, it remains to show that the term on the right hand side of the last inequality is equal to

$$\sup_{\lambda>0} \left\{G(u,\lambda) - J^1_\beta(\lambda)\right\} .$$

It is enough to show that we may restrict the supremum on the right hand side of the inequality to a compact subset of $[0,\infty)$ because for each compact there exists a real A_0 such that G and G_{A_0} coincide in this compact for every $A > A_0$.

By assumption (1.8), we have that

$$G(u,\lambda) - J^1_\beta(\lambda) \leq C_0 + C_1\lambda - J^1_\beta(\lambda) .$$

Taking the derivative of the rate function $J^1_\beta(\cdot)$ and keeping in mind identity (1.3) and that $J^1_\beta(\beta) = 0$, we deduce that

$$J^1_\beta(\lambda) = \int_\beta^\lambda \log\left(\frac{\Phi(\sigma)}{\Phi(\beta)}\right) d\sigma .$$

From the choice of the constant C_1,

$$\lim_{\lambda\to\infty} \frac{1}{\lambda} J^1_\beta(\lambda) = \log \frac{\varphi^*}{\Phi(\beta)} > C_1 .$$

In consequence,

$$\limsup_{\lambda\to\infty}\left[\sup_A G_A(u,\lambda) - J^1_\beta(\lambda)\right] \le \lim_{\lambda\to\infty}\left\{C_0 + C_1\lambda - J^1_\beta(\lambda)\right\} = -\infty$$

and we may restrict the supremum to a compact subset of $[0,\infty)$.

A lower bound for the expression in (1.9) can be proved with similar arguments:

$$\liminf_{\ell\to\infty}\frac{1}{\ell}\log E_\beta\left[\exp\{\ell G(u,\bar X_\ell)\}\right] \ge \sup_{\lambda>0}\{G(u,\lambda) - J^1_\beta(\lambda)\}\,.$$

It remains to justify the exchange of the limit and the integral. From assumption (1.8) and from the definition of the constant C_1, we deduce a bound, uniform over ℓ and u, of

$$\frac{1}{(2\ell+1)^d}\log E_{\rho(u)}\left[\exp(2\ell+1)^d G(u,\bar X_{(2\ell+1)^d})\right]\,.$$

The theorem of the dominated convergence permits to conclude the proof of the lemma. □

Applying the previous lemma to the function

$$G(u,\lambda) = \gamma F(s,u)\left\{\Phi(\lambda) - \Phi(\rho(s,u)) - \Phi'(\rho(s,u))[\lambda - \rho(s,u)]\right\}$$

we conclude the first step of the proof of Proposition 1.6. We summarize the conclusions in the next corollary. To state it notice that the just defined function G is such that

$$\sup_{u\in\mathbb{T}^d}\left|G(u,\lambda)\right| \le \gamma\,\|F\|_\infty\left\{2g^*\lambda + \sup_{\beta\in[0,K_2]}\Phi(\beta) + \sup_{\beta\in[0,K_2]}\beta\Phi'(\beta)\right\}$$

because $\Phi(0) = 0$ and, by Corollary 2.3.6, $0 \le \Phi(\beta) - \Phi(\alpha) \le g^*(\beta - \alpha)$ for $\alpha \le \beta$ so that $\Phi(\lambda) \le g^*\lambda$ and $\Phi'(\lambda) \le g^*$. In this formula, $\|F\|_\infty$ stands for the $L^\infty([0,t]\times\mathbb{T}^d)$ norm of F:

$$\|F\|_\infty = \sup_{(s,u)\in[0,t]\times\mathbb{T}^d}|F(s,u)|\,.$$

Corollary 1.9 *Recall that K_2 stands for the upper bound for the initial profile ρ_0. Let*

$$\gamma_1 = \frac{1}{2\|F\|_\infty g^*}\log\frac{\varphi^*}{\Phi(K_2)}\,.$$

Then, for all $\gamma < \gamma_1$ and all $0 \le s \le t$,

$$\limsup_{\ell\to\infty}\limsup_{N\to\infty}$$
$$\frac{1}{N^d}\log E_{\nu^N_{\rho(s,\cdot)}}\left[\exp\left\{\gamma\sum_{x\in\mathbb{T}^d_N}F(s,x/N)M(\eta^\ell(x),\rho(s,x/N))\right\}\right]$$
$$\le \int_{\mathbb{T}^d}du\,\sup_{\lambda>0}\left\{\gamma F(s,u)M(\lambda,\rho(s,u)) - J^1_{\rho(s,u)}(\lambda)\right\}.$$

To conclude the proof of Proposition 1.6, we have to show that the right hand side of the previous inequality is non positive for all γ sufficiently small. This result follows from the next lemma.

Lemma 1.10 *For every* $0 < K_1 < K_2 < \infty$,

$$C_2 := \sup_{\substack{\beta\in[K_1,K_2]\\ \lambda\geq 0}} \frac{|M(\lambda,\beta)|}{J^1_\beta(\lambda)} < \infty .$$

Proof. We first choose $0 < \epsilon < K_1/2$. Throughout this proof, K_1^ϵ and K_2^ϵ stand respectively for $K_1 - \epsilon$ and $K_2 + \epsilon$. We decompose the set $[K_1, K_2] \times \mathbb{R}_+$ in three disjoint subsets ($\lambda << \beta$, $\lambda \sim \beta$ and $\lambda >> \beta$) and prove the result in each of these subsets by different ways. We start with the region $\lambda \sim \beta$.

Consider the set

$$\mathcal{E}_1 = \left\{(\lambda,\beta) \in \mathbb{R}_+ \times [K_1,K_2] \; ; \; K_1^\epsilon \leq \lambda \leq K_2^\epsilon\right\} .$$

Let A be the constant defined by

$$A = \sup_{K_1^\epsilon\leq\sigma\leq K_2^\epsilon} |\Phi''(\sigma)| .$$

Taylor expansion permit to bound M on $\mathcal{E}_1$:

$$|M(\lambda,\beta)| \leq \frac{A}{2}(\beta-\lambda)^2 \qquad \text{for} \quad (\lambda,\beta)\in\mathcal{E}_1 .$$

On the other hand, a simple computation taking advantage of the relation (2.3.5), permits to compute the first two derivatives of the rate function J^1_β:

$$\partial_\lambda J^1_\beta(\lambda) = \log\left(\frac{\Phi(\lambda)}{\Phi(\beta)}\right) \quad , \quad \partial^2_\lambda J^1_\beta(\lambda) = \frac{\Phi'(\lambda)}{\Phi(\lambda)} .$$

In particular, both J^1_β and its derivative vanish at β. Let B be the constant defined by

$$B = \inf_{K_1^\epsilon\leq\sigma\leq K_2^\epsilon} (\partial^2_\sigma J^1_\beta)(\sigma) = \inf_{K_1^\epsilon\leq\sigma\leq K_2^\epsilon} \left(\frac{\Phi'(\sigma)}{\Phi(\sigma)}\right) .$$

B is strictly positive because Φ and Φ' are smooth functions strictly positive on $(0,\infty)$. J^1_β and its derivative vanishing at β, by Taylor expansion and the definition of B,

$$J^1_\beta(\lambda) \geq \frac{B}{2}(\lambda-\beta)^2 \qquad \text{for} \quad (\lambda,\beta)\in\mathcal{E}_1 .$$

In conclusion, for $(\lambda,\beta)\in\mathcal{E}_1$

$$\frac{|M(\lambda,\beta)|}{J^1_\beta(\lambda)} \leq \frac{A}{B} =: C_3 < \infty .$$

We turn now to the set

$$\mathcal{E}_2 = \left\{(\lambda, \beta) \in \mathbb{R}_+ \times [K_1, K_2] \,;\, \lambda \geq K_2^\epsilon\right\} .$$

Notice that on this set $\lambda \geq \beta + \epsilon$. On the one hand, since $\Phi(\beta) \leq g^* \beta$ and $\lambda \geq K_2^\epsilon > \beta$,

$$\begin{aligned} \frac{1}{\lambda} |M(\lambda, \beta)| \; &\leq \; \frac{1}{\lambda} \Phi(\lambda) + \Phi'(\beta) + \frac{1}{\lambda} [\Phi(\beta) + \beta \Phi'(\beta)] \\ &\leq \; 2 [g^* + \Phi'(\beta)] \, . \end{aligned}$$

On the other hand, since $J_\beta^1(\beta) = 0$, $(\partial_\lambda J_\beta^1)(\cdot) = \log(\Phi(\cdot)/\Phi(\beta))$ and $\lambda \geq \epsilon + \beta$, by an integration by parts,

$$\begin{aligned} \frac{1}{\lambda} J_\beta^1(\lambda) \; &= \; \frac{1}{\lambda} \int_\beta^\lambda [\lambda - \sigma] \frac{\Phi'(\sigma)}{\Phi(\sigma)} \, d\sigma \\ &\geq \; \frac{\epsilon}{2(\epsilon + \beta)} \int_\beta^{\beta + \epsilon/2} \frac{\Phi'(\sigma)}{\Phi(\sigma)} \, d\sigma \, . \end{aligned}$$

This last expression is denoted by $C(\beta)$. Thus,

$$\sup_{(\lambda,\beta) \in \mathcal{E}_2} \frac{|M(\lambda, \beta)|}{J_\beta^1(\lambda)} \; \leq \; \sup_{\beta \in [K_1, K_2]} \frac{2[g^* + \Phi'(\beta)]}{C(\beta)} \; =: \; C_4 < \infty$$

because $\Phi'(\cdot)$ et $C(\cdot)$ are continuous and positive.

Finally, for the set $\mathcal{E}_3 = \left\{(\lambda, \beta) \in \mathbb{R}_+ \times [K_1, K_2] \,;\, \lambda \leq K_1^\epsilon\right\}$ we proceed in the following way. Since on this set $\lambda \leq K_1 - \epsilon \leq \beta - \epsilon$, by Taylor expansion

$$|M(\lambda, \beta)| \; \leq \; \frac{1}{2} (\beta - \lambda)^2 \sup_{\sigma \leq \beta} |\Phi''(\sigma)| \; \leq \; \frac{1}{2} \beta^2 \sup_{\sigma \leq K_2} |\Phi''(\sigma)| \, .$$

On the other hand, repeating the arguments presented for the set $\mathcal{E}_2$,

$$J_\beta^1(\lambda) \; = \; \int_\lambda^\beta [\sigma - \lambda] \frac{\Phi'(\sigma)}{\Phi(\sigma)} \, d\sigma \; \geq \; \frac{\epsilon}{2} \int_{\beta - \epsilon/2}^\beta \frac{\Phi'(\sigma)}{\Phi(\sigma)} \, d\sigma \, .$$

We denote by $2^{-1} C^*(\beta)$ this last function. From these two inequalities, we deduce an estimate on the set $\mathcal{E}_3$:

$$\sup_{(\lambda,\beta) \in \mathcal{E}_3} \frac{|M(\lambda, \beta)|}{J_\beta^1(\lambda)} \; \leq \; \sup_{\beta \in [K_1, K_2]} \frac{K_2^2 \sup_{\sigma \leq K_2} |\Phi''(\sigma)|}{C^*(\beta)} \; =: \; C_5 < \infty \, .$$

It remains to set $C_2 = C_3 \vee C_4 \vee C_5$ to conclude the proof of the lemma. □

Corollary 1.11 *There exists $\gamma_0 > 0$ such that for all $\gamma < \gamma_0$*

$$\sup_{(s,u) \in [0,t] \times \mathbb{T}^d} \left\{\gamma F(s, u) M(\lambda, \rho(s, u)) - J_{\rho(s,u)}^1(\lambda)\right\} \; \leq \; 0 \, .$$

Proof. Straightforward from Lemma 1.10 because F is bounded in $[0,t] \times \mathbb{T}^d$ and the range of $\rho(\cdot,\cdot)$ is contained in $[K_1, K_2]$. □

Remark 1.12 Without the assumption concerning the order of magnitude of the entropy at time 0, the same proof gives an upper bound for the entropy production. Indeed, if for $t \geq 0$ $H(t)$ stands for the limit of the specific entropy:

$$H(t) \;:=\; \limsup_{N\to\infty} N^{-d} H_N(t) \;;$$

the arguments presented above show that there exists $\gamma > 0$ such that

$$H(t) \;\leq\; H(0) \;+\; \frac{1}{\gamma}\int_0^t H(s)\, ds \;.$$

Remark 1.13 The special form of the hydrodynamic equation (1.1) played no special role. We just needed the existence of a smooth solution of the hydrodynamic equation. In particular, the relative entropy method extends to a large class of interacting particle systems that includes conservative asymmetric dynamics, described by first order quasi–linear hyperbolic equations, up to the appearance of the first shock.

Remark 1.14 As noticed in Chapter 5, for attractive processes either one of the assumptions (**SLG**) or (**FEM**) is fulfilled. Theorem 1.1 applies therefore to attractive zero range processes.

2. Comments and References

The method presented in this chapter is due to Yau (1991). It was extensively used to investigate the first order correction to the hydrodynamic equation. This topic is discussed in the last section of the next chapter.
Euler equations. Olla, Varadhan and Yau (1993) considered a superposition of an Hamiltonian dynamics with an infinite range stochastic noise on the velocities that exchanges momenta and preserves the conserved quantities (the density, the momentum and the energy). Adapting the relative entropy method to this context, they proved that the conserved quantities evolve according to the Euler equations

$$\begin{cases} \partial_t \rho + \sum_{j=1}^{3} \partial_{u_j}\{\rho\pi^j\} = 0\,, \\[2ex] \partial_t(\rho\pi^i) + \sum_{j=1}^{3} \partial_{u_j}\{\rho\pi^i\pi^j + \delta_{i,j}P\} = 0\,, \\[2ex] \partial_t(\rho e) + \sum_{j=1}^{3} \partial_{u_j}\{\rho e\pi^j - \pi^j P\} = 0\,, \end{cases}$$

in the time interval where the solutions of these equations are smooth. In this formula ρ stands for the density, π for the velocity per particle, e for the energy per particle and P is the pressure, a function of ρ, π, e.

One of the main ingredients in this derivation is the proof of the ergodicity of the dynamics. Liverani and Olla (1996) proved ergodicity for Hamiltonian systems superposed to finite range stochastic interactions on the velocities: they proved that translation invariant measures that are stationary for the deterministic Hamiltonian dynamics, reversible for the stochastic dynamics and have finite specific entropy are convex combinations of Gibbs states. Fritz, Liverani and Olla (1997) removed the requirement of reversibility proving that all translation invariant stationary states of finite specific entropy are reversible with respect to the stochastic evolution. For lattice systems this question has been solved by Fritz, Funaki and Lebowitz (1994): They proved that all translation invariant stationary states with finite local entropy are microcanonical Gibbs states in the case of Hamiltonian systems with a local random perturbation that conserves the energy.

Cahn–Hilliard equations. Bertini, Landim and Olla (1997) deduced the Cahn–Hilliard equation

$$\partial_t \rho = \Delta(F'(\rho)\Delta\rho)$$

from a stochastic microscopic Ginzburg–Landau dynamics. Giacomin and Lebowitz (1997a) examined an interacting particle system evolving according to a local mean field Kawasaki dynamics and showed that the hydrodynamic equation is given by

$$\partial_t \rho = \nabla \cdot \left\{ A(\rho) \nabla \frac{\delta}{\delta \rho} F(\rho) \right\} ,$$

where $A(\rho) = \beta\rho(1-\rho)$, β is the inverse of the temperature and

$$F(\rho) \;=\; -\frac{1}{\beta}\int_{\mathbb{T}^d} du\, s(\rho(u)) \;-\; \frac{1}{2}\int_{\mathbb{T}^d} du \int_{\mathbb{T}^d} dv\, J(u-v)\rho(u)\rho(v)\, .$$

In this formula $s(\alpha) = -\alpha\log\alpha - (1-\alpha)\log(1-\alpha)$ and $J(\cdot)$ is the mean field interaction. Giacomin and Lebowitz (1997b) compared the solution of these equations with the behavior of solutions of Cahn–Hilliard equations.

Reaction–diffusion equations. De Masi, Ferrari and Lebowitz (1986) considered a superposition of Glauber and speeded up Kawasaki dynamics to obtain reaction–diffusion equations. In these models at most one particle is allowed per site. To describe the stochastic evolution, fix a positive cylinder function $c(\eta)$. For each site x, at rate N^2 the occupation variables $\eta(x)$ and $\eta(x+e_i)$ are exchanged and at rate $c(\tau_x\eta)$ the occupation variable $\eta(x)$ is flipped. De Masi, Ferrari and Lebowitz (1986) proved that the hydrodynamic behavior of the system is given by the solution of the reaction–diffusion equation

$$\partial_t \rho \;=\; \Delta\rho \;+\; F(\rho)\, ,$$

where $F(\alpha) = E_{\nu_\alpha}[(1-2\eta(0))c(\eta)]$ and ν_α is the Bernoulli product measure of density α. Mourragui (1996) extended this analysis to zero range processes with creation and annihilation of particles applying the relative entropy method. Nappo, Orlandi (1988) and Nappo, Orlandi and Rost (1989) deduce a non linear reaction–diffusion equation for Brownian particles moving on $\mathbb{R}^d$ with an interaction that kills a particle at some rate which depends on its distance to the others.

Noble (1992) and Durrett and Neuhauser (1994) investigate the behavior of a superposition of an attractive Glauber dynamics with a speeded up Kawasaki dynamics. Using the result of De Masi, Ferrari and Lebowitz (1986) they prove the existence of non trivial stationary states for a wide class of examples provided the stirring rate is large enough.

Bramson and Lebowitz (1991) consider a system with two types of particles that evolve according to independent random walks. When two particles of different type meet, they annihilate each other. They investigate the limit density of each type of particles and they examine the spatial structure of the process. In this model the critical dimension is 4 and while in dimension $d < 4$ there is segregation of types of particles, in dimension $d > 4$ there is coexistence of the two types.

Stefan problems. Chayes and Swindle (1996) analyze a one-dimensional exclusion process with two types of particles. The first type of particle evolves as an usual exclusion process and the other type is kept frozen. Superposed to this evolution there is an annihilation mechanism that either eliminates one particle of each type when they meet or that transforms a free particle in a frozen particle when their distance reaches 1. The hydrodynamic behavior of these systems for a class of initial states is shown to be described by the solution of a Stefan problem with one free boundary:

$$\begin{cases} \partial_t \rho = \Delta \rho \,, \\ \rho(0,u) = \rho_0(u) \quad \text{for } 0 \le u \le B_0 \,, \\ \rho(t,0) = a(t) \,; \end{cases} \qquad \begin{cases} B(0) = B_0 \,, \\ \rho(t, B(t)) = 0 \,, \\ (dB/dt)(t) = \pm(\partial_u \rho)(t, B(t)) \,. \end{cases}$$

Landim, Olla and Volchan (1997), (1998) considered a nearest neighbor one-dimensional symmetric simple exclusion process with an asymmetric tagged particle. The hydrodynamic behavior is given by the solution of the Stefan problem

$$\begin{cases} \partial_t \rho = (1/2)\Delta \rho \,, \\ -\, v_t = (\partial_u \log \rho)(t, v_t+) = (\partial_u \log \rho)(t, v_t-) \,, \\ p\{1 - \rho(t, v_t+)\} = q\{1 - \rho(t, v_t-)\} \,, \\ \rho(0, \cdot) = \rho_0(\cdot) \,. \end{cases}$$

Gravner and Quastel (1998) derived the hydrodynamic equation of a system where particles are created at a finite number of fixed sites and then perform zero range random walks. Each particle jumping to a site occupied by less than κ particles is kept frozen at this site. They showed that the macroscopic behavior of this process is described by the solution of a Stefan problem.

Carleman and Broadwell equation. The Carleman equation is a special case of a discrete Boltzmann equation. It describes the evolution of two types of particles on $\mathbb{R}$ whose density $\rho_0(t,\cdot)$, $\rho_1(t,\cdot)$ evolves according to

$$\begin{cases} \partial_t \rho_a + m\, \partial_u \rho_a = (\rho_{1-a})^2 - (\rho_a)^2 \, , \\ \rho_a(0,\cdot) = \rho_{a,0}(\cdot) \, , \quad a = 0, 1 \, . \end{cases}$$

De Masi and Presutti (1991) deduce this equation by the method of truncated correlation functions from a microscopic model where two type of particles evolve on the discrete torus $\mathbb{T}_N^d$ according to independent asymmetric random walks speeded up by N. Particles of type 0 jump only to the right nearest neighbor while particles of type 1 jump only to the left nearest neighbor. Superposed to this displacement there is a collision dynamics whose generator is

$$(L_c f)(\eta) = \sum_{a=0}^{1} \sum_{x \in \mathbb{T}_N^d} \eta(x,a)[\eta(x,a) - 1][f(\eta - 2\mathfrak{d}_{x,a} + 2\mathfrak{d}_{x,1-a}) - f(\eta)] \, .$$

In this formula, $\eta(x,a)$ stands for the total number of a-particles at site x and $\mathfrak{d}_{x,a}$ is the configuration with no particles but one a-particle at site x.

Carleman equation is derived by Caprino, De Masi, Presutti and Pulvirenti (1989, 1990) from diffusion processes evolving on $\mathbb{R}$. The two-dimensional version of the system gives rise to the Broadwell equation and is derived in Caprino, De Masi, Presutti and Pulvirenti (1991) in the time interval where the equation admits smooth solutions.

Boltzmann equations. Rezakhanlou (1996a) considered an interacting particles system from which he deduced a discrete Boltzmann equation. Fix a finite set $\mathcal{I}$ of labels. Particles evolve on $\mathbb{Z}$, each one with a label a in $\mathcal{I}$. A particle with label a evolves independently according to a random walk with finite range transition probability $p_a(\cdot)$ with mean drift $q_a = \sum_x x p_a(x)$. Two particles at the same site with labels a, b collide with probability N^{-1}, the rescaled interdistance between sites. If they collide they gain new labels a', b' at rate $K(a,b;a',b')$. These rates are chosen symmetric and vanish when the conservation of momentum is violated: $K(a,b;a',b') = K(b,a;a',b') = K(a,b;b',a')$ and $K(a,b;a',b') = 0$ when $q_a + q_b \neq q'_a + q'_b$ or when $q_a = q_b$.

Starting from a product measure associated to a bounded integrable profile $\rho_0(\cdot) = \{\rho_0^a(\cdot),\, a \in \mathcal{I}\}$, Rezakhanlou (1996a) proved that the macroscopic evolution of the empirical measure is described by the discrete Boltzmann equation

$$\partial_t \rho^a + q_a \partial_u \rho^a = \sum_{b,c,d} \left\{ K(c,d;a,b)\rho^c \rho^d - K(a,b;c,d)\rho^a \rho^b \right\} . \tag{2.1}$$

Rezakhanlou (1996b) proved the propagation of chaos (cf. Chapter 8 for the terminology) for this model. Rezakhanlou and Tarver (1997) deduced the hydrodynamic equation (2.1) for a one-dimensional, continuous version of the previous model. Here, instead of moving according to random walks, each particle moves deterministically with a velocity determined by its label. The collision rules are similar but the assumption on conservation of momentum is dropped. The proof of the above results in higher dimension remains an open problem.

Rezakhanlou (1997) proved the equilibrium fluctuations in any dimension for the model introduced by Rezakhanlou and Tarver (1997). He showed that the rescaled fluctuation field converges to an Ornstein–Uhlenbeck process with a drift given by the linearized Boltzmann equation.

Caprino and Pulvirenti (1995) consider a system of N identical particles moving freely on $\mathbb{R}$ until they collide. Particles collide independently with probability ε. They prove that in the Boltzmann–Grad limit, i.e., as $N \uparrow \infty$ and $\varepsilon N \to \lambda$, the density profile converges to the solution of a Boltzmann equation, globally in time. Caprino and Pulvirenti (1996) consider the same evolution with stochastic reflection at the boundary of the interval $[0, 1]$. They show that the density profile of the unique invariant measure converges in the Boltzmann–Grad limit to the solution of the Boltzmann stationary equation.

Degenerate diffusions. Rezakhanlou (1990) considers Ginzburg–Landau models where the equilibrium states are canonical Gibbs measures for a finite range interaction. He deduces the hydrodynamic behavior of the system under assumptions on the interaction that do not exclude the possibility of phase transition, in which case the diffusion coefficient might vanish. Carmona and Xu (1997) extend this result to the case of random finite range interactions.

Lebowitz, Orlandi and Presutti (1991) consider a class of one-dimensional, infinite volume exclusion processes with a small drift toward the region of higher density. They deduce a non–linear parabolic hydrodynamic equation of type $\partial_t \rho = \partial_u(D(\rho)\partial_u \rho)$. The diffusion coefficient might be negative on an interval (a, b). In this case the hydrodynamic equation is proved only for initial data taking values on $[0, a) \cup (b, 1]$. Giacomin (1991) extends this investigation to reversible models. Computer simulations suggest that the system undergoes phase segregation on the scale of the interaction and that the system does not change on the macroscopic scale, indicating that the diffusion coefficient vanishes in this region. There are, however, no rigorous results.

Carlson, Grannan, Swindle and Tour (1993) prove the hydrodynamic behavior of a one-dimensional symmetric exclusion process in which a particle at x jumps to $x+y$ at rate $c(|y|)$ provided all sites between x and y are occupied. If $c(\cdot)$ decays slowly, the diffusion coefficient $D(\cdot)$ has a singularity at $\alpha = 1$: $\lim_{\alpha\to 1} D(\alpha) = \infty$.

Interface motion. In dimension 1 the hydrodynamic behavior of an interacting particle system can be interpreted as the motion of an interface (cf. for instance De Masi, Ferrari and Vares (1989)). Indeed, consider to fix ideas a nearest neighbor symmetric zero range process on $\{0, \dots, N\}$ with reflexive boundary conditions. The generator of this process is given by $L_N = \sum_{0\le x\le N-1} L_{x,x+1}$, where

$$(L_{x,x+1} f)(\eta) \;=\; g(\eta(x))[f(\eta^{x,x+1}) - f(\eta)] \;+\; g(\eta(x+1))[f(\eta^{x+1,x}) - f(\eta)]\;.$$

Denote by χ the "integral" of the configuration η: $\chi(x) = \sum_{0\le y\le x} \eta(y)$. χ_t is itself a Markov process with one conserved quantity $\chi(N)$ and with generator

$\mathcal{L}_N = \sum_{0 \le x \le N-1} \mathcal{L}_{x,x+1}$, where

$$(\mathcal{L}_{x,x+1} f)(\chi) = g(\chi(x) - \chi(x-1))[f(\chi - \mathfrak{d}_x) - f(\chi)] + g(\chi(x+1) - \chi(x))[f(\chi + \mathfrak{d}_x) - f(\chi)] \,.$$

In this formula $\mathfrak{d}_x$ stands for a configuration with no particles but one at x.

Denote by $\chi_N(t, \cdot) : [0,1] \to \mathbb{R}_+$ the profile $\chi_N(t,u) = N^{-1}\chi_{tN^2}([uN])$. It is easy to check that $\chi_N(t,\cdot)$ converges to some nondecreasing function $h(t,\cdot)$ if and only if the empirical measure associated to the η_t process converges to an absolutely continuous measure with density $\rho(t,u) = (\partial_u h)(t,u)$. In particular, it follows from the hydrodynamic behavior of the symmetric zero range process that $\chi_N(t,\cdot)$ converges to $h(t,\cdot)$, as $N \uparrow \infty$, where $h(t,\cdot)$ is the solution of the equation

$$\begin{cases} \partial_t h = \partial_u \Phi(\partial_u h)\,, \\ h(0,\cdot) = h_0(\cdot)\,, \\ h(\cdot,0) = 0\,, \quad h(\cdot,1) = a_0\,. \end{cases}$$

Marchand and Martin (1986) used similar ideas to investigate the macroscopic behavior of a droplet evolving according to a Glauber dynamics in $\mathbb{Z}^2$.

The whole problem is to extend these ideas to higher dimensions. At the moment there are very few rigorous results. Naddaf and Spencer (1997) and Funaki and Spohn (1997) considered d-dimensional, continuous spins Ginzburg–Landau models on a cube with periodic boundary conditions. The spins $\{\phi_t(x),\, |x| \le N\}$ evolve according to the differential equations

$$d\phi_t(x) = -\sum_{|x-y|=1} V'(\phi_t(x) - \phi_t(y))\, dt + \sqrt{2} dW_t(x)\,,$$

where $W_x(t)$ is a collection of independent Brownian motions and V is a strictly convex, smooth, symmetric potential. Naddaf and Spencer (1997) examined the fluctuations of density field at a fixed time.

Defining $\phi^N(t,u)$ by $\phi^N(t,u) = N^{-1}\phi_t([uN])$, Funaki and Spohn (1997) proved that starting from a measure μ^N associated to some profile $h_0 \colon \mathbb{T}^d \to \mathbb{R}$, $\phi^N(t,\cdot)$ converges in L^2 to the solution of

$$\begin{cases} \partial_t h = \sum_{j=1}^{d} \partial_{u_j} \{\sigma_j(\nabla h)\}\,, \\ h(0,\cdot) = h_0(\cdot)\,, \end{cases} \tag{2.2}$$

where $\sigma_j = \partial_{u_j}\sigma$ and σ is the surface tension. Giacomin, Olla and Spohn (1998) derived the equilibrium fluctuations for this model.

A second possible way to derive a macroscopic interface motion from microscopic local dynamics is to examine zero temperature Glauber dynamics. Consider, for instance, the Ising model starting from a configuration in which the + domain is separated from the – domain by a single contour Γ without self intersections.

The zero temperature Glauber dynamics forbids flips that increase the energy. We modify slightly this dynamics excluding also flips that create a second contour in the separation of the + and – domains. Such a model has been considered by Spohn (1993) who proved the hydrodynamic behavior of a two-dimensional system for some special initial configurations. The macroscopic evolution is shown to be described by solutions of the equation (2.2). Landim, Olla and Volchan (1997) investigated the interface motion obtained by the zero temperature dynamics of a Potts model.

A third possible approach would consist in studying the evolution of a + region in a see of – for a reversible Ising model without external field at low enough temperature to have phase coexistence. There are no rigorous results in this direction and we refer to Spohn (1993) for an overview on the problems and on the available techniques. Recently Evans and Rezakhanlou (1997) derived the hydrodynamic equation of a sandpile model.

Motion by mean curvature, Ising models with long range interactions
Kac potential. Fix a smooth potential $J:\mathbb{R} \to \mathbb{R}$ symmetric and with compact support, $\gamma > 0$, that will represent the inverse of the range of the interaction, and an external field $h > 0$. Define the Kac potential $J_\gamma : \mathbb{Z}^d \times \mathbb{Z}^d \to \mathbb{R}_+$ by $J_\gamma(x,y) = \gamma^d J(\gamma \|y-x\|)$ and the formal energy $H_\gamma : \{-1,1\}^{\mathbb{Z}^d} \to \mathbb{R}$ of a spin configuration σ by

$$H_\gamma(\sigma) \;=\; -\frac{1}{2} \sum_{x,y\in\mathbb{Z}^d} J_\gamma(x,y)\sigma(x)\sigma(y) \;-\; h \sum_{x\in\mathbb{Z}^d} \sigma(x)\,.$$

Fix $\beta > 0$, the inverse of the temperature, and consider the Glauber dynamics associated to the energy H_γ at temperature β^{-1}. This is the Markov process on $\{-1,1\}^{\mathbb{Z}^d}$ whose generator L_γ acts on cylinder functions as

$$(L_\gamma f)(\sigma) \;=\; \sum_{x\in\mathbb{Z}^d} c(x,\sigma)[f(\sigma^x) - f(\sigma)]\,.$$

Here σ^x is the spin configuration obtained form σ by flipping the spin at x:

$$(\sigma^x)(y) = \begin{cases} \sigma(y) & \text{if } y \neq x\,, \\ -\sigma(x) & \text{if } y = x\,, \end{cases}$$

and $c(x,\sigma)$ is the jump rate given by

$$c(x,\sigma) \;=\; \frac{\exp\{-(\beta/2)(\Delta_x H_\gamma)(\sigma)\}}{\exp\{-(\beta/2)\sigma(x)(\Delta_x H_\gamma)(\sigma)\} + \exp\{(\beta/2)\sigma(x)(\Delta_x H_\gamma)(\sigma)\}}\,,$$

and $(\Delta_x H_\gamma)(\sigma) = H_\gamma(\sigma^x) - H_\gamma(\sigma)$. Notice that jump rates $\{c(x,\sigma),\, x \in \mathbb{Z}^d\}$ satisfy the detailed balance condition:

$$\frac{c_x(\sigma^x)}{c_x(\sigma)} \;=\; e^{\beta(\Delta_x H_\gamma)(\sigma)}\,.$$

Lebowitz-Penrose limits A basic question in the theory of Ising models with long range interactions is the investigation of the behavior of the system as the range of the interaction increases to infinity, i.e., as $\gamma \downarrow 0$. This limit is known as the Lebowitz–Penrose limit (Lebowitz and Penrose (1966)).

De Masi, Orlandi, Presutti and Triolo (1994) considered this Ising model starting from a product measure μ_γ with marginals given by

$$E_{\mu_\gamma}[\sigma(x)] = m_0(\gamma x) \; , \quad x \in \mathbb{Z}^d \; ,$$

where m_0 is a profile in $C^1(\mathbb{R}^d)$ with bounded derivatives. They proved that for each $n \geq 1$,

$$\lim_{\gamma \to 0} \sup \Big| \mathbb{E}_{\mu_\gamma} \Big[\prod_{i=1}^n \sigma_t(x_i) \Big] - \prod_{i=1}^n m(t, \gamma x_i) \Big| \; = \; 0 \; ,$$

provided m is the unique solution of

$$\begin{cases} \partial_t m + m - \tanh\{\beta(J * m + h)\} = 0 \; , \\ m(0, \cdot) = m_0(\cdot) \; . \end{cases} \tag{2.3}$$

In the previous formulas, σ_t is the state of the Markov process at time t, the supremum is taken over all $(x_1, \dots, x_n)$ in $(\mathbb{Z}^d)^n$ such that $x_i \neq x_j$ for $i \neq j$ and $J * m$ stands for the convolution of J and m:

$$(J * m)(t, u) \; = \; \int_{\mathbb{R}^d} dv \; J(\|u - v\|) m(t, v) \; .$$

Notice that time is not rescaled so that each spin undergoes a finite number of flips in this regime. The deterministic limit is obtained by means of a law of large numbers since for γ small a large numbers of spins feel the effect of the same potential. This limit is called by the authors a mesoscopic limit to differentiate it from asymptotic behaviors where time is rescaled.

Motion by mean curvature. Assume now that the potential is nonnegative and normalized, that the external field vanishes and that the temperature is below 1: $J \geq 0$, $\int_{\mathbb{R}^d} J(\|u\|) du = 1$, $h = 0$, $\beta > 1$. $\beta = 1$ is the inverse critical temperature in the Lebowitz–Penrose limit. Denote by m_β the strictly positive solution of the equation

$$m \; = \; \tanh(\beta m) \; .$$

$\pm m_\beta$ are the magnetization of the two extremal Gibbs states in the limit $\gamma \to 0$.

To define the macroscopic time and space variables, let

$$\lambda = \frac{1}{\sqrt{\log \gamma^{-1}}} \; , \qquad \varepsilon \; = \; \lambda \gamma \; = \; \frac{\gamma}{\sqrt{\log \gamma^{-1}}} \; .$$

We shall rescale space by ε and time by λ^{-2} so that the macroscopic space variable ξ is equal to $x\varepsilon$ and the macroscopic time variable τ is equal to $t\lambda^{-2}$.

Fix a compact domain Λ_0 of $\mathbb{R}^d$ whose boundary Σ_0 is a C^∞ connected surface. Let Σ_t evolve according to the motion by mean curvature with normal velocity θ (θ depends only on the potential J and on the instanton solution of (2.3). Its explicit form is given in De Masi, Orlandi, Presutti and Triolo (1994), equation (5.2a)). We refer to Evans and Spruck (1991) for the terminology of motion by mean curvature. Assume that Σ_τ is smooth for $0 \le \tau \le \tau^*$ and denote by Λ_τ the compact domain whose boundary is Σ_τ.

Let μ_ε, be a product measure on $\{-1, 1\}^{\mathbb{Z}^d}$ such that

$$E_{\mu_\varepsilon}[\sigma(x)] \;=\; \begin{cases} m_\beta & \text{if } x \in \varepsilon\Lambda_0\,, \\ -m_\beta & \text{otherwise}\,. \end{cases}$$

De Masi, Orlandi, Presutti and Triolo (1994) proved that the Ising process σ_t with space rescaled by ε and time rescaled by λ^{-2} ($\sigma_t^\gamma = \varepsilon\sigma_{t\lambda^{-2}}$) converges, as $\gamma \downarrow 0$, to $m_\beta \mathbf{1}\{\Lambda_t\} - m_\beta \mathbf{1}\{\Lambda_t^c\}$. It is not a problem to generalize this result to the case of several initial interfaces. Buttà (1994) extended this result to all times in the two-dimensional case, where the unique singularity is the shrinking of a surface to a point. Katsoulakis and Souganidis (1995) extend this convergence beyond the appearance of the first geometric singularity of the interface.

De Masi, Orlandi, Presutti and Triolo (1996a,b) investigate the fluctuations around the mesoscopic limit showing that they converge to a generalized Ornstein–Uhlenbeck process. They prove also the existence of a deterministic time scale ($\log \gamma^{-1}$) where phase separation occurs for a system starting from a Bernoulli product measure with zero average.

Interface dynamics and reaction–diffusion equations. Bonaventura (1995) considers a spin system on $\mathbb{Z}^d$, where a Glauber dynamics is superposed to a Kawasaki dynamics. Speeding up the Kawasaki dynamics by ε^{-2} and rescaling space by ε, one obtains an interacting particle system whose hydrodynamic equation is a reaction–diffusion equation of type

$$\partial_t m \;=\; \Delta m \;-\; V'(m)\,.$$

Assume that V is a symmetric double well potential. Exploiting the connection between motion by mean curvature and reaction–diffusion equations with vanishing viscosity (cf. Evans, Soner and Souganidis (1992)) and the scaling properties of the reaction–diffusion equation (the fact that $n(t,u)$ defined by $n(t,u) = m(t, \lambda u)$ is the solution of $\partial_t n = \lambda^{-2}\Delta n - V'(n)$), Bonaventura proves the existence of $a > 0$ for which the macroscopic behavior of the process when the Kawasaki dynamics is speeded up by N^{2-a}, is given by the motion of an interface evolving by mean curvature up to the appearance of the first singularity. Katsoulakis and Souganidis (1994) prove the same result for processes evolving on a torus without the assumption of smoothness of the interface.

Interacting diffusions. Varadhan (1991) investigated the evolution of reversible and repulsive interacting Brownian motions on the one-dimensional torus. He proved that starting from an initial state associated to some profile $\rho_0 \colon \mathbb{T}^d \to \mathbb{R}$

whose entropy is of order N, the empirical measure converges to a Lebesgue absolutely continuous measure whose density is the solution of the nonlinear parabolic equation

$$\partial_t \rho = (1/2)\Delta P(\rho) ,$$

where $P(\cdot)$ is the pressure. Olla and Varadhan (1991) extended this investigation to interacting Ornstein–Uhlenbeck processes, whose dynamics is not reversible. Uchiyama (1994) removed the assumption made in Varadhan (1991) concerning the initial entropy bound.

Weakly interacting diffusions. Dittrich (1987) consider a system of independent Brownian motions on $\mathbb{R}^d$, $d \geq 3$, that are replaced after exponential random times by a random number of particles with finite second moment and first moment equal to 1. He proves that the hydrodynamic behavior is described by the heat equation and deduces the non equilibrium fluctuations. Dittrich (1988a,b) proves the conservation of local equilibrium and non equilibrium fluctuations for independent Brownian motions evolving in the interval $[0, 1]$ with reflexion at the boundary and local destruction of particles. The macroscopic behavior of the process is shown to be governed by reaction–diffusion equations.

Oelschläger (1984) considers a finite number of diffusion processes evolving on $\mathbb{R}^d$ according to a mean–field type interaction and shows that the empirical distribution converges, as the number of particles increases to infinity, to a deterministic measure–valued process. Oelschläger (1985) investigates the evolution of finitely many weakly interacting diffusions and deduces a non linear parabolic equation as hydrodynamic limit for the process. The nonequilibrium fluctuations are studied in Oelschläger (1987) and a reaction–diffusion equation is obtained in Oelschläger (1989) for a model where, besides the space evolution, particles are created and destroyed.

7. Hydrodynamic Limit of Reversible Nongradient Systems

We investigate in this chapter the hydrodynamic behavior of reversible nongradient systems. To fix ideas we consider one of the simplest examples, the so called symmetric generalized exclusion process. This is the Markov process introduced in section 2.4 that describes the evolution of particles on a lattice with an exclusion rule that allows at most κ particles per site. Here κ is a fixed positive integer greater or equal than 2. The generator of this Markov process acts on cylinder functions as

$$(L_N f)(\eta) = (1/2) \sum_{\substack{x,y\in\mathbb{T}_N^d \\ |x-y|=1}} r(\eta(x),\eta(y))[f(\eta^{x,y}) - f(\eta)] , \tag{0.1}$$

where $r(a,b) = \mathbf{1}\{a > 0, b < \kappa\}$ and $\eta^{x,y}$ is the configuration obtained from η moving a particle from x to y.

Before proceeding we explain the terminology. For a site x and $1 \le i \le d$, denote by $W_{x,x+e_i}$ the instantaneous current from x to $x+e_i$, i.e., the rate at which a particle jumps from x to $x+e_i$ minus the rate at which a particle jumps from $x+e_i$ to x. With this definition, for nearest neighbor interacting particle systems, we have that

$$L_N\eta(x) = \sum_{i=1}^{d}\{W_{x-e_i,x} - W_{x,x+e_i}\} .$$

For the generalized exclusion process considered in this chapter the current $W_{x,x+e_i}$ writes

$$W_{x,x+e_i} = (1/2)\Big\{\mathbf{1}\{\eta(x) > 0,\ \eta(x+e_i) < \kappa\} - \mathbf{1}\{\eta(x+e_i) > 0,\ \eta(x) < \kappa\}\Big\}.$$

In contrast with zero range processes, where $W_{x,x+e_i} = g(\eta(x)) - g(\eta(x+e_i))$, the current can not be written as a difference $\tau_x h - \tau_{x+e_i} h$ for some cylinder function h. This characteristic of nongradient systems adds a major difficulty in the derivation of the hydrodynamic behavior of the process.

Fix a sequence of initial probability measures μ^N on $\{0,\dots,\kappa\}^{\mathbb{T}_N^d}$ and denote by $\mathbb{P}_{\mu^N}$ the probability measure on $D(\mathbb{R}_+, \{0,\dots,\kappa\}^{\mathbb{T}_N^d})$ induced by the Markov process with generator L_N defined by (0.1) speeded up by N^2 and the measure μ^N. Hereafter $\mathbb{E}_{\mu^N}$ stands for expectation with respect to $\mathbb{P}_{\mu^N}$.

We have presented in Chapters 4 and 5 a general method to deduce the hydrodynamic behavior of interacting particle systems. We started considering a class of martingales associated to the empirical measure: for each smooth function $H : [0,T] \times \mathbb{T}^d \to \mathbb{R}$, let $M^{H,N}(t) = M^H(t)$ be the martingale defined by

$$M^H(t) = < \pi_t^N, H_t > - < \pi_0^N, H_0 > - \int_0^t (\partial_s + N^2 L_N) < \pi_s^N, H_s > ds \, . \tag{0.2}$$

A simple computation shows that its quadratic variation is of order $O(N^{-d})$ so that by Doob inequality, for every $T > 0$ and $\delta > 0$,

$$\lim_{N\to\infty} \mathbb{P}_{\mu^N}\Big[\sup_{0\le t\le T} |M^H(t)| \ge \delta \Big] = 0 \, . \tag{0.3}$$

Since $L_N \eta(x) = \sum_{1\le i\le d} \{W_{x-e_i,x} - W_{x,x+e_i}\}$, and $W_{x,x+e_i} = \tau_x W_{0,e_i}$, a spatial summation by parts permits to rewrite the martingale $M^H(t)$ as

$$\begin{aligned} M^H(t) &= < \pi_t^N, H_t > - < \pi_0^N, H_0 > - \int_0^t < \pi_s^N, \partial_s H_s > ds \\ &- \sum_{i=1}^d \int_0^t N^{1-d} \sum_{x\in\mathbb{T}_N^d} (\partial_{u_i}^N H)(s, x/N) \tau_x W_{0,e_i}(s) \, ds \, , \end{aligned} \tag{0.4}$$

where $\partial_{u_i}^N$ represents the discrete derivative in the i–th direction:

$$(\partial_{u_i}^N H)(x/N) = N\Big[H((x+e_i)/N) - H(x/N) \Big] \, .$$

At this point we face the additional difficulty in the proof of the hydrodynamic behavior of nongradient systems. For gradient systems, like zero range processes, the current $W_{x,x+e_i}$ is itself equal to the difference $\tau_x h - \tau_{x+e_i} h$ of a cylinder function and its translation. This property permits a second summation by parts in the integral term of the martingale. After this second summation by parts, the integral term becomes

$$\int_0^t N^{-d} \sum_{x\in\mathbb{T}_N^d} (\Delta H)(s, x/N) \tau_x h(\eta_s) \, ds + O(N^{-1}) \, .$$

The proof is thus reduced to the replacement of the cylinder function h by a function of the empirical measure.

In contrast with the gradient case the current W_{0,e_i} in the generalized exclusion process can no longer be written as a difference $h - \tau_{e_i} h$ for some cylinder function h so that a second summation by parts is impossible. We have thus not only to replace a cylinder function but a cylinder function multiplied by N.

The main theorem of this chapter asserts that there exists a collection of continuously differentiable increasing functions $\{d_{i,j}: [0,\kappa] \to \mathbb{R}_+, 1 \le i,j \le d\}$ such that

$$\limsup_{\varepsilon\to 0}\limsup_{N\to\infty}\mathbb{E}_{\mu^N}\Big[\Big|\int_0^t N^{-d}\sum_{x\in\mathbb{T}_N^d} H(s,x/N)\times$$

$$\times\; N\Big\{W_{x,x+e_i}(s)+\sum_{j=1}^d\Big[d_{i,j}(\eta_s^{\varepsilon N}(x+e_j))-d_{i,j}(\eta_s^{\varepsilon N}(x))\Big]\Big\}\,ds\Big|\Big] \;=\; 0 \tag{0.5}$$

for all $1\le i\le d$, smooth function H and $t>0$.

A sketch of the proof of this result is presented at the end of this introduction. This proof relies on the characterization of the closed forms of $\{0,\dots,\kappa\}^{\mathbb{Z}^d}$. This characterization, presented in section A3.4, requires a sharp estimate for the second eigenvalue of the generator of the process restricted to finite cubes: for a positive integer ℓ and $0\le K\le\kappa|\Lambda_\ell|$, denote by L_{Λ_ℓ} the restriction to Λ_ℓ of the generator of the symmetric generalized exclusion process and by $\lambda_{\ell,K}$ the largest, strictly negative eigenvalue of the symmetric operator L_{Λ_ℓ} on $\Sigma^{\kappa}_{\Lambda_\ell,K}$. The proof of the characterization of germs of closed forms requires that $|\lambda_{\ell,K}|$ shrinks at a rate slower than ℓ^{-2}, i.e., that there exists a universal constant C such that $|\lambda_{\ell,K}|\ge C\ell^{-2}$ for all $\ell\ge 2$ and $0\le K\le\kappa|\Lambda_\ell|$. This estimate on the spectrum of L_{Λ_ℓ} is proved in section A3.2. Therefore, though hidden in Appendix 3.4, the proof presented in this chapter of the hydrodynamic behavior of nongradient reversible systems relies on a sharp estimate of the spectral gap for the generator of the process restricted to finite cubes.

In possession of (0.5), the conclusion of the proof is rather simple. Statement (0.5) permits a second summation by parts in the integral term of the martingale. $M^H(t)$ can thus be written as

$$M^H(t) \;=\; <\pi_t^N,H_t> \;-\; <\pi_0^N,H_0> \;-\; \int_0^t <\pi_s^N,\partial_s H_s>\,ds$$

$$-\;\sum_{i,j=1}^d\int_0^t N^{-d}\sum_{x\in\mathbb{T}_N^d}(\partial^2_{u_i,u_j}H)(s,x/N)d_{i,j}(\eta_s^{\varepsilon N}(x))\,ds \;+\; o_N(1)\,,$$

where $o_N(1)$ indicates a random variable that converges to 0 in $L^1(\mu^N)$ as $N\uparrow\infty$. It follows from (0.3) that in the limit as $N\uparrow\infty$, the empirical measure satisfies the nonlinear parabolic differential equation:

$$\partial_t\rho \;=\; \sum_{i,j=1}^d\partial^2_{u_i,u_j}d_{i,j}(\rho) \;=\; \sum_{i,j=1}^d\partial_{u_i}\{D_{i,j}(\rho)\partial_{u_j}\rho\}\,,$$

where $D_{i,j}(\alpha)=(d/d\alpha)d_{i,j}(\alpha)$.

To state the hydrodynamic behavior of the generalized exclusion process it remains to present an explicit formula for the diffusion coefficient $D_{i,j}(\alpha)$ introduced above. Hereafter, by an oriented bond b, we understand a pair of nearest neighbor sites, most of the time denoted by $b=(b_1,b_2)$. Notice that with this convention $(b_1,b_2)\ne(b_2,b_1)$. For an oriented bond (b_1,b_2) and a cylinder function g, define the gradient $(\nabla_{b_1,b_2}g)$ by

$$(\nabla_{b_1,b_2} g)(\eta) = r_{b_1,b_2}(\eta)[g(\eta^{b_1,b_2}) - g(\eta)] \, . \tag{0.6}$$

Set $\Sigma^\kappa = \{0, \dots, \kappa\}^{\mathbb{Z}^d}$. For every cylinder function $g \colon \Sigma^\kappa \to \mathbb{R}$, denote by $\Gamma_g(\eta)$ the formal sum

$$\Gamma_g = \sum_{x \in \mathbb{Z}^d} \tau_x g$$

which does not make sense but for which the "gradient"

$$\nabla \Gamma_g = \Big(\nabla_{0,e_1} \Gamma_g, \dots, \nabla_{0,e_d} \Gamma_g \Big)$$

is well defined since it involves only a finite number of non zero differences. For each α, the diffusion coefficient of the hydrodynamic equation for the generalized symmetric exclusion process is the unique symmetric matrix $\{D_{i,j}(\alpha), 1 \le i, j \le d\}$ such that

$$a^* D(\alpha) a = \frac{1}{2\chi(\alpha)} \inf_{h \in \mathcal{C}_0} \sum_{i=1}^d \Big\langle \Big(\sum_{j=1}^d a_j (\mathfrak{A}^j)_i + \nabla_{0,e_i} \Gamma_h \Big)^2 \Big\rangle_\alpha \tag{0.7}$$

for every vector a in $\mathbb{R}^d$. In this formula a^* represents the transposition of a, χ stands for the static compressibility, which in our case is equal to

$$\chi(\alpha) = < \eta(0)^2 >_\alpha - < \eta(0) >_\alpha^2 \, ,$$

$\mathfrak{A}^j$ stands for the vector defined by

$$(\mathfrak{A}^j)_i(\eta) = r_{0,e_j}(\eta) \delta_{j,i} \, ,$$

$< \cdot >_\alpha$ for the expectation with respect to ν_α and $\mathcal{C}_0$ for the space of cylinder functions on Σ^κ with mean zero with respect to all canonical measures, i.e., the space of cylinder functions g such that $E_{\nu_{\Lambda_\ell, K}}[g] = 0$ for all large enough ℓ so that Λ_ℓ contains the support of g and all $0 \le K \le (2\ell+1)^d \kappa$. Examples of functions in $\mathcal{C}_0$ are the currents $\{W_{0,e_i}, 1 \le i \le d\}$, the gradients $\{\eta(e_i) - \eta(0), 1 \le i \le d\}$ and the range of the generator L: $L\mathcal{C}_0 = \{Lg, g \in \mathcal{C}_0\}$. Here L sands for the generator extended to $\mathbb{Z}^d$.

The reader can find in Spohn (1991) (Proposition II.2.2) the equivalence of the variational formula (0.7) for the diffusion coefficient with the Green–Kubo formula based on the current–current correlation functions:

$$D_{i,j}(\alpha) = \frac{1}{2\chi(\alpha)} \Big\{ \sum_{x \in \mathbb{Z}^d} (x \cdot e_i) < \eta(x) W_{0,e_j} >_\alpha + \int_0^\infty dt \sum_{x \in \mathbb{Z}^d} < e^{tL} W_{0,e_j}, W_{x,x+e_i} >_\alpha \Big\} \, .$$

Here $(a \cdot b)$ stands for the usual inner product of $\mathbb{R}^d$.

We are now ready to state the hydrodynamic behavior of the symmetric generalized exclusion process.

Theorem 0.1 *Assume $d = 1$. Let $\rho_0 \colon \mathbb{T}^d \to \mathbb{R}_+$ be a bounded function. Let $\{\mu^N, N \geq 1\}$ be a sequence of probabilities on $\Sigma^{\kappa}_{\mathbb{T}^d_N}$ associated to the profile ρ_0:*

$$\lim_{N\to\infty} \mu^N \Big[\Big| \frac{1}{N^d} \sum_{x\in\mathbb{T}^d_N} G(x/N)\eta(x) - \int G(u)\rho_0(u)\, du \Big| > \delta \Big] = 0 ,$$

for every continuous function $G \colon \mathbb{T}^d \to \mathbb{R}$ and for every δ strictly positive. Then, for every $t \geq 0$,

$$\lim_{N\to\infty} \mathbb{P}_{\mu^N} \left[\left| \frac{1}{N^d} \sum_{x\in\mathbb{T}^d_N} G(x/N)\eta_t(x) - \int G(u)\rho(t,u)\, du \right| > \delta \right] = 0 ,$$

for every continuous function $G \colon \mathbb{T}^d \to \mathbb{R}$ and for every δ strictly positive, where $\rho(t,u)$ is the unique weak solution of the parabolic equation

$$\begin{cases} \partial_t \rho = \displaystyle\sum_{1\leq i,j\leq d} \partial_{u_i} \Big\{ D_{i,j}(\rho) \partial_{u_j} \rho \Big\} \\ \rho(0,\cdot) = \rho_0(\cdot) . \end{cases} \tag{0.8}$$

Remark 0.2 We assumed the dimension to be equal to 1. The reason is that we know only the diffusion coefficient given by the variational formula (0.7) to be continuous and there is no uniqueness result for weak solutions of equation (0.8) in dimension $d \geq 2$ under such weak assumptions on $D(\cdot)$. Proofs are presented in general dimension to stress that we need nowhere else the dimension to be equal to 1.

We conclude this section presenting the strategy of the proof of the replacement of the current by a gradient. Recall that $\mathcal{C}_0$ stands for the space of mean-zero cylinder functions and that L_{Λ_ℓ} stands for the restriction of the generator L_N to the cube Λ_ℓ. In section 4 we prove that for each pair f, g of cylinder functions in $\mathcal{C}_0$, and each sequence K_ℓ such that $K_\ell/(2\ell)^d$ converges to α,

$$\frac{1}{|\Lambda_\ell|} \Big\langle \sum_{|x|\leq \ell_f} \tau_x f, (-L_{\Lambda_\ell})^{-1} \sum_{|y|\leq \ell_g} \tau_y g \Big\rangle_{\nu_{\Lambda_\ell, K_\ell}} \tag{0.9}$$

converges as $\ell \uparrow \infty$. In this formula, for a cylinder function f, ℓ_f stands for $\ell - s_f$, where s_f is the smallest integer k such that Λ_k contains the support of f. In this way the support of $\tau_x f$ is contained in Λ_ℓ for all $|x| \leq \ell_f$. The proof of this convergence requires the characterization of all closed forms of $\{0, \dots, \kappa\}^{\mathbb{Z}^d}$. It is here that a sharp estimate on the spectral gap of L_{Λ_ℓ} is needed.

The limit of (0.9), denoted by $\ll f, g \gg_\alpha$, defines a semi-inner product on the space $\mathcal{C}_0$. The Hilbert space induced by $\mathcal{C}_0$ and this semi-inner product is denoted by $\mathcal{H}_\alpha$.

In section 3, with the type of arguments used in the proof of the one block estimate (the entropy inequality, Feynman–Kac formula, a variational formula for the largest eigenvalue of a symmetric operator), we prove that for each $t > 0$, smooth function $H: \mathbb{T}^d \to \mathbb{R}$ and h in $\mathcal{C}_0$,

$$\limsup_{N\to\infty} \mathbb{E}_{\mu^N}\Big[\Big|\int_0^t N^{1-d} \sum_{x\in\mathbb{T}_N^d} H(s,x/N)(\tau_x h)(\eta_s)\, ds\Big|\Big]$$
$$\le C(t,H) \limsup_{\ell\to\infty} \sup_K \frac{1}{|\Lambda_\ell|}\Big\langle \sum_{|x|\le \ell_h} \tau_x h, (-L_{\Lambda_\ell})^{-1} \sum_{|y|\le\ell_h} \tau_y h\Big\rangle_{\nu_{\Lambda_\ell,K}}$$

for some finite constant $C(t,H)$. This inequality together with (0.9) shows that for each $t > 0$, smooth function $H: \mathbb{T}^d \to \mathbb{R}$ and h in $\mathcal{C}_0$,

$$\limsup_{N\to\infty} \mathbb{E}_{\mu^N}\Big[\Big|\int_0^t N^{1-d} \sum_{x\in\mathbb{T}_N^d} H(s,x/N)(\tau_x h)(\eta_s)\, ds\Big|\Big] \le C(t,H) \sup_{0\le\alpha\le\kappa} \ll h,h \gg_\alpha \tag{0.10}$$

for some finite constant $C(t,H)$.

The structure of $\mathcal{H}_\alpha$ is quite simple and is examined in details in section 5. The gradients $\{\eta(e_i) - \eta(0),\ 1 \le i \le d\}$ are linearly independent vectors orthogonal to the space $L\mathcal{C}_0$ and $\mathcal{H}_\alpha$ is generated by these two subspaces. There exists, in particular, a matrix $\{D_{i,j}(\alpha), 1 \le i,j \le d\}$, that depends on the density because the inner product depends on the density, such that

$$W_{0,e_i} + \sum_{j=1}^d D_{i,j}(\alpha)[\eta(e_j) - \eta(0)] \in \overline{L\mathcal{C}_0}$$

for $1 \le i \le d$. The matrix $D_{i,j}$ that appears in the above formula is the diffusion coefficient of the hydrodynamic equation (0.8). Thus the diffusion coefficient $D_{i,j}(\alpha)$ is just the coefficient of the projection of the current W_{0,e_i} on the gradient $\eta(e_j) - \eta(0)$ in the Hilbert space $\mathcal{H}_\alpha$. Moreover, for each fixed $\delta > 0$, there exists f in $\mathcal{C}_0$ such that $\ll W_{0,e_i} + \sum_{j=1}^d D_{i,j}(\alpha)[\eta(e_j) - \eta(0)] - Lf \gg_\alpha \le \delta$.

It follows from this observation, (0.10) and some two blocks type argument (to replace $D_{i,j}(\alpha)[\eta(e_j) - \eta(0)]$ by $D_{i,j}(\eta^{\varepsilon N}(0))[\eta^{\varepsilon N}(e_j) - \eta^{\varepsilon N}(0)]$ that

$$\inf_{f\in\mathcal{C}_0} \limsup_{\varepsilon\to 0} \limsup_{N\to\infty} \mathbb{E}_{\mu^N}\Big[\Big|\int_0^t N^{1-d} \sum_{x\in\mathbb{T}_N^d} H(s,x/N) \times$$
$$\tau_x\Big\{W_{0,e_i}(s) + \sum_{j=1}^d D_{i,j}(\eta_s^{\varepsilon N}(0))[\eta_s^{\varepsilon N}(e_j) - \eta_s^{\varepsilon N}(0)] - (\tau_x Lf)(\eta_s)\Big\}\, ds\Big|\Big] = 0.$$

By Taylor's expansion and the continuity of the diffusion coefficient $D(\cdot)$, statement (0.5) follows from the previous limit and from the easy to prove identity (cf. the proof of Corollary 1.2)

$$\limsup_{N\to\infty} \mathbb{E}_{\mu^N}\Big[\Big|\int_0^t N^{1-d}\sum_{x\in\mathbb{T}_N^d} H(x/N)Lf(\eta_s)\,ds\Big|\Big] = 0 \tag{0.11}$$

for each f in $\mathcal{C}_0$.

1. Replacing Currents by Gradients

We show in this section that the current W_{0,e_i} may be decomposed as a linear combination of the gradients $\{\eta(e_j)-\eta(0),\ 1\le j\le d\}$ with a function in the range of the generator L_N: $W_{0,e_i}+\sum_{1\le j\le d} D_{i,j}(\alpha)\{\eta(e_j)-\eta(0)\} = L_N f$ for some matrix $D_{i,j}(\alpha)$ that depend on the density and a cylinder function f. The gradient part in the decomposition permits a second summation by parts, while the $L_N f$ term turns out to be negligible. This is the content of Theorem 1.1 below.

For positive integers ℓ, N, a smooth function H in $C^2(\mathbb{T}^d)$ and a cylinder function $\mathfrak{f}$, let

$$X_{N,\ell}^{\mathfrak{f},i}(H,\eta) := N^{1-d}\sum_{x\in\mathbb{T}_N^d} H(x/N)\tau_x V_i^{\mathfrak{f},\ell}(\eta)$$

where,

$$V_i^{\mathfrak{f},\ell} = W_{0,e_i}(\eta) + \sum_{j=1}^d D_{i,j}\big(\eta^\ell(0)\big)\Big\{\eta^\ell(e_j)-\eta^\ell(0)\Big\} - L_N\mathfrak{f}(\eta)\,.$$

Theorem 1.1 *For every smooth function H in $C^{1,2}([0,T]\times\mathbb{T}^d)$, $1\le i\le d$ and $T>0$,*

$$\inf_{\mathfrak{f}\in\mathcal{C}}\limsup_{\varepsilon\to0}\limsup_{N\to\infty}\frac{1}{N^d}\log\mathbb{E}_{\nu_\alpha^N}\left[\exp N^d\Big|\int_0^T X_{N,\varepsilon N}^{\mathfrak{f},i}(H_s,\eta_s)\,ds\Big|\right] = 0\,.$$

The proof of this result is postponed to section 3. We first conclude the proof of the hydrodynamic behavior of the generalized symmetric exclusion process. We start showing that the $L_N\mathfrak{f}$ term is negligible. For a positive integer ℓ and a smooth function H in $C^2(\mathbb{T}^d)$, let

$$Y_{N,\ell}^i(H,\eta) :=$$

$$N^{1-d}\sum_{x\in\mathbb{T}_N^d} H(x/N)\Big\{W_{x,x+e_i} + \sum_{j=1}^d D_{i,j}\big(\eta^\ell(x)\big)\Big\{\eta^\ell(x+e_j)-\eta^\ell(x)\Big\}\Big\}\,.$$

Corollary 1.2 *For every smooth function H in $C^{1,2}([0,T]\times\mathbb{T}^d)$, $1\le i\le d$ and $T>0$,*

$$\limsup_{\varepsilon\to 0}\limsup_{N\to\infty}\mathbb{E}_{\mu^N}\left[\left|\int_0^T Y^i_{N,N\varepsilon}(H_s,\eta_s)\,ds\right|\right] = 0\,.$$

Proof. Fix a smooth function H, $T>0$ and $1\le i\le d$. For a cylinder function $\mathfrak{f}$ write

$$Y^i_{N,\ell}(H,\eta) \;=\; X^{\mathfrak{f},i}_{N,\ell}(H,\eta) \;+\; N^{1-d}\sum_{x\in\mathbb{T}^d_N} H(x/N)\tau_x L_N\mathfrak{f}(\eta)\,.$$

To prove the corollary, we have just to show that

$$\inf_{\mathfrak{f}\in\mathcal{C}}\limsup_{\varepsilon\to 0}\limsup_{N\to\infty}\mathbb{E}_{\mu^N}\left[\left|\int_0^T X^{\mathfrak{f},i}_{N,N\varepsilon}(H_s,\eta_s)\,ds\right|\right] = 0 \tag{1.1}$$

and that for every cylinder function $\mathfrak{f}$,

$$\lim_{N\to\infty}\mathbb{E}_{\mu^N}\left[\left|\int_0^T N^{1-d}\sum_{x\in\mathbb{T}^d_N} H(t,x/N)\tau_x L_N\mathfrak{f}(\eta_t)\,dt\right|\right] = 0\,. \tag{1.2}$$

To prove (1.1) apply the entropy inequality to obtain that

$$\begin{aligned}
&\mathbb{E}_{\mu^N}\left[\left|\int_0^T X^{\mathfrak{f},i}_{N,N\varepsilon}(H_s,\eta_s)\,ds\right|\right]\\
&\quad\le\; \frac{1}{\gamma N^d}H(\mu^N\,|\nu^N_\alpha) \;+\; \frac{1}{\gamma N^d}\log\mathbb{E}_{\nu^N_\alpha}\left[\exp\gamma N^d\left|\int_0^T X^{\mathfrak{f},i}_{N,\varepsilon N}(H_s,\eta_s)\,ds\right|\right]
\end{aligned}$$

for every positive γ. Since there are at most κ particles per site, the entropy $H(\mu^N\,|\nu^N_\alpha)$ is bounded by $C(\alpha,\kappa)N^d$ (cf. discussion preceding the statement of Lemma 5.5.7). In particular, (1.1) follows from Theorem 1.1 and the arbitrariness of γ.

We turn now to (1.2). Since time is speeded up by N^2, for any smooth function $H:[0,T]\times\mathbb{T}^d\to\mathbb{R}$ and any cylinder (and thus bounded) function $\mathfrak{f}$,

$$\begin{aligned}
M^{H,\mathfrak{f}}(t) \;=\;& N^{-(d+1)}\sum_{x\in\mathbb{T}^d_N} H(t,x/N)\tau_x\mathfrak{f}(\eta_t)\\
&-\; N^{-(d+1)}\sum_{x\in\mathbb{T}^d_N} H(0,x/N)\tau_x\mathfrak{f}(\eta_0)\\
&-\;\int_0^t N^{-(d+1)}\sum_{x\in\mathbb{T}^d_N}(\partial_s H)(s,x/N)\tau_x\mathfrak{f}(\eta_s)\,ds\\
&-\;\int_0^t N^{1-d}\sum_{x\in\mathbb{T}^d_N} H(s,x/N)\tau_x L_N\mathfrak{f}(\eta_s)\,ds
\end{aligned}$$

is a martingale. The first three terms on the right hand side are of order N^{-1} because $\mathfrak{f}$ is bounded. A simple computation of the quadratic variation of the martingale shows that $\mathbb{E}_{\mu^N}[M^{H,\mathfrak{f}}(t)^2]$ is bounded above by

$$\frac{N^{-2d}}{2}\sum_{|x-y|=1}\mathbb{E}_{\mu^N}\Big[\int_0^t r_{x,y}(\eta_s)\Big\{\sum_{z\in\mathbb{T}_N^d}H(s,z/N)[(\tau_z\mathfrak{f})(\eta_s^{x,y})-(\tau_z\mathfrak{f})(\eta_s)]\Big\}^2\Big]\,.$$

Since $\mathfrak{f}$ is a cylinder function, the difference $(\tau_z\mathfrak{f})(\eta^{x,y})-(\tau_z\mathfrak{f})(\eta)$ vanishes for all but a finite number of sites z. This expectation is therefore of order N^{-d}, which proves (1.2). □

We have now all elements to prove the hydrodynamic behavior of a nongradient system.

Proof of Theorem 0.1. Recall that we denote by π_t^N the empirical measure defined by

$$\pi_t^N(\eta) \;=\; N^{-d}\sum_{x\in\mathbb{T}_N^d}\eta_t(x)\delta_{x/N}\,.$$

Fix $T>0$ and denote by Q_{μ^N} the probability measure on the path space $D([0,T],\mathcal{M}_{+,\kappa}(\mathbb{T}^d))$ corresponding to the process π_t^N with generator L_N speeded up by N^2 starting from μ^N. We have already seen in the proof of the hydrodynamic behavior of symmetric simple exclusion processes that a law of large numbers for the empirical measure π_t^N follows from the weak convergence of the sequence Q_{μ^N} to the probability measure concentrated on the deterministic, absolutely continuous trajectory $\pi(t,du)=\pi(t,u)du$ whose density is the weak solution of equation (0.8).

In section 6 we prove that the sequence $\{Q_{\mu^N}, N\ge 1\}$ is weakly relatively compact and that all limit points Q^* are concentrated on weakly continuous paths π_t that are absolutely continuous with respect to the Lebesgue measure with density bounded by κ:

$$\pi(t,du) \;=\; \pi(t,u)du \;\text{ and }\; \pi(t,u)\le\kappa \;\text{ for all } 0\le t\le T,\ Q^* \text{ almost surely.}$$

From Brezis and Crandall (1979), in dimension 1, there exists a unique weak solution of (0.8). Therefore, to conclude the proof of the theorem, it remains to show that all limit points of the sequence $\{Q_{\mu^N}, N\ge 1\}$ are concentrated on absolutely continuous trajectories $\pi(t,du)=\pi(t,u)du$ whose density are weak solutions of equation (0.8).

Fix a smooth function $H\colon[0,T]\times\mathbb{T}^d\to\mathbb{R}$ and recall from (0.2) the definition of the martingale $M^H(t)$. Since $\mathbb{E}_{\mu^N}[(M^H(t))^2]$ vanishes as $N\uparrow\infty$, by Doob's inequality, for every $\delta>0$,

$$\lim_{N\to\infty}\mathbb{P}_{\mu^N}\Big[\sup_{0\le t\le T}\big|M^H(t)\big|>\delta\Big] \;=\; 0\,. \tag{1.3}$$

Applying Corollary 1.2 to the second integral term in the explicit formula (0.4) of $M^H(t)$, we get that for every $\delta>0$,

$$\limsup_{\varepsilon\to 0}\limsup_{N\to\infty} \mathbb{P}_{\mu^N}\Big[\Big| <\pi_T, H_T> \ - \ <\pi_0, H_0> \ - \int_0^T <\pi_s, \partial_s H_s> ds$$
$$+ \sum_{i,j=1}^d \int_0^T N^{1-d} \sum_{x\in\mathbb{T}_N^d} (\partial_{u_i}^N H)(s,x/N)\tau_x V_{i,j,\varepsilon N}(\eta_s)\, ds\Big| > \delta\Big] = 0\,,$$

where

$$V_{i,j,\varepsilon N}(\eta) = D_{i,j}\big(\eta^{\varepsilon N}(0)\big)\Big[\eta^{\varepsilon N}(e_j) - \eta^{\varepsilon N}(0)\Big]\,.$$

Denote by $d_{i,j}$ the integral of $D_{i,j}$: $d_{i,j}(\alpha) = \int_0^\alpha D_{i,j}(\beta)d\beta$. Since by Theorem 5.8 $D_{i,j}$ is continuous,

$$D_{i,j}(\eta^{\varepsilon N}(0))\Big\{\eta^{\varepsilon N}(e_j) \ - \ \eta^{\varepsilon N}(0)\Big\}$$
$$= \ d_{i,j}(\eta^{\varepsilon N}(e_j)) \ - \ d_{i,j}(\eta^{\varepsilon N}(0)) \ + \ N^{-1}o_N(1)\,,$$

where $o_N(1)$ represents a term that converges uniformly to 0 as $N\uparrow\infty$. Summing by parts, for each $1\le i,j\le d$, we obtain that

$$N^{1-d}\sum_{x\in\mathbb{T}_N^d}(\partial_{u_i}^N H)(s,x/N)\tau_x D_{i,j}\big(\eta^{\varepsilon N}(0)\big)\Big[\eta^{\varepsilon N}(e_j) - \eta^{\varepsilon N}(0)\Big]$$
$$= \ -N^{-d}\sum_{x\in\mathbb{T}_N^d}(\partial_{u_i,u_j}^2 H)(s,x/N)d_{i,j}\big(\eta^{\varepsilon N}(x)\big) \ + \ o_N(1)\,.$$

Therefore,

$$\lim_{\varepsilon\to 0}\lim_{N\to\infty} \mathbb{P}_{\mu^N}\Big[\Big| <\pi_T, H_T> \ - \ <\pi_0, H_0> \ - \int_0^T <\pi_s, \partial_s H_s> ds$$
$$- \sum_{i,j=1}^d \int_0^T N^{-d}\sum_{x\in\mathbb{T}_N^d}(\partial_{u_i,u_j}^2 H)(s,x/N)d_{i,j}(\eta_s^{\varepsilon N}(x))\, ds\Big| > \delta\Big] = 0\,.$$

Recall the definition of the approximation of the identity ι_ε. By continuity, for every limit point Q^* of the sequence Q_{μ^N},

$$\lim_{\varepsilon\to 0} Q^*\Big[\Big| <\pi_T, H_T> \ - \ <\pi_0, H_0> \ - \int_0^T <\pi_s, \partial_s H_s> ds$$
$$- \sum_{i,j=1}^d \int_0^T ds \int_{\mathbb{T}^d} du\, (\partial_{u_i,u_j}^2)H(s,u)d_{i,j}(<\pi_s, \iota_\varepsilon(u-\cdot)>)\Big| > \delta\Big] = 0\,.$$

Since each limit point Q^* is concentrated on absolutely continuous paths $\pi_t = \pi(t,u)du$ with density $\pi(t,u)$ bounded by κ, for each fixed $0\le s\le T$, as $\varepsilon\downarrow 0$, $<\pi_s, \iota_\varepsilon(u-\cdot)>$ converges to $\pi(s,u)$ for almost all u in $\mathbb{T}^d$. From this remark and the continuity of $\{d_{i,j},\ 1\le i,j\le d\}$, we obtain that

$$Q^*\Big[\Big| <\pi_T, H_T> \; - \; <\pi_0, H_0> \; - \int_0^T <\pi_s, \partial_s H_s> ds$$
$$- \sum_{i,j=1}^d \int_0^T ds \int du\, (\partial^2_{u_i,u_j}) H(s,u) d_{i,j}(\pi(s,u))\Big| > \delta\Big] = 0$$

for all H in $C^{1,2}([0,T], \mathbb{T}^d)$. In particular, Q^* is concentrated on weak solution of (0.8) what concludes the proof of the theorem. □

It may seem odd to define $X^{\mathfrak{f},i}_{N,\ell}$ as we do since it involves the term

$$N^{1-d} \sum_{x \in \mathbb{T}^d_N} H(x/N) \tau_x L_N \mathfrak{f}$$

which vanishes and gives no contribution to the equation in the limit. As a matter of fact this term is important in order to establish Theorem 1.1 where the current is not only multiplied by N but also exponentiated.

2. An Integration by Parts Formula

For a cylinder function ψ, denote by Λ_ψ the smallest d-dimensional rectangle that contains the support of ψ and by s_ψ the the smallest positive integer s such that $\Lambda_\psi \subset \Lambda_s$. If ψ is the gradient $\eta(e_i) - \eta(0)$, for instance, $\Lambda_\psi = \{0, e_i\}$ and $s_\psi = 1$. in this example Λ_{s_ψ} and Λ_ψ do not coincide. Let $\mathcal{C}_0$ be the space of cylinder functions with mean zero with respect to all canonical invariant measures:

$$\mathcal{C}_0 \; = \; \Big\{ g \in \mathcal{C}\, ; \; <g>_{\Lambda_g, K} \; = \; 0 \text{ for all } 0 \le K \le \kappa|\Lambda_g| \Big\} \, .$$

Here, for a finite subset Λ of $\mathbb{Z}^d$ and $0 \le K \le \kappa|\Lambda|$, $< \cdot >_{\Lambda,K}$ stands for the expectation with respect to the canonical measure $\nu_{\Lambda,K}$. Examples of functions in $\mathcal{C}_0$ are $L_\Lambda g$ for finite subsets Λ and cylinder functions g with support contained in Λ, currents $\{W_{0,e_i},\ 1 \le i \le d\}$ and gradients $\{\eta(e_i) - \eta(0),\ 1 \le i \le d\}$.

For each cylinder g in $\mathcal{C}_0$, $<g>_{\Lambda,K} = 0$ for all $\Lambda \supset \Lambda_g$, $0 \le K \le \kappa|\Lambda|$,

$$<g>_\alpha = 0 \qquad \text{and} \qquad \sum_{x \in \mathbb{Z}^d} <g, \eta(x)>_\alpha \; = \; 0 \quad \text{for all } 0 \le \alpha \le \kappa \, .$$

To prove the first assertion, fix a set $\Lambda \supset \Lambda_g$ and $0 \le K \le \kappa|\Lambda|$. $<g>_{\Lambda,K}$ is equal to

$$E_{\nu_{\Lambda,K}}\Big[E_{\nu_{\Lambda,K}}\Big[g \,\big|\, \sum_{x \in \Lambda_g} \eta(x) \Big]\Big] \, .$$

The conditional expectation $E_{\nu_{\Lambda,K}}[g \mid \sum_{x \in \Lambda_g} \eta(x)]$ is equal to the expectation of g with respect to the canonical measure $\nu_{\Lambda_g, \sum_{x\in\Lambda_g} \eta(x)}$, that vanishes because g

belongs to $\mathcal{C}_0$. The second identity follows immediately from the previous result and the convergence of the finite marginals of $\nu_{\ell,K}$ to ν_α, as $\ell \uparrow \infty$ and $K/(2\ell)^d \to \alpha$ (cf. Section A2.2). The third identity is also easy to check. First observe that $< g, \eta(x) >_\alpha$ does not vanish only for a finite number of sites x since g has mean zero and ν_α is a product measure. In particular, $\sum_x < g, \eta(x) >_\alpha$ is well defined and equal to $< g, \sum_{x\in\Lambda_g} \eta(x) >_\alpha$. Taking conditional expectation with respect to $\sum_{x\in\Lambda_g} \eta(x)$, this last expectation writes $E_{\nu_\alpha}[E_{\nu_\alpha}[g|\sum_{x\in\Lambda_g}\eta(x)]\sum_{x\in\Lambda_g}\eta(x)]$ that vanishes because the conditional expectation $E_{\nu_\alpha}[\,\cdot\,|\sum_{x\in\Lambda_g}\eta(x)]$ reduces to the expectation with respect to the canonical measure $\nu_{\Lambda_g,\sum_{x\in\Lambda_g}\eta(x)}$ and g belongs to $\mathcal{C}_0$.

For a finite subset Λ of $\mathbb{Z}^d$, denote by $\mathcal{F}_\Lambda$ the σ-algebra generated by $\{\eta(x);\ x \in \Lambda\}$:

$$\mathcal{F}_\Lambda \;=\; \sigma\{\eta(x);\ x \in \Lambda\}$$

and abbreviate $\mathcal{F}_{\Lambda_\ell}$ by $\mathcal{F}_\ell$.

For a rectangle Λ and a canonical measure $\nu_{\Lambda,K}$ on $\Sigma^\kappa_{\Lambda,K}$, denote by $<\cdot,\cdot>_{\ell,K}$ (resp. $<\cdot,\cdot>_\alpha$) the inner product in $L^2(\nu_{\ell,K})$ (resp. $L^2(\nu_\alpha)$) and by $\mathfrak{D}(\nu_{\Lambda,K},\cdot)$ the Dirichlet form relative to $\nu_{\Lambda,K}$ defined by

$$\mathfrak{D}(\nu_{\Lambda,K}, f) \;=\; \Big\langle -L_\Lambda f, f\Big\rangle_{\Lambda,K} \;=\; \sum_{b\in\Lambda} \mathfrak{D}_b(\nu_{\Lambda,K}, f)\,,$$

where, for each bond $b = (b_1, b_2)$,

$$\mathfrak{D}_b(\nu_{\Lambda,K}, f) \;=\; \frac{1}{4}\int r_{b_1,b_2}(\eta)\Big[f(\eta^{b_1,b_2}) - f(\eta)\Big]^2 \nu_{\Lambda,K}(d\eta)\,.$$

In this formula and below summation is carried out over all oriented bonds b in Λ (a bond $b = (b_1, b_2)$ is said to be in Λ if both end points are in Λ). Notice that $(b_1, b_2) \neq (b_2, b_1)$ so that both (b_1, b_2) and (b_2, b_1) appear in the above summation.

Consider a cylinder function ψ in $\mathcal{C}_0$. We claim that ψ is in the range of L_{Λ_ψ}. To prove this statement fix $0 \le K \le \kappa|\Lambda_\psi|$ and consider the generalized symmetric exclusion process on $\Sigma_{\Lambda_\psi,K}$. The kernel of L_{Λ_ψ} in $L^2(\nu_{\Lambda_\psi,K})$ has dimension 1: assume that $L_{\Lambda_\psi} f = 0$. Multiply both sides of the identity by f and integrate with respect to $\nu_{\Lambda_\psi,K}$ to obtain that f must be constant. Since the kernel of L_{Λ_ψ} in $L^2(\nu_{\Lambda_\psi,K})$ has dimension 1, the range of L_{Λ_ψ} has codimension 1. Since the range is included in the subspace of mean-zero functions that has codimension 1, both spaces are equal, i.e., all $\nu_{\Lambda_\psi,K}$-mean-zero functions are in the range of L_{Λ_ψ}.

We may therefore write the cylinder function ψ as $\psi = (-L_{\Lambda_\psi})(-L_{\Lambda_\psi})^{-1}\psi$ for some mean-zero function $(-L_{\Lambda_\psi})^{-1}\psi$, measurable with respect to the variables $\{\eta(z),\ z \in \Lambda_\psi\}$. Fix a rectangle Λ that contains Λ_ψ and $0 \le K \le \kappa|\Lambda|$. Since the canonical measure $\nu_{\Lambda,K}$ is reversible, we have that

$$\begin{aligned}\Big\langle \psi, h\Big\rangle_{\Lambda,K} &= \Big\langle (-L_{\Lambda_\psi})(-L_{\Lambda_\psi})^{-1}\psi, h\Big\rangle_{\Lambda,K} \\ &= (1/4)\sum_{b\in\Lambda_\psi}\Big\langle \nabla_b(-L_{\Lambda_\psi})^{-1}\psi, \nabla_b h\Big\rangle_{\Lambda,K} \;=\; \sum_{b\in\Lambda_\psi}\Big\langle \Phi^\psi_b, \nabla_b h\Big\rangle_{\Lambda,K}\end{aligned}$$

provided we set $\Phi_b^\psi = (1/4)\nabla_b(-L_{\Lambda_\psi})^{-1}\psi$ for b in Λ_ψ. In this formula summation is carried out over all oriented bonds b in Λ_ψ and ∇_b is defined in (0.6). We may extend the definition of Φ_b^ψ for bonds not in Λ_ψ setting $\Phi_b^\psi = 0$ if $b \notin \Lambda_\psi$. From this explicit formula for Φ_b^ψ, it is easy to check that $\tau_y\Phi_b^\psi = \Phi_{b+y}^{\tau_y\psi}$ for all y in $\mathbb{Z}^d$. Moreover, by reversibility, we have that

$$\sum_{b\in\Lambda_\psi}\Big\langle(\Phi_b^\psi)^2\Big\rangle_{\Lambda,K} = (1/16)\sum_{b\in\Lambda_\psi}\Big\langle\nabla_b(-L_{\Lambda_\psi})^{-1}\psi, \nabla_b(-L_{\Lambda_\psi})^{-1}\psi\Big\rangle_{\Lambda,K}$$

$$= (1/4)\Big\langle\psi, (-L_{\Lambda_\psi})^{-1}\psi\Big\rangle_{\Lambda,K} = C(\psi)$$

for some finite constant that depends only on ψ (and not on Λ nor on K). We have thus proved the integration by parts formula:

Lemma 2.1 (Integration by parts formula) *Let ψ be a cylinder function in $\mathcal{C}_0$. There exists a family of cylinder functions $\{\Phi_b^\psi;\ b \in \Lambda_\psi\}$, measurable with respect to $\mathcal{F}_{\Lambda_\psi}$, such that*

$$\Big\langle\psi, h\Big\rangle_{\Lambda,K} = \sum_{b\in\Lambda_\psi}\Big\langle\Phi_b^\psi, \nabla_b h\Big\rangle_{\Lambda,K}$$

for all rectangles $\Lambda \supset \Lambda_\psi$, $0 \le K \le \kappa|\Lambda|$ and functions h in $L^2(\nu_{\Lambda,K})$. Moreover,

$$\sum_{b\in\Lambda_\psi}\Big\langle(\Phi_b^\psi)^2\Big\rangle_{\Lambda,K} \le C(\psi)$$

for some finite constant $C(\psi)$ and $\tau_y\Phi_b^\psi = \Phi_{b+y}^{\tau_y\psi}$ for all d-dimensional integer y. The same result may be restated with canonical measures $\nu_{\Lambda,K}$ replaced by grand canonical measures ν_α.

The integration by parts formula assumes a particularly simple form for three types of functions. First of all, if $\psi = L_\Lambda h$, for some finite rectangle Λ and some $\mathcal{F}_\Lambda$-measurable cylinder function h, then $\Phi_b^\psi = -(1/4)\nabla_b h$ for $b \in \Lambda$. If ψ is a current, $\psi = W_{(b_1,b_2)}$, we obtain that

$$\Phi_{b'}^\psi = \begin{cases} -(1/4) & if b' = b = (b_1, b_2)\,, \\ (1/4) & \text{if } b' = (b_2, b_1)\,. \end{cases} \tag{2.1}$$

For each $1 \le k \le d$, an elementary computation shows that $L_{0,e_k}^{-1}[\eta(e_k) - \eta(0)] = [\eta(e_k) - \eta(0)]\{-(1/3)\eta(0)\eta(e_k) + F(\eta(0) + \eta(e_k))\}$, where

$$\begin{aligned} F(A) &= F_\kappa(A) \\ &= \begin{cases} -(1/6)(A+1)(A+2) & \text{for } 0 \le A \le \kappa\,, \\ -(1/6)A^2 + [\kappa + (1/2)]A - [\kappa^2 + \kappa + (1/3)] & \text{for } \kappa \le A \le 2\kappa\,. \end{cases} \end{aligned}$$

Both definitions coincide at $A = \kappa$. Let $\Psi_{0,e_k} = -(1/4)\nabla_{(0,e_k)}L_{0,e_k}^{-1}[\eta(e_k) - \eta(0)]$ and $\Psi_{e_k,0} = -(1/4)\nabla_{(e_k,0)}L_{0,e_k}^{-1}[\eta(e_k) - \eta(0)]$. From the explicit formula for $L_{0,e_k}^{-1}[\eta(e_k) - \eta(0)]$, we get that

$$\Psi_{0,e_k} = \frac{1}{4} r_{0,e_k}(\eta)\Big\{ 2\eta(0)\eta(e_k) - [\eta(e_k) - \eta(0)] - G(\eta(0) + \eta(e_k))\Big\}$$
$$\Psi_{e_k,0} = \frac{1}{4} r_{0,e_k}(\eta)\Big\{ - 2\eta(0)\eta(e_k) - [\eta(e_k) - \eta(0)] + G(\eta(0) + \eta(e_k))\Big\},$$

where,

$$G(A) = G_\kappa(A) = \begin{cases} -A & \text{for } 0 \le A \le \kappa, \\ (2\kappa + 1)A - 2(\kappa^2 + \kappa) & \text{for } \kappa \le A \le 2\kappa. \end{cases}$$

We leave the reader to check that $2\Psi_{0,e_k} + 2\Psi_{e_k,0} = -[\eta(e_k) - \eta(0)]$ and that

$$< \eta(e_k) - \eta(0)\,,\, h >_\alpha = < \Psi_{0,e_k}\,,\, \nabla_{0,e_k} h >_\alpha + < \Psi_{e_k,0}\,,\, \nabla_{e_k,0} h >_\alpha \qquad (2.2)$$

for every $0 \le \alpha \le \kappa$ and for all cylinder functions h in $L^2(\nu_\alpha)$.

Remark 2.2 A change of variables shows that

$$< \nabla_{b_1,b_2} f\,,\, \nabla_{b_1,b_2} h >_\alpha = < \nabla_{b_2,b_1} f\,,\, \nabla_{b_2,b_1} h >_\alpha$$

for every bond (b_1, b_2), $0 \le \alpha \le \kappa$ and cylinder functions f, h in $L^2(\nu_\alpha)$. Since for ψ in $\mathcal{C}_0$, $\Phi_b^\psi = (1/4)\nabla_b(-L_{\Lambda_\psi})^{-1}\psi$, we have

$$\sum_{b \in \Lambda_\psi} < \Phi_b^\psi\,,\, \nabla_{b_1,b_2} h >_\alpha = 2\sum_{i=1}^{d} \sum_{\substack{z;\, z\in\Lambda_\psi \\ z+e_i\in\Lambda_\psi}} < \Phi^\psi_{(z,z+e_i)}\,,\, \nabla_{z,z+e_i} h >_\alpha$$

and the integration by parts formula becomes

$$< \psi\,,\, h >_\alpha = 2\sum_{i=1}^{d} \sum_{\substack{z;\, z\in\Lambda_\psi \\ z+e_i\in\Lambda_\psi}} < \Phi^\psi_{(z,z+e_i)}\,,\, \nabla_{z,z+e_i} h >_\alpha .$$

3. Nongradient Large Deviations Estimates

We prove in this section Theorem 1.1. To detach the main arguments of the proof we divide it in several steps.

Step 1: Reduction to an eigenvalue problem. Our purpose in this first step is to reduce the dynamic problem stated in the theorem to a static problem involving the largest eigenvalue of a small perturbation of the generator $N^2 L_N$. This reduction relies mainly on Feynman–Kac formula and on a variational formula for the largest eigenvalue of a symmetric operator.

Since $e^{|x|} \le e^x + e^{-x}$ and since

$$\limsup_{N\to\infty} N^{-d} \log\{a_N + b_N\} \le \limsup_{N\to\infty} N^{-d} \log a_N \vee \limsup_{N\to\infty} N^{-d} \log b_N ,$$

it is enough to show that

$$\inf_{\mathfrak{f}\in\mathcal{C}} \limsup_{\varepsilon\to 0} \limsup_{N\to\infty} \frac{1}{N^d} \log \mathbb{E}_{\nu_\alpha^N}\left[\exp\left\{N^d \int_0^T X_{N,\varepsilon N}^{\mathfrak{f},i}(H_s,\eta_s)\,ds\right\}\right] \le 0$$

for every smooth function H in $C^{1,2}([0,T]\times\mathbb{T}^d)$, $1\le i\le d$ and $T>0$.

Fix such smooth function H, $1\le i\le d$ and $T>0$. By Feynman-Kac formula (A1.7.5),

$$\mathbb{E}_{\nu_\alpha^N}\left[\exp\left\{N^d \int_0^T X_{N,\varepsilon N}^{\mathfrak{f},i}(H_s,\eta_s)\,ds\right\}\right] \le \exp\left\{\int_0^T \lambda_N(s)\,ds\right\},$$

where $\lambda_N(s)$ is the largest eigenvalue of the symmetric operator $N^2L_N + N^dX_{N,\varepsilon N}^{\mathfrak{f},i}(H_s,\eta)$. From the variational formula for the largest eigenvalue of an operator in a Hilbert space (A3.1.1),

$$\lambda_N(s) \le \sup_f \left\{\left\langle N^d X_{N,\varepsilon N}^{\mathfrak{f},i}(H_s,\eta) f(\eta)\right\rangle_\alpha - N^2 D_N(f)\right\},$$

where the supremum is taken over all densities f with respect to ν_α^N and $<\cdot>_\alpha$ denotes the expectation with respect to ν_α. In particular,

$$\begin{aligned} N^{-d}\log \mathbb{E}_{\nu_\alpha^N}&\left[\exp\left\{\int_0^T N^d X_{N,\varepsilon N}^{\mathfrak{f},i}(H_s,\eta_s)\,ds\right\}\right] \\ &\le \int_0^T ds \sup_f \left\{\left\langle X_{N,\varepsilon N}^{\mathfrak{f},i}(H_s,\eta) f(\eta)\right\rangle_\alpha - N^{2-d} D_N(f)\right\}. \end{aligned}$$

Therefore, to prove Theorem 1.1 we have to show that

$$\inf_{\mathfrak{f}\in\mathcal{C}} \limsup_{\varepsilon\to 0} \limsup_{N\to\infty} \sup_f \left\{\left\langle X_{N,\varepsilon N}^{\mathfrak{f},i}(H,\eta) f(\eta)\right\rangle_\alpha - N^{2-d} D_N(f)\right\} \le 0$$

uniformly over the set of continuous functions H in $C^2(\mathbb{T}^d)$ that are bounded as well as their first and second derivatives by some fixed constant.

The proof of this inequality is divided in three steps. The strategy follows closely the one adopted in the proof of the replacement lemma in Chapter 5. Recall that $X_{N,\varepsilon N}^{\mathfrak{f},i}$ is equal to $N^{1-d}\sum_x H(x/N)\tau_x V_i^{\mathfrak{f},\varepsilon N}(\eta)$. We first reduce the problem on a small macroscopic cube to the same problem on a large microscopic block. In our context this corresponds to replace $V_i^{\mathfrak{f},\varepsilon N}(\eta)$ by $V_i^{\mathfrak{f},\ell}(\eta)$ and constitutes the goal of step 2 below. We will then follow the arguments presented in the proof of the one block estimate to bound the largest eigenvalue of $N^2L_N + \gamma X_{N,\ell}^{\mathfrak{f},i}$ by the largest eigenvalue of $L_{\Lambda_\ell} + \beta_N V_i^{\mathfrak{f},\ell}$, where, for a finite subset Λ of $\mathbb{Z}^d$, L_Λ represents the restriction of the generator L_N to Λ and β_N is a small constant. In order to estimate the largest eigenvalue of $L_{\Lambda_\ell} + \beta_N V_i^{\mathfrak{f},\ell}$ we use a perturbation method that relies on the existence of a spectral gap for the generator restricted to finite boxes. This argument provides a bound on the largest eigenvalue in terms of the variance of $V_i^{\mathfrak{f},\ell}$. To conclude the proof it will remain to compute the variance and to show that it vanishes for some cylinder function $\mathfrak{f}$.

We start localizing the eigenvalue problem, a rather technical step.

Step 2. Reduction to microscopic blocks. Notice that there is no spatial average of the current W_{0,e_i} in the definition of $X_{N,\varepsilon N}^{\mathfrak{f},i}$ and recall from the proof of the one block estimate in Chapter 5 that such a spatial average is crucial. It can easily be inserted because

$$N^{1-d} \sum_{x\in\mathbb{T}_N^d} H(x/N)\Big\{\tau_x W_{0,e_i}(\eta) \;-\; (2\ell'+1)^{-d} \sum_{|y-x|\le\ell'} \tau_y W_{0,e_i}(\eta)\Big\}$$

is of order ℓ^2/N as one can see after performing a summation by parts and from the presence of a discrete Laplacian. Here and below ℓ denotes a positive integer independent of N and ε that increases to infinity after $N\uparrow\infty$, $\varepsilon\downarrow 0$ and $\ell' = \ell-1$. We averaged over $|y-x|\le\ell'$ so that $\tau_y W_{0,e_i}$ is measurable with respect to $\{\eta(z),\ z\in\Lambda_\ell\}$ for y in $\Lambda_{\ell'}$.

Denote by $s_{\mathfrak{f}}$ the linear size of the support of the cylinder function $\mathfrak{f}$. In the definition of $X_{N,\ell}^{\mathfrak{f},i}$, we may replace L_N by L_Λ for some cube Λ large enough to contain $\Lambda_{s_{\mathfrak{f}}+1}$. Furthermore, since $\mathfrak{f}$ is a cylinder function, a summation by parts shows that

$$N^{1-d}\sum_{x\in\mathbb{T}_N^d} H(x/N)\tau_x(L_\Lambda\mathfrak{f})(\eta) \;-\; N^{1-d}\sum_{x\in\mathbb{T}_N^d} H(x/N)\frac{1}{|\Lambda_\ell|}\sum_{|y|\le\ell}\tau_{x+y}(L_\Lambda\mathfrak{f})(\eta)$$

is of order ℓ^2/N. This justifies the replacement of $\tau_x L_N\mathfrak{f}$ by the average $|\Lambda_\ell|^{-1}\sum_{|y|\le\ell}\tau_{x+y}L_\Lambda\mathfrak{f}$ in $X_{N,\varepsilon N}^{\mathfrak{f},i}$. We have thus averaged in space the terms W_{0,e_i} and $L_N\mathfrak{f}$ in the definition of $X_{N,\varepsilon N}^{\mathfrak{f},i}$. We turn now to the substitution of $D_{i,j}(\eta^{\varepsilon N}(0))\{\eta^{\varepsilon N}(e_j) \;-\; \eta^{\varepsilon N}(0)\}$ by a local function.

Lemma 3.1 *For each $1\le i,j\le d$, $\varepsilon>0$ and positive integers N, ℓ, define $V_{\ell,N\varepsilon}^{i,j}(\eta)$ by*

$$V_{\ell,N\varepsilon}^{i,j}(\eta)$$
$$= D_{i,j}(\eta^{\varepsilon N}(0))\Big\{\eta^{\varepsilon N}(e_j) \;-\; \eta^{\varepsilon N}(0)\Big\} \;-\; D_{i,j}(\eta^{\ell}(0))\Big\{\eta^{\ell'}(e_j) \;-\; \eta^{\ell'}(0)\Big\}\,.$$

For every $\delta>0$ and $1\le i,j\le d$,

$$\limsup_{\ell\to\infty}\limsup_{\varepsilon\to 0}\limsup_{N\to\infty}\sup_{f}$$
$$\Big\{N^{1-d}\sum_{x\in\mathbb{T}_N^d} H(x/N)\Big\langle \tau_x V_{\ell,N\varepsilon}^{i,j}(\eta)f(\eta)\Big\rangle_\alpha \;-\; \delta N^{2-d}D_N(f)\Big\} \;\le\; 0\,.$$

We averaged the density over $|y|\le\ell'=\ell-1$ for $\eta^{\ell'}(e_j)$ and $\eta^{\ell'}(0)$ to be measurable with respect to $\{\eta(x);\ x\in\Lambda_\ell\}$.

Proof. We first rewrite the difference $V^{i,j}_{\ell,N\varepsilon}(\eta)$ as

$$
\begin{aligned}
& D_{i,j}(\eta^{\varepsilon N}(0))\Big\{\Big[\eta^{\varepsilon N}(e_j) - \eta^{\varepsilon N}(0)\Big] - \Big[\eta^{\ell'}(e_j) - \eta^{\ell'}(0)\Big]\Big\} \\
& \quad + \Big\{D_{i,j}(\eta^{\varepsilon N}(0)) - D_{i,j}(\eta^{\ell}(0))\Big\}\Big\{\eta^{\ell'}(e_j) - \eta^{\ell'}(0)\Big\}
\end{aligned}
\tag{3.1}
$$

and consider the two lines separately. A summation by parts shows that the first one translated by x, multiplied by $H(x/N)$ and summed over x is equal to

$$
\sum_{x\in\mathbb{T}^d_N} b_x(\eta)[\eta(x+e_j) - \eta(x)] \;,
$$

where $b_x(\eta)$ is given by

$$
\begin{aligned}
b_x(\eta) := \; & \frac{1}{|\Lambda_{\varepsilon N}|} \sum_{|y-x|\le \varepsilon N} D_{i,j}(\eta^{\varepsilon N}(y))H(y/N) \\
& - \frac{1}{|\Lambda_{\ell'}|} \sum_{|y-x|\le \ell'} D_{i,j}(\eta^{\ell}(y))H(y/N) \,.
\end{aligned}
$$

We shall prove that

$$
\limsup_{\ell\to\infty}\limsup_{\varepsilon\to 0}\limsup_{N\to\infty}\sup_{f}
\Big\{N^{1-d}\sum_{x\in\mathbb{T}^d_N}\Big\langle b_x(\eta)[\eta(x+e_j)-\eta(x)]f(\eta)\Big\rangle_\alpha - \delta N^{2-d}D_N(f)\Big\} \le 0
$$

for each $\delta > 0$.

The main difficulty in obtaining an estimate for $N^{1-d}\sum_x < b_x(\eta)[\eta(x+e_j) - \eta(x)]f(\eta) >_\rho$ in terms of the Dirichlet form comes from the factor N that is multiplying the space average (otherwise the problem would reduce to the replacement lemma of Chapter 5 because D is a continuous function in virtue of Theorem 5.8). The extra factor N^{-1} shall be obtained integrating by parts the function $[\eta(x) - \eta(0)]$ and applying Schwarz inequality.

Recall from (2.2) the definition of the cylinder functions Ψ_{0,e_k}, $\Psi_{e_k,0}$ for $1 \le k \le d$. By formula (2.2) and Remark 2.2, for all cylinder functions h,

$$
< \eta(e_k) - \eta(0)\,,\, h >_\alpha \,=\, 2 < \Psi_{0,e_k}, \nabla_{0,e_k} h >_\alpha \;.
$$

Applying this identity to $h(\eta) = b_x(\eta)f(\eta)$, we get that $\sum_x < b_x(\eta)[\eta(x+e_j) - \eta(x)]f(\eta) >_\alpha$ may be rewritten as

$$
2\sum_{x\in\mathbb{T}^d_N}\Big\langle(\tau_x\Psi_{0,e_j})(\eta)r_{x,x+e_j}(\eta)\Big\{b_x(\eta^{x,x+e_j})f(\eta^{x,x+e_j}) - b_x(\eta)f(\eta)\Big\}\Big\rangle_\alpha \;. \tag{3.2}
$$

Note that the difference $b_x(\eta^{x,x+e_j}) - b_x(\eta)$ is equal to

$$\frac{1}{|\Lambda_{\varepsilon N}|}\sum_{\substack{|y-x|\le\varepsilon N\\ y_j=x_j-\varepsilon N}} H(y/N)\Big\{D_{i,j}\Big(\eta^{\varepsilon N}(y)-(2\varepsilon N+1)^{-d}\Big)-D_{i,j}(\eta^{\varepsilon N}(y))\Big\}$$

which is of order $(\varepsilon N)^{-1}o_N(1)$ because $D_{i,j}(\cdot)$ is continuous. Here $o_N(1)$ stands for a constant that vanishes in the limit as $N\uparrow\infty$. In particular, (3.2) is equal to

$$2\sum_{x\in\mathbb{T}_N^d}\Big\langle(\tau_x\Psi_{0,e_j})(\eta)b_x(\eta)r_{x,x+e_j}(\eta)\Big\{f(\eta^{x,x+e_j})-f(\eta)\Big\}\Big\rangle_\alpha + \varepsilon^{-1}N^{d-1}o_N(1)\,.$$

To estimate the first term of this expression, rewrite $|f(\eta^{x,x+e_j})-f(\eta)|$ as

$$\Big|\sqrt{f(\eta^{x,x+e_j})}-\sqrt{f(\eta)}\Big|\;\Big|\sqrt{f(\eta^{x,x+e_j})}+\sqrt{f(\eta)}\Big|\,.$$

By the elementary inequality $2cd\le Ac^2+A^{-1}d^2$ that holds for any positive A, the expectation $<(\tau_x\Psi_{0,e_j})b_x\nabla_{x,x+e_j}f>_\alpha$ is bounded by

$$\frac{1}{2A}<(\tau_x\Psi_{0,e_j}(\eta))^2b_x(\eta)^2r_{x,x+e_j}(\eta)\Big[\sqrt{f(\eta^{x,x+e_j})}+\sqrt{f(\eta)}\Big]^2\Big\rangle_\alpha$$
$$+\frac{A}{2}\Big\langle r_{x,x+e_j}(\eta)\Big[\sqrt{f(\eta^{x,x+e_j})}-\sqrt{f(\eta)}\Big]^2\Big\rangle_\alpha$$

for all $A>0$. The second term of this expression is just $2AI_{x,x+e_j}(f)$. Since $(c+d)^2\le 2c^2+2d^2$, since Ψ_{0,e_j} is bounded by a finite constant that depends only on κ and since by reversibility $<r_{x+e_j,x}(\eta)h(\eta^{x+e_j,x})>_\alpha=<r_{x,x+e_j}(\eta)h(\eta)>_\alpha$ for every h in $L^2(\nu_\alpha)$, the first term is bounded above by

$$\frac{C(\kappa)}{A}\Big\langle b_x(\eta)^2f(\eta)\Big\rangle_\alpha+\frac{C(\kappa)}{A}\Big\langle b_x(\eta^{x+e_j,x})^2f(\eta)\Big\rangle_\alpha\,.$$

Since $b_x(\eta^{x+e_j,x})=b_x(\eta)+(\varepsilon N)^{-1}o_N(1)$, summing over x and diving by N^{d-1}, we obtain that (3.2) divided by N^{d-1} is bounded above by

$$\frac{C(\kappa)}{A}N^{1-d}\sum_{x\in\mathbb{T}_N^d}\Big\langle b_x(\eta)^2f(\eta)\Big\rangle_\alpha+2AN^{1-d}D_N(f)+\varepsilon^{-1}\Big(1+\frac{1}{\varepsilon NA}\Big)o_N(1)$$

for every $A>0$. Choosing $A=\delta/4N$, we get that

$$\sup_f\Big\{N^{1-d}\sum_{x\in\mathbb{T}_N^d}\Big\langle[\eta(x+1)-\eta(x)]b_x(\eta)f(\eta)\Big\rangle_\alpha-\delta N^{2-d}D_N(f)\Big\}$$
$$\le\sup_f\Big\{C\delta^{-1}N^{-d}\sum_{x\in\mathbb{T}_N^d}\Big\langle[b_x(\eta)]^2f(\eta)\Big\rangle_\alpha-(\delta/2)N^{2-d}D_N(f)\Big\}+o_N(1)\,.$$

The proof that the limit of this expression as $N\uparrow\infty$, $\varepsilon\downarrow 0$ and $\ell\uparrow\infty$ is nonpositive follows from the usual two blocks estimate since $D_{i,j}(\cdot)$ is continuous by Theorem 5.8 below.

The second expression in (3.1) is handled in the same way. This concludes the proof of the lemma. □

For every positive integers N, ℓ, $1 \le i \le d$, smooth function $H \in C^2(\mathbb{T}^d)$ and cylinder function $\mathfrak{f}$, let

$$\tilde{X}^{\mathfrak{f},i}_{N,\ell}(H,\eta) := N^{1-d} \sum_{x\in\mathbb{T}^d_N} H(x/N)\tau_x \tilde{V}^{\mathfrak{f},\ell}_i(\eta)$$

where,

$$\begin{aligned}\tilde{V}^{\mathfrak{f},\ell}_i &= (2\ell'+1)^{-d} \sum_{|y|\le \ell'} \tau_y W_{0,e_i}(\eta) + \sum_{j=1}^{d} D_{i,j}\big(\eta^{\ell}(0)\big)\Big\{\eta^{\ell'}(e_j) - \eta^{\ell'}(0)\Big\}\\ &\quad - \frac{1}{(2\ell_{\mathfrak{f}}+1)^d} \sum_{y\in\Lambda_{\ell_{\mathfrak{f}}}} (\tau_y L_N \mathfrak{f})(\eta)\end{aligned}$$

and $\ell_{\mathfrak{f}} = \ell - s_{\mathfrak{f}} - 1$ so that $\tau_y L\mathfrak{f}$ is $\mathcal{F}_\ell$-measurable for every y in $\Lambda_{\ell_{\mathfrak{f}}}$. Up to this point we showed that in order to prove Theorem 1.1 it is enough to prove the same statement with $\tilde{X}^{\mathfrak{f},i}_{N,\ell}(\eta)$ instead of $X^{\mathfrak{f},i}_{N,\varepsilon N}$, i.e., to show that for all $\delta > 0$,

$$\inf_{\mathfrak{f}\in C} \limsup_{\ell\to\infty} \limsup_{N\to\infty} \sup_{f} \Big\{ \Big\langle \tilde{X}^{\mathfrak{f},i}_{N,\ell}(\eta) f(\eta) \Big\rangle_\alpha - \delta N^{2-d} D_N(f) \Big\} \le 0\,. \qquad (3.3)$$

Step 3. Estimate on small perturbations of a reversible generator. Since ν_α is translation invariant, we may rewrite $< \tilde{X}^{\mathfrak{f},i}_{N,\ell}(\eta) f(\eta) >_\alpha$ as

$$N^{1-d} \sum_{x\in\mathbb{T}^d_N} \Big\langle H(x/N) \tilde{V}^{\mathfrak{f},\ell}_i(\eta) \tau_{-x} f(\eta) \Big\rangle_\alpha .$$

We now repeat the usual procedure of the proof of the one block estimate. We first project the density $\tau_{-x} f$ on the finite hyperplanes with fixed total number of particles $\Sigma^{\kappa}_{\Lambda_\ell,K}$. Recall that for $0 \le K \le \kappa|\Lambda_\ell|$ and a density f with respect to ν_α, we denote by $\nu_{\ell,K}$ the measure ν_α conditioned on the hyperplane $\Sigma^{\kappa}_{\Lambda_\ell,K}$:

$$\nu_{\ell,K}(\cdot) = \nu_\alpha\Big(\cdot \mid \sum_{x\in\Lambda_\ell} \eta(x) = K \Big)$$

and by $f_{\ell,K}$ the projection of f on $\Sigma^{\kappa}_{\Lambda_\ell,K}$:

$$f_{\ell,K}(\xi) = \frac{E_{\nu_\alpha}\Big[f \,\Big|\, \eta(x) = \xi(x)\,,\, x\in\Lambda_\ell \Big]}{E_{\nu_\alpha}\Big[f \,\Big|\, \sum_{x\in\Lambda_\ell} \eta(x) = K \Big]}$$

for all configurations ξ of $\Sigma^{\kappa}_{\Lambda_\ell,K}$. Since $\tilde{V}^{\mathfrak{f},\ell}_i(\eta)$ depends on η only through $\{\eta(z);\ z\in\Lambda_\ell\}$, we have that

$$\left\langle \tilde V_i^{\mathfrak f,\ell}(\eta)(\tau_{-x}f)(\eta)\right\rangle_\alpha = \sum_K c(x,f,K)\left\langle \tilde V_i^{\mathfrak f,\ell}(\eta)(\tau_{-x}f)_{\ell,K}(\eta)\right\rangle_{\ell,K},$$

where $c(x,f,K)$ is given by

$$c(x,f,K) = \int \mathbf{1}\{\eta;\ \sum_{x\in\Lambda_\ell}\eta(x)=K\}(\tau_{-x}f)(\eta)\nu_\alpha(d\eta)$$

and $<\cdot>_{\ell,K}$ stands for expectation with respect to the canonical measure $\nu_{\ell,K} = \nu_{\Lambda_\ell,K}$. Notice that summation over K of $C(x,f,K)$ is equal to 1 for all x. Moreover, with respect to $\nu_{\ell,K}$, $\eta^\ell(0)$ is a constant equal to $K/(2\ell+1)^d$.

Denote by $D_{\ell,K}$ the Dirichlet form on $\Sigma^{\kappa}_{\Lambda_\ell,K}$:

$$D_{\ell,K}(f) = (1/4)\sum_{\substack{x,y\in\Lambda_\ell\\|x-y|=1}}\left\langle r_{x,y}(\eta)\left[\sqrt{f(\eta^{x,y})}-\sqrt{f(\eta)}\right]^2\right\rangle_{\ell,K}.$$

By convexity of the Dirichlet form, we have that

$$\sum_{x\in\mathbb{T}_N^d}\sum_K c(x,f,K)D_{\ell,K}\Big((\tau_{-x}f)_{\ell,K}\Big) \le (2\ell+1)^{-d}D_N(f).$$

In conclusion, the expression inside braces in (3.3) is bounded by

$$\sum_{x\in\mathbb{T}_N^d}\sum_K c(x,f,K)\Big\{N^{1-d}\left\langle H(x/N)\tilde V_i^{\mathfrak f,\ell}(\eta)(\tau_{-x}f)_{\ell,K}(\eta)\right\rangle_{\ell,K}$$
$$- C_\ell\delta N^{2-d}(2\ell)^{-d}D_{\ell,K}((\tau_{-x}f)_{\ell,K})\Big\}$$

for some constant C_ℓ that converges to 1 as $\ell\uparrow\infty$. Since $(\tau_{-x}f)_{\ell,K}$ is a density with respect to $\nu_{\ell,K}$, and since summation over K of $c(x,f,K)$ is equal to 1, this expression is bounded above by

$$C_\ell\delta N^{2-d}(2\ell)^{-d}\times$$
$$\times\sum_{x\in\mathbb{T}_N^d}\sup_K\sup_h\Big\{(C_\ell\delta N)^{-1}(2\ell)^dH(x/N)\left\langle\tilde V_i^{\mathfrak f,\ell}(\eta)h(\eta)\right\rangle_{\ell,K} - D_{\ell,K}(h)\Big\}, \tag{3.4}$$

where the supremum is carried over all densities h with respect to $\nu_{\ell,K}$. Let $\beta=\beta(x,H,\delta,N,\ell)=(C_\ell\delta N)^{-1}(2\ell)^dH(x/N)$. The expression

$$\sup_h\Big\{\beta\left\langle\tilde V_i^{\mathfrak f,\ell}(\eta)h(\eta)\right\rangle_{\ell,K} - D_{\ell,K}(h)\Big\}$$

is a variational formula for the largest eigenvalue of a small perturbation of the generator L_N restricted to a cube of length $2\ell+1$.

Recall that L_{Λ_ℓ} stands for the restriction of the generator L_N to Λ_ℓ. Since the generalized symmetric exclusion process on a finite cube is ergodic, L_{Λ_ℓ} has a strictly positive spectral gap σ_ℓ. Denote by λ_β be the largest eigenvalue of $L_{\Lambda_\ell} + \beta \tilde{V}_i^{\mathfrak{f},\ell}(\cdot)$. Theorem A3.1.1 asserts that

$$\lambda_\beta \le \frac{\beta^2}{1 - 2\|\tilde{V}_i^{\mathfrak{f},\ell}\|_\infty |\beta| \sigma_\ell} \Big\langle (-L_{\Lambda_\ell})^{-1} \tilde{V}_i^{\mathfrak{f},\ell}, \tilde{V}_i^{\mathfrak{f},\ell} \Big\rangle_{\ell,K}$$

uniformly in K. Since β vanishes in the limit as $N \uparrow \infty$, the right hand side of the last inequality is bounded by $2\beta^2 < (-L_{\Lambda_\ell})^{-1} \tilde{V}_i^{\mathfrak{f},\ell}, \tilde{V}_i^{\mathfrak{f},\ell} >_{\ell,K}$ for sufficiently large N. Therefore, (3.4) is bounded above by

$$\sup_K \frac{2\|H\|_\infty^2}{\delta C_\ell} (2\ell)^d \Big\langle (-L_{\Lambda_\ell})^{-1} \tilde{V}_i^{\mathfrak{f},\ell}, \tilde{V}_i^{\mathfrak{f},\ell} \Big\rangle_{\ell,K} .$$

To conclude the proof of the proposition, it remains to show that

$$\inf_{\mathfrak{f} \in \mathcal{C}} \lim_{\ell \to \infty} \sup_K (2\ell)^d \Big\langle (-L_{\Lambda_\ell})^{-1} \tilde{V}_i^{\mathfrak{f},\ell}, \tilde{V}_i^{\mathfrak{f},\ell} \Big\rangle_{\ell,K} = 0 . \tag{3.5}$$

This follows from Theorem 4.6 and Corollary 5.9 below.

4. Central Limit Theorem Variances

We assume in this section that the reader is acquainted with the concept of closed and exact forms in the context of interacting particle systems. The main ideas and all results needed below are presented in section A3.4.

In last section we reduced the proof of the hydrodynamic limit of nongradient systems to the computation of a central limit theorem variance. The purpose of this section is to obtain a variational formula for this variance. We start introducing a semi–norm on $\mathcal{C}_0$ closely related to the central limit theorem variance. For $1 \le k \le d$ denote by $\mathfrak{A}^k = (\mathfrak{A}_1^k, \dots, \mathfrak{A}_d^k)$ the d-dimensional cylinder function with coordinates defined by

$$(\mathfrak{A}^k)_i(\eta) = \delta_{i,k}\, r_{0,e_k}(\eta) \quad \text{for} \quad 1 \le i \le d . \tag{4.1}$$

Here $\delta_{i,j}$ stands for the delta of Kronecker, equal to 1 if $i = j$ and 0 otherwise. For cylinder functions g and h in $\mathcal{C}_0$ and $1 \le j \le d$, let

$$\ll g, h \gg_{\alpha,0} = \sum_{x \in \mathbb{Z}^d} \Big\langle g, \tau_x h \Big\rangle_\alpha \quad \text{and} \quad \ll g \gg_{\alpha,j} = \sum_{x \in \mathbb{Z}^d} x_j \Big\langle g, \eta(x) \Big\rangle_\alpha ,$$

where x_j stands for the j-th coordinate of $x \in \mathbb{Z}^d$. Both $\ll g, h \gg_{\alpha,0}$ and $\ll g \gg_{\alpha,j}$ are well defined because g and h belongs to $\mathcal{C}_0$ and therefore all but a finite number of terms vanish. For h in $\mathcal{C}_0$, define the semi–norm $\ll h \gg_\alpha^{1/2}$ by

$$\ll h \gg_\alpha = \sup_{\substack{g\in\mathcal{C}_0\\ a\in\mathbb{R}^d}} \Big\{ 2 \ll g, h \gg_{\alpha,0} + 2\sum_{i=1}^d a_i \ll h \gg_{\alpha,i} - (1/2)\sum_{i=1}^d \Big\langle \Big(\sum_{1\le j\le d} a_j(\mathfrak{A}^j)_i + \nabla_{(0,e_i)}\Gamma_g \Big)^2 \Big\rangle_\alpha \Big\} .$$

In this formula Γ_g stands for the formal sum $\sum_{x\in\mathbb{Z}^d} \tau_x g$. To keep notation simple, denote by $\nabla\Gamma_g$ the vector $(\nabla_{(0,e_1)}\Gamma_g, \dots, \nabla_{(0,e_d)}\Gamma_g)$ and by $\|a\|$ the Euclidean norm of a d-dimensional vector a: $\|a\|^2 = \sum_{1\le i\le d}(a_i)^2$. With this notation, the semi–norm $\ll \cdot \gg_\alpha$ writes

$$\ll h \gg_\alpha = \sup_{\substack{g\in\mathcal{C}_0\\ a\in\mathbb{R}^d}} \Big\{ 2 \ll g, h \gg_{\alpha,0} + 2\sum_{i=1}^d a_i \ll h \gg_{\alpha,i} - (1/2)\Big\langle \Big\| \sum_{1\le j\le d} a_j\mathfrak{A}^j + \nabla\Gamma_g \Big\|^2 \Big\rangle_\alpha \Big\} .$$

We investigate in the next section the main properties of the semi–norm $\ll \cdot \gg_\alpha^{1/2}$, while in this section we prove that the variance

$$(2\ell)^{-d} < (-L_{\Lambda_\ell})^{-1} \sum_{|x|\le\ell_\psi} \tau_x\psi \, , \sum_{|x|\le\ell_\psi} \tau_x\psi >_{\ell,K_\ell}$$

of any cylinder function ψ in $\mathcal{C}_0$ converges to $\ll \psi \gg_\alpha$, as $\ell\uparrow\infty$ and $K_\ell/(2\ell)^d \to \alpha$. Here ℓ_ψ stands for $\ell - s_\psi$ so that the support of $\tau_x\psi$ is included in Λ_ℓ for every $|x|\le\ell_\psi$. To prove this result we need an alternative formula for $\ll \cdot \gg_\alpha$. By elementary computations relying on an adequate change of variables, we obtain that the quadratic term in last formula writes $-(1/2)\|a\|^2 < r_{0,e_1} >_\alpha$ $+2\sum_{1\le j\le d} a_j \ll W_{0,e_j}, g \gg_{\alpha,0} -(1/2) < \|\nabla\Gamma_g\|^2 >_\alpha$. The norm $\ll \cdot \gg_\alpha$ may thus be rewritten as

$$\ll h \gg_\alpha = \sup_{\substack{g\in\mathcal{C}_0\\ a\in\mathbb{R}^d}} \Big\{ 2 \ll g, h \gg_{\alpha,0} + 2\sum_{i=1}^d a_i \ll h \gg_{\alpha,i} + 2\sum_{i=1}^d a_i \ll W_{0,e_i}, g \gg_{\alpha,0} - \frac{\|a\|^2}{2} < r_{0,e_1} >_\alpha - \frac{1}{2}\Big\langle \|\nabla\Gamma_g\|^2 \Big\rangle_\alpha \Big\} \tag{4.2}$$

We are now in a position to state the main result of this section.

Theorem 4.1 *Consider a cylinder function ψ in $\mathcal{C}_0$ and a sequence of positive integers K_ℓ such that $0 \le K_\ell \le \kappa(2\ell+1)^d$ and $\lim_{\ell\to\infty} K_\ell/(2\ell)^d = \alpha$. Then,*

$$\lim_{\ell\to\infty} (2\ell)^{-d}\Big\langle (-L_{\Lambda_\ell})^{-1} \sum_{|x|\le\ell_\psi} \tau_x\psi \, , \sum_{|x|\le\ell_\psi} \tau_x\psi \Big\rangle_{\ell,K_\ell} = \ll \psi \gg_\alpha .$$

We first prove that the left hand side of this identity is bounded above by the right hand side. This is done in two steps that we state as separate lemmas in sake of clarity. In the first lemma we estimate the variances with respect to canonical measures by variances with respect to grand canonical measures. We then recall that the space of germs of closed forms is a direct sum of the germs $\{\mathfrak{A}^k, 1 \le k \le d\}$ introduced in the beginning of this section and the gradients $\{\nabla \Gamma_g, g \in \mathcal{C}_0\}$. This permits to bound the variance, with respect to the grand canonical measure ν_α, of a cylinder function ψ in $\mathcal{C}_0$ by $\ll \psi \gg_\alpha$. The necessity of a sharp estimate for the spectral gap of the generator restricted to finite cubes is hidden in this second step, for such an estimate is crucial in the description of the structure of the space of germs of closed forms (cf. Theorem A3.4.14).

We start with a corollary of the integration by parts formula. For a subset Λ of $\mathbb{Z}^d$, a site $x \in \mathbb{Z}^d$ and a positive integer k, denote by $d(x, \Lambda)$ the distance from x to Λ for the norm $|\cdot|$ and by $\Omega_k(\Lambda)$ the collection of sites at a distance less than or equal to k from Λ:

$$\Omega_k(\Lambda) = \{y \in \mathbb{Z}^d;\ d(y, \Lambda) \le k\} .$$

Lemma 4.2 *Fix a cylinder function ψ in $\mathcal{C}_0$. There exist constants $C_1(\psi)$, $C_2(\psi)$ depending only on ψ such that for every positive integer ℓ, $0 \le K \le (2\ell+1)^d \kappa$, subset Λ of Λ_{ℓ_ψ} and $\mathcal{F}_{\Omega_{s_\psi}(\Lambda)}$-measurable function h in $L^2(\nu_{\ell,K})$,*

$$2\Big\langle \sum_{x\in\Lambda} \tau_x\psi\, ,\, h\Big\rangle_{\ell,K} \le \frac{1}{A} C_1(\psi)|\Lambda| + A C_2(\psi) \sum_{b\in\Omega_{s_\psi}(\Lambda)} \mathfrak{D}_b(\nu_{\ell,K}, h)$$

for every $A > 0$. The statement remains in force if the canonical measure $\nu_{\ell,K}$ is replaced by the grand canonical measure ν_α.

In this lemma Λ shall be thought as a cube $x + \Lambda_k$ much smaller than Λ_ℓ: $k \ll \ell$. The lemma is just saying that the left hand side can be estimated by $|\Lambda|$ and the Dirichlet form of h on a set slightly larger than $x + \Lambda_k$.

Proof. By the integration by parts formula,

$$2\sum_{x\in\Lambda} \Big\langle \tau_x\psi\, ,\, h\Big\rangle_{\ell,K} = 2\sum_{x\in\Lambda}\sum_{b\in x+\Lambda_\psi} \Big\langle \Phi_b^{\tau_x\psi}\, ,\, \nabla_b h\Big\rangle_{\ell,K}$$

because $\Lambda_{\tau_x\psi} = x + \Lambda_\psi$. For each oriented bond b, denote by $A^\psi_{b,\Lambda}$ the set of sites x in Λ such that b belongs to $x + \Lambda_\psi$: $A^\psi_{b,\Lambda} = \{x \in \Lambda;\ b \in x + \Lambda_\psi\}$. By Schwarz inequality $2ad \le A^{-1}a^2 + Ad^2$, the right hand side of the last identity is bounded above by

$$\frac{1}{A}\sum_{x\in\Lambda}\sum_{b\in x+\Lambda_\psi} \Big\langle \Big(\Phi_b^{\tau_x\psi}\Big)^2\Big\rangle_{\ell,K} + A\sum_{b\in\Omega_{s_\psi}(\Lambda)} |A^\psi_{b,\Lambda}| \Big\langle \Big(\nabla_b h\Big)^2\Big\rangle_{\ell,K} \tag{4.3}$$

for all $A > 0$. The explicit formula for $\Phi_b^{\tau_x \psi}$ derived in the proof of Lemma 2.1 shows that $\sum_{b \in x + \Lambda_\psi} < (\Phi_b^{\tau_x\psi})^2 >_{\ell,K}$ is bounded by a finite constant $C_1(\psi)$ that depends only on ψ and not on x because under $\nu_{\ell,K}$ the distribution of the collection $(\eta(z),\, z \in x + \Lambda_\psi)$ does not depend on x. To conclude the proof of the lemma it remains to observe that $|A_{b,\Lambda}^\psi| \le C_2(\psi)$. □

Lemma 4.3 *Under the assumptions of Theorem 4.1,*

$$\limsup_{\ell\to\infty} (2\ell)^{-d} \Big\langle (-L_{\Lambda_\ell})^{-1} \sum_{|x|\le \ell_\psi} \tau_x\psi \,,\, \sum_{|x|\le \ell_\psi} \tau_x \psi \Big\rangle_{\ell, K_\ell}$$
$$\le \limsup_{k\to\infty} (2k)^{-d} \Big\langle (-L_{\Lambda_k})^{-1} \sum_{|x|\le k_\psi} \tau_x\psi \,,\, \sum_{|x|\le k_\psi} \tau_x \psi \Big\rangle_{\alpha} \cdot$$

Proof. By the variational formula for the variance,

$$\begin{aligned} &(2\ell)^{-d} < (-L_{\Lambda_\ell})^{-1} \sum_x \tau_x\psi, \sum_x \tau_x\psi >_{\ell,K_\ell} \\ &\quad = (2\ell)^{-d} \sup_h \Big\{ 2 \Big\langle \sum_{|x|\le \ell_\psi} \tau_x\psi \,,\, h \Big\rangle_{\ell,K_\ell} - \Big\langle -L_{\Lambda_\ell} h \,,\, h \Big\rangle_{\ell,K_\ell} \Big\} , \end{aligned} \tag{4.4}$$

where the supremum is taken over all functions in $L^2(\nu_{\ell,K_\ell})$. By Lemma 4.2 with $\Lambda = \Lambda_{\ell_\psi}$, $K = K_\ell$ and $A = (1/2)C_2(\psi)^{-1}$, the expression inside braces in (4.4) is bounded by

$$(2\ell)^d C(\psi) \;-\; (1/2)\mathfrak{D}(\nu_{\ell,K_\ell}, h)$$

which is negative if $\mathfrak{D}(\nu_{\ell,K_\ell}, h) \ge C(\psi)(2\ell)^d$. Here $C(\psi)$ is a constant depending only on ψ that may change from line to line. On the other hand, since ψ has mean zero with respect to all canonical measures, for a constant function h the difference $2\sum_{x\in\Lambda_{\ell_\psi}} < \tau_x\psi \,,\, h >_{\ell,K_\ell} - \mathfrak{D}(\nu_{\ell,K_\ell}, h)$ vanishes. We may therefore restrict the supremum to functions h with Dirichlet form bounded by $C(\psi)(2\ell)^d$.

Fix a positive integer k larger than $s_\psi + 1$ and that shall converge to infinity after ℓ. Divide the hypercube Λ_ℓ in cubes of length $2k+1$. Denote these subcubes by B_a, $1 \le a \le p = [(2\ell+1)/(2k+1)]^d$. Here $[r]$ stands for the integer part of r. Since $2\ell + 1$ might not be divisible by $2k+1$, let B_{p+1} denote the set of sites that do not belong to any of the cubes B_a and notice that $|B_{p+1}| \le Ck\ell^{d-1}$ for some universal constant C. Thus

$$\Lambda_\ell \;=\; \bigcup_{a=1}^{p+1} B_a \quad \text{and} \quad B_{a_1} \cap B_{a_2} = \phi \quad \text{for } a_1 \neq a_2 \,.$$

For each fixed $1 \le a \le p$, denote by B_a^o the "interior" of B_a, that is, the set of points in B_a that are at a distance at least $s_\psi + 1$ of the boundary of B_a and by Λ_ℓ^o the union of all interior points of Λ_ℓ:

$$B_a^o = \{x \in B_a;\ d(x, B_a^c) \ge s_\psi + 1\}\ , \quad \Lambda_\ell^o = \bigcup_{a=1}^{p} B_a^o \ \text{ and } \ \Lambda_\ell^1 = \Lambda_\ell - \Lambda_\ell^o\ ,$$

where B_a^c stands for the complement of B_a. Notice that Λ_ℓ^1 contains B_{p+1} and that $|\Lambda_\ell^1| \le C(\psi)\ell^d\{k\ell^{-1} + k^{-1}\}$.

For each fixed h in $L^2(\nu_{\ell,K_\ell})$ with Dirichlet form bounded by $C(\psi)\ell^d$, rewrite the expression inside braces in (4.4) as

$$2\sum_a \sum_{x\in B_a^o} \Big\langle \tau_x\psi\, ,\, h\Big\rangle_{\ell,K_\ell} + 2 \sum_{x\in \Lambda_\ell^1 \cap \Lambda_{\ell_\psi}} \Big\langle \tau_x\psi\, ,\, h\Big\rangle_{\ell,K_\ell} - \mathfrak{D}(\nu_{\ell,K_\ell}, h)\ . \qquad (4.5)$$

By Lemma 4.2, the second term in this formula is bounded above by $C(\psi)\min_{A>0}\{A^{-1}|\Lambda_\ell^1| + A\ell^d\} = C(\psi)\sqrt{\ell^d|\Lambda_\ell^1|}$ because the supremum is restricted to functions with Dirichlet form bounded by $C(\psi)\ell^d$. The second term in (4.5) is thus bounded above by $C(\psi)\ell^d\{k\ell^{-1} + k^{-1}\}^{1/2}$ since $|\Lambda_\ell^1| \le C(\psi)\ell^d\{k\ell^{-1} + k^{-1}\}$.

We turn now to the other two terms in (4.5). By construction, for x in B_a^o, $\tau_x\psi$ is measurable with respect to $\{\eta(x);\ x \in B_a\}$. Denote by h_a the conditional expectation of h with respect to this σ-algebra: $h_a = E_{\ell,K_\ell}[h\,|\,\eta(x),\ x \in B_a]$ so that $< \tau_x\psi\, ,\, h >_{\ell,K_\ell} = < \tau_x\psi\, ,\, h_a >_{\ell,K_\ell}$ for x in B_a. Let $\mathfrak{D}_a$ be the restriction of the Dirichlet form to B_a:

$$\mathfrak{D}_a(\nu_{\ell,K_\ell}, h) = \sum_{b\in B_a} \mathfrak{D}_b(\nu_{\ell,K_\ell}, h)\ .$$

By convexity, $\mathfrak{D}_a(\nu_{\ell,K_\ell}, h_a) \le \mathfrak{D}_a(\nu_{\ell,K_\ell}, h)$. On the other hand, we have that $\sum_a \mathfrak{D}_a(\nu_{\ell,K_\ell}, \cdot\,) \le \mathfrak{D}(\nu_{\ell,K_\ell}, \cdot\,)$ because in the first Dirichlet form bonds that links different cubes B_a do not appear. Therefore, expression (4.5) is bounded above by

$$\sum_a \Big\{ 2\sum_{x\in B_a^o} \Big\langle \tau_x\psi\, ,\, h_a\Big\rangle_{\ell,K_\ell} - \mathfrak{D}_a(\nu_{\ell,K_\ell}, h_a)\Big\} + C(\psi)\ell^d\sqrt{\frac{k}{\ell} + \frac{1}{k}}\ . \qquad (4.6)$$

Fix $1 \le a \le p$ and denote by $\mathcal{F}_{B_a}$ the σ-algebra generated by $\{\eta(z),\ z \in B_a\}$. The expression inside braces in last formula is bounded above by $\sup_g\{2\sum_{x\in B_a^o} < \tau_x\psi\, ,\, g >_{\ell,K_\ell} - \mathfrak{D}_a(\nu_{\ell,K_\ell}, g)\}$ where the supremum is taken over all $\sigma\{\eta(z),\ z \in B_a\}$-measurable functions g in $L^2(\nu_{\ell,K_\ell})$. Since under ν_{ℓ,K_ℓ} the distribution of the collection $(\eta(z),\ z \in x + \Lambda_\psi)$ does not depend on x,

$$\begin{aligned} &\sup_{g\in\mathcal{F}_{B_a}} \Big\{ 2\sum_{x\in B_a^o} \Big\langle \tau_x\psi\, ,\, g\Big\rangle_{\ell,K_\ell} - \mathfrak{D}_a(\nu_{\ell,K_\ell}, g)\Big\} \\ &\quad = \sup_{g\in\mathcal{F}_k} \Big\{ 2\sum_{x\in\Lambda_{k_\psi}} \Big\langle \tau_x\psi\, ,\, g\Big\rangle_{\ell,K_\ell} - \mathfrak{D}_{\Lambda_k}(\nu_{\ell,K_\ell}, g)\Big\} \end{aligned}$$

where on the right hand side the supremum is taken over all $\mathcal{F}_k$-measurable functions in $L^2(\nu_{\ell,K_\ell})$ and $\mathfrak{D}_{\Lambda_k}(\nu_{\ell,K_\ell}, \cdot\,)$ is the Dirichlet form $\mathfrak{D}(\nu_{\ell,K_\ell}, \cdot\,)$ restricted to

Λ_k: $\mathfrak{D}_{\Lambda_k}(\nu_{\ell,K_\ell}, \cdot) = \sum_{b\in\Lambda_k} \mathfrak{D}_b(\nu_{\ell,K_\ell}, \cdot)$. In particular, for each h in $L^2(\nu_{\ell,K_\ell})$, formula (4.6) and thus (4.5), is bounded above by

$$(2\ell+1)^d(2k)^{-d} \sup_{g\in\mathcal{F}_k} \Big\{2 \sum_{x\in\Lambda_{k_\psi}} \Big\langle \tau_x\psi , g\Big\rangle_{\ell,K_\ell} - \mathfrak{D}_{\Lambda_k}(\nu_{\ell,K_\ell}, g)\Big\}$$
$$+ \; C(\psi)(2\ell+1)^d\sqrt{\frac{k}{\ell}+\frac{1}{k}} \; .$$

Since the expression inside braces in (4.4) is just (4.5), to prove the lemma it is enough to show that for each fixed k,

$$\limsup_{\ell\to\infty} \sup_{g\in\mathcal{F}_k} \Big\{2 \sum_{x\in\Lambda_{k_\psi}} \Big\langle \tau_x\psi , g\Big\rangle_{\ell,K_\ell} - \mathfrak{D}_{\Lambda_k}(\nu_{\ell,K_\ell}, g)\Big\}$$
$$= \sup_{g\in\mathcal{F}_k} \Big\{2 \sum_{x\in\Lambda_{k_\psi}} \Big\langle \tau_x\psi , g\Big\rangle_{\alpha} - \mathfrak{D}_{\Lambda_k}(\nu_{\alpha}, g)\Big\} \; . \tag{4.7}$$

By the variational formula for the variance, the supremum on the left hand side of this identity is equal to $< (-L_{\Lambda_k})^{-1}\sum_{x\in\Lambda_{k_\psi}} \tau_x\psi \, , \, \sum_{x\in\Lambda_{k_\psi}} \tau_x\psi >_{\ell,K_\ell}$ while the right hand side is equal to the variance

$$< (-L_{\Lambda_k})^{-1} \sum_{x\in\Lambda_{k_\psi}} \tau_x\psi \, , \, \sum_{x\in\Lambda_{k_\psi}} \tau_x\psi >_\alpha \; .$$

Since both $(-L_{\Lambda_k})^{-1}\sum_{x\in\Lambda_{k_\psi}} \tau_x\psi$ and $\sum_{x\in\Lambda_{k_\psi}} \tau_x\psi$ are cylinder functions, identity (4.7) follows from the equivalence of ensembles stated in Lemma A2.2.2. □

We now conclude the proof of the upper bound for the central limit theorem variances.

Lemma 4.4 *Under the assumptions of Theorem 4.1*

$$\limsup_{k\to\infty} (2k)^{-d}\Big\langle(-L_{\Lambda_k})^{-1} \sum_{|x|\le k_\psi} \tau_x\psi \, , \, \sum_{|x|\le k_\psi} \tau_x\psi\Big\rangle_\alpha \le \ll \psi \gg_\alpha \; . \tag{4.8}$$

Proof. By the variational formula for the variance, for each fixed k, the expression on the left hand side of (4.8) is equal to

$$(2k)^{-d} \sup_h \Big\{2 \sum_{x\in\Lambda_{k_\psi}} \Big\langle \tau_x\psi , h\Big\rangle_{\alpha} - \mathfrak{D}_{\Lambda_k}(\nu_{\alpha}, h)\Big\} \; .$$

In this formula the supremum is taken over all $\mathcal{F}_k$-measurable functions h in $L^2(\nu_\alpha)$. Lemma 4.2 and arguments similar to the ones presented at the beginning of the proof of the previous lemma permit to restrict the supremum to functions

h with Dirichlet form $\mathfrak{D}_{\Lambda_k}(\nu_\alpha, h)$ bounded by $C(\psi)k^d$ for some finite constant $C(\psi)$.

By the integration by parts formula, $2\sum_{x\in\Lambda_{k_\psi}} < \tau_x\psi\,,\, h >_\alpha$ is equal to

$$2\sum_{x\in\Lambda_{k_\psi}}\sum_{b\in x+\Lambda_\psi}\Big\langle \Phi_b^{\tau_x\psi}\,,\,\nabla_b h\Big\rangle_\alpha = 2\sum_{b\in\Lambda_k}\Big\langle \sum_{x\in A^\psi_{b,k}} \Phi_b^{\tau_x\psi}\,,\,\nabla_b h\Big\rangle_\alpha\,,$$

where $A^\psi_{b,k}$ stands for the set of sites x in Λ_{k_ψ} such that $b\in x+\Lambda_\psi$. For each fixed bond b in $\Lambda_{k_\psi - s_\psi}$, $\sum_{x\in A^\psi_{b,k}}\Phi_b^{\tau_x\psi} = \sum_{x;\, b\in x+\Lambda_\psi}\Phi_b^{\tau_x\psi}$. Let $\hat\Phi_b^\psi = \sum_{x;\, b\in x+\Lambda_\psi}\Phi_b^{\tau_x\psi}$. With this notation the last sum writes

$$2\sum_{b\in\Lambda_k}\Big\langle \hat\Phi_b^\psi\,,\,\nabla_b h\Big\rangle_\alpha + 2\sum_{b\in\Lambda_k-\Lambda_{k_\psi-s_\psi}}\Big\langle\sum_{x\in A^\psi_{b,k}}\Phi_b^{\tau_x\psi}\,,\,\nabla_b h\Big\rangle_\alpha$$
$$-\,2\sum_{b\in\Lambda_k-\Lambda_{k_\psi-s_\psi}}\Big\langle \hat\Phi_b^\psi\,,\,\nabla_b h\Big\rangle_\alpha\,.$$

By Schwarz inequality, the a priori bound on the Dirichlet form of h and the estimates of the L^2 norm of $\Phi_b^{\tau_x\psi}$, the second and third terms are bounded by $C(\psi)k^{d-(1/2)}$ for some constant $C(\psi)$ that depends only on ψ. In particular, the left hand side of (4.8) is bounded above by

$$\begin{aligned}&\limsup_{k\to\infty}(2k)^{-d}\sup_h\Big\{2\sum_{b\in\Lambda_k}\Big\langle\hat\Phi_b^\psi\,,\,\nabla_b h\Big\rangle_\alpha - \mathfrak{D}_{\Lambda_k}(\nu_\alpha, h)\Big\}\\ &\quad = \lim_{k\to\infty}(2k)^{-d}\Big\{\sum_{b\in\Lambda_k}2\Big\langle\hat\Phi_b^\psi\,,\,\nabla_b h_k\Big\rangle_\alpha - \sum_{b\in\Lambda_k}\mathfrak{D}_b(\nu_\alpha, h_k)\Big\}\end{aligned} \tag{4.9}$$

for some sequence of $\mathcal{F}_k$-measurable functions h_k in $L^2(\nu_\alpha)$.

Since by Lemma 2.1 $\tau_y\Phi_b^\psi = \Phi_{b+y}^{\tau_y\psi}$, $\hat\Phi_b^\psi$ is translation covariant in the sense that $\tau_y\hat\Phi_b^\psi = \hat\Phi_{b+y}^\psi$. Therefore, recalling that $b=(b_1,b_2)$, that ν_α is translation invariant and that $\tau_y\nabla_b = \nabla_{b+y}\tau_y$

$$\Big\langle\hat\Phi_b^\psi\,,\,\nabla_b h\Big\rangle_\alpha = \Big\langle\tau_{-b_1}\hat\Phi_b^\psi\,,\,\tau_{-b_1}\nabla_b h\Big\rangle_\alpha = \Big\langle\hat\Phi^\psi_{(0,b_2-b_1)}\,,\,\nabla_{(0,b_2-b_1)}\tau_{-b_1}h\Big\rangle_\alpha\,.$$

On the other hand, for every oriented bond $b=(b_1,b_2)$, a change of variables shows that $<\nabla_{(b_1,b_2)}f, \nabla_{(b_1,b_2)}g>_\alpha$ is equal to $<\nabla_{(b_2,b_1)}f, \nabla_{(b_2,b_1)}g>_\alpha$. In particular, since $\Phi_b^\psi = (1/4)\nabla_b(-L_{\Lambda_\psi})^{-1}\psi$, for $1\le i\le d$, we have that $<\hat\Phi^\psi_{(e_i,0)}, \nabla_{(e_i,0)}h>_\alpha = <\hat\Phi^\psi_{(0,e_i)}, \nabla_{(0,e_i)}h>_\alpha$. Thus,

$$\begin{aligned}&2(2k)^{-d}\sum_{b\in\Lambda_k}\Big\langle\hat\Phi_b^\psi\,,\,\nabla_b h_k\Big\rangle_\alpha\\ &\quad = 4\sum_{i=1}^d\Big\langle\hat\Phi^\psi_{(0,e_i)}\,,\,\nabla_{(0,e_i)}(2k)^{-d}\sum_{\substack{x;\,x\in\Lambda_k\\ x+e_i\in\Lambda_k}}\tau_{-x}h_k\Big\rangle_\alpha\,.\end{aligned}$$

On the other hand, by Schwarz inequality and the bound on the Dirichlet form of h,

$$\sum_{i=1}^{d} \Big\langle \Big(\nabla_{(0,e_i)} (2k)^{-d} \sum_{\substack{x;\, x\in\Lambda_k \\ x+e_i\in\Lambda_k}} \tau_{-x} h_k \Big)^2 \Big\rangle_\alpha$$

$$\le \sum_{i=1}^{d} (2k)^{-d} \sum_{\substack{x;\, x\in\Lambda_k \\ x+e_i\in\Lambda_k}} \Big\langle \Big(\nabla_{x,x+e_i} h_k \Big)^2 \Big\rangle_\alpha + O(k^{-1})$$

$$= 2(2k)^{-d} \sum_{b\in\Lambda_k} \mathfrak{D}_b(\nu_\alpha, h_k) + O(k^{-1}) .$$

The remainder $O(1/k)$ appeared because last summation is carried over $(2k+1)^{d-1}(2k)$ bonds, while we are dividing only by $(2k)^d$ and the Dirichlet form is bounded by k^d.

Therefore, if we denote by R_i^k the cylinder function

$$\nabla_{(0,e_i)} (2k)^{-d} \sum_{x;\, x, x+e_i\in\Lambda_k} \tau_{-x} h_k ,$$

for $1 \le i \le d$, the right hand side of identity (4.9) is bounded above by

$$\lim_{k\to\infty} \sum_{i=1}^{d} \Big\{ 4\Big\langle \hat{\Phi}^{\psi}_{(0,e_i)}, R_i^k \Big\rangle_\alpha - (1/2)\Big\langle (R_i^k)^2 \Big\rangle_\alpha \Big\} .$$

The upper bound on the Dirichlet form of h_k implies that the sequence of vectors $\{R_i^k,\, k\ge 1\}$ is bounded in $L^2(\nu_\alpha)$. There exists therefore a weakly converging subsequence. Denote by R_i a weak limit and assume, without loss of generality, that the sequence R^k converges weakly to $R = (R_1, \dots, R_d)$. Since the L^2 norm may only decrease along weakly converging subsequences, the limit of last sum, as $k \uparrow \infty$, is bounded by

$$\sum_{i=1}^{d} \Big\{ 4\Big\langle \hat{\Phi}^{\psi}_{(0,e_i)}, R_i \Big\rangle_\alpha - (1/2)\Big\langle (R_i)^2 \Big\rangle_\alpha \Big\} .$$

It is not difficult to check that R_i is a germ of a closed form in the terminology of Definition A3.4.12. Therefore, according to Theorem A3.4.14, R can be decomposed as a sum of the germs $\{\mathfrak{A}^j, 1 \le j \le d\}$ defined in (4.1) and of a gradient: $R = \sum_{1\le i\le d} a_j \mathfrak{A}^j + \nabla \Gamma_g$ for some cylinder function g in $\mathcal{C}_0$. Therefore, the right hand side of (4.9) is bounded above by

$$\sup_{\substack{a\in\mathbb{R}^d \\ g\in\mathcal{C}_0}} \Big\{ \sum_{i=1}^{d} 4\Big\langle \hat{\Phi}^{\psi}_{(0,e_i)}, \sum_{1\le j\le d} a_j (\mathfrak{A}^j)_i + \nabla_{(0,e_i)} \Gamma_g \Big\rangle_\alpha$$

$$- (1/2)\Big\langle \Big\| \sum_{1\le j\le d} a_j \mathfrak{A}^j + \nabla \Gamma_g \Big\|^2 \Big\rangle_\alpha \Big\} .$$

To conclude the proof of the lemma it remains to check that

$$4\sum_{i=1}^{d}\Big\langle \hat{\Phi}^{\psi}_{(0,e_i)}\,,\,(\mathfrak{A}^j)_i\Big\rangle_\alpha = 2\sum_{x\in\mathbb{Z}^d} x_j\Big\langle \psi\,,\,\eta(x)\Big\rangle_\alpha = 2 \ll \psi \gg_{\alpha,j}$$

for $1 \le j \le d$ and

$$4\sum_{i=1}^{d}\Big\langle \hat{\Phi}^{\psi}_{(0,e_i)}\,,\,\nabla_{(0,e_i)}\Gamma_g\Big\rangle_\alpha = 2\sum_{x\in\mathbb{Z}^d} \Big\langle \psi\,,\,\tau_x g\Big\rangle_\alpha = 2 \ll \psi, g \gg_{\alpha,0}$$

for each cylinder function g in $\mathcal{C}_0$. The first identity relies on the integration by parts formula for the current: by definition of the germ $\mathfrak{A}^j$, summation over i is equal to $4 < \hat{\Phi}^{\psi}_{(0,e_j)} >$. By definition of $\hat{\Phi}^{\psi}_b$ and a change of variables, this expectation is equal to

$$\sum_{\substack{x;\,0\in x+\Lambda_\psi\\ e_j\in x+\Lambda_\psi}} \Big\langle \nabla_{0,e_j}\tau_x(-L_{\Lambda_\psi})^{-1}\psi\Big\rangle_\alpha$$

$$= -2\sum_{\substack{x;\,0\in x+\Lambda_\psi\\ e_j\in x+\Lambda_\psi}} \Big\langle (-L_{\Lambda_\psi})^{-1}\psi\,,\,W_{-x,-x+e_j}\Big\rangle_\alpha\ .$$

A simple computation shows that $L_{\Lambda_\psi}\sum_{x\in\Lambda_\psi} x_j\eta(x) = \sum_{z,z+e_j\in\Lambda_\psi} W_{z,z+e_j}$ from what the first identity follows. The second identity is elementary to check. One has just to replace $\hat{\Phi}^{\psi}_{(0,e_i)}$ by its value and use the integration by parts formula after performing change of variables of type $\xi = \tau_{-x}\eta$. □

To conclude the proof of Theorem 4.1 , it remains to obtain a lower bound for the variance. This is much easier since the variational formula for the variance is expressed as a supremum.

Lemma 4.5 *Under the assumptions of Theorem 4.1 ,*

$$\lim_{\ell\to\infty}(2\ell)^{-d}\Big\langle (-L_{\Lambda_\ell})^{-1}\sum_{|x|\le\ell_\psi}\tau_x\psi\,,\,\sum_{|x|\le\ell_\psi}\tau_x\psi\Big\rangle_{\ell,K_\ell} \ge \ll \psi \gg_\alpha\ .$$

Proof. We need to obtain a lower bound for the variance given by the variational formula (4.4). Consider a cylinder function g. For $\ell \ge s_g + 1$, take h in (4.4) as $h = \sum_{|x|\le\ell_g}\tau_x g + \sum_{1\le i\le d} a_i\sum_{x\in\Lambda_\ell} x_i\eta(x)$.

On the one hand, for $1 \le i \le d$, the equivalence of ensembles gives that

$$\lim_{\ell\to\infty}(2\ell)^{-d}\Big\langle \sum_{|x|\le\ell_\psi}\tau_x\psi\,,\,\sum_{|y|\le\ell_g}\tau_y g\Big\rangle_{\ell,K_\ell} = \ll \psi, g \gg_{\alpha,0}\ ,$$

$$\lim_{\ell\to\infty}(2\ell)^{-d}\Big\langle \sum_{|x|\le\ell_\psi}\tau_x\psi\,,\,\sum_{y\in\Lambda_\ell} y_i\eta(y)\Big\rangle_{\ell,K_\ell} = \ll \psi \gg_{\alpha,i}$$

because ψ belongs to $\mathcal{C}_0$ and, by assumption, $K_\ell/(2\ell)^d \to \alpha$.

On the other hand, to compute the limit as $\ell \uparrow \infty$ of the Dirichlet form $(2\ell)^{-d}\mathfrak{D}(\nu_{\ell,K_\ell}, \sum_{|x|\le \ell_g} \tau_x g + \sum_{1\le i\le d} a_i \sum_{x\in\Lambda_\ell} x_i\eta(x))$, recall the elementary identity $L_{\Lambda_\ell}\sum_{x\in\Lambda_\ell} x_i\eta(x) = \sum_{x,x+e_i\in\Lambda_\ell} W_{x,x+e_i}$. We may therefore decompose the Dirichlet form as a sum of terms of three kinds:

$$- (2\ell)^{-d}\Big\langle L_{\Lambda_\ell}\sum_{|x|\le \ell_g}\tau_x g\,,\, \sum_{|y|\le\ell_g}\tau_y g\Big\rangle_{\ell,K_\ell}\,,$$

$$- 2(2\ell)^{-d}\Big\langle \sum_{x,x+e_i\in\Lambda_\ell} W_{x,x+e_i}\,,\, \sum_{|y|\le\ell_g}\tau_y g\Big\rangle_{\ell,K_\ell}$$

$$\text{and}\quad - (2\ell)^{-d}\Big\langle \sum_{x,x+e_i\in\Lambda_\ell} W_{x,x+e_i}\,,\, \sum_{y\in\Lambda_\ell} y_j\eta(y)\Big\rangle_{\ell,K_\ell}\,.$$

By the equivalence of ensembles, as $\ell \uparrow \infty$, the first term converges to $-\ll Lg, g\gg_{\alpha,0}$ that may be rewritten as $(1/2)< \|\nabla\Gamma_g\|^2 >_\alpha$. By similar reasons the second expression converges to $-2\ll W_{0,e_i}, g\gg_{\alpha,0}$ and the third one to $-< W_{0,e_i}, \sum_y y_j\eta(y) >_\alpha = -\delta_{i,j} < W_{0,e_i}, \eta(e_i) >_\alpha = (1/2)\delta_{i,j} < r_{0,e_i} >_\alpha$.

In conclusion, the limit, as $\ell \uparrow \infty$, of (4.4) is bounded below by

$$2\ll \psi, g\gg_{\alpha,0} + 2\sum_{i=1}^d a_i \ll \psi \gg_{\alpha,i} + 2\sum_{i=1}^d a_i \ll W_{0,e_i}, g\gg_{\alpha,0}$$
$$- \frac{\|a\|^2}{2} < r_{0,e_i} >_\alpha - \frac{1}{2}\Big\langle \|\nabla\Gamma_g\|^2\Big\rangle_\alpha\,.$$

To conclude the proof of the lemma it remains to take a supremum over $a \in \mathbb{R}^d$ and g in $\mathcal{C}_0$. and recall formula (4.2) □

We conclude this section proving that for each ψ in $\mathcal{C}_0$ the function $\ll \psi \gg : [0,\kappa] \to \mathbb{R}_+$ that associates to each density α the value $\ll \psi \gg_\alpha$ is continuous and that the convergence of the finite volume variances to $\ll \cdot \gg_\alpha$ is uniform on $[0,\kappa]$. For each ℓ in $\mathbb{N}$ and $0 \le K \le (2\ell+1)^d\kappa$, denote by $V_\ell^\psi(K/(2\ell+1)^d)$ the variance of $(2\ell+1)^{-d}\sum_{|x|\le\ell_\psi}\tau_x\psi$ with respect to $\nu_{\ell,K}$:

$$V_\ell^\psi(K/(2\ell+1)^d) \;=\; (2\ell)^{-d}\Big\langle (-L_{\Lambda_\ell})^{-1}\sum_{|x|\le\ell_\psi}\tau_x\psi\,,\, \sum_{|x|\le\ell_\psi}\tau_x\psi\Big\rangle_{\ell,K}\,.$$

We may interpolate linearly to extend the definition of V_ℓ^ψ to the all interval $[0,\kappa]$. With this definition V_ℓ^ψ is continuous. Theorem 4.1 asserts that $V_\ell^\psi(K_\ell/(2\ell+1)^d)$ converges, as $\ell \uparrow \infty$, to $\ll \psi \gg_\alpha$, for any sequence K_ℓ such that $K_\ell/(2\ell+1)^d \to \alpha$. In particular, $\lim_{\ell\to\infty} V_\ell^\psi(\alpha_\ell) = \ll \psi \gg_\alpha$ for any sequence $\alpha_\ell \to \alpha$. This implies that $\ll \psi \gg_\alpha$ is continuous and that $V_\ell^\psi(\cdot)$ converges uniformly to $\ll \psi \gg$. as $\ell \uparrow \infty$. We have thus proved the following theorem.

Theorem 4.6 *For each fixed h in $\mathcal{C}_0$, $\ll h \gg_\alpha$ is continuous as a function of the density α on $[0, \kappa]$. Moreover, the variance*

$$(2\ell)^{-d} < (-L_{\Lambda_\ell})^{-1} \sum_{|x|\le \ell_h} \tau_x h\, , \sum_{|x|\le \ell_h} \tau_x h >_{\ell,K}$$

converges uniformly to $\ll h \gg_\alpha$ as $\ell \uparrow \infty$ and $K_\ell/(2\ell)^d \to \alpha$. In particular,

$$\lim_{\ell\to\infty} \sup_{0\le K\le (2\ell+1)^d\kappa} (2\ell)^{-d}\Big\langle (-L_{\Lambda_\ell})^{-1} \sum_{|x|\le \ell_h} \tau_x h\, , \sum_{|x|\le \ell_h} \tau_x h \Big\rangle_{\ell,K}$$
$$= \sup_{0\le\alpha\le\kappa} \ll h \gg_\alpha \, .$$

5. The Diffusion Coefficient

We investigate here the main properties of the semi norm $\ll \cdot \gg_\alpha$ introduced in the previous section and of the diffusion coefficient defined in the beginning of the chapter. We first show that we may define from $\ll \cdot \gg_\alpha$ a semi–inner product on $\mathcal{C}_0$ through polarization.

Lemma 5.1 *For every g, h in $\mathcal{C}_0$ and $\lambda \in \mathbb{R}$*

(a) $\ll g \gg_\alpha \ge 0$,
(b) $\ll \lambda g \gg_\alpha = \lambda^2 \ll g \gg_\alpha$ *and*
(c) (parallelogram identity) $\ll g+h \gg_\alpha + \ll g-h \gg_\alpha = 2\{\ll g \gg_\alpha + \ll h \gg_\alpha \}$.

The proof of this lemma is elementary. On $\mathcal{C}_0 \times \mathcal{C}_0$ let $\ll \cdot, \cdot \gg_\alpha$ be defined by

$$\ll g, h \gg_\alpha = \frac{1}{4}\Big\{ \ll g+h \gg_\alpha \; - \; \ll g-h \gg_\alpha \Big\} . \tag{5.1}$$

It is easy to check that (5.1) defines a semi–inner product on $\mathcal{C}_0$: for all g_1, g_2, h in $\mathcal{C}_0$ and $\lambda \in \mathbb{R}$ we have that

(a) (symmetry) $\ll g_1, g_2 \gg_\alpha = \ll g_2, g_1 \gg_\alpha$,
(b) (linearity) $\ll \lambda g_1 + g_2, h \gg_\alpha = \lambda \ll g_1, h \gg_\alpha + \ll g_2, h \gg_\alpha$ and
(c) (positiveness) $\ll h \gg_\alpha \ge 0$.

Linearity is a simple consequence of the parallelogram identity in Lemma 5.1 and its proof can be found in any standard text on functional analysis.

Denote by $\mathcal{N}_\alpha$ the kernel of the semi–norm $\ll \cdot \gg_\alpha^{1/2}$ on $\mathcal{C}_0$. Since $\ll \cdot, \cdot \gg_\alpha$ is a semi–inner product on $\mathcal{C}_0$, the completion of $\mathcal{C}_0|_{\mathcal{N}_\alpha}$, denoted by $\mathcal{H}_\alpha$, is a Hilbert space.

Recall that the linear space generated by the currents $\{W_{0,e_i},\ 1\le i\le d\}$ and $L\mathcal{C}_0=\{Lg,\ g\in\mathcal{C}_0\}$ are subsets of $\mathcal{C}_0$. The first main result of this section consists in showing that $\mathcal{H}_\alpha$ is the completion of $L\mathcal{C}_0|\mathcal{N}_\alpha+\{W_{0,e_i},\ 1\le i\le d\}$, in other words, that all elements of $\mathcal{H}_\alpha$ can be approximated by $\sum_{1\le i\le d} a_iW_{0,e_i}+Lg$ for some a in $\mathbb{R}^d$ and g in $\mathcal{C}_0$. To prove this result we derive two elementary identities:

$$\ll h, Lg\gg_\alpha = -\ll h,g\gg_{\alpha,0} \qquad\text{and}\qquad \ll h, W_{0,e_i}\gg_\alpha = -\ll h\gg_{\alpha,i} \tag{5.2}$$

for all h, g in $\mathcal{C}_0$ and $1\le i\le d$.

These identities are easily explained. By Theorem 4.1 and (5.1), the semi–inner product $\ll h,g\gg_\alpha$ is the limit of the covariance $(2\ell)^{-d} < (-L_{\Lambda_\ell})^{-1}\sum_{|x|\le \ell_g}\tau_x g$, $\sum_{|x|\le\ell_h}\tau_x h >_{\ell,K_\ell}$, as $\ell\uparrow\infty$ and $K_\ell/(2\ell)^d\to\alpha$. In particular, if $g=Lg_0$, for some cylinder function g_0, since $\tau_x Lg_0=L_{\Lambda_\ell}\tau_x g_0$ for $|x|\le \ell_{g_0}$, we have that $\ll h, Lg\gg_\alpha=\lim_{\ell\to\infty}(2\ell)^{-d} < (-L_{\Lambda_\ell})^{-1}\sum_{|x|\le\ell_{g_0}} L_{\Lambda_\ell}\tau_x g_0$, $\sum_{|x|\le \ell_h}\tau_x h>_{\ell,K_\ell}$ for some sequence K_ℓ such that $K_\ell/(2\ell)^d\to\alpha$. The inverse of the generator cancels with the generator. Therefore, $\ll h,Lg\gg_\alpha$ is equal to

$$-\lim_{\ell\to\infty}(2\ell)^{-d} < \sum_{|x|\le\ell_g}\tau_x g,\ \sum_{|x|\le\ell_h}\tau_x h>_{\ell,K_\ell} = -\ll g,h\gg_{\alpha,0}\ .$$

The second identity is proved in a similar way, we just need to recall the elementary relation $L_{\Lambda_\ell}\sum_{x\in\Lambda_\ell}x_j\eta(x)=\sum_{x;\,x,x+e_j\in\Lambda_\ell}W_{x,x+e_j}$. It is also possible to prove both identities directly from the definition of the semi–norm $\ll\cdot\gg_\alpha^{1/2}$ through the variational formula. We leave the second proof to the reader as an exercise.

It follows from the first identity that the gradients $\{\eta(e_i)-\eta(0),\ 1\le i\le d\}$ are orthogonal to the space $L\mathcal{C}_0$, while the second identity permits to compute inner product of cylinder functions with the current:

$$\begin{aligned}\ll \eta(e_i)-\eta(0), Lh\gg_\alpha &= 0 \quad \text{for all } 1\le i\le d \text{ and all } h \text{ in } \mathcal{C}_0.\\ \ll \eta(e_i)-\eta(0), W_{0,e_j}\gg_\alpha &= -\chi(\alpha)\delta_{i,j}\\ \text{and}\quad \ll W_{0,e_i}, W_{0,e_j}\gg_\alpha &= (1/2) < r_{0,e_1}>_\alpha \delta_{i,j}\end{aligned} \tag{5.3}$$

for $1\le i,j\le d$. In this formula $\chi(\alpha)$ stands for the static compressibility and is equal to $<\eta(0)^2>_\alpha - <\eta(0)>_\alpha^2$. Furthermore,

$$\ll \sum_{1\le j\le d} a_jW_{0,e_j}+Lg\gg_\alpha = (1/2)\sum_{i=1}^d\Big\langle\Big\{\sum_{1\le j\le d}a_j(\mathfrak{A}^j)_i+\nabla_{(0,e_i)}\Gamma_g\Big\}^2\Big\rangle_\alpha \tag{5.4}$$

for a in $\mathbb{R}^d$ and g in $\mathcal{C}_0$. In particular, by (5.2), the variational formula for $\ll h\gg_\alpha$ writes

$$\ll h\gg_\alpha = \sup_{\substack{a\in\mathbb{R}^d\\ g\in\mathcal{C}_0}}\Big\{-2\ll h,\sum_{1\le j\le d}a_jW_{0,e_j}+Lg\gg_\alpha - \ll\sum_{1\le j\le d}a_jW_{0,e_j}+Lg\gg_\alpha\Big\}. \tag{5.5}$$

We may now prove that in $\mathcal{H}_\alpha$ a function can be approximated by

$$\sum_{1\le j\le d} a_j W_{0,e_j} \; + \; Lg$$

for some a in $\mathbb{R}^d$ and g in $\mathcal{C}_0$.

Proposition 5.2 *Recall that we denote by $L\mathcal{C}_0$ the space $\{Lg;\; g \in \mathcal{C}_0\}$. For each $0 \le \alpha \le \kappa$,*

$$\mathcal{H}_\alpha \;=\; \overline{L\mathcal{C}_0}\big|_{\mathcal{N}_\alpha} \oplus \{W_{0,e_i},\; 1 \le i \le d\}\;.$$

Proof. Let us first show that the sum is direct. Suppose $\sum_{1\le j\le d} a_j W_{0,e_j}$ belongs to $\overline{L\mathcal{C}_0}$ for some vector a. There exists, therefore, a sequence of functions g_k in $\mathcal{C}_0$ such that Lg_k converges to $\sum_{1\le j\le d} a_j W_{0,e_j}$. Take the inner product with respect to $\eta(e_i) - \eta(0)$. By (5.3),

$$\begin{aligned} -\,\chi(\alpha) a_i \;&= \ll \sum_{1\le j\le d} a_j W_{0,e_j}\,,\, \eta(e_i) - \eta(0) \gg_\alpha \\ &= \lim_{k\to\infty} \ll Lg_k\,,\, \eta(e_i) - \eta(0) \gg_\alpha \,=\, 0\;. \end{aligned}$$

Thus $a_j = 0$ for $1 \le j \le d$ proving that the sum is direct.

We now turn to the proof that $\mathcal{H}_\alpha$ is generated by $L\mathcal{C}_0$ and the currents. Since $\{W_{0,e_i},\; 1 \le i \le d\}$ and $L\mathcal{C}_0$ are contained in $\mathcal{C}_0$, by definition, $\mathcal{H}_\alpha$ contains the right hand space. To prove the converse inclusion, let $h \in \mathcal{C}_0$ so that $\ll h, W_{0,e_i} \gg_\alpha = 0$ for $1 \le i \le d$ and $\ll h, Lg \gg_\alpha = 0$ for g in $\mathcal{C}_0$. From (5.5) it follows that $\ll h \gg_\alpha = 0$. Thus, $\mathcal{C}_0|_{\mathcal{N}_\alpha} \subset (L\mathcal{C}_0 + \{W_{0,e_i},\; 1 \le i \le d\})|_{\mathcal{N}_\alpha}$. □

Corollary 5.3 *For each g in $\mathcal{C}_0$, there exists a unique vector a in $\mathbb{R}^d$ such that*

$$g \;-\; \sum_{j=1}^{d} a_j W_{0,e_j} \;\in\; \overline{L\mathcal{C}_0} \quad \textit{in} \quad \mathcal{H}_\alpha\;.$$

From (5.3) we know that the space generated by $\{\eta(e_i) - \eta(0),\; 1 \le i \le d\}$ is orthogonal to $L\mathcal{C}_0$. Thus $\mathcal{H}_\alpha$ is a Hilbert space with a quite simple structure. It is generated by $L\mathcal{C}_0$ and the current $\{W_{0,e_i},\; 1 \le i \le d\}$ and has inner product defined by (5.1). Moreover, $\{\eta(e_i) - \eta(0),\; 1 \le i \le d\}$ and $L\mathcal{C}_0$ are orthogonal in $\mathcal{H}_\alpha$.

We shall now start to describe the diffusion coefficient D of the hydrodynamic equation. From Corollary 5.3, there exists a matrix $\{Q_{i,j},\; 1 \le i, j \le d\}$ such that

$$\eta(e_i) - \eta(0) \;+\; \sum_{j=1}^{d} Q_{i,j} W_{0,e_j} \;\in\; \overline{L\mathcal{C}_0} \quad \text{in} \quad \mathcal{H}_\alpha\;. \tag{5.6}$$

We replaced the minus sign by a plus for the matrix Q to be positive as we shall see in the next lemma. Notice that the matrix $Q = Q(\alpha)$ depends on the density α because the inner product depends on α. We claim that Q is symmetric, strictly positive and has all eigenvalues bounded below by a finite constant.

Lemma 5.4 *Consider the matrix Q defined by (5.6). Recall that $\chi(\alpha)$ denotes the static compressibility and is equal to $<\eta(0)^2>_\alpha - <\eta(0)>_\alpha^2$. Q is a symmetric, strictly positive matrix with eigenvalues bounded below by $\{2\chi(\alpha)/<r_{0,e_i}>_\alpha\}$.*

Proof. The proof of this lemma is quite simple. Since the vectors $\{\eta(e_i) - \eta(0),\ 1 \le i \le d\}$ are orthogonal to LC_0,

$$\begin{aligned}\ll \eta(e_i) - \eta(0), \eta(e_k) - \eta(0) \gg_\alpha &= -\sum_{j=1}^{d} Q_{i,j} \ll W_{0,e_j}, \eta(e_k) - \eta(0) \gg_\alpha \\ &= \chi(\alpha) Q_{i,k}\end{aligned} \tag{5.7}$$

because, by (5.3), $\ll W_{0,e_j}, \eta(e_k) - \eta(0) \gg_\alpha = -\chi(\alpha)\delta_{j,k}$. Q is therefore symmetric and strictly positive. It remains to show that all eigenvalues of Q are bounded below by $\{2\chi(\alpha)/<r_{0,e_i}>_\alpha\}$.

To keep notation simple, assume that there exist functions H_i in $\mathcal{C}_0$, $1 \le i \le d$, so that $\eta(e_i) - \eta(0) + \sum_{j=1}^d Q_{i,j} W_{0,e_j} = LH_i$. Otherwise, we approximate $\eta(e_i) - \eta(0) + \sum_{j=1}^d Q_{i,j} W_{0,e_j}$ by a strongly convergent sequence $\{LH_i^n;\ n \ge 1\}$ with H_i^n in $\mathcal{C}_0$ for each $1 \le i \le d$ and $n \ge 1$.

Taking inner product on both sides with respect to W_{0,e_k}, by (5.3), we obtain that $-\chi(\alpha)\delta_{i,k} + (1/2) <r_{0,e_1}>_\alpha Q_{i,k} = \ll LH_i, W_{0,e_k} \gg_\alpha$. Denote by M the matrix with entries $M_{i,k} = \ll LH_i, W_{0,e_k} \gg_\alpha$. Thus,

$$-\chi(\alpha) I + (1/2) <r_{0,e_1}>_\alpha Q = M\ ,$$

if I stands for the identity. Taking now inner product with respect to LH_k, since $\eta(e_i) - \eta(0)$ is orthogonal to LC_0, we get that $\sum_{1 \le j \le d} Q_{i,j} \ll W_{0,e_j}, LH_k \gg_\alpha = \ll LH_i, LH_k \gg_\alpha$. Notice that the matrix with entries $\ll LH_i, LH_k \gg_\alpha$ is positive definite. Therefore, $QM^* \ge 0$ in the matrix sense, if M^* denotes the adjoint of M. Since, by the first part of the proof, $M = -\chi(\alpha) I + (1/2) <r_{0,e_1}>_\alpha Q$,

$$Q\Big\{\chi(\alpha) I - (1/2) <r_{0,e_1}>_\alpha Q\Big\}^* \le 0$$

in the sense of matrices. Let λ be an eigenvalue of Q and denote by v an associated eigenvector. λ is positive because Q is a positive definite matrix. The previous inequality asserts that

$$\begin{aligned} 0 &\ge \Big\langle [\chi(\alpha) I - (1/2) <r_{0,e_1}>_\alpha Q] v, Qv \Big\rangle \\ &= \lambda \Big\{\chi(\alpha) - (1/2) <r_{0,e_1}>_\alpha \lambda\Big\} \|v\|^2\ .\end{aligned}$$

Thus, $\lambda(\chi(\alpha) - (1/2) < r_{0,e_1} >_\alpha \lambda) \le 0$. Since λ is positive, λ is bounded below by $\{2\chi(\alpha)/ < r_{0,e_i} >_\alpha\}$. □

Denote by $D = D(\alpha)$ the inverse of Q. We shall see below that $D(\alpha)$ is the diffusion coefficient of the hydrodynamic equation (0.8). From the previous lemma, D is symmetric, positive definite, with eigenvalues bounded above by $\{< r_{0,e_i} >_\alpha /2\chi(\alpha)\}$:

$$D \le \frac{< r_{0,e_i} >_\alpha}{2\chi(\alpha)} I \tag{5.8}$$

in the sense of matrices. Our purpose now is to obtain an explicit formula for D and then prove that D is continuous on $[0,\kappa]$ and nonlinear. Since D is the inverse of Q, we have that

$$W_{0,e_i} + \sum_{j=1}^{d} D_{i,j}[\eta(e_j) - \eta(0)] \in \overline{LC_0} \quad \text{in} \quad \mathcal{H}_\alpha$$

for $1 \le i \le d$. This relation provides a variational characterization of the diffusion coefficient D. Indeed, for all vectors a in $\mathbb{R}^d$,

$$\inf_{g\in C_0} \Big\{ \ll \sum_{i=1}^{d} a_i W_{0,e_i} + \sum_{i,j=1}^{d} a_i D_{i,j}[\eta(e_j) - \eta(0)] - Lg \gg_\alpha \Big\} = 0 \, .$$

Since gradients are orthogonal to the space LC_0, since

$$\ll \eta(e_j) - \eta(0), W_{0,e_i} \gg_\alpha = -\chi(\alpha)\delta_{i,j}$$

and since, by (5.7),

$$\ll \eta(e_j) - \eta(0), \eta(e_k) - \eta(0) \gg_\alpha = \chi(\alpha) Q_{j,k} = \chi(\alpha)[D^{-1}]_{j,k} \, ,$$

the last identity reduces to

$$\inf_{g\in C_0} \Big\{ -\chi(\alpha) a^* D a + \ll \sum_{i=1}^{d} a_i W_{0,e_i} - Lg \gg_\alpha \Big\} = 0 \, ,$$

where a^* stands for the transposition of a. We have thus obtained a variational formula for $D(\alpha)$.

Theorem 5.5 *The diffusion coefficient $D(\alpha)$ is such that*

$$\begin{aligned} a^* D(\alpha) a &= \frac{1}{\chi(\alpha)} \inf_{g\in C_0} \ll \sum_{j=1}^{d} a_j W_{0,e_j} - Lg \gg_\alpha \\ &= \frac{1}{2\chi(\alpha)} \inf_{g\in C_0} \sum_{i=1}^{d} \Big\langle \Big(\sum_{j=1}^{d} a_j (\mathfrak{A}^j)_i - \nabla_{(0,e_i)} \Gamma_g \Big)^2 \Big\rangle_\alpha \end{aligned}$$

for all a in $\mathbb{R}^d$.

The second identity follows from equation (5.4). Moreover, this formula determines the matrix D since D is symmetric by Lemma 5.4.

We now prove that the diffusion coefficient D is continuous. The proof is divided in three steps. We first show that D is continuous on the open interval $(0, \kappa)$. Then, taking advantage of the integration by parts formula for $\eta(e_i) - \eta(0)$, we prove a lower bound for D. This lower bound in addition to the upper bound (5.8) shall prove that D is continuous at the boundary of $[0, \kappa]$.

The following functional space plays a key role in the proof of the continuity of the diffusion coefficient. Denote by $\mathfrak{F}$ the space of functions $\mathfrak{f} : [0, \kappa] \times \Sigma_{\mathbb{Z}^d}^{\kappa} \to \mathbb{R}$ such that

(i) For each $\alpha \in [0, \kappa]$, $\mathfrak{f}(\alpha, \cdot)$ is a mean-zero cylinder function with uniform support: there exist a finite set $\Lambda \subset \mathbb{Z}^d$ that contains the support of each $\mathfrak{f}(\alpha, \cdot)$ and the expected value of $\mathfrak{f}(\alpha, \cdot)$ with respect to all canonical measures $\nu_{\Lambda,K}$ vanishes:

$$E_{\nu_{\Lambda,K}}[\mathfrak{f}(\alpha, \cdot)] = 0 \quad \text{for all } 0 \le K \le \kappa|\Lambda| .$$

(ii) For each configuration η, $\mathfrak{f}(\cdot, \eta)$ is a smooth function of class $C^2([0, \kappa])$.

Theorem 5.6 *The diffusion coefficient* $D_{i,j}(\,\cdot\,)$ *is continuous on* $(0, \kappa)$.

Proof. Fix $\varepsilon > 0$ and α in $[0, \kappa]$. Since $W_{0,e_i} + \sum_{j=1}^d D_{i,j}(\alpha)[\eta(e_j) - \eta(0)]$ belongs to $\overline{L\mathcal{C}_0}$, there exists a cylinder function $H_i(\alpha, \eta)$ in $\mathcal{C}_0$ such that

$$\ll W_{0,e_i} - \sum_{j=1}^{d} D_{i,j}(\alpha)[\eta(e_j) - \eta(0)] - LH_i(\alpha, \eta) \gg_\alpha \le \varepsilon .$$

Since by Theorem 4.6 $\ll h \gg_\alpha$ is continuous in α for all h in $\mathcal{C}_0$, for each α_0 in $[0, \kappa]$, there exists a neighborhood O_{α_0} of α_0 such that $\ll W_{0,e_i} - \sum_{j=1}^d D_{i,j}(\alpha_0)[\eta(e_j) - \eta(0)] - LH_i(\alpha_0, \eta) \gg_\alpha \le 2\varepsilon$ for α in O_{α_0}. The family $\{O_\alpha,\ \alpha \in [0, \kappa]\}$ forms an open covering of the compact set $[0, \kappa]$. There exists therefore a finite subcovering $\{O_{\alpha_k},\ 1 \le k \le n\}$.

From $\{D_{i,j}(\alpha_k),\ 1 \le k \le n\}$ and $\{H(\alpha_k, \eta),\ 1 \le k \le n\}$ it is possible to define by interpolation continuous functions $D_{i,j}^\varepsilon : [0, \kappa] \to \mathbb{R}$, $1 \le j \le d$, and a function $H(\alpha, \eta)$ in $\mathfrak{F}$ so that

$$\sup_{0 \le \alpha \le \kappa} \ll W_{0,e_i} + \sum_{j=1}^{d} D_{i,j}^\varepsilon(\alpha)[\eta(e_j) - \eta(0)] - LH_i^\varepsilon(\alpha, \eta) \gg_\alpha \le 4\varepsilon .$$

We now prove that the continuous functions $D_{i,j}^\varepsilon$ uniformly approximate $D_{i,j}$ on compact sets of $(0, \kappa)$. On the one hand, by Schwarz inequality, $\ll \sum_j [D_{i,j}^\varepsilon - D_{i,j}][\eta(e_j) - \eta(0)] - L\{H_i^\varepsilon(\alpha, \eta) - H_i(\alpha, \eta)\} \gg_\alpha$ is bounded above by $2 \ll W_{0,e_i} -$

$\sum_j D^\varepsilon_{i,j}(\alpha)[\eta(e_j) - \eta(0)] - LH^\varepsilon_i(\alpha,\eta) \gg_\alpha +2 \ll W_{0,e_i} - \sum_j D_{i,j}(\alpha)[\eta(e_j) - \eta(0)] - LH_i(\alpha,\eta) \gg_\alpha$. By the previous estimate, this last sum is bounded above by 10ε. On the other hand, since the vectors $\{\eta(e_j)-\eta(0),\ 1 \le j \le d\}$ are orthogonal to the space $L\mathcal{C}_0$, $\ll \sum_j [D^\varepsilon_{i,j} - D_{i,j}][\eta(e_j) - \eta(0)] - L[H^\varepsilon_i(\alpha,\eta) - H_i(\alpha,\eta)] \gg_\alpha$ is bounded below by $\ll \sum_j [D^\varepsilon_{i,j} - D_{i,j}][\eta(e_j) - \eta(0)] \gg_\alpha$. In conclusion, we have that

$$\ll \sum_{j=1}^{d} [D^\varepsilon_{i,j} - D_{i,j}][\eta(e_j) - \eta(0)] \gg_\alpha \le 10\varepsilon .$$

Recall the definition of the matrix Q defined in (5.6) and keep in mind that $Q(\alpha)$ is the inverse of the diffusion coefficient D. By (5.7), last sum thus writes $\chi(\alpha) \sum_{j,k} B^\varepsilon_{i,j} Q_{j,k} B^\varepsilon_{i,k}$. Here, to keep notation simple, we denoted the difference $D^\varepsilon_{i,j} - D_{i,j}$ by $B^\varepsilon_{i,j}$. Since Q is bounded below by $2\chi(\alpha)/ < r_{0,e_1} >_\alpha$, We obtain that

$$\frac{2\chi(\alpha)^2}{< r_{0,e_1} >_\alpha} \sum_{1 \le j \le d} \Big[D^\varepsilon_{i,j} - D_{i,j} \Big]^2 \le 10\varepsilon .$$

This proves that D is uniformly approximated by continuous functions on any compact set of $(0,\kappa)$. In particular, D is continuous on $(0,\kappa)$. □

We are now ready to prove a lower bound for the diffusion coefficient.

Lemma 5.7 *For every a in $\mathbb{R}^d$, we have that*

$$a^* D a \ \ge\ \frac{\chi(\alpha)}{8 < \Psi^2_{0,e_1}\ r_{0,e_1} >_\alpha} \|a\|^2 .$$

Here Ψ_{0,e_i} is the cylinder function defined in (2.2) and related to the gradients by the integration by parts formula.

Proof. By definition of the static compressibility $\chi(\alpha) = (1/2) < \eta(e_i) - \eta(0), \eta(e_i) - \eta(0) >_\alpha$. Since $\nabla_{0,e_i}[\eta(e_i) - \eta(0)] = 2r_{0,e_i}$, by the integration by parts formula (2.2) and Remark 2.2, for all a in $\mathbb{R}^d$

$$\|a\|^2 \chi(\alpha) = 2 \sum_{i=1}^{d} a_i^2 \Big\langle \Psi_{0,e_i}\,,\, r_{0,e_i} \Big\rangle_\alpha .$$

Up to the end of this proof we denote by $\cdot$ the inner product on $\mathbb{R}^d$. Recall the definition of the germs $\{\mathfrak{A}^i, 1 \le i \le d\}$, as vectors of $\mathbb{R}^d$. We may rewrite this last identity as

$$(1/2)\|a\|^2 \chi(\alpha) = \Big\langle \Big[\sum_{i=1}^{d} a_i \Psi_{0,e_i} \mathfrak{A}^i \Big] \cdot \Big[\sum_{i=1}^{d} a_i \mathfrak{A}^i \Big] \Big\rangle_\alpha .$$

In contrast, since gradients are orthogonal to the space $L\mathcal{C}_0$, by the integration by parts formula, for every g in $\mathcal{C}_0$,

$$
\begin{aligned}
0 &= -(1/2) \ll \sum_{i=1}^{d} a_i[\eta(e_i) - \eta(0)]\,,\, Lg \gg_\alpha \\
&= (1/2)\Big\langle \sum_{i=1}^{d} a_i[\eta(e_i) - \eta(0)]\,,\, \Gamma_g \Big\rangle_\alpha \\
&= \Big\langle \Big[\sum_{i=1}^{d} a_i \Psi_{0,e_i} \mathfrak{A}^i\Big] \cdot \nabla \Gamma_g \Big\rangle_\alpha \,.
\end{aligned}
$$

Adding the two previous identities and applying Schwarz inequality, we obtain that

$$
\begin{aligned}
(1/4)\|a\|^4 \chi(\alpha)^2 &= \Big\{ \Big\langle \Big[\sum_{i=1}^{d} a_i \Psi_{0,e_i} \mathfrak{A}^i\Big] \cdot \Big[\sum_{i=1}^{d} a_i \mathfrak{A}^i + \nabla \Gamma_g\Big] \Big\rangle_\alpha \Big\}^2 \\
&\le \Big\langle \Big\| \sum_{i=1}^{d} a_i \Psi_{0,e_i} \mathfrak{A}^i \Big\|^2 \Big\rangle_\alpha \Big\langle \Big\| \sum_{j=1}^{d} a_j \mathfrak{A}^j + \nabla \Gamma_g \Big\|^2 \Big\rangle_\alpha
\end{aligned}
$$

for every g in $\mathcal{C}_0$. Minimizing over all g in $\mathcal{C}_0$, by the variational characterization of the diffusion coefficient presented in Theorem 5.5, we obtain that

$$
a^* D a \ \ge\ \frac{\chi(\alpha)\|a\|^4}{8\Big\langle \Big\| \sum_{i=1}^{d} a_i \Psi_{0,e_i} \mathfrak{A}^i \Big\|^2 \Big\rangle_\alpha} \,. \tag{5.9}
$$

Since the vectors $\mathfrak{A}^j$ are orthogonal, the denominator is equal to $8\|a\|^2 < \Psi^2_{0,e_1} r_{0,e_1} >_\alpha$. □

It is now easy to prove that the diffusion coefficient is continuous at the boundary of $[0,\kappa]$. By duality among particles and holes we need only to check the continuity at one of the boundary points, say the origin. From the explicit formulas for $\chi(\alpha)$, $< r_{0,e_1} >_\alpha$ and $< \Psi_{0,e_1} r_{0,e_1} >_\alpha$, we have that $\chi(R(\varphi)) = \varphi + O(\varphi^2)$, $< r_{0,e_1} >_{R(\varphi)} = \varphi + O(\varphi^2)$ and $< \Psi^2_{0,e_1} r_{0,e_1} >_{R(\varphi)} = (1/4)\varphi + O(\varphi^2)$. Thus, by the previous lemma, $a^* D(R(\varphi))a \ge (1/2)\|a\|^2 + O(\varphi)$. In contrast, by the lower bound (5.8) for the diffusion coefficient, $a^* D(R(\varphi))a \le (1/2)\|a\|^2 + O(\varphi)$. Therefore,

Theorem 5.8 *The diffusion coefficient $D(\alpha)$ is continuous on $[0,\kappa]$. Moreover it converges to $(1/2)I$ as $\alpha \downarrow 0$ or $\alpha \uparrow \kappa$.*

From the continuity of the diffusion coefficient and the proof of Theorem 5.6 we have

Corollary 5.9 *Let D be the matrix defined in Theorem 5.5. Then, for each $1 \le i \le d$,*

$$\inf_{\mathfrak{f}\in\mathcal{C}_0} \sup_{0\le\alpha\le\kappa} \ll W_{0,e_i} + \sum_{j=1}^{d} D_{i,j}(\alpha)[\eta(e_j) - \eta(0)] - L\mathfrak{f}(\eta) \gg_\alpha = 0 .$$

This result together with (3.5), the definition of $\tilde{V}_i^{\mathfrak{f},\ell}$ and Theorem 4.6 concludes the proof of Theorem 1.1.

Proof of Corollary 5.9. Fix $1 \le i \le d$ and $\varepsilon > 0$. From the proof of Theorem 5.6, there exists $H(\alpha, \eta)$ in $\mathfrak{F}$ such that

$$\sup_{0\le\alpha\le\kappa} \ll W_{0,e_i} + \sum_{j=1}^{d} D_{i,j}^{\varepsilon}(\alpha)[\eta(e_j) - \eta(0)] - LH(\alpha,\eta) \gg_\alpha \le \varepsilon .$$

Fix a positive integer ℓ and set $\mathfrak{f}(\eta) = H(\eta^\ell(0), \eta)$. Notice that, for sufficiently large ℓ, $\mathfrak{f}$ belongs to $\mathcal{C}_0$ because H is in $\mathfrak{F}$. By the triangle inequality,

$$\begin{aligned} &\sup_{0\le\alpha\le\kappa} \ll W_{0,e_i} + \sum_{j=1}^{d} D_{i,j}^{\varepsilon}(\alpha)[\eta(e_j) - \eta(0)] - L\mathfrak{f}(\eta) \gg_\alpha \\ &\qquad \le 2 \sup_{0\le\alpha\le\kappa} \ll L\mathfrak{f} - LH(\alpha,\eta) \gg_\alpha + 2\varepsilon . \end{aligned} \tag{5.10}$$

By identity (5.4) with $a = 0$,

$$\ll L\{\mathfrak{f} - H(\alpha,\eta)\} \gg_\alpha = \frac{1}{2}\sum_{i=1}^{d} \Big\langle \Big\{ \nabla_{(0,e_i)} \sum_{x\in\mathbb{Z}^d} \tau_x \Big[H(\eta^\ell(0),\eta) - H(\alpha,\eta) \Big] \Big\}^2 \Big\rangle_\alpha .$$

Since $\nabla_{(0,e_i)}\tau_x = \tau_x \nabla_{(-x,-x+e_i)}$ and since ν_α is translation invariant, the previous expression is equal to

$$\frac{1}{2}\sum_{i=1}^{d} \Big\langle \Big\{ \sum_{x\in\mathbb{Z}^d} \nabla_{(x,x+e_i)} \Big[H(\eta^\ell(0),\eta) - H(\alpha,\eta) \Big] \Big\}^2 \Big\rangle_\alpha .$$

Since H belongs to $\mathfrak{F}$, there exists a cube Λ such that

$$\begin{aligned} &\sum_{x\in\mathbb{Z}^d} \nabla_{(x,x+e_i)} \Big\{ H(\eta^\ell(0),\eta) - H(\alpha,\eta) \Big\} \\ &\quad = \sum_{x\in\Lambda} \nabla_{(x,x+e_i)} \Big\{ H(\eta^\ell(0),\eta) - H(\alpha,\eta) \Big\} + O(\ell^{-1}) . \end{aligned} \tag{5.11}$$

The second term on the right hand side comes from a jump of a particle from Λ_ℓ to Λ_ℓ^c or from a jump in the opposite direction. For ℓ large enough the contribution of this jump is $H(\eta^\ell(0) \pm (2\ell+1)^{-d}, \eta) - H(\eta^\ell(0), \eta)$. Since H belongs to $\mathfrak{F}$, this difference is bounded by $C(H)\ell^{-d}$. Summing over all sites at the boundary of Λ_ℓ, we obtain (5.11).

From identity (5.11) and since for every bond b and every $L^2(\nu_\alpha)$ function h, $< (\nabla_b h)^2 >_\alpha \le 4 < h^2 >_\alpha$, we obtain that the first term in (5.10) is bounded above by

$$\sup_\alpha \Big\{ C(H) \Big\langle \Big\{ H(\eta^\ell(0), \eta) - H(\alpha, \eta) \Big\}^2 \Big\rangle_\alpha + O(\ell^{-2}) \Big\}$$

that vanishes as $\ell \uparrow \infty$ by the law of large numbers. This concludes the proof of the corollary. □

We conclude this section proving that the diffusion coefficient is nonlinear.

Proposition 5.10 *$D(\cdot)$ is a nonlinear function and*

$$\frac{\chi(\alpha)\|a\|^4}{8\Big\langle \Big\| \sum_{i=1}^d a_i \Psi_{0,e_i} \mathfrak{A}^i \Big\|^2 \Big\rangle_\alpha} \le a^* D(\alpha) a \le \frac{< r_{0,e_i} >_\alpha}{2\chi(\alpha)} \|a\|^2$$

for every a in $\mathbb{R}^d$.

Proof. The inequalities were proved in (5.8) and (5.9) so that $D(0) = D(\kappa) = (1/2)I$. By the upper bound for D, that may also be obtained setting $g = 1$ in Theorem 5.5, we have

$$a^* D(R(\varphi)) a \le \frac{1}{2} \{1 - 2\varphi + O(\varphi^2)\} \|a\|^2$$

for φ small. This proves that D is nonlinear. □

6. Compactness

We prove in this section that the sequence Q_{μ^N} is compact.

Theorem 6.1 *The sequence of probability measures Q_{μ^N} is relatively compact. Moreover, every limit point Q^* is concentrated on absolutely continuous paths $\pi(t, du)$ with density bounded above by κ: $\pi(t, du) = \pi(t, u) du$, $\pi(t, u) \le \kappa$.*

The proof of this theorem relies on the following exponential estimate.

Lemma 6.2 *For any smooth function $H \colon \mathbb{T}^d \to [0, 1]$, $1 \le i \le d$ and $s < t$:*

$$\mathbb{E}_{\nu_\alpha^N} \Big[\exp \Big\{ N^d \Big| \int_s^t N^{1-d} \sum_{x \in \mathbb{T}_N^d} H(x/N) W_{x, x+e_i}(r) dr \Big| \Big\} \Big] \le 2 \exp \Big\{ (1/2)(t - s) \sum_{x \in \mathbb{T}_N^d} H(x/N)^2 \Big\} .$$

Proof. Since $e^{|x|} \le e^x + e^{-x}$, it is enough to prove that

$$\mathbb{E}_{\nu_\alpha^N}\Big[\exp\Big\{N^d\int_s^t N^{1-d}\sum_{x\in\mathbb{T}_N^d} H(x/N)W_{x,x+e_i}(r)dr\Big\}\Big] \le \exp\Big\{(1/2)(t-s)\sum_{x\in\mathbb{T}_N^d} H(x/N)^2\Big\}$$

for every smooth function $H\colon \mathbb{T}^d \to [0,1]$, $1 \le i \le d$ and $s < t$.

Fix a smooth function $H\colon \mathbb{T}^d \to \mathbb{R}_+$. By Feynman–Kac formula (A1.7.5) and by stationarity of ν_α^N,

$$\mathbb{E}_{\nu_\alpha^N}\Big[\exp\Big\{\int_s^t N\sum_{x\in\mathbb{T}_N^d} H(x/N)W_{x,x+e_i}(r)dr\Big\}\Big] = \mathbb{E}_{\nu_\alpha^N}\Big[\exp\Big\{\int_0^{t-s} N\sum_{x\in\mathbb{T}_N^d} H(x/N)W_{x,x+e_i}(r)dr\Big\}\Big] \le e^{(t-s)\lambda_N(H)},$$

where $\lambda_N(H)$ is the largest eigenvalue of $N^2L_N + N\sum_{x\in\mathbb{T}_N^d} H(x/N)W_{x,x+e_i}$ and has the variational expression:

$$\lambda_N(H) = \sup_f\Big\{N\sum_{x\in\mathbb{T}_N^d} H(x/N) < W_{x,x+e_i}f >_\alpha -(1/2)N^2D_N(f)\Big\}.$$

By the integration by parts formula (2.1) for the current $W_{x,x+e_i}$ and a computation similar to the one performed just after (3.2), $H(x/N) < W_{x,x+e_i}f >_\alpha$ is bounded above by $(A/8)H(x/N)^2 < (r_{x,x+e_i} + r_{x+e_i,x})f >_\alpha + (2/A)\mathfrak{D}_{x,x+e_i}(f)$ for every $A > 0$. Since the jump rate is bounded by 1, setting $A = (2/N)$, we obtain that the expression inside braces in last formula is bounded above by

$$(1/2)\sum_{x\in\mathbb{T}_N^d} H(x/N)^2,$$

what concludes the proof of the lemma. □

Corollary 6.3 *Assume that $T > 1/2$. For any smooth function $H\colon \mathbb{T}^d \to \mathbb{R}_+$ and $1 \le i \le d$, there is a constant $C = C(H,T)$ depending only on H and T such that for any N and any small enough δ:*

$$\mathbb{E}_{\mu^N}\Big[\sup_{\substack{|t-s|<\delta\\ 0\le s<t\le T}}\Big|\int_s^t N^{1-d}\sum_{x\in\mathbb{T}_N^d} H(x/N)W_{x,x+e_i}(r)dr\Big|\Big] \le C\sqrt{\delta}\log\delta^{-1}.$$

Proof. Denote the time integral $\int_0^t N^{1-d}\sum_{x\in\mathbb{T}_N^d} H(x/N)W_{x,x+e_i}(r)dr$ by $g(t)$. By the entropy inequality, $\mathbb{E}_{\mu^N}[\sup|g(t)-g(s)|]$ is bounded above by

$$\frac{C_1(\delta)}{N^d} H(\mu^N | \nu_\alpha^N) + \frac{C_1(\delta)}{N^d} \log \mathbb{E}_{\nu_\alpha^N} \Big[\exp \Big\{ N^d C_1(\delta)^{-1} \sup |g(t) - g(s)| \Big\} \Big] \quad (6.1)$$

where $C_1(\delta)$ is a function of δ to be determined later. Since there are at most κ particles per site, the relative entropy $H(\mu^N|\nu_\alpha^N)$ is bounded by $C(\kappa, \alpha)N^d$.

Recall the Garsia–Rodemich–Rumsey inequality (cf. Stroock and Varadhan (1979)) that can be stated as follows. Let $g(t)$ a continuous function and $\psi(u)$, $p(u)$ strictly increasing functions such that $\psi(0) = 0$, $p(0) = 0$ and $\lim_{u\to\infty} \psi(u) = \infty$, and define B as

$$B = \int_0^T ds \int_0^T dt\, \psi\Big(\frac{|g(t) - g(s)|}{p(|t-s|)} \Big) .$$

Then,

$$\sup_{\substack{|t-s|<\delta \\ 0\le t,s\le T}} |g(t) - g(s)| \le 8 \int_0^\delta \psi^{-1}\Big(\frac{4B}{u^2}\Big) p(du) .$$

Set $p(u) = \sqrt{u}$ and $\psi(u) = \exp\{N^d u\} - 1$ so that $\psi^{-1}(u) = N^{-d} \log(1 + u)$. Integrating by parts we get that

$$\int_0^\delta \log\Big\{1 + \frac{4B}{u^2}\Big\} \frac{1}{\sqrt{u}}\, du \le 2\sqrt{\delta} \log\Big\{1 + \frac{4B}{\delta^2}\Big\} + 8\sqrt{\delta}$$

because $B/(4B+u^2) \le 1/4$. For $\delta < e^{-2}$, the right hand side of the last inequality is bounded above by $8\sqrt{\delta}\log\delta^{-1}\{1 + \log^+(4B + \delta^2)\}$, where $\log^+ u = (\log u) \vee 0$. Therefore, choosing $C_1(\delta) = 32\sqrt{\delta}\log\delta^{-1}$, we get that

$$N^d C_1(\delta)^{-1} \sup |g(t) - g(s)| \le 1 + \log^+(4B + \delta^2) .$$

In particular, the second term of (6.1) is bounded above by

$$\frac{32\sqrt{\delta}\log\delta^{-1}}{N^d} \Big\{ 1 + \log(4\mathbb{E}_{\nu_\alpha^N}[B] + \delta^2 + 1) \Big\} .$$

Recalling the definitions of B and $g(t)$, for $\delta^2 < 4T^2 - 1$, which is possible because we assumed $T > 1/2$, by Lemma 6.2, $4\mathbb{E}_{\nu_\alpha^N}[B] + \delta^2 + 1$ is bounded above by

$$\begin{aligned}
& 4\int_0^T dt \int_0^T ds\, \mathbb{E}_{\nu_\alpha^N}\Big[\exp\Big| \int_s^t N \sum_{x\in\mathbb{T}_N^d} |t-s|^{-1/2} H(x/N) W_{x,x+e_i}(r) dr \Big| \Big] \\
& \le 8 \int_0^T dt \int_0^T ds \exp\Big\{ (1/2) \sum_{x\in\mathbb{T}_N^d} H(x/N)^2 \Big\} \\
& = 8T^2 \exp\Big\{ (1/2) \sum_{x\in\mathbb{T}_N^d} H(x/N)^2 \Big\} .
\end{aligned}$$

This estimate together with the bound on the relative entropy obtained earlier in the proof shows that (6.1) is bounded above by

$$32\sqrt{\delta}\log\delta^{-1}\Big\{C(\kappa,\alpha) + N^{-d}\Big(1+\log(8T^2)+(1/2)\sum_{x\in\mathbb{T}_N^d} H(x/N)^2\Big)\Big\}.$$

This concluded the proof of the corollary. □

An estimate of the modulus of continuity of the trajectories follows immediately:

Corollary 6.4 *For any smooth function $H:\mathbb{T}^d \to \mathbb{R}_+$, there exists a constant $C = C(H,T)$ depending only H and T such that*

$$\limsup_{N\to\infty} \mathbb{E}_{\mu^N}\Bigg[\sup_{\substack{|t-s|<\delta\\ 0\le t,s\le T}} \Big| < H,\pi_t > - < H,\pi_s > \Big|\Bigg] \le C(H,T)d\sqrt{\delta}\log\delta^{-1}.$$

Proof. Fix a smooth function H and consider the martingale $M^H(t) = M^{H,N}(t)$ defined by $M^H(t) =< H,\pi_t^N > - < H,\pi_0^N > - \int_0^t N^2 L_N < H,\pi_s^N > ds$. A simple computation shows that the quadratic variation of this martingale is bounded by $C(d,H)N^{-d}$. In particular, by Doob's inequality,

$$\mathbb{E}_{\mu^N}\Bigg[\sup_{\substack{|t-s|<\delta\\ 0\le t,s\le T}} \Big|M_t^H - M_s^H\Big|\Bigg] \le 2\mathbb{E}_{\mu^N}\Big[\sup_{0\le t\le T}\Big|M_t^H\Big|\Big] \le C(d,H)N^{-d/2}.$$

On the other hand,

$$\int_0^t N^2 L_N < H,\pi_s^N > ds = \int_0^t N^{1-d}\sum_{i,x}(\partial_{u_i}^N H)(x/N) W_{x,x+e_i}(r)dr .$$

To prove the corollary, it is therefore enough to show for each $1\le i\le d$ that

$$\limsup_{N\to\infty}\mathbb{E}_{\mu^N}\Bigg[\sup_{\substack{|t-s|<\delta\\ 0\le s<t\le T}} \Big|\int_s^t N^{1-d}\sum_{x\in\mathbb{T}_N^d}(\partial_{u_i}^N H)(x/N)W_{x,x+e_i}(r)dr\Big|\Bigg]$$
$$\le C(T,H)\sqrt{\delta}\log\delta^{-1}$$

which is the content of the previous corollary. □

It is now easy to conclude the proof of Theorem 6.1.

Proof of Theorem 6.1. Since there are at most κ particles per site,

$$\mathbb{P}_{\mu^N}\Big[\sup_{t\ge 0} < \pi_t, 1 > \ge A\Big] = 0$$

for all $A > \kappa$. By Theorem 4.1.3, Remark 4.1.4 and Proposition 4.1.7, the tightness of the sequence Q_{μ^N} follows from this identity and the previous corollary.

Furthermore, since there are at most κ particles per site, for any continuous function $H: \mathbb{T}^d \to \mathbb{R}$, $| < \pi_t^N, H > |$ is bounded above by $\kappa N^{-d} \sum_x |H(x/N)|$ that converges to $\kappa \int |H(u)| du$ as $N \uparrow \infty$. All limit points Q^* of the sequence Q_{μ^N} are thus concentrated on paths $\pi(t, du)$ such that $\sup_{t \ge 0} | < \pi_t, H > | \le \kappa \int |H(u)| du$. The trajectories are therefore absolutely continuous with density bounded by κ Q^*-almost surely. □

7. Comments and References

The nongradient method just presented is due to Varadhan (1994a) and Quastel (1992). It permitted to extend to reversible nongradient systems the entropy method presented in Chapter 5, provided the generator of the system restricted to a cube o linear size ℓ has a spectral gap that shrinks as ℓ^{-2}. The integration by parts formula for the current W_{0,e_i} is presented in Varadhan (1994a) and Quastel (1992). It was extended to mean-zero functions by Esposito, Marra and Yau (1994). The proof proposed here is taken from this latter article, as well as the one of Theorem 4.1. Section 5 is a mixture of Esposito, Marra and Yau (1994) and Landim, Olla and Yau (1997). The proof of the continuity of the diffusion coefficient, Theorem 5.6, is taken from Landim, Olla and Yau (1997) while the proof of Corollary 5.9 is taken from Funaki, Uchiyama and Yau (1995). The continuity of the diffusion coefficient was already present in Varadhan (1994a) and Quastel (1992).

Wick (1989) proved the hydrodynamic behavior of a one-dimensional nongradient model in which the current can be written as the sum of a gradient $h - \tau h$ and a term of type Lf, for cylinder functions h and f. Kipnis, Landim and Olla (1994) applied Varadhan (1994a) and Quastel (1992) ideas to derive the hydrodynamic behavior of the symmetric generalized exclusion process. Xu (1993) extended the nongradient approach to the non reversible setting by considering mean-zero asymmetric simple exclusion processes. Spohn and Yau (1995), based on the variational formula presented in Theorem 5.5 for the diffusion coefficient, obtained a lower and an upper bound for the diffusion matrix of lattice gases that are valid close to the critical temperature. They showed that $d_- \alpha(1-\alpha)\chi(\alpha)^{-1} \le D(\alpha) \le d_+ \alpha(1-\alpha)\chi(\alpha)^{-1}$, where d_-, d_+ are universal constants and $\chi(\alpha)$ is the static compressibility. Funaki, Uchiyama and Yau (1995), assuming that the diffusion coefficient is smooth, applied the relative entropy method to derive the hydrodynamic equation of nongradient lattice gases that are reversible with respect to Bernoulli product measures. Komoriya (1997) extended these ideas to asymmetric mean-zero exclusion processes with speed change. Varadhan and Yau (1997) prove the hydrodynamic limit of Kawasaki dynamics satisfying mixing conditions.

In the nongradient context, there are two problems that deserve to be studied. The first one consists in proving the hydrodynamic behavior of a nongradient system without using any information on the size of the spectral gap. In another direction, it would be interesting to derive the hydrodynamic behavior of nongradient interacting Brownian particles.

Navier–Stokes equations. A fundamental question in mathematical physics is the derivation and the interpretation of the Navier–Stokes equations. One of difficulties in the interpretation of this equation is that it is not scaling invariant and thus cannot be obtained by a scaling limit. Although this problem is still out of reach for Hamiltonian systems, important progress has been made recently in the context of interacting particle systems.

To fix ideas consider an asymmetric zero range process evolving on the lattice $\mathbb{T}_N^d$. The macroscopic evolution of the process under Euler scaling is described by the first order quasi–linear hyperbolic equation

$$\partial_t \rho \ + \ \gamma \cdot \nabla \Phi(\rho) \ = \ 0 \, , \tag{7.1}$$

where γ stands for the mean drift: $\gamma = \sum_{x \in \mathbb{Z}^d} x p(x)$ and $\Phi(\alpha)$ for the expected value of the jump rate $g(\eta(0))$ under the invariant measure with density α. Assume that the system starts from a product measure with slowly varying parameter associated to a profile $\rho_0 \colon \mathbb{T}^d \to \mathbb{R}_+$. We shall see in Chapter 8 that under Euler scaling (times of order tN) the density has still a slowly varying profile $q^N(t,u) = \mathbb{E}_{\nu^N_{\rho_0(\cdot)}}[\eta_{tN}([uN])]$ that converges weakly (in fact pointwisely at every continuity point according to Theorem 9.0.2) to the entropy solution of equation (7.1) with initial data ρ_0.

In the context of interacting particle systems with one conserved quantity the Navier–Stokes equations takes the form

$$\partial_t \rho^N \ + \ \gamma \cdot \nabla \Phi(\rho^N) \ = \ N^{-1} \sum_{i,j=1}^{d} \partial_{u_i} \Big(D_{i,j}(\rho^N) \partial_{u_j} \rho^N \Big) \, , \tag{7.2}$$

where D is a diffusion matrix. Three different interpretations have been proposed for the Navier–Stokes corrections:

(a) **The incompressible limit**: Consider a small perturbation of a constant profile θ: $\rho_0^N(u) \ = \ \theta + N^{-1} a_0(u)$. Assuming that this form persists at latter times ($\rho^N(t,u) = \theta + N^{-1} a(t,u)$) we obtain from (7.2) the following equation for $a_N(t,u) = a(tN, u)$

$$\begin{aligned} \partial_t a_N \ + \ N\Phi'(\theta)\gamma \cdot \nabla a_N \ + \ (1/2)\Phi''(\theta)\gamma \cdot \nabla a_N^2 \\ = \sum_{i,j=1}^{d} D_{i,j}(\theta) \partial^2_{u_i,u_j} a_N \ + \ O(N^{-1}) \, . \end{aligned}$$

A Galilean transformation $m_N(t,u) = a_N(t, u + Nt\Phi'(\theta)\gamma)$ permits to remove the diverging term of the last differential equation and to get a limit equation for $m = \lim_{N \to \infty} m_N$:

$$\partial_t m \; + \; (1/2)\Phi''(\theta)\gamma \cdot \nabla m^2 \; = \; \sum_{i,j=1}^{d} D_{i,j}(\theta)\partial^2_{u_i,u_j} m \; . \tag{7.3}$$

(b) **First order correction to the hydrodynamic equation**: Fix a smooth profile $\rho_0 \colon \mathbb{T}^d \to \mathbb{R}_+$ and consider a process starting from a product measure with slowly varying parameter associated to the profile $\rho_0(\cdot)$. We have seen that under Euler scaling the expected density $q^N(t,u) = \mathbb{E}_{\nu^N_{\rho_0(\cdot)}}[\eta_{tN}([uN])]$ has still a slowly varying profile that converges weakly to the entropy solution of equation (7.1) with initial data ρ_0. This second interpretation asserts that the solution of equation (7.2) with initial profile ρ_0 approximates q^N up to the order N^{-1}:

$$\lim_{N\to\infty} N\big(q^N - \rho^N\big) \; = \; 0$$

in a weak sensc as $N \uparrow \infty$.

(c) **Long time behavior**: The third interpretation relies on the following observation. Denote by $\rho(t,u)$ the solution of the hyperbolic equation (7.1). In the interacting case in finite volume with periodic boundary conditions, asymptotically as $t \uparrow \infty$, the entropy solution $\rho(t,u)$ converges to a stationary solution which is constant along the drift:

$$\lim_{t\to\infty} \rho(t,\cdot) \; = \; \rho_\infty(u) \; = \; \int_0^1 \rho_0(u + r\gamma)\, dr \; ,$$

provided ρ_0 stands for the initial data. In particular, if we consider the asymptotic process under diffusive scaling, we expect it to become immediately constant along the drift direction:

$$\gamma \cdot \nabla \lim_{N\to\infty} \mathbb{E}_{\nu^N_{\rho_0(\cdot)}}[\eta_{tN^2}([uN])] \; = \; 0$$

for every $t > 0$ and for any initial profile. In contrast, on the hyperplane orthogonal to the drift the profile should evolve smoothly in time according to a parabolic equation.

The third interpretation consists therefore in analyzing the behavior of the solution of equation (7.2) in time scales of order tN on the hyperplane orthogonal to the drift direction. Let $b_N(t,u) = \rho(tN,u)$. From (7.2) we obtain the following equation for b_N:

$$\partial_t b_N \; + \; N\gamma \cdot \nabla\Phi(b_N) \; = \; \sum_{i,j=1}^{d} \partial_{u_i}\Big(D_{i,j}(b_N)\partial_{u_j} b_N\Big) \; .$$

To eliminate the diverging term $N\gamma \cdot \nabla\Phi(b_N)$, assume that the initial data (and therefore the solution at any fixed time) is constant along the drift direction: $\gamma \cdot \nabla\rho_0 = 0$. In this case we get the parabolic equation

$$\partial_t b = \sum_{i,j=1}^{d} \partial_{u_i}\Big(D_{i,j}(b)\partial_{u_j} b\Big) \tag{7.4}$$

that describes the evolution of the system in the hyperplane orthogonal to the drift.

Notice that while the first and the third interpretation concern the behavior of the system under diffusive scaling, the second one is a statement on the process under Euler scaling.

Dobrushin (1989) was the first to investigate the corrections to the hydrodynamic equations. He considered the evolution of independent Markov processes and proposed a systematic approach to deduce the corrections of all orders to the hydrodynamics equations of interacting particle systems. The method has been successfully applied to harmonic random oscillators in Dobrushin, Pellegrinotti, Suhov and Triolo (1988) and in Dobrushin, Pellegrinotti and Suhov (1990).

Esposito and Marra (1994) deduced formally the Navier–Stokes equations

$$\begin{cases} \text{div } A = 0 \\ \partial_t A + K_0 A \cdot \nabla A = \nabla P + K_1 \nabla \cdot D \nabla A \end{cases} \tag{7.5}$$

from Hamiltonian dynamics. Here P stands for the pressure and D for the diffusion matrix. In the sequel Esposito, Marra and Yau (1994) proved the incompressible limit for asymmetric simple exclusion processes in dimension $d \geq 3$: They considered an asymmetric simple exclusion process evolving on the torus $\mathbb{T}_N^d$ starting form a product measure with slowly varying parameter associated to a profile $\rho_0^N(u) = \theta + N^{-1}a_0(u)$, where θ is a fixed parameter in $(0,1)$. Recall that in the context of exclusion processes $\Phi(\theta) = \theta(1-\theta)$. Denote by $\Pi_t^N(du)$ the corrected empirical measure defined by

$$\Pi_t^N = \frac{1}{N^{d-1}} \sum_{x \in \mathbb{T}_N^d} \big\{\eta_{tN^2}(x) - \theta\big\}\delta_{x/N}\,.$$

Notice the diffusive scaling of time and that the sum is divided by N^{d-1} instead of N^d. Esposito, Marra and Yau (1994) proved the following result:

Theorem 7.1 *In dimension $d \geq 3$, as $N \uparrow \infty$, $\tau_{[tN\Phi'(\theta)\gamma]}\Pi_t^N$ converges weakly in probability to an absolutely continuous measure whose density is the solution of the equation (7.3) with initial data a_0. Moreover, the diffusion coefficient D is given by a variational formula.*

Landim, Olla and Yau (1997) examined the question of the Navier–Stokes equations from the point of view of the first order corrections. They considered an asymmetric simple exclusion process evolving on the torus $\mathbb{T}_N^d$ starting from a product measure with slowly varying parameter associated to a profile ρ_0. Denote by $q^N(t,u)$ the expected density of particles at time t around u: $q^N(t,u) = \mathbb{E}_{\nu^N_{\rho_0(\cdot)}}[\eta_{tN}([uN])]$ and denote by ρ^N the solution of equation (7.2) with $\Phi(\theta) = \theta(1-\theta)$. Landim, Olla and Yau (1997) proved

Theorem 7.2 *For the asymmetric simple exclusion process in dimension $d \geq 3$, as $N \uparrow \infty$, $N(q^N - \rho^N)$ converges weakly to 0 in some appropriate $\mathcal{H}_{-1}$ space.*

Landim, Olla and Yau (1996) proved regularity properties of the diffusion matrix D of the Navier–Stokes equation (7.2), (7.3). Landim and Yau (1997) filled a gap left in the previous works showing that in the exclusion models, each cylinder function h such that $E_{\nu_\alpha}[h] = (d/d\alpha)E_{\nu_\alpha}[h] = 0$ for all α can be approximate in some $\mathcal{H}_{-1}$ space by functions in the range of the generator.

Benois, Koukkous and Landim studied an asymmetric zero range process evolving on $\mathbb{T}_N^d$ with drift $\gamma = \sum_x xp(x)$ along the first direction: $\gamma = ce_1$ for some constant $c \neq 0$. Fix a profile $\rho_0 \colon \mathbb{T}^d \to \mathbb{R}_+$ constant along the drift direction: $\partial_{u_1}\rho_0 = 0$ and consider as initial state a product measure with slowly varying parameter associated to a profile ρ_0. Benois, Koukkous and Landim (1997) proved that in dimension $d \geq 2$ the empirical measure diffusively rescaled converges to an absolutely continuous measure whose density is the solution of

$$\partial_t \rho = \sum_{i,j=1}^{d} \sigma_{i,j} \partial_{u_i} \Big(\Phi'(\rho) \partial_{u_j} \rho \Big)$$

with initial condition ρ_0. In this formula $\sigma_{i,j}$ stands for the covariance matrix of the transition probability $p(\cdot)$: $\sigma_{i,j} = \sum_x x_i x_j p(x)$. Janvresse (1997) obtained the first order corrections to the parabolic hydrodynamic equations of Bernoulli reversible speed change exclusion processes in dimension $d \geq 3$.

The proofs of Theorems 7.1 and 7.2 rely on the relative entropy method present in Chapter 6 and require a logarithmic Sobolev inequality, which at the moment where this book has been concluded has been proved only for reversible generalized exclusion processes and lattice gases with mixing conditions in Yau (1996), (1997).

More recently exclusion processes in which particles have velocities have been considered. The dynamics can be briefly described as follows. Fix a finite set $\mathcal{Q} \subset \mathbb{R}^d$ of possible velocities. Each particle has a velocity q and evolves on the lattice $\mathbb{T}_N^d$. An exclusion rule forbids the presence of two particles with the same velocity at some site. The state space is thus $\{\{0,1\}^{\mathcal{Q}}\}^{\mathbb{T}_N^d}$.

The evolution can be decomposed in two pieces. Particles evolve according to random walks on $\mathbb{T}_N^d$ obeying the exclusion rule described above. Thus a particle with velocity q at site x, independently from the others, waits a mean-one exponential time at the end of which it jumps to $x + y$ with probability $p(y, q)$. If the chosen site is already occupied by a particle with velocity q the jump is suppressed. Here the transition probabilities $p(\cdot, q)$ are such that their mean drift are q: $\sum_x xp(x, q) = q$.

Superposed to this jump dynamics there is a collision process that exchanges the velocity of a pair of particles sitting on the same site. More precisely, if two particles with velocities q_1 and q_2 are at x and there are no particles on this site with velocities q_1', q_2', then simultaneously at rate 1 the particle with velocity q_i

assumes velocity q_i', $i = 1, 2$, provided $q_1 + q_2 = q_1' + q_2'$. This latter assumption is imposed to guarantee the conservation of momentum.

Notice that both the total number of particles and the total momentum are conserved by the dynamics. For x in $\mathbb{T}_N^d$, denote by $\eta(x, q)$ the total number of particles with velocity q at site x, by $I_\ell(x, \eta)$ the total momentum at site x for the configuration η in the ℓ-th direction: $I_\ell(x, \eta) = \sum_{q \in \mathcal{Q}} (q \cdot e_\ell)\, \eta(x, q)$ and by $\Pi_t^{N,\ell}(du)$ the corrected empirical measure associated to the moment in the ℓ-th direction:

$$\Pi_t^{N,\ell} = \frac{1}{N^{d-1}} \sum_{x \in \mathbb{T}_N^d} I_\ell(x, \eta_t) \delta_{x/N} .$$

Starting from a product measure μ^N with slowly varying parameter associated to the profile $(\rho_0, A_0^1, \dots, A_0^d)$ with density $\rho_0(u) = r + N^{-2} a_0(u)$ and momentum in the ℓ-th direction $A_0^\ell(u) = N^{-1} a_0^\ell(u)$, Esposito, Marra and Yau (1996) proved, under some assumptions on the set $\mathcal{Q}$, that the corrected empirical measure $\Pi_t^{N,\ell}$ diffusively scaled converges in probability to a Navier–Stokes equation of type (7.5), provided the solution of this equation is smooth, and obtained a variational formula for the diffusion matrix D.

Smooth solutions of the incompressible Navier–Stokes equations are only known to exist for short period of times. To avoid this problem, Quastel and Yau (1997) proved that the sequence of probability measures on the path space induced by the process $(\Pi_t^{N,1}, \dots, \Pi_t^{N,d})$ and the initial state μ^N is tight and that all its limit points are concentrated on weak solutions of the incompressible Navier–Stokes equation (7.5).

8. Hydrodynamic Limit of Asymmetric Attractive Processes

We examine in this chapter an alternative method to prove the hydrodynamic behavior of asymmetric interacting particle systems. This approach has the advantage over the one presented in Chapter 6 that it does not require the solution of the hydrodynamic equation to be smooth. On the other hand its main inconvenience is that it assumes the process to be attractive to permit the use of coupling arguments and the initial state to be a product measure. To illustrate this approach we consider an asymmetric attractive zero range process on the discrete d–dimensional torus $\mathbb{T}_N^d$. The generator of this Markov process, denoted by L_N, is given by

$$(L_N f)(\eta) \;=\; \sum_{\substack{x\in\mathbb{T}_N^d\\ y\in\mathbb{Z}^d}} p(y)g(\eta(x))\Big[f(\eta^{x,x+y}) - f(\eta)\Big]\;, \tag{0.1}$$

where $p(\cdot)$ is a finite range irreducible transition probability on $\mathbb{Z}^d$. Irreducible means here that for every z in $\mathbb{Z}^d$, there exists a positive integer m and a sequence $0 = x_0, x_1, \dots, x_m = z$ such that $p(x_{i+1} - x_i) + p(x_i - x_{i+1}) > 0$ for $0 \le i \le m-1$. Throughout this chapter we assume the process to be attractive. We have seen in Chapter 2 that this hypothesis corresponds to assume that the rate at which a particle leaves a site is a non decreasing function of the total number of particles at that site:

(H) $g(\cdot)$ is a non decreasing function.

To avoid minor technical difficulties we shall assume also that $g(\cdot)$ is bounded. This hypothesis is not a crucial one and can be removed without problems.

We have seen in Chapter 1 that to study the hydrodynamic behavior of asymmetric processes we have to speed up the process by a factor N. We denote therefore by η_t the Markov process on $\mathbb{N}^{\mathbb{T}_N^d}$ associated to the generator L_N defined by (0.1) speeded up by N.

A formal argument, assuming conservation of local equilibrium, permits to foresee the hydrodynamic equation of asymmetric zero range processes. Indeed, assume the initial state μ^N to be a local equilibrium of profile ρ_0. For a smooth function $H : \mathbb{T}^d \to \mathbb{R}$, M_t^H defined by

$$
\begin{aligned}
M_t^H \;=\; & N^{-d} \sum_{x \in \mathbb{T}_N^d} H(x/N)\eta_t(x) \;-\; N^{-d} \sum_{x \in \mathbb{T}_N^d} H(x/N)\eta_0(x) \\
& - \int_0^t N^{-d} \sum_{x \in \mathbb{T}_N^d} H(x/N) N L_N \eta_s(x)\, ds
\end{aligned}
\tag{0.2}
$$

is a martingale vanishing at 0. The additional factor N in the integral term comes from the time renormalization. A discrete integration by parts shows that the integral term is equal to

$$
\sum_{j=1}^d \int_0^t N^{-d} \sum_{x \in \mathbb{T}_N^d} \Big(\sum_y y_j p(y) \Big) (\partial_{u_j} H)(x/N) g(\eta_s(x))\, ds \;+\; O(N^{-1})\, .
$$

For $1 \le j \le d$, denote by γ_j the drift in the j-th direction of the evolution of each elementary particle:

$$
\gamma_j \;=\; \sum_y y_j p(y)\, .
$$

Since the martingale vanishes at 0, if $\rho^N(t, x/N)$ denotes the expected value of the number of particles at time t and at site x,

$$
\begin{aligned}
& N^{-d} \sum_{x \in \mathbb{T}_N^d} H(x/N) \rho^N(t, x/N) \;-\; N^{-d} \sum_{x \in \mathbb{T}_N^d} H(x/N) \rho^N(0, x/N) \\
& \quad = \sum_{j=1}^d N^{-d} \sum_{x \in \mathbb{T}_N^d} \int_0^t \gamma_j (\partial_{u_j} H)(x/N) \mathbb{E}_{\mu^N} \Big[g(\eta_s(x)) \Big]\, ds\, .
\end{aligned}
$$

Recall from Chapter 2 that we denoted by $\Phi(\alpha)$ the expected value of the jump rate $g(\eta(0))$ with respect to the invariant measure ν_α^N. If there is conservation of local equilibrium, around site x at time s the state of the process should be close to some equilibrium. Since the equilibrium states are parametrized by the density of particles and since we denote by $\rho^N(s, x/N)$ the density at time s at site x, the state of the process should be close to $\nu^N_{\rho^N(s,x/N)}$. In consequence the expected value of $g(\eta_s(x))$ should be close to $\Phi(\rho^N(s, x/N))$. From these considerations we see that $\rho^N(t, x/N)$ should be the solution of

$$
\begin{aligned}
& N^{-d} \sum_{x \in \mathbb{T}_N^d} H(x/N) \rho^N(t, x/N) \;-\; N^{-d} \sum_{x \in \mathbb{T}_N^d} H(x/N) \rho^N(0, x/N) \\
& \quad = \sum_{j=1}^d N^{-d} \sum_{x \in \mathbb{T}_N^d} \int_0^t \gamma_j (\partial_{u_j} H)(x/N) \Phi(\rho^N(s, x/N))\, ds\, .
\end{aligned}
$$

Thus, we expect the macroscopic behavior of asymmetric zero range processes to be described by solutions of the first order quasi–linear hyperbolic partial differential equation

$$\begin{cases} \partial_t \rho + \sum_{i=1}^{d} \gamma_i \partial_{u_i} \Phi(\rho) = 0 , \\ \rho(0, \cdot) = \rho_0(\cdot) . \end{cases} \tag{0.3}$$

This equation is well known by now. We refer to Smoller (1983) for a background on properties of solutions of this equation. We just point out here and review in Appendix 2 some of the main features.

First of all, even if the initial profile $\rho_0(\cdot)$ is smooth, solutions of equation (0.3) may develop shocks. Therefore, there may not exist classical solutions for this equation and we are forced to consider weak solutions in the sense of Definition 4.2.2. Enlarging, however, the set of admissible solutions in this way we loose uniqueness and we have to introduce a criterion to pick among all weak solutions the physically significant one. We shall consider the entropy condition proposed by Kružkov.

Definition 0.1 (Entropy solutions) A bounded function $\rho: \mathbb{R}_+ \times \mathbb{T}^d \to \mathbb{R}$ is an entropy solution of equation (0.3) if:

(a) (Entropy inequality) For every $c \in \mathbb{R}$,

$$\partial_t \big|\rho - c\big| + \sum_{i=1}^{d} \gamma_i \partial_{u_i} \big|\Phi(\rho) - \Phi(c)\big| \le 0$$

in the sense of distributions on $(0, \infty) \times \mathbb{T}^d$, that is, if for every smooth function $H: (0, \infty) \times \mathbb{T}^d \to \mathbb{R}$ with compact support,

$$\int_{(0,\infty)\times\mathbb{T}^d} dt\, du \Big\{ (\partial_t H) \big|\rho - c\big| + \sum_{i=1}^{d} \gamma_i (\partial_{u_i} H) \big|\Phi(\rho) - \Phi(c)\big| \Big\} \ge 0 .$$

(b) ($L^1(\mathbb{T}^d)$ convergence at $t = 0$) $\rho(t, \cdot)$ converges in $L^1(\mathbb{T}^d)$ to $\rho_0(\cdot)$ as t decreases to 0:

$$\lim_{t\to 0} \int_{\mathbb{T}^d} |\rho(t, u) - \rho_0(u)| du = 0 .$$

Notice that the initial data does not play any role in part (a) of this definition since the smooth functions are assumed to have compact support on $(0, \infty) \times \mathbb{T}^d$. It is condition (b) that connects the solution to the initial data.

Kružkov (1970) proved the existence of a unique entropy solution of equation (0.3) if the initial data is bounded:

Theorem 0.2 *Assume that $\rho_0 \in L^\infty(\mathbb{T}^d)$. There exists a unique entropy solution $\rho(t, u)$ of equation (0.3).*

In dimension 1, if $\Phi(\cdot)$ is concave and $\gamma = \gamma_1 > 0$, the entropy inequality is equivalent to require the solution not to have decreasing jumps:

$$\liminf_{u\downarrow u_0} \rho(t,u) \;\ge\; \limsup_{u\uparrow u_0} \rho(t,u) \qquad \text{for every } u_0 \in \mathbb{T}^d\,.$$

In view of this discussion, our natural goal in this chapter is to prove that the empirical measure converges in probability to the absolutely continuous measure whose density is the entropy solution of the first order quasi-linear equation (0.3). This is the content of the main theorem of this chapter. To state this result we have to describe the initial measures considered. Let $(\mu^N)_{N\ge 1}$ be a sequence of probability measures on $\mathbb{N}^{\mathbb{T}_N^d}$. We shall assume that

(M1) μ^N is a product measure.

(M2) There exists an equilibrium state $\nu^N_{\alpha^*}$ that bounds above μ^N:

$$\mu^N \;\le\; \nu^N_{\alpha^*} \qquad \text{for all } N\ge 1\,.$$

(M3) Define $\rho_0^N : \mathbb{T}^d \to \mathbb{R}_+$ as the profile associated to μ^N: on an hypercube of length $1/N$ centered at x/N, ρ_0^N is equal to the expected value of the total number of particles for μ^N:

$$\rho_0^N(u) \;=\; \sum_{x\in\mathbb{T}_N^d} E_{\mu^N}[\eta(x)]\mathbf{1}\{u\in \square_N(x)\}\,,$$

where $\square_N(x)$ is a hypercube of length $1/N$ centered at x:

$$\square_N(x) \;=\; \Big\{y\in\mathbb{R}^d;\; x_i/N - 1/2N \le y_i < x_i/N + 1/2N \text{ for } 1\le i\le d\Big\}\,.$$

We assume that ρ_0^N converges in $L^1(\mathbb{T}^d)$ to the initial profile ρ_0:

$$\lim_{N\to\infty}\int_{\mathbb{T}^d} |\rho_0^N(u) - \rho_0(u)|du \;=\; 0\,.$$

Of course such initial measures exist. Fix for example ρ_0 in $L^\infty(\mathbb{T}^d)$ and define ρ_0^N as the profile given by

$$\rho_0^N(u) \;=\; N^d \int_{\square_N(x)} \rho_0(v)\,dv \qquad \text{if } u\in\square_N(x)$$

and μ^N as the product measure with marginals

$$\mu^N\{\eta;\; \eta(x)=k\} \;=\; \nu^N_{\rho_0^N(x/N)}\{\eta;\; \eta(0)=k\} \qquad \text{for } x \text{ in } \mathbb{T}_N^d.$$

It is easy to check that this sequence of initial states satisfies hypotheses **(M1)**–**(M3)**.

We are now ready to state the main theorem of this chapter.

Theorem 0.3 *Fix a bounded profile $\rho_0\colon \mathbb{T}^d\to\mathbb{R}_+$. Let $(\mu^N)_{N\ge 1}$ be a sequence of probability measures on $\mathbb{N}^{\mathbb{T}_N^d}$ satisfying assumptions* **(M1)**–**(M3)**. *For each $t\ge 0$ the empirical measure π_t^N converges in probability to the absolutely continuous* *measure $\pi(t,du) = \rho(t,u)du$ whose density $\rho(t,u)$ is the entropy solution of equation (0.3).*

We have presented in Chapter 4 an approach to prove this type of result in which the first step is to show that the empirical measure $\pi^N_\cdot$ converges in distribution to a Dirac measure. Thus, for a fixed time $T > 0$ and for a sequence of probabilities $(\mu^N)_{N\geq 1}$ on $\mathbb{N}^{\mathbb{T}^d_N}$ denote by $Q^N_{\mu^N}$ the probability measure on the path space $D([0,T], \mathcal{M}_+(\mathbb{T}^d))$ corresponding to the process π^N_t with generator (0.1) speeded up by N and starting from μ^N. To keep notation simple we will often omit the subscript of $Q^N_{\mu^N}$.

We would like to show that Q^N converges to a Dirac measure concentrated on the path $\pi(t,du)$ whose density is the entropy solution of (0.3). In view of Chapter 5 the first natural idea is to repeat in this context the entropy approach. Consider, therefore, for a smooth function $H\colon \mathbb{T}^d \to \mathbb{R}$, the martingale M^H_t defined by (0.2). We showed that after a summation by parts this martingale is equal to

$$N^{-d} \sum_{x\in\mathbb{T}^d_N} H(x/N)\eta_t(x) \;-\; N^{-d} \sum_{x\in\mathbb{T}^d_N} H(x/N)\eta_0(x)$$
$$-\sum_{j=1}^{d} \int_0^t N^{-d} \sum_{x\in\mathbb{T}^d_N} \gamma_j (\partial_{u_j} H)(x/N) g(\eta_s(x))\, ds \;+\; O(N^{-1})\,.$$

At this point to close the equation we would like to replace the expression

$$N^{-d} \sum_{x\in\mathbb{T}^d_N} (\partial_{u_j} H)(x/N) g(\eta_s(x))$$

by $N^{-d}\sum_{x\in\mathbb{T}^d_N} (\partial_{u_j} H)(x/N)\Phi(\eta^{N\varepsilon}_s(x))$. Notice that once the replacement is performed the equation is indeed closed since $\eta^{N\varepsilon}(x)$ is a function of the empirical measure. In fact,

$$\eta^{N\varepsilon}(x) \;=\; <\pi^N, \iota_\varepsilon(x/N - \cdot)>$$

if $\iota_\varepsilon = \iota_{\varepsilon,N}$ is the approximation of the identity defined by

$$\iota_\varepsilon(u) \;=\; \frac{N^d}{(2[\varepsilon N]+1)^d} \mathbf{1}\{[-\varepsilon,\varepsilon]\}(u)\,.$$

Once the equation is closed we would proceed as in the proof of the hydrodynamic equation for symmetric simple exclusion processes to show that Q^N converges to a Dirac measure.

There are however two serious problems in this approach. First of all the generator L_N is speeded up by N and not by N^2. This factor N^2 was crucial in the proof of the two blocks estimate. Thus the replacement lemma allows only to replace $g(\eta(x))$ by $\Phi(\eta^\ell(x))$ where ℓ is an integer independent of N and that increases to infinity after N. But $\eta^\ell(x)$ is not a function of the empirical measure and hence the equation is not closed.

The second problem comes from the previously mentioned existence of several weak solution of the hyperbolic equation (0.3). It is therefore not enough to close the equation and to show that all limit point of the sequence Q^N are concentrated on weak solutions of (0.3), we still have to guaranty that they are concentrated on *entropy* solutions.

The second problem is solved proving an entropy inequality at a microscopic level. This result is stated at Corollary 2.2 of the next section and relies on coupling techniques allowed by the attractiveness assumption. It consists in showing that for every $c \in \mathbb{R}$,

$$\partial_t \big| \eta_t^\ell(\cdot) - c \big| \; + \; \sum_{i=1}^d \gamma_i \partial_{u_i} \big| \Phi(\eta_t^\ell(\cdot)) - \Phi(c) \big| \; \le \; 0 \qquad (0.4)$$

in the sense of distributions on $(0, \infty) \times \mathbb{T}^d$ and for $N \uparrow \infty$ and then $\ell \uparrow \infty$. Notice that here also we need a two blocks estimate to replace $\eta_t^\ell(\cdot)$ by $\eta_t^{N\varepsilon}(\cdot)$ in order to obtain functions of the empirical measure. If the replacement was possible it would follow from this entropy inequality that all limit points of the sequence Q^N are concentrated on entropy solutions of (0.3) and to conclude the proof of Theorem 0.3 it would remain to show the relative compactness and some $L^1(\mathbb{T}^d)$ continuity at $t = 0$ to ensure condition (b) of Definition 0.1.

Unfortunately there is a proof of the two blocks estimate in the asymmetric case only for attractive nearest neighbor processes in dimension 1 (cf. Rezakhanlou (1991)). A new approach is thus needed to close the equation. The idea here is to introduce the measure valued solutions of equation (0.3). For a background on this subject we refer to Di Perna (1985).

Definition 0.4 (Measure valued solutions) Denote by $\mathcal{P}(\mathbb{R}_+)$ the set of probability measures on $\mathbb{R}_+$. A measurable map $\mu : (0, \infty) \times \mathbb{T}^d \to \mathcal{P}(\mathbb{R}_+)$ is a measure valued solution of (0.3) if

$$\partial_t < \mu(t, u), \lambda > \; + \; \sum_{i=1}^d \gamma_i \partial_{u_i} < \mu(t, u), \Phi(\lambda) > \; = \; 0$$

in the sense of distributions, that is, if for every smooth function H in $C_K^{1,1}(\mathbb{R}_+ \times \mathbb{T}^d)$,

$$\int dt \int du \Big\{ \partial_t H(t, u) < \mu(t, u), \lambda > + \sum_{i=1}^d \gamma_i \, (\partial_{u_i} H)(t, u) < \mu(t, u), \Phi(\lambda) > \Big\}$$
$$+ \; \int H(0, u) \rho_0(u) \, du \; = \; 0 \, .$$

Here and below, for a continuous function $\Psi : \mathbb{R}_+ \to \mathbb{R}_+$, $< \mu(t, u), \Psi(\lambda) >$ stands for the expected value of Ψ with respect to the probability $\mu(t, u)$ and $C_K^{m,n}(\mathbb{R}_+ \times \mathbb{T}^d)$ for the space of continuous functions with compact support and with m continuous derivatives in time and n continuous derivatives in space.

This concept is clearly weaker than the one of weak solution since to each weak solutions $\rho(t,u)$ we may associate the measure valued solution defined by $\mu(t,u)(d\lambda) = \delta_{\rho(t,u)}(d\lambda)$. Since this concept is weaker it will be easier to prove existence of measure valued solution of (0.3). In fact such solutions can easily be obtained by the viscosity method which consists in adding a small diffusion term to the hyperbolic equation (0.3) and in obtaining a measure valued solution as a weak limit of classical solutions for the second order equation as the diffusion coefficient vanishes.

The real problem is therefore to prove uniqueness of measure valued solutions. To consider this problem we need some terminology. A measure valued solution $\mu(t,u)(d\lambda)$ is said to be a Dirac solution if there exists a bounded measurable function $\rho(t,u)$ such that $\mu(t,u)(d\lambda) = \delta_{\rho(t,u)}(d\lambda)$.

From the previous discussion we know that to guarantee uniqueness we have at least to impose an entropy condition and some $L^1(\mathbb{T}^d)$ continuity at $t = 0$.

Definition 0.5 (Entropy measure valued solutions) A measure valued solution $\mu(t,u)(d\lambda)$ is said to be an entropy measure valued solution if

(a) (Entropy inequality) For every $c \in \mathbb{R}$

$$\partial_t < \mu(t,u), |\lambda - c| > \, + \sum_{i=1}^{d} \gamma_i \, \partial_{u_i} < \mu(t,u), |\Phi(\lambda) - \Phi(c)| > \, \le \, 0$$

in the sense of distributions on $(0,\infty) \times \mathbb{T}^d$.

(b) ($L^1(\mathbb{T}^d)$ continuity at $t = 0$)

$$\lim_{t\to 0} \int_{\mathbb{T}^d} du \, < \mu(t,u), |\lambda - \rho_0(u)| > \, = \, 0 \, .$$

All entropy measure valued solutions $\mu(t,u)(d\lambda)$ that are Dirac solutions $\delta_{\rho(t,u)}(d\lambda)$ are such that the associated profile $\rho(t,u)$ is a weak entropy solution of (0.3). Therefore the proof of uniqueness of entropy measure valued solutions is reduced to the proof that all entropy measure valued solutions are Dirac solutions. Moreover, condition (b) in Definitions 0.1 and 0.5 can be relaxed. It is enough to prove that

$$\lim_{t\to 0} \frac{1}{t} \int_0^t ds \int_{\mathbb{T}^d} du \, < \mu(s,u), |\lambda - \rho_0(u)| > \, = \, 0 \, . \tag{0.5}$$

The following result solves the question of uniqueness (cf. Di Perna (1985)).

Theorem 0.6 *Assume that the initial profile ρ_0 is in $L^\infty(\mathbb{T}^d)$. An entropy measure valued solution satisfying (0.5) instead of condition (b) is a Dirac solution if there exists λ_0 such that*

$$\sup_{t,u} \mu(t,u)\Big([-\lambda_0, \lambda_0]^c\Big) \, = \, 0 \, .$$

In view of this discussion on measure valued solutions of quasi-linear hyperbolic equations, for positive integers N and ℓ, associate to each configuration η the Young measure

$$\pi^{N,\ell}(\eta) = \pi^{N,\ell}(du, d\lambda) = N^{-d} \sum_{x \in \mathbb{T}_N^d} \delta_{x/N}(du)\, \delta_{\eta^\ell(x)}(d\lambda) .$$

denote by $\pi_t^{N,\ell}$ the Young measure at time t: $\pi_t^{N,\ell} = \pi^{N,\ell}(\eta_t)$.

Notice first that for a continuous function $H: \mathbb{T}^d \to \mathbb{R}$ and a smooth function $\Psi: \mathbb{R}_+ \to \mathbb{R}$, the expression

$$N^{-d} \sum_{x \in \mathbb{T}_N^d} H(x/N) \Psi(\eta_t^\ell(x))$$

is a function of the Young measure since it is equal to

$$\int H(u)\Psi(\lambda)\pi_t^{N,\ell}(du, d\lambda) =: < \pi_t^{N,\ell}, H(u)\Psi(\lambda) > \cdot$$

In particular, for the martingale M_t^H defined in (0.2), the one block estimate closes the equation in terms of the Young measure. More precisely, after replacement of $N^{-d} \sum_x H(x/N) g(\eta_s(x))$ by $N^{-d} \sum_x H(x/N) \Phi(\eta_s^\ell(x))$, we obtain that M_t^H is equal to

$$\begin{aligned} &< \pi_t^{N,\ell}, H\lambda > - < \pi_0^{N,\ell}, H\lambda > \\ &\quad - \sum_{i=1}^d \gamma_i \int_0^t ds < \pi_s^{N,\ell}, (\partial_{u_i} H)\Phi(\lambda) > + o_N(1) . \end{aligned} \tag{0.6}$$

Here we performed a discrete integration by parts and used the smoothness of H to get that

$$\begin{aligned} N^{-d} \sum_{x \in \mathbb{T}_N^d} H(x/N)\eta_t(x) &= N^{-d} \sum_{x \in \mathbb{T}_N^d} H(x/N)\eta_t^\ell(x) + O(\ell/N^2) \\ &= < \pi_t^{N,\ell}, H\lambda > + O(\ell/N^2) . \end{aligned}$$

Notice that the expression (0.6) is closed in $\pi^{N,\ell}$. Thus the introduction of the Young measure $\pi^{N,\ell}$ solves the first objection raised in the beginning of this section. To prove that the Young measures converges in probability to the entropy measure valued solution of (0.3), we first show that it converges in distribution. Thus for integers N and ℓ and a sequence μ^N of probability measures on $\mathbb{N}^{\mathbb{T}_N^d}$, denote by $Q_{\mu^N}^{N,\ell} = Q^{N,\ell}$ the probability measure on the path space $D([0,T], \mathcal{M}_+(\mathbb{T}^d \times \mathbb{R}_+))$ corresponding to the process $\pi_t^{N,\ell} = \pi^{N,\ell}(\eta_t)$ with generator (0.1) speeded up by N and starting from μ^N.

To prove that all limit points are concentrated on measure valued entropy solutions we need to prove that they are concentrated on absolutely continuous measures $\pi_t(du, d\lambda) = \pi_t(u, d\lambda)du$ that satisfy the conditions of Definition 0.5 and Theorem 0.6. It is easy to prove that the measures are absolutely continuous in the first coordinate (cf. section 3). To show that the limit points are concentrated on measures that satisfy the entropy inequality, notice that inequality (0.4) may be rewritten in terms of the Young measure as

$$\partial_t < \pi_t^{N,\ell}, |\lambda - c| > + \sum_{i=1}^{d} \gamma_i \partial_{u_i} < \pi_t^{N,\ell}, |\Phi(\lambda) - \Phi(c)| > \leq 0 .$$

From this result it will be easy to prove that all limit points of $Q^{N,\ell}$ are concentrated on measure valued solutions satisfying condition (a) of Definition 0.5. The proof of condition (b) relies on coupling techniques and is the unique point in the proof where we need the initial measures to be product. At this point, we are left to show that the condition of Theorem 0.6 is satisfied and that the sequence is relatively compact. These problems are solved in the third section of this chapter. From this result we obtain the convergence in probability by standard arguments already used in Chapter 4 for symmetric simple exclusion process.

Finally, we claim that the convergence in probability of the empirical measure to an absolutely continuous measure whose density is the entropy solution of (0.3) follows from the convergence of the Young measure $\pi^{N,\ell}(du, d\lambda)$ to the entropy measure valued solution $\pi_t(u, d\lambda)du$ of (0.3). Indeed, for every smooth function H, we have that

$$\begin{aligned} < \pi_t^N, H > &= N^{-d} \sum_{x \in \mathbb{T}_N^d} H(x/N)\eta_t(x) \\ &= N^{-d} \sum_{x \in \mathbb{T}_N^d} H(x/N)\eta_t^\ell(x) + O(\ell/N^2) . \end{aligned}$$

Since, by assumption, the Young measure $\pi^{N,\ell}(du, d\lambda)$ converges in probability, the first term on the right hand side, equal to $< \pi_t^{N,\ell}, H\lambda >$, converges to

$$\int < \pi_t(u, d\lambda), \lambda > H(u)\, du .$$

Since $\pi_t(u, d\lambda)$ is the measure valued entropy solution, $\pi_t(u, d\lambda) = \delta_{\rho(t,u)}(d\lambda)$ and last integral is equal to

$$\int H(u)\rho(t, u)\, du .$$

This shows that π_t^N converges in probability to the entropy solution of (0.3).

1. Young Measures

We prove in this section, under restrictive assumptions regarding the initial profile ρ_0 and the initial state μ^N, that the Young measures $\pi^{N,\ell}$ converges in distribution to the entropy measure valued solution of equation (0.3).

Recall from section 1 that for positive integers N and ℓ, to each configuration η we associate the Young measure $\pi^{N,\ell}(du, d\lambda)$ on $\mathbb{T}^d \times \mathbb{R}_+$ defined by

$$\pi^{N,\ell}(\eta) \;=\; \pi^{N,\ell}(du, d\lambda) \;=\; N^{-d} \sum_{x \in \mathbb{T}_N^d} \delta_{x/N}(du)\, \delta_{\eta^\ell(x)}(d\lambda)\,.$$

Thus, if $H: \mathbb{T}^d \times \mathbb{R}_+ \to \mathbb{R}$ is a smooth function, the integral of H with respect to $\pi^{N,\ell}(du, d\lambda)$, denoted by $< \pi^{N,\ell}, H >$, is equal to

$$< \pi^{N,\ell}, H > = \; N^{-d} \sum_{x \in \mathbb{T}_N^d} H(x/N, \eta^\ell(x))\,.$$

Denote by $\mathcal{M}_+(\mathbb{T}^d \times \mathbb{R}_+)$ the space of positive Radon measures on $\mathbb{T}^d \times \mathbb{R}_+$ endowed with the weak topology and fix a time $T \ge 0$. Denote by $L^\infty([0,T])$ the space of bounded functions from $[0,T]$ to $\mathbb{R}$ endowed with the weak topology. This topology is generated by the metric

$$d(f,g) \;=\; \sum_{k \ge 1} \frac{1}{2^k} \frac{\Big| \int_0^T dt\, h_k(t) g(t) - \int_0^T dt\, h_k(t) f(t) \Big|}{1 + \Big| \int_0^T dt\, h_k(t) g(t) - \int_0^T dt\, h_k(t) f(t) \Big|}\,,$$

where $\{h_k,\, k \ge 1\}$ is a dense sequence of functions in $L^1([0,T])$. Denote, furthermore, by $L^\infty([0,T], \mathcal{M}_+(\mathbb{T}^d \times \mathbb{R}_+))$ the space of bounded functions $\pi: [0,T] \to \mathcal{M}_+(\mathbb{T}^d \times \mathbb{R}_+)$ such that for each function F in $C_K(\mathbb{T}^d \times \mathbb{R}_+)$, $< \pi, F >$ belongs to $L^\infty([0,T])$. This space is endowed with the weak topology generated by the metric

$$d_{MV}(\pi, \tilde\pi) \;=\; \sum_{k \ge 1} \frac{1}{2^k} \frac{d(< \pi, F_k >, < \tilde\pi, F_k >)}{1 + d(< \pi, F_k >, < \tilde\pi, F_k >)}\,,$$

where $\{F_k,\, k \ge 1\}$ is a dense, with respect to the uniform topology, sequence of functions in $C_K(\mathbb{T}^d \times \mathbb{R}_+)$. In particular, a sequence π^n converges to some measure valued function π if $< \pi^n, F >$ converges weakly in $L^\infty([0,T])$ to $< \pi, F >$ for every F in $C_K(\mathbb{T}^d \times \mathbb{R}_+)$. For a sequence μ^N of probability measures on $\mathbb{N}^{\mathbb{T}_N^d}$ satisfying assumptions of Theorem 0.3, denote by $Q^{N,\ell} = Q^{N,\ell}_{\mu^N}$ the measure on the space $L^\infty([0,T], \mathcal{M}_+(\mathbb{T}^d \times \mathbb{R}_+))$ induced by the process $\pi_t^{N,\ell} := \pi^{N,\ell}(\eta_t)$ with generator (0.1) speeded up by N and starting from μ^N. Recall also that $\mathbb{P}^N_{\mu^N} = \mathbb{P}^N$ stands for the probability on $D([0,T], \mathbb{N}^{\mathbb{T}_N^d})$ corresponding to the process η_t with generator (0.1) speeded up by N and starting from μ^N. Expectation with respect to $\mathbb{P}^N$ is denoted by $\mathbb{E}_N$.

The following theorem is the main result of this section.

Theorem 1.1 *Let $\rho_0: \mathbb{T}^d \to \mathbb{R}_+$ be a Lipschitz continuous profile. For each $N \geq 1$, define μ^N as the product measure with marginals given by*

$$\mu^N\{\eta;\ \eta(x) = k\} \;=\; \nu^N_{\rho_0(x/N)}\{\eta;\ \eta(0) = k\} \qquad x \in \mathbb{T}^d_N\,,\ k \in \mathbb{N}\,.$$

Then, the sequence $Q^{N,\ell}$ converges, as $N \uparrow \infty$ and then $\ell \uparrow \infty$, to the probability measure concentrated on the entropy measure valued solution of equation (0.3).

We have seen in Chapter 4 that the proof of the hydrodynamic behavior of interacting particle systems comprehend essentially two steps. We first show that the sequence of probability measures is weakly relatively compact and then prove uniqueness of limit points, that is, that all converging subsequences converge to the same limit. We start with the tightness.

Lemma 1.2 *The sequence $Q^{N,\ell}$ is weakly relatively compact.*

Proof. It is well known that the unit ball of $L^\infty([0,T])$ for the strong topology is compact for the weak topology. In particular, for each $b > 0$, the sets $A_b = \{f \in L^\infty([0,T]),\ \|f\| \leq b\}$ are weakly compact.

Fix a countable, dense subset $\{G_k,\ k \geq 1\}$ of $C_K(\mathbb{T}^d \times \mathbb{R}_+)$. It is not difficult to check that for any sequence of finite numbers $\{C_k,\ k \geq 1\}$, the set

$$\Big\{\pi,\ \| < \pi, G_k > \|_\infty \leq C_k \text{ for all } k \geq 1\Big\}$$

is compact in $L^\infty([0,T], \mathcal{M}_+(\mathbb{T}^d \times \mathbb{R}_+))$. In particular, the tightness of the sequence $Q^{N,\ell}$ follows from the existence of finite constants $\{C(F),\ F \in C_K(\mathbb{T}^d \times \mathbb{R}_+)\}$ such that

$$\sup_{N,\ell} E_{Q^{N,\ell}}\Big[\Big\| < \pi, F > \Big\|_\infty\Big] \;\leq\; C(F)$$

for every F. This estimate in our case is trivial because, by definition, $\Big\| < \pi, F > \Big\|_\infty$ is trivially bounded by $\|F\|_\infty$. □

It remains to investigate the uniqueness of limit points.

Proof of Theorem 1.1. In view of Definition 0.5 and of Theorem 0.6 of section 1, to prove that a probability Q^* is concentrated on the entropy measure valued solution of equation (0.3) we have to show that Q^* almost surely

(i) For every $0 \leq t \leq T$, π_t is absolutely continuous on $\mathbb{T}^d$:

$$\pi_t(du, d\lambda) \;=\; \pi_t(u, d\lambda)\, du\,.$$

(ii) For every $0 \leq t \leq T$, π_t is concentrated on a compact set: there exists $\lambda_0 > 0$ such that

$$\pi_t(u, [0, \lambda_0]^c) \;=\; 0$$

for all (t, u) in $[0,T] \times \mathbb{T}^d$.

(iii) Entropy inequality: for every $c \in \mathbb{R}$

$$\partial_t < \pi_t, |\lambda - c| > + \sum_{i=1}^{d} \gamma_i \partial_{u_i} < \pi_t, |\Phi(\lambda) - \Phi(c)| > \leq 0$$

on $(0, \infty) \times \mathbb{T}^d$ in the sense of distributions.

(iv) Convergence in $L^1(\mathbb{T}^d)$ to the initial data:

$$\liminf_{t \to 0} \frac{1}{t} \int_0^t ds \int du < \pi_s, |\lambda - \rho_0(u)| > = 0 .$$

For each fixed integer ℓ, denote by $\mathcal{Q}^{*,\ell}$ the set of probability measures on $L^\infty([0,T], \mathcal{M}_+(\mathbb{T}^d \times \mathbb{R}_+))$ that are limit points of the sequence $Q^{N,\ell}$ with $N \uparrow \infty$. Denote also by $\mathcal{Q}^*$ the set of probability measures that are limit points of sequences $\{Q^{*,\ell}, \ell \geq 1\}$ such that, $Q^{*,\ell}$ belongs to $\mathcal{Q}^{*,\ell}$ for each ℓ. Fix Q^* in $\mathcal{Q}^*$. We shall prove that Q^* is concentrated on paths satisfying (i)-(iv).

We start with property (i). Fix $G: \mathbb{R}_+ \to \mathbb{R}$ and $H : \mathbb{T}^d \to \mathbb{R}$ two bounded continuous functions. From the definition of $\pi_t^{N,\ell}$,

$$\Big| < \pi_t^{N,\ell}, G(\lambda)H(u) > \Big| = \Big| N^{-d} \sum_{x \in \mathbb{T}_N^d} H(x/N) G(\eta_t^\ell(x)) \Big|$$
$$\leq \|G\|_\infty N^{-d} \sum_{x \in \mathbb{T}_N^d} |H(x/N)| .$$

Since H is continuous, the right hand side converges to $\|G\|_\infty \int |H(u)|\, du$, as $N \uparrow \infty$. In particular,

$$\lim_{N \to \infty} Q^{N,\ell} \Big[\sup_{0 \leq t \leq T} \big| < \pi_t, G(\lambda)H(u) > \big| \leq \|G\|_\infty \int |H(u)|\, du \Big] = 1 .$$

Since $\|\cdot\|_\infty$ is lower semicontinuous for the weak topology of $L^\infty([0,T])$, the above set is closed. Therefore, Q^* a.s.

$$\sup_{0 \leq t \leq T} \big| < \pi_t, G(\lambda)H(u) > \big| \leq \|G\|_\infty \int |H(u)|\, du .$$

This shows that Q^* a.s. the first coordinate of π_t is absolutely continuous:

$$\pi_t(du, d\lambda) = \pi_t(u, d\lambda) du .$$

Statement (ii) is a simple consequence of attractiveness and the boundness of the initial profile. Indeed, for every $A > 0$,

$$E_{Q^{N,\ell}} \Big[\int_0^T < \pi_t, \mathbf{1}\{|\lambda| \geq A\} > dt \Big] = \mathbb{E}_N \Big[\int_0^T N^{-d} \sum_{x \in \mathbb{T}_N^d} \mathbf{1}\{\eta_t^\ell(x) \geq A\} \Big] .$$

To avoid minor technical difficulties we assumed in this chapter the jump rate to be bounded. It follows from the boundness of g that the equilibrium measures do not have all exponential moments finite ($E_{\nu_\alpha^N}[\exp\{\theta\eta(0)\}] = \infty$ for $\theta > \log\{\sup_k g(k)\} - \log\Phi(\alpha)$). This characteristic of bounded jump rate zero range processes prevent us in using the entropy inequality in order to show that the last expectation is small for A large. Coupling arguments, available due to the attractiveness assumption, replace the entropy inequality.

Recall that by assumption $\mu^N \le \nu_{\alpha^*}^N$. Denote by S_t^N the semigroup of the Markov process (η_t) with generator L_N speeded up by N. By the attractiveness assumption and since $\nu_{\alpha^*}^N$ is invariant $\mu^N S_t^N \le \nu_{\alpha^*}^N$. Since $\mathbf{1}\{\eta^\ell(x) \ge A\}$ is an increasing function last expectation is bounded above by

$$T\,\nu_{\alpha^*}^N[\eta^\ell(0) \ge A]\ .$$

For $A > \alpha^*$, by the law of large numbers, this expression converges to 0 as ℓ increases to ∞. We proved therefore that for A sufficiently large,

$$\limsup_{\ell\to\infty}\limsup_{N\to\infty} E_{Q^{N,\ell}}\Big[\int_0^T < \pi_t, \mathbf{1}\{|\lambda| \ge A\} > dt\Big] \;=\; 0\ .$$

Since the application $\pi_\cdot \to \int_0^T < \pi_t, \mathbf{1}\{|\lambda| > A\} > dt$ is lower semicontinuous,

$$E_{Q^*}\Big[\int_0^T < \pi_t, \mathbf{1}\{|\lambda| > A\} > dt\Big] \;=\; 0$$

for A sufficiently large. Thus

$$\int_0^T dt \int du\, \pi_t(u, [0,A]^c) \;=\; 0$$

Q^* a.s. Redefining, if necessary, $\pi_t(u, d\lambda)$ in a subset of measure 0 of $[0,T]\times\mathbb{T}^d$ we conclude the proof of (ii).

We now turn to property (iii). In the next section we prove the following entropy inequality at the microscopic level:

Theorem 1.3 *Let μ^N be a sequence of probability measures bounded by some equilibrium state $\nu_{\alpha^*}^N$: $\mu^N \le \nu_{\alpha^*}^N$. For every smooth positive function H with compact support in $(0,\infty)\times\mathbb{T}^d$, every constant $c \in \mathbb{R}$ and every $\varepsilon > 0$,*

$$\lim_{\ell\to\infty}\lim_{N\to\infty} \mathbb{P}_{\mu^N}^N\Big[\int_0^\infty dt\, N^{-d} \sum_{x\in\mathbb{T}_N^d} \Big\{\partial_t H(t,x/N)\Big|\eta_t^\ell(x) - c\Big| + \sum_{i=1}^d \gamma_i\Big(\partial_{u_i} H\Big)(t,x/N)\Big|\Phi(\eta_t^\ell(x)) - \Phi(c)\Big|\Big\} \ge -\varepsilon\Big] \;=\; 1\ .$$

Notice that in this theorem, which is the main step toward the proof of Theorem 1.1 we only require the initial state to be bounded above.

We claim that (iii) follows from this result. Indeed, in terms of the Young measure this result can be restated as

$$\lim_{\ell\to\infty}\lim_{N\to\infty} Q^{N,\ell}\Big[\int_0^T dt\,\Big\langle \pi_t, (\partial_t H)(t,u)|\lambda - c|\Big\rangle + \sum_{i=1}^d \gamma_i\Big\langle \pi_t, (\partial_{u_i} H)(t,u)|\Phi(\lambda)-\Phi(c)|\Big\rangle \ \geq\ -\varepsilon\Big] \;=\; 1$$

for all positive smooth functions $H : (0,T)\times\mathbb{T}^d \to \mathbb{R}_+$ with compact support, all $c \in \mathbb{R}$ and all $\varepsilon > 0$. Since Q^* is a weak limit point concentrated on absolutely continuous measures, from this result and property (ii) already proved, we obtain that

$$Q^*\Big[\int_0^T dt \int_{\mathbb{T}^d} du\,\Big\{(\partial_t H)(t,u) < \pi_t(u,d\lambda), |\lambda - c| > + \sum_{i=1}^d \gamma_i\,(\partial_{u_i} H)(t,u) < \pi_t(u,d\lambda), |\Phi(\lambda)-\Phi(c)| > \Big\} \ \geq\ -\varepsilon\Big] \;=\; 1\,.$$

Letting $\varepsilon \downarrow 0$ and since the statement is valid for every smooth function H we have that Q^* a.s.

$$\partial_t < \pi_t, |\lambda - c| > \, + \sum_{i=1}^d \gamma_i \partial_{u_i} < \pi_t, |\Phi(\lambda)-\Phi(c)| > \,\leq\, 0$$

on $(0,T)\times\mathbb{T}^d$ in the sense of distributions for every $c \in \mathbb{R}$.

It remains to prove (iv). Here is the unique point in the proof of Theorem 1.1 where we need the initial measure μ^N to be product. We claim that in order to prove (iv) it is enough to show that

$$\lim_{t\to 0}\limsup_{\ell\to\infty}\limsup_{N\to\infty} E_{Q^{N,\ell}}\Big[< \pi_t, |\lambda - \rho_0(u)| > \Big] \;=\; 0\,. \tag{1.1}$$

Indeed, it follows from (1.1) that

$$\lim_{t\to 0}\limsup_{\ell\to\infty}\limsup_{N\to\infty} E_{Q^{N,\ell}}\Big[\frac{1}{t}\int_0^t ds\, < \pi_s, |\lambda - \rho_0(u)| > \Big] \;=\; 0.$$

Since Q^* is concentrated on paths π_t such that $\pi_t(u, [0,\lambda_0]^c) = 0$ for large enough λ_0 and $\rho_0(\cdot)$ is a bounded profile, by weak convergence, for every $t > 0$,

$$E_{Q^*}\Big[\int_0^t ds\, < \pi_s, |\lambda - \rho_0(u)| > \Big] \leq \limsup_{\ell\to\infty}\limsup_{N\to\infty} E_{Q^{N,\ell}}\Big[\int_0^t ds\, < \pi_s, |\lambda - \rho_0(u)| > \Big]\,.$$

In particular,

$$\lim_{t\to 0} Q^*\Big[\frac{1}{t}\int_0^t ds < \pi_s, |\lambda - \rho_0(u)| > \Big] = 0$$

and, by the dominated convergence theorem,

$$Q^*\Big[\limsup_{t\to 0} \frac{1}{t}\int_0^t ds < \pi_s, |\lambda - \rho_0(u)| > \Big] = 0 .$$

We now turn to the proof of (1.1). Recall that we denoted by $\rho_0^N(x/N)$ the expected value of particles at site x for the measure μ^N: $\rho_0^N(x/N) = E_{\mu^N}[\eta(x)]$. From the definition of the empirical measure $\pi^{N,\ell}$, we have that

$$E_{Q^{N,\ell}}\Big[< \pi_t, |\lambda - \rho_0(u)| > \Big] = \mathbb{E}_N\Big[N^{-d}\sum_{x\in\mathbb{T}_N^d} |\eta_t^\ell(x) - \rho_0(x/N)|\Big] . \qquad (1.2)$$

We shall need to couple two and three coordinate processes starting from product measures. We first define the initial measures and the evolution. For each fixed N and sites y_1 and y_2 of $\mathbb{T}_N^d$, consider the product measure $\bar\mu_{y_1}^N$ (resp. $\bar\mu_{y_1,y_2}^N$) on $\big(\mathbb{N}^{\mathbb{T}_N^d}\big)^2$ (resp. $\big(\mathbb{N}^{\mathbb{T}_N^d}\big)^3$) with first marginal equal to μ^N, second marginal equal to $\nu_{\rho_0(y_1/N)}^N$ (resp. third marginal equal to $\nu_{\rho_0(y_2/N)}^N$) and ordered at each site z:

$$\bar\mu_{y_1}^N\{(\eta,\xi^{y_1});\ \eta(z)\le\xi^{y_1}(z)\} = \begin{cases} 1 & \text{if } \rho_0^N(z/N)\le\rho_0(y_1/N)\,, \\ 0 & \text{otherwise}\,; \end{cases}$$

$$\bar\mu_{y_1,y_2}^N\{(\eta,\xi^{y_1},\xi^{y_2});\ \eta(z)\le\xi^{y_1}(z)\} = \begin{cases} 1 & \text{if } \rho_0^N(z/N)\le\rho_0(y_1/N)\,, \\ 0 & \text{otherwise}\,; \end{cases}$$

an analogous inequality with ξ^{y_2} replacing ξ^{y_1} and

$$\begin{cases} \bar\mu_{y_1,y_2}^N\{(\eta,\xi^{y_1},\xi^{y_2});\ \xi^{y_1}\le\xi^{y_2}\} = 1 & \text{if } \rho_0(y_1/N)\le\rho_0(y_2/N)\,, \\ \bar\mu_{y_1,y_2}^N\{(\eta,\xi^{y_1},\xi^{y_2});\ \xi^{y_1}\ge\xi^{y_2}\} = 1 & \text{otherwise}\,. \end{cases}$$

In these last formulas and below ξ^{y_1} stands for the second coordinate and ξ^{y_2} for the third one in the three coordinates coupled process.

We let processes $(\eta_t, \xi_t^{y_1}, \xi_t^{y_2})$ and $(\eta_t, \xi_t^{y_1})$ evolve according to the basic coupling defined in section 2.5 speeded up by N. Though we defined it only for two coordinate models, it is straightforward to define a generator for a three coordinates processes that preserve the order. Denote by $\mathbb{P}_{\bar\mu_{y_1}^N}$ (resp. $\mathbb{P}_{\bar\mu_{y_1,y_2}^N}$) the probability measure induced by the basic coupling speeded up by N and the probability measure $\bar\mu_{y_1}^N$ (resp. $\bar\mu_{y_1,y_2}^N$). Expectation with respect to $\mathbb{P}_{\bar\mu_{y_1}^N}$ (resp. $\mathbb{P}_{\bar\mu_{y_1,y_2}^N}$) is denote by $\mathbb{E}_{\bar\mu_{y_1}^N}$ (resp. $\mathbb{E}_{\bar\mu_{y_1,y_2}^N}$).

Expectation (1.2) is bounded above by

$$N^{-d}\sum_{x\in\mathbb{T}_N^d}\mathbb{E}_{\bar\mu_x^N}\Big[|\eta_t^\ell(x) - \xi_t^{x,\ell}(x)|\Big] + N^{-d}\sum_{x\in\mathbb{T}_N^d}\mathbb{E}_{\bar\mu_x^N}\Big[|\xi_t^{x,\ell}(x) - \rho_0(x/N)|\Big] . \qquad (1.3)$$

In this formula, $\xi^{x,\ell}(y)$ stands for $(2\ell+1)^{-d}\sum_{|z-y|\le\ell}\xi^x(z)$. It is easy to show that the second sum vanishes in the limit, as $N\uparrow\infty$ and then $\ell\uparrow\infty$, since the second coordinate starts from the equilibrium state $\nu^N_{\rho_0(x/N)}$. More precisely, recall that the initial state μ^N is bounded above by $\nu^N_{\alpha^*}$. In particular $\rho_0(\cdot)$ is bounded above by α^*. Since ξ^x starts from the equilibrium state $\nu^N_{\rho_0(x/N)}$, the second sum is equal to

$$N^{-d} \sum_{x \in \mathbb{T}^d_N} E_{\nu^N_{\rho_0(x/N)}} \Big[|\eta^\ell(x) - \rho_0(x/N)| \Big] \;\le\; \sup_{\alpha \le \alpha^*} E_{\nu^N_\alpha} \Big[|\eta^\ell(0) - \alpha| \Big] \,.$$

Schwarz inequality shows that the right hand side is bounded above by

$$\ell^{-1/2} \sup_{\alpha \le \alpha^*} \Big\{ E_{\nu^N_\alpha} \big[(\eta(0) - \alpha)^2 \big] \Big\}^{1/2}$$

and thus converges to 0 as ℓ increases to ∞.

On the other hand, the first term of (1.3) is bounded by

$$N^{-d} \sum_{x \in \mathbb{T}^d_N} (2\ell + 1)^{-d} \sum_{|z-x| \le \ell} \mathbb{E}_{\bar\mu^N_x} \Big[|\eta_t(z) - \xi^x_t(z)| \Big] \,.$$

Using now the three coordinates coupling we may bound this expression by

$$\begin{aligned} & N^{-d} \sum_{x \in \mathbb{T}^d_N} \mathbb{E}_{\bar\mu^N_x} \Big[|\eta_t(x) - \xi^x_t(x)| \Big] \\ & \quad + N^{-d} \sum_{x \in \mathbb{T}^d_N} (2\ell + 1)^{-d} \sum_{|z-x| \le \ell} \mathbb{E}_{\bar\mu^N_{x,z}} \Big[|\xi^x_t(z) - \xi^z_t(z)| \Big] \,. \end{aligned} \tag{1.4}$$

By construction of the coupling measure $\bar\mu^N_{x,z}$, either $\xi^x_0 \le \xi^z_0$ or $\xi^x_0 \ge \xi^z_0$. Since the order is preserved by the dynamics, we may move the absolute value to outside the expectation. Since both marginal start from equilibrium, this term is equal to

$$N^{-d} \sum_{x \in \mathbb{T}^d_N} (2\ell + 1)^{-d} \sum_{|z-x| \le \ell} |\rho_0(x/N) - \rho_0(z/N)| \,.$$

Since the initial profile ρ_0 is Lipschitz continuous, this expression converges to 0 as $N \uparrow \infty$.

It remains to prove that the first term of (1.4) converges to 0. It is equal to

$$N^{-d} \sum_{x \in \mathbb{T}^d_N} \mathbb{E}_{\bar\mu^N_x} \Big[|\eta_0(x) - \xi^x_0(x)| \Big] + N^{-d} \sum_{x \in \mathbb{T}^d_N} \mathbb{E}_{\bar\mu^N_x} \Big[\int_0^t ds\, N \bar L_N |\eta_s(x) - \xi^x_s(x)| \Big] \,. \tag{1.5}$$

Here $\bar L_N$ stands for the generator of the basic coupling introduced in section 2.5. Since at each site the measure $\bar\mu^N_x$ is ordered, in the first expression we may exchange the absolute value and the expectation to obtain that this expected value is identically 0. On the other hand, since by a straightforward computation involving the generator of the coupled process, we have that

$$\bar{L}_N|\eta(x) - \xi^x(x)|$$
$$\leq -|g(\eta(x)) - g(\xi^x(x))| + \sum_y p(-y)|g(\eta(x+y)) - g(\xi^x(x+y))| \, ,$$

the second expression is bounded above by the time integral of

$$\begin{aligned} N^{1-d} \sum_{x\in\mathbb{T}_N^d} \sum_y p(-y)\mathbb{E}_{\bar\mu_x^N}\Big[|g(\eta_s(x+y)) - g(\xi_s^x(x+y))|\Big] \\ - N^{1-d} \sum_{x\in\mathbb{T}_N^d} \mathbb{E}_{\bar\mu_x^N}\Big[|g(\eta_s(x)) - g(\xi_s^x(x))|\Big] \, . \end{aligned} \tag{1.6}$$

Using once more the three coordinates coupled process, the first line is bounded above by

$$\begin{aligned} N^{1-d} \sum_{x\in\mathbb{T}_N^d} \sum_y p(-y)\mathbb{E}_{\bar\mu_{x,x+y}^N}\Big[|g(\xi_s^x(x+y)) - g(\xi_s^{x+y}(x+y))|\Big] \\ + N^{1-d} \sum_{x\in\mathbb{T}_N^d} \mathbb{E}_{\bar\mu_x^N}\Big[|g(\eta_s(x)) - g(\xi_s^x(x))|\Big] \, . \end{aligned}$$

The second line is compensated by the negative part of (1.6). Therefore, (1.5) is bounded above by

$$\int_0^t ds\, N^{1-d} \sum_{x\in\mathbb{T}_N^d} \sum_y p(-y)\mathbb{E}_{\bar\mu_{x,x+y}^N}\Big[|g(\xi_s^x(x+y)) - g(\xi_s^{x+y}(x+y))|\Big] \, .$$

Since the jump rate g is an increasing function and ξ^x and ξ^{x+y} are ordered, as we did before, we may move the absolute value outside the expectation and obtain, since both marginals start from equilibrium, that this integral is equal to

$$tN^{-d} \sum_{x\in\mathbb{T}_N^d} \sum_y p(-y)N|\Phi(\rho_0((x+y)/N)) - \Phi(\rho_0(x/N))| \, .$$

Since the transition probability $p(\cdot)$ is of finite range and ρ_0 is Lipschitz continuous, this expression is bounded above by $tC(\rho_0)$ and converges to 0 as t decreases to 0. This proves (iv). □

Remark 1.4 It is known that the entropy solution starting from a smooth initial profile remains smooth in a finite time interval. One may therefore use the relative entropy method, presented in Chapter 6, to prove statement (iv) for processes starting from smooth initial profiles and avoid all computations performed above.

2. An Entropy Inequality at Microscopic Level

Recall from Chapter 2 that we denote by $\overline{L}_N$ the generator of the basic coupling of two copies of zero range processes. This section is devoted to the proof of an entropy inequality at a microscopic level. In order to state this result for a probability $\overline{\mu}^N$ on the configuration space $\mathbb{N}^{\mathbb{T}_N^d} \times \mathbb{N}^{\mathbb{T}_N^d} = \mathcal{X}_N^2$ denote by $\overline{\mathbb{P}}_{\overline{\mu}^N}^N$ the probability measure on the path space $D([0,\infty), \mathcal{X}_N^2)$ corresponding to the Markov process (η_t, ξ_t) evolving according to the generator $\overline{L}_N$ defined above speeded up by N and starting from $\overline{\mu}^N$. Expectation with respect to $\overline{\mathbb{P}}_{\overline{\mu}^N}^N$ is denoted by $\overline{\mathbb{E}}_{\overline{\mu}^N}^N$. Furthermore, for a measure $\overline{\mu}^N$ on the product space $\mathcal{X}_N^2$, we denote by $\overline{\mu}_i^N$ its i–th marginal.

Theorem 2.1 *Let $\overline{\mu}^N$ be a measure with both marginals bounded by an invariant product measure $\nu_{\alpha_0}^N$:*

$$\overline{\mu}_i^N \;\le\; \nu_{\alpha_0}^N \tag{2.3}$$

for $i = 1, 2$ and some density α_0. Recall the definition of $\eta^\ell(x)$ given in (5.1.10). For every smooth positive function H in $C_K^{1,1}((0,\infty) \times \mathbb{T}^d)$ and every positive ε,

$$\lim_{\ell\to\infty} \lim_{N\to\infty} \overline{\mathbb{P}}_{\overline{\mu}^N}^N \Bigg[\int_0^\infty dt\, N^{-d} \sum_{x\in\mathbb{T}_N^d} \Big\{ \partial_t H(t,x/N) \Big|\eta_t^\ell(x) - \xi_t^\ell(x)\Big| + \sum_{i=1}^d \gamma_i (\partial_{u_i} H)(t,x/N) \Big| \Phi\big(\eta_t^\ell(x)\big) - \Phi\big(\xi_t^\ell(x)\big)\Big| \Big\} \ge -\varepsilon \Bigg] \;=\; 1\,.$$

Theorem 1.3 follows straightforwardly from this result:

Proof of Theorem 1.3. Fix a constant $c \ge 0$. Define the measure $\overline{\mu}_c^N$ on the product space $\mathcal{X}_N^2$ by

$$\overline{\mu}_c^N \;=\; \mu^N \otimes \nu_c^N\,.$$

Since ν_c^N is a translation invariant stationary measure, for every positive t_0

$$\overline{\mathbb{E}}_{\overline{\mu}^N} \Bigg[\int_0^{t_0} dt\, N^{-d} \sum_{x\in\mathbb{T}_N^d} \Big|\Phi(\xi_t^\ell(x)) - \Phi(c)\Big| \Bigg] \;=\; t_0 \nu_c^N \Big[\Big|\Phi(\xi^\ell(0)) - \Phi(c)\Big| \Big]\,.$$

This last expression converges to 0 as ℓ increases to infinity by the law of large numbers because $\Phi(\cdot)$ is Lipschitz continuous in virtue of Corollary 2.3.6. By the same reasons $\overline{\mathbb{E}}_{\overline{\mu}^N}[\int_0^{t_0} dt\, N^{-d} \sum_{x\in\mathbb{T}_N^d} |\xi_t^\ell(x) - c|]$ vanishes as $N \uparrow \infty$ and $\ell \uparrow \infty$. □

The proof of Theorem 2.1 is divided in several lemmas. We first prove that in the limit as $N \uparrow \infty$ the configurations η and ξ are ordered. Before stating this result, notice that assumption (2.3) on the initial measure $\overline{\mu}^N$ implies that the expected value of the mean density of particles is bounded:

$$\limsup_{N\to\infty} \mathbb{E}_{\mu^N}\left[N^{-d}\sum_{x\in\mathbb{T}_N^d}\Big\{\eta_0(x)+\xi_0(x)\Big\}\right] < \infty\,. \tag{2.1}$$

Next lemma requires only this weaker assumption on the sequence of initial measures.

Lemma 2.2 *Assume that the sequence of initial measure $\overline{\mu}^N$ satisfies hypothesis (2.1). Then, for every positive time T and every d–dimensional integer y,*

$$\lim_{N\to\infty} \overline{\mathbb{E}}^N_{\overline{\mu}^N}\left[\int_0^T dt\, N^{-d}\sum_{x\in\mathbb{T}_N^d} G_{x,x+y}(\eta_t,\xi_t)\right] = 0\,,$$

where, for two sites x and y in $\mathbb{Z}^d$, $G_{x,y}(\eta,\xi)$ is an indicator function equal to one if the configurations η and ξ are not ordered at sites x and y:

$$G_{x,y}(\eta,\xi) \;=\; \mathbf{1}\{\eta(x)<\xi(x)\,,\,\eta(y)>\xi(y)\} \;+\; \mathbf{1}\{\eta(x)>\xi(x)\,,\,\eta(y)<\xi(y)\}\,.$$

Proof. The proof is divided in 3 steps.

Step 1. We first derive from the fact that the total number of uncoupled particles decreases in time that

$$\lim_{N\to\infty} \overline{\mathbb{E}}^N_{\overline{\mu}^N}\Bigg[\int_0^T dt$$
$$N^{-d}\sum_{x,y} p(-y)\,\Big|g(\eta_t(x+y))-g(\xi_t(x+y))\Big|G_{x,x+y}(\eta_t,\xi_t)\Bigg] = 0\,.$$

Consider the martingale $M(t)$ vanishing at 0 given by

$$M(t) \;=\; N^{-d}\sum_{x\in\mathbb{T}_N^d}\Big|\eta_t(x)-\xi_t(x)\Big| \;-\; N^{-d}\sum_{x\in\mathbb{T}_N^d}\Big|\eta_0(x)-\xi_0(x)\Big|$$
$$-\int_0^t ds\,N^{-d}\sum_{x\in\mathbb{T}_N^d} N\overline{L}_N\Big|\eta_s(x)-\xi_s(x)\Big|\,.$$

Since the total number of uncoupled particles decreases in time, the time integral of the last formula has to be negative. In fact, a careful computation relying on the explicit form of the generator $N\overline{L}_N$ shows that the integrand is equal to

$$
\begin{aligned}
&- N^{-(d-1)} \sum_{x \in \mathbb{T}_N^d} \Big| g(\eta_t(x)) - g(\xi_t(x)) \Big| \\
&- N^{-(d-1)} \sum_{x,y} p(-y) G_{x,x+y}(\eta_t, \xi_t) \Big| g(\eta_t(x+y)) - g(\xi_t(x+y)) \Big| \\
&+ N^{-(d-1)} \sum_{x,y} p(-y) \Big[1 - G_{x,x+y}(\eta_t, \xi_t) \Big] \Big| g(\eta_t(x+y)) - g(\xi_t(x+y)) \Big| \\
&= -2N^{-(d-1)} \sum_{x,y} p(-y) G_{x,x+y}(\eta_t, \xi_t) \Big| g(\eta_t(x+y)) - g(\xi_t(x+y)) \Big| \, .
\end{aligned}
$$

Since the total number of particles is conserved by the dynamics, by assumption (2.1) on the initial measure, the expected value of the martingale $M(t)$ is well defined. Therefore, since $M(t)$ vanishes at time 0, we have that

$$
\begin{aligned}
&\overline{\mathbb{E}}^N_{\overline{\mu}^N} \Big[\int_0^T dt\, N^{-d} \sum_{x,y} p(-y) G_{x,x+y}(\eta_t, \xi_t) \Big| g(\eta_t(x+y)) - g(\xi_t(x+y)) \Big| \Big] \\
&\quad \le N^{-1} E_{\overline{\mu}^N} \Big[N^{-d} \sum_{x \in \mathbb{T}_N^d} \Big| \eta(x) - \xi(x) \Big| \Big] \, .
\end{aligned}
$$

A factor N^{-1} coming from the time renormalization of the dynamics appeared on the right hand side of the last inequality. This factor and assumption (2.1) permits to conclude the first step of the proof.

Step 2. We deduce from the previous step that the configurations η and ξ are ordered on sites x and $x + y$ such that $p(y) + p(-y) > 0$.

From Step 1 and since by assumption $g(\cdot)$ is bounded below by a strictly positive constant ($g(k) \ge g(1) > 0$), it follows that

$$
\lim_{N \to \infty} \overline{\mathbb{E}}^N_{\overline{\mu}^N} \Big[\int_0^T dt\, N^{-d} \sum_{x,y} p(-y) \Big\{ I_0^{x,y}(\eta_t, \xi_t) + J_0^{x,y}(\eta_t, \xi_t) \Big\} \Big] = 0 \, ,
$$

where, for a positive integer m,

$$
\begin{aligned}
I_m^{x,y}(\eta, \xi) &= \mathbf{1}\{ m = \eta(x+y) < \xi(x+y),\ \eta(x) > \xi(x) \} \\
J_m^{x,y}(\eta, \xi) &= \mathbf{1}\{ \eta(x+y) > \xi(x+y) = m,\ \eta(x) < \xi(x) \} \, .
\end{aligned}
$$

We now remove the condition $\eta_t(x+y) \wedge \xi_t(x+y) = 0$ in the left hand side of the last equality. For a positive integer m, let

$$
I_m(\eta, \xi) \;=\; N^{-d} \sum_{x,y} p(-y) \mathbf{1}\{ m = \eta(x+y) < \xi(x+y) ;\ \eta(x) > \xi(x) \} \, .
$$

We shall prove by induction that

$$
\lim_{N \to \infty} \overline{\mathbb{E}}^N_{\overline{\mu}^N} \Big[\int_0^T dt\, I_m(\eta_t, \xi_t) \Big] = 0
$$

for every positive m. We just proved this equality for $m = 0$. Consider now for two fixed d–dimensional integers x and y the mean-zero martingale $M_m^{x,y}(t)$ defined by

$$M_m^{x,y}(t) \;=\; I_m^{x,y}(\eta_t, \xi_t) - I_m^{x,y}(\eta_0, \xi_0) - \int_0^t N\overline{L}_N I_m^{x,y}(\eta_s, \xi_s)\, ds \,.$$

If we compute carefully the expression $\overline{L}_N I_m^{x,y}$ we obtain two different kinds of terms. The first ones are negative and correspond to jumps of $I_m^{x,y}(\eta_t, \xi_t)$ from 1 to 0 and the second ones are positive and correspond to jumps from 0 to 1. Notice that while the negative terms are equal to $I_m^{x,y}\overline{L}_N I_m^{x,y}$, the positive terms are given by $(1 - I_m^{x,y})\overline{L}_N I_m^{x,y}$. Among the positive terms there is one which comes from a jump of two coupled particles from site $x+y$ to another site, when site $x+y$ is occupied by $m+1$ η–particles and by at least $m+2$ ξ–particles and site x is occupied by more η–particles than ξ–particles. This jump happens at rate $g(m+1)N$. Therefore

$$\begin{aligned}(1 - I_m^{x,y})\overline{L}_N I_m^{x,y} \;&\ge\; g(m+1)\mathbf{1}\{m+1 = \eta(x+y) < \xi(x+y)\,;\; \eta(x) > \xi(x)\} \\ &=\; g(m+1) I_{m+1}^{x,y} \,.\end{aligned}$$

In particular, since $M_m^{x,y}(t)$ is a mean-zero martingale,

$$\begin{aligned} & g(m+1)N\overline{\mathbb{E}}_{\overline{\mu}^N}^N\Big[\int_0^T dt\, I_{m+1}^{x,y}(\eta_t, \xi_t)\Big] \\ & \quad \le\; \overline{\mathbb{E}}_{\overline{\mu}^N}^N\Big[I_m^{x,y}(\eta_T, \xi_T)\Big] \;-\; N\int_0^T dt\, \overline{\mathbb{E}}_{\overline{\mu}^N}^N\Big[I_m^{x,y} N\overline{L}_N I_m^{x,y}(\eta_t, \xi_t)\Big] \,. \end{aligned} \tag{2.2}$$

A simple computation shows that $-I_m^{x,y}\overline{L}_N I_m^{x,y}(\eta, \xi)$ is equal to $I_m^{x,y}(\eta, \xi)$ times

$$\begin{aligned} g(m) \;&+\; \Big[g(m+1) - g(m)\Big]\mathbf{1}\{\xi(x+y) = m+1\} \\ &+\; \Big[g(\eta(x)) - g(\xi(x))\Big]\mathbf{1}\{\eta(x) = \xi(x)+1\} \\ &\quad+\; \sum_z p(x-z)\Big[g(\xi(z)) - g(\eta(z))\Big]\mathbf{1}\{\eta(z) < \xi(z);\; \eta(x) = \xi(x)+1\} \\ &\qquad+\; \sum_z p(x+y-z) g(\eta(z)) \end{aligned}$$

Therefore, multiplying inequality (2.2) by $p(-y)N^{-(d+1)}$, summing over all x and y and keeping in mind that $g(\cdot)$ is bounded by $g(\infty)$, we obtain that

$$\begin{aligned} & g(m+1)\int_0^T dt\, \overline{\mathbb{E}}_{\overline{\mu}^N}^N\Big[I_{m+1}(\eta_t, \xi_t)\Big] \\ & \quad \le\; N^{-1}\overline{\mathbb{E}}_{\overline{\mu}^N}^N\Big[I_m(\eta_T, \xi_T)\Big] \;+\; 4g(\infty)\int_0^T dt\, \overline{\mathbb{E}}_{\overline{\mu}^N}^N\Big[I_m(\eta_t, \xi_t)\Big] \,. \end{aligned}$$

Since $I_m(\eta, \xi)$ is bounded by 1 it follows by induction that

$$\lim_{N\to\infty} \int_0^T dt\, \overline{\mathbb{E}}_{\overline{\mu}^N}^N \Big[I_m(\eta_t, \xi_t) \Big] = 0 \tag{2.3}$$

for every positive integer m. We prove In the same way this result with the roles of η and ξ interchanged.

We are now ready to prove that

$$\lim_{N\to\infty} \int_0^T dt\, \overline{\mathbb{E}}_{\overline{\mu}^N}^N \Big[N^{-d} \sum_{x\in\mathbb{T}_N^d} G_{x,x+y}(\eta_t, \xi_t) \Big] = 0$$

for every y such that $p(y)+p(-y) > 0$. Fix such integer y and assume without loss of generality that $p(y) > 0$. The time integral of last formula is bounded above by

$$p(y)^{-1} \int_0^T dt\, \overline{\mathbb{E}}_{\overline{\mu}^N}^N \Big[N^{-d} \sum_{x,z} p(z) G_{x,x+z}(\eta_t, \xi_t) \Big]$$

$$= p(y)^{-1} \int_0^T dt\, \overline{\mathbb{E}}_{\overline{\mu}^N}^N \Big[N^{-d} \sum_{x,z} p(-z) G_{x,x+z}(\eta_t, \xi_t) \Big] .$$

For every integer m, $N^{-d} \sum_{x,z} p(-z) G_{x,x+z}(\eta, \xi)$ is bounded above by

$$\sum_{n=0}^{m} \Big\{ I_m(\eta, \xi) + I_m(\xi, \eta) \Big\} + \frac{1}{m} N^{-d} \sum_{x,z} p(-z) \Big\{ \eta(x+z) + \xi(x+z) \Big\} .$$

The claim follows therefore from (2.3), from conservation of total number of particles and from assumption (2.1) on the initial measure.

Step 3 It remains now to remove the assumption $p(y) + p(-y) > 0$. Fix z in $\mathbb{Z}^d$. Since $p(\cdot)$ is irreducible, there exists $m \ge 1$ and a sequence $0 = x_0, x_1, \dots, x_m = z$ such that $p(x_{i+1} - x_i) + p(x_i - x_{i+1}) > 0$ for $0 \le i \le m-1$. Denote by $M(z)$ the length of the smallest path linking the origin to z. We shall prove that

$$\lim_{N\to\infty} \overline{\mathbb{E}}_{\overline{\mu}^N}^N \Big[\int_0^T dt\, N^{-d} \sum_{x\in\mathbb{T}_N^d} G_{x,x+z}(\eta_t, \xi_t) \Big] = 0 \tag{2.4}$$

by induction in M. In Step 2 we proved the above statement for paths of length $m = 1$. Assume that this statement is true for all $n \le m$. Fix z in $\mathbb{Z}^d$ such that $M(z) = m+1$ and a path $0 = x_0, x_1, \dots, x_{m+1} = z$ from the origin to z of length $m+1$. Assume, without loss of generality, that $p(x_m - z) > 0$. If $p(x_m - z) = 0$, we would consider the path $x_m - z, 0, x_1, \dots, x_m$ and replace in the proof below x_m by x_0, z by $x_m - z$ and x_0 by x_m.

By the martingale argument presented in steps 1 and 2,

$$\overline{\mathbb{E}}_{\overline{\mu}^N}^N \Big[N^{-(d+1)} \sum_{x\in\mathbb{T}_N^d} I^{x,x_m}(\eta_T, \xi_T) \Big] - \overline{\mathbb{E}}_{\overline{\mu}^N}^N \Big[N^{-(d+1)} \sum_{x\in\mathbb{T}_N^d} I^{x,x_m}(\eta_0, \xi_0) \Big]$$

$$= \overline{\mathbb{E}}_{\overline{\mu}^N}^N \Big[\int_0^T dt\, N^{-d} \sum_{x\in\mathbb{T}_N^d} \overline{L}_N I^{x,x_m}(\eta_t, \xi_t) \Big] ,$$

where

$$I^{x,y}(\eta_t, \xi_t) = \mathbf{1}\{\eta(x) < \xi(x);\ \eta(x+y) > \xi(x+y)\} .$$

The left hand side of the previous identity vanishes as $N \uparrow \infty$ and therefore the right hand side. We may decompose $\overline{L}_N I^{x,x_m}$ in positive and negative terms. The positive terms are $(1 - I^{x,x_m})\overline{L}_N I^{x,x_m}$ and the negative terms are $I^{x,x_m}\overline{L}_N I^{x,x_m}$. A simple computation similar to the one performed in the second step shows that the negative terms are bounded in absolute value by a constant depending only on g multiplied by I^{x,x_m}. In particular, by the induction hypothesis, the expectation of the negative terms vanishes as $N \uparrow \infty$. Therefore,

$$\lim_{N\to\infty} \overline{\mathbb{E}}^N_{\overline{\mu}^N}\Big[\int_0^T dt\, N^{-d} \sum_{x\in\mathbb{T}_N^d} [1 - I^{x,x_m}(\eta_t,\xi_t)]\overline{L}_N I^{x,x_m}(\eta_t,\xi_t)\Big] = 0 .$$

Since $p(x_m - z)$ is positive, the positive expression $[1 - I^{x,x_m}]\overline{L}_N I^{x,x_m}$ is bounded below by $p(x_m - z)[g(\eta(x+z)) - g(\xi(x+z))]\mathbf{1}\{\eta(x) < \xi(x), \eta(x+x_m) = \xi(x+x_m), \eta(x+z) > \xi(x+z)\}$. This indicator function can be written as

$$\begin{aligned}
&\mathbf{1}\{\eta(x) < \xi(x), \eta(x+z) > \xi(x+z)\}\\
&\quad - \mathbf{1}\{\eta(x) < \xi(x), \eta(x+x_m) > \xi(x+x_m), \eta(x+z) > \xi(x+z)\}\\
&\quad\quad - \mathbf{1}\{\eta(x) < \xi(x), \eta(x+x_m) < \xi(x+x_m), \eta(x+z) > \xi(x+z)\} .
\end{aligned}$$

By the induction assumption, the expectation of the average of the second and third terms vanishes as $N \uparrow \infty$. Therefore, since $g(k) \geq g(1) > 0$ for $k \geq 1$,

$$\lim_{N\to\infty} \overline{\mathbb{E}}^N_{\overline{\mu}^N}\Big[\int_0^T dt\, N^{-d} \sum_{x\in\mathbb{T}_N^d} \mathbf{1}\{\eta_t(x) < \xi_t(x), \eta_t(x+z) > \xi_t(x+z) = 0\}\Big] = 0 .$$

We now repeat the arguments presented in the previous step to remove the condition $\xi(x+z) = 0$. We may of course repeat the proof with the roles of η and ξ interchanged to conclude the proof of the lemma. □

This ordering of coordinates made by the process permits to replace averages of absolute values of differences of monotone functions by absolute values of averages. This statement is made clear in the next lemma.

Recall from Chapter 2 that a cylinder function is said to be Lipschitz if there exists a finite subset Λ of $\mathbb{Z}^d$ and a constant $C(\Psi)$ such that

$$\big|\Psi(\eta) - \Psi(\xi)\big| \leq C(\Psi) \sum_{x\in\Lambda} \big|\eta(x) - \xi(x)\big|$$

for all configurations η and ξ. Notice that for all Lipschitz cylinder functions Ψ there exists a constant $C'(\Psi)$ and a finite subset Λ of $\mathbb{Z}^d$ such that

$$\big|\Psi(\eta)\big| \leq C'(\Psi)\Big(1 + \sum_{x\in\Lambda} \eta(x)\Big)$$

for all configurations η. On the other hand, every monotone cylinder function is bounded below by $\Psi(\underline{0})$, if $\underline{0}$ represents the configuration without particle: $\underline{0}(x) = 0$ for every site x.

Lemma 2.3 *Let $\overline{\mu}^N$ be a sequence of measures satisfying the assumption stated in Theorem 2.1. Let Ψ be a monotone Lipschitz function. Then, for every positive integer ℓ and every positive time T,*

$$\lim_{N\to\infty} \overline{\mathbb{E}}^N_{\overline{\mu}^N} \Big[\int_0^T dt\, N^{-d} \sum_{x\in\mathbb{T}^d_N} \tau_x V_\Psi(\eta_t, \xi_t) \Big] = 0 ,$$

where $V_\Psi(\eta, \xi)$ is the positive function defined by

$$V_\Psi(\eta,\xi) = (2\ell+1)^{-d} \sum_{|y|\le\ell} \big|\tau_y\Psi(\eta) - \tau_y\Psi(\xi)\big| - \Big|(2\ell+1)^{-d} \sum_{|y|\le\ell} \big[\tau_y\Psi(\eta) - \tau_y\Psi(\xi)\big]\Big| .$$

Proof. In the case where Ψ is a bounded function this result is an immediate consequence of the previous lemma. Indeed, for a finite subset Λ of $\mathbb{Z}^d$, let $\Lambda_{D,\ell}$ be the subset of $\mathbb{Z}^d$ consisting of all integers at a distance smaller than ℓ from Λ: $\Lambda_{D,\ell} = \{y \in \mathbb{Z}^d;\ \exists x \in \Lambda,\ |x-y| \le \ell\}$. Define $G_{\Lambda,\ell}(\eta,\xi)$ as the indicator function equal to 1 if η and ξ are not ordered at $\Lambda_{D,\ell}$:

$$G_{\Lambda,\ell}(\eta,\xi) = 1 - \prod_{x,y\in\Lambda_{D,\ell}} \Big(1 - G_{x,y}(\eta,\xi)\Big) .$$

If the configurations η and ξ are ordered on the set $\Lambda_{D,\ell}$ translated by x $\tau_x V_\Psi(\eta,\xi)$ vanishes because Ψ is monotone. Therefore to prove the lemma for bounded functions Ψ it is enough to show that

$$\lim_{N\to\infty} \overline{\mathbb{E}}^N_{\overline{\mu}^N} \Big[\int_0^T dt\, N^{-d} \sum_{x\in\mathbb{T}^d_N} \tau_x G_{\Lambda,\ell}(\eta_t, \xi_t) \Big] = 0$$

and this follows from the previous lemma since $G_{\Lambda,\ell}(\eta,\xi)$ is bounded by

$$\sum_{x,y\in\Lambda_{D,\ell}} G_{x,y}(\eta,\xi) .$$

We now turn to the general case. Recall that a monotone Lipschitz function is bounded below (by $\Psi(\underline{0})$). The idea is to reduce the general case to the bounded case by means of a cutoff. Thus, for a real positive A, let Ψ_A be the cutoff of Ψ at level A:

$$\Psi_A(\eta) = \Psi(\eta) \wedge A .$$

The expected value which appears in the statement of the lemma is bounded above by

$$\overline{\mathbb{E}}^N_{\overline{\mu}^N}\Big[\int_0^T dt\, N^{-d}\sum_{x\in\mathbb{T}^d_N}\tau_x V_{\Psi_A}(\eta_t,\xi_t)\Big]$$
$$+\ 2\overline{\mathbb{E}}^N_{\overline{\mu}^N}\Big[\int_0^T dt\, N^{-d}\sum_{x\in\mathbb{T}^d_N}\Big\{\Big(\tau_x\Psi(\eta_t)-A\Big)^+ + \Big(\tau_x\Psi(\xi_t)-A\Big)^+\Big\}\Big]\,.$$

For every A the first term converges to 0 as N increases to ∞ by the first part of the proof. On the other hand, since both marginal of the initial measure $\overline{\mu}^N$ are bounded by the translation invariant measure $\nu^N_{\alpha_0}$ and since $\Big(\Psi(\eta)-A\Big)^+$ is an increasing function, the second term is bounded by

$$4E_{\nu_{\alpha_0}}\Big[N^{-d}\sum_{x\in\mathbb{T}^d_N}\Big(\tau_x\Psi(\eta)-A\Big)^+\Big]\ \le\ 4E_{\nu_{\alpha_0}}\Big[\Psi(\eta)-\Psi(\eta)\wedge A\Big]\,.$$

This expected value converges to 0 as A increases to ∞ by the dominated convergence theorem since Ψ is Lipschitz and therefore ν_{α_0}–integrable. □

The third result towards the proof of Theorem 2.1 is a one block estimate for the uncoupled process. This result is stated in section 5.4. Notice that all zero range processes considered in this chapter satisfy assumption (**SLG**) because $g(\cdot)$ is bounded.

We are now ready to prove the entropy inequality at the microscopic level.

Proof of Theorem 2.1. Fix a smooth positive function H with compact support in $(0,\infty)\times\mathbb{T}^d$. Let M^H_t be the martingale vanishing at 0 defined by

$$\begin{aligned} M^H_t &= N^{-d}\sum_{x\in\mathbb{T}^d_N} H(t,x/N)\Big|\eta_t(x)-\xi_t(x)\Big| \\ &- \int_0^t \Big(\partial_s + N\overline{L}_N\Big)N^{-d}\sum_{x\in\mathbb{T}^d_N} H(s,x/N)\Big|\eta_s(x)-\xi_s(x)\Big|\,ds\,. \end{aligned}\tag{2.5}$$

We will show in Lemma 2.4 below that under assumption (2.1) the expected value of the square of the martingale converges to 0:

$$\lim_{N\to\infty}\overline{\mathbb{E}}^N_{\overline{\mu}^N}\Big[\Big(M^H_t\Big)^2\Big] = 0\,.$$

From Chebychev inequality we obtain that for every $t\ge 0$ and every positive ε,

$$\lim_{N\to\infty}\overline{\mathbb{P}}^N_{\overline{\mu}^N}\Big[\Big|M^H_t\Big|>\varepsilon\Big] = 0\,. \tag{2.6}$$

On the other hand, since H has compact support, for sufficiently large t, the martingale M_t^H is equal to

$$-\int_0^\infty \Big(\partial_s + N\overline{L}_N\Big) N^{-d} \sum_{x\in\mathbb{T}_N^d} H(s,x/N)\Big|\eta_s(x) - \xi_s(x)\Big|\, ds \,.$$

A straightforward computation shows that this integral is equal to

$$\begin{aligned}-\int_0^\infty N^{-d} \sum_{x\in\mathbb{T}_N^d} \Big\{ &\partial_s H(s,x/N)\Big|\eta_s(x) - \xi_s(x)\Big| \\ &+ \sum_{i=1}^d \gamma_i \big(\partial_{u_i} H\big)(s,x/N)\Big|g(\eta_s(x)) - g(\xi_s(x))\Big|\Big\}\, ds \\ &+ R(H,\eta_\cdot,\xi_\cdot) + O(1/N)\,.\end{aligned}$$

Here R is a positive term and a remainder of order $O(1/N)$ appeared when we replaced the discrete partial derivative by the usual one. Therefore, for sufficiently large t, the martingale M_t^H is bounded below by

$$\begin{aligned}-\int_0^\infty N^{-d} \sum_{x\in\mathbb{T}_N^d} \Big\{ &\partial_s H(s,x/N)\Big|\eta_s(x) - \xi_s(x)\Big| \\ &+ \sum_{i=1}^d \gamma_i \big(\partial_{u_i} H\big)(s,x/N)\Big|g(\eta_s(x)) - g(\xi_s(x))\Big|\Big\}\, ds \; - \; O(1/N)\,.\end{aligned}$$

By assumption (2.1), for every continuous function $G\colon \mathbb{R}_+ \times \mathbb{T}^d \to \mathbb{R}$ with compact support,

$$\begin{aligned}\lim_{N\to\infty} \overline{\mathbb{E}}^N_{\overline{\mu}^N} \Bigg[\int_0^\infty dt\, N^{-d} \sum_{x\in\mathbb{T}_N^d} \Big[G(t,x/N) - (2\ell+1)^{-d} \sum_{|y-x|\le\ell} G(t,y/N)\Big] \times \\ \times \Big\{ \Big|\eta_t(x) - \xi_t(x)\Big| + \Big|g(\eta_t(x)) - g(\xi_t(x))\Big|\Big\}\Bigg] = 0\,.\end{aligned}$$

Applying this result to the functions $\partial_t H$ and $\partial_{u_i} H$ and making a discrete integration by parts we obtain that for t sufficiently large, the martingale is bounded below by

$$\begin{aligned}-\int_0^\infty N^{-d} \sum_{x\in\mathbb{T}_N^d} \Big\{ &\partial_s H(s,x/N)(2\ell+1)^{-d} \sum_{|y-x|\le\ell} \Big|\eta_s(y) - \xi_s(y)\Big| \\ &+ \sum_{i=1}^d \gamma_i \big(\partial_{u_i} H\big)(s,x/N)(2\ell+1)^{-d} \sum_{|y-x|\le\ell} \Big|g(\eta_s(y)) - g(\xi_s(y))\Big|\Big\}\, ds \\ &- o_N(1)\,.\end{aligned}$$

Therefore, from (2.6) and Lemma 2.3, we obtain that for every $\varepsilon > 0$,

$$\lim_{N\to\infty} \overline{\mathbb{P}}^N_{\overline{\mu}^N} \left[\int_0^\infty N^{-d} \sum_{x\in\mathbb{T}^d_N} \left\{ \partial_s H(s,x/N) \Big| (2\ell+1)^{-d} \sum_{|y-x|\le\ell} [\eta_s(y) - \xi_s(y)] \Big| \right.\right.$$
$$\left.\left. + \sum_{i=1}^d \gamma_i (\partial_{u_i} H)(s,x/N)\tau_x V_\ell(\eta_s) \right\} ds < -\varepsilon \right] = 0 ,$$

where

$$V_\ell(\eta) = \Big| (2\ell+1)^{-d} \sum_{|y|\le\ell} [g(\eta_s(y)) - g(\xi_s(y))] \Big| .$$

Finally applying the one block estimate to the function $g(\eta(0))$ and recalling that the expectation of this function with respect to the product measure ν^N_α is $\Phi(\alpha)$, we conclude the proof of the theorem. □

Lemma 2.4 *For a smooth positive function $H: (0,\infty)\times\mathbb{T}^d \to \mathbb{R}$, let M^H_t be the martingale defined by (2.5). Then,*

$$\lim_{N\to\infty} \overline{\mathbb{E}}^N_{\overline{\mu}^N} \left[\left(M^H_t \right)^2 \right] = 0 .$$

Proof. From Lemma A1.5.1 $N^H(t)$ given by

$$N^H(t) = (M^H(t))^2 - \int_0^t ds \left\{ N\overline{L}_N \left(A^H(s) \right)^2 - 2A^H(s) N\overline{L}_N \left(A^H(s) \right) \right\}$$

is a martingale provided

$$A^H(t) = N^{-d} \sum_{x\in\mathbb{T}^d_N} H(t,x/N) \Big| \eta_t(x) - \xi_t(x) \Big| .$$

A straightforward computation shows that the expression inside braces is equal to

$$N^{1-2d} \sum_{x,y} |g(\eta_s(x)) - g(\xi_s(x))| p(y)\, G_{x,x+y}(\eta_s,\xi_s) [\nabla^+_{N,y} H_s(x/N)]^2$$
$$+ N^{1-2d} \sum_{x,y} |g(\eta_s(x)) - g(\xi_s(x))| p(y) \left[1 - G_{x,x+y} \right] [\nabla^-_{N,y} H_s(x/N)]^2 ,$$

where $\nabla^\pm_{N,y} H_s(x/N) = H(s,(x+y)/N) \pm H(s,x/N)$. The second line is of order $O(N^{-d-1})$ and therefore converges to 0 as N increases to ∞. The first line is of order $O(N^{1-d})$. Its expected value integrated in time converges to 0 in virtue of Lemma 2.2. □

3. Law of Large Numbers for the Empirical Measure

In this section we prove Theorem 0.3. The proof relies on the law of large numbers for the Young measure proved in section 1.

We first show that if the initial profile $\rho_0(\cdot)$ is Lipschitz continuous and the initial measure is product with marginals given by

$$\mu^N\{\eta;\ \eta(x)=k\} \;=\; \nu^N_{\rho_0(x/N)}\{\eta;\ \eta(0)=k\} \qquad \text{for} \quad x\in\mathbb{T}^d_N\,, \quad k\in\mathbb{N}\,,$$

then a law of large numbers for the empirical measure follows from Theorem 1.1.

For each $N\ge 1$, denote by $Q^N = Q^N_{\mu^N}$ the probability measure on $D([0,T],\mathcal{M}_+(\mathbb{T}^d))$ induced by the empirical measure π^N_t speeded up by N and the sequence of initial measures μ^N.

Lemma 3.1 *The sequence Q^N is tight. Moreover, all limit points are concentrated on weakly continuous paths $\pi(t,du)$ that are absolutely continuous with respect to the Lebesgue measure: $\pi(t,du)=\rho(t,u)du$.*

This lemma is proved essentially in the same way as the tightness is proved in the diffusive case presented in Chapter 5. We leave the details to the reader.

We now claim that it follows from Theorem 1.1 that all limit points of the sequence Q^N are concentrated on entropy solutions of the hyperbolic equation (0.3). Indeed, denote by $\rho(t,u)$ the entropy solution of (0.3), fix a smooth function $H:[0,T]\times\mathbb{T}^d\to\mathbb{R}$ of class $C^{1,2}([0,T]\times\mathbb{T}^d)$ and $\varepsilon>0$. By definition of the empirical measure,

$$Q^N\Big[\Big|\int_0^T dt\ <\pi_t,H_t> - \int_0^T dt\int du\, H(t,u)\rho(t,u)\Big| > \varepsilon\Big] \;=$$
$$\mathbb{P}_N\Big[\Big|\int_0^T dt\, N^{-d}\sum_{x\in\mathbb{T}^d_N} H(t,x/N)\eta_t(x) - \int_0^T dt\int du\, H(t,u)\rho(t,u)du\Big| > \varepsilon\Big]\,.$$

By a discrete integration by parts,

$$N^{-d}\sum_{x\in\mathbb{T}^d_N} H(t,x/N)\eta_t(x) \;=\; N^{-d}\sum_{x\in\mathbb{T}^d_N} H(t,x/N)\eta^\ell_t(x) \;+\; r_{N,\ell}(\eta_t)$$

where the remainder $r_{N,\ell}(\eta_t)$ has absolute value bounded above by

$$\frac{\ell^2}{N^2}C(H)N^{-d}\sum_{x\in\mathbb{T}^d_N}\eta_t(x)\,.$$

Since the total number of particles is conserved, by Chebychev inequality,

$$\mathbb{P}_N\Big[|r_{N,\varepsilon}| > \varepsilon/2\Big] \;\le\; \frac{2\ell^2}{\varepsilon N^2}C(H)\alpha^*$$

if α^* is an upper bound for the initial profile ρ_0. Therefore,

$$\limsup_{N\to\infty} Q^N \Big[\Big| \int_0^T dt \, < \pi_t, H_t > - \int_0^T dt \int du \, H(u)\rho(t,u) \Big| > \varepsilon \Big]$$

$$\le \limsup_{N\to\infty} \mathbb{P}_N \Big[\Big| \int_0^T dt \, N^{-d} \sum_{x\in\mathbb{T}_N^d} H(t,x/N)\eta_t^\ell(x) \tag{3.1}$$

$$- \int_0^T dt \int du \, H(t,u)\rho(t,u) \Big| > \varepsilon/2 \Big] .$$

for every ℓ in $\mathbb{N}$. Notice that the expressions inside the absolute value in the last probability may respectively be rewritten as the time integral of $< \pi_t^{N,\ell}, H(t,u)\lambda >$ and the time integral of $< \tilde\pi_t, H(t,u)\lambda >$, if $\tilde\pi_t = \delta_{\rho(t,u)}(d\lambda)du$ stands for the entropy measure valued solution of equation (0.3). Thus, with the notation introduced in section 1, last probability may be rewritten as

$$Q^{N,\ell} \Big[\Big| \int_0^T dt \, < \pi_t, H(t,u)\lambda > \, - \, \int_0^T dt \, < \tilde\pi_t, H(t,u)\lambda > \Big| > \varepsilon/2 \Big] .$$

By Theorem 1.1, this expression converges to 0 as $N \uparrow \infty$ and then $\ell \uparrow \infty$. In particular, we have from (3.1) that

$$\limsup_{N\to\infty} Q^N \Big[\Big| \int_0^T dt \, < \pi_t, H_t > - \int_0^T dt \int du \, H(u)\rho(t,u) \Big| > \varepsilon \Big] \; = \; 0 ,$$

what concludes the proof of the claim.

Finally, since by Lemma 3.1, all limit points are concentrated on weakly continuous paths, we have that for all $0 \le t \le T$, all continuous functions $H\colon \mathbb{T}^d \to \mathbb{R}$ and all $\varepsilon > 0$,

$$\limsup_{N\to\infty} Q^N \Big[\Big| < \pi_t, H > - \int du \, H(u)\rho(t,u) \Big| > \varepsilon \Big] \; = \; 0 .$$

To prove Theorem 0.3, it remains to extend this result to profiles in $L^\infty(\mathbb{T}^d)$. Fix $\rho_0 \in L^\infty(\mathbb{T}^d)$ and μ^N a sequence of product measures satisfying assumptions of Theorem 0.3.

Consider a sequence $\rho_\varepsilon : \mathbb{T}^d \to \mathbb{R}_+$ of Lipschitz continuous functions converging in $L^1(\mathbb{T}^d)$ to ρ_0. Denote by $\rho_\varepsilon(t,u)$ the entropy solution of equation (0.3) with initial data ρ_ε. It is known (cf. Theorem A2.5.11) that the $L^1(\mathbb{T}^d)$ norm of the difference of two entropy solutions of equation (0.3) decreases in time:

$$\int_{\mathbb{T}^d} du \, |\rho_\varepsilon(t,u) - \rho(t,u)| \; \le \; \int_{\mathbb{T}^d} du \, |\rho_\varepsilon(u) - \rho_0(u)| \, .$$

To each profile ρ_ε associate the product measure $\mu^{\varepsilon,N}$ with marginals given by

$$\mu^{\varepsilon,N}\{\eta;\ \eta(x) = k\} \; = \; \nu^N_{\rho_\varepsilon(x/N)}\{\eta;\ \eta(0) = k\} \, ,$$

for x in $\mathbb{T}_N^d$ and k in $\mathbb{N}$.

For each $\varepsilon > 0$, define the measure $\bar\mu^{\varepsilon,N}$ on $\mathbb{N}^{\mathbb{T}_N^d} \times \mathbb{N}^{\mathbb{T}_N^d}$ with first marginal equal to μ^N, second marginal equal to $\mu^{\varepsilon,N}$ and such that

$$\bar\mu^{\varepsilon,N}\{(\eta,\xi);\ \eta(x) \le \xi(x)\} \;=\; \begin{cases} 1 & \text{if } \rho_0^N(x/N) \le \rho_\varepsilon(x/N)\,, \\ 0 & \text{otherwise}\,. \end{cases}$$

Let (η_t, ξ_t) evolve according to the basic coupling defined in section 2.5 and denote by $\bar{\mathbb{P}}_{N,\varepsilon} = \bar{\mathbb{P}}_{\bar\mu^{\varepsilon,N}}$ the probability on the path space induced by the Markov process with generator $N\bar L_N$ and by the initial measure $\bar\mu^{\varepsilon,N}$. Fix $\delta > 0$ and a continuous function $H : \mathbb{T}^d \to \mathbb{R}$. From the definition of the empirical measure, we have

$$Q^N\Big[\Big| < \pi_t, H > - \int H(u)\rho(t,u)du\Big| > \delta\Big]$$
$$= \bar{\mathbb{P}}_{N,\varepsilon}\Big[\Big|N^{-d}\sum_{x\in\mathbb{T}_N^d} H(x/N)\eta_t(x) - \int H(u)\rho(t,u)du\Big| > \delta\Big]\,.$$

For ε such that $\int_{\mathbb{T}^d} du\,|\rho_\varepsilon(u) - \rho_0(u)| < (3\|H\|_\infty)^{-1}\delta$, the last probability is bounded above by

$$\bar{\mathbb{P}}_{N,\varepsilon}\Big[\Big|N^{-d}\sum_{x\in\mathbb{T}_N^d} H(x/N)[\eta_t(x) - \xi_t(x)]\Big| > \delta/3\Big]$$
$$+\ \bar{\mathbb{P}}_{N,\varepsilon}\Big[\Big|N^{-d}\sum_{x\in\mathbb{T}_N^d} H(x/N)\xi_t(x) - \int H(u)\rho_\varepsilon(t,u)du\Big| > \delta/3\Big]$$

because the $L^1(\mathbb{T}^d)$ norm of the difference of two entropy solutions decreases in time. By the first claim of this section and since $\rho_\varepsilon(u)$ is Lipschitz continuous, the second term converges to 0 as N increases to ∞ for each $\varepsilon > 0$. On the other hand, since the total number of uncoupled particles decreases in time, the first term is bounded above by

$$3\|H\|_\infty\delta^{-1}N^{-d}\sum_{x\in\mathbb{T}_N^d} E_{\bar\mu_{N,\varepsilon}}[|\eta(x) - \xi(x)|]\,.$$

Since at each site the measure $\bar\mu_{N,\varepsilon}$ is ordered, we may move the absolute value outside the expectation and obtain that this last sum is equal to

$$3\|H\|_\infty\delta^{-1}N^{-d}\sum_{x\in\mathbb{T}_N^d} |E_{\mu^N}[\eta(x)] - \rho_\varepsilon(x/N)|\,.$$

Recall from section 1 that we denoted by $\rho_0^N : \mathbb{T}^d \to \mathbb{R}_+$ the profile associated to the sequence μ^N:

$$\rho_0^N(x/N) \;=\; E_{\mu^N}[\eta(x)]$$

for x in $\mathbb{T}_N^d$. With this notation the last line is bounded above by

$$3\|H\|_\infty \delta^{-1}\Big\{\int du\,|\rho_0^N(u) - \rho_0(u)| + \int du\,|\rho_0(u) - \rho_\varepsilon(u)|\Big\}$$
$$+\ 3\|H\|_\infty \delta^{-1} N^{-d} \sum_{x\in\mathbb{T}_N^d} \Big|\rho_\varepsilon(x/N) - N^d \int_{\square_N(x)} du\, N^d \rho_\varepsilon(u)\Big| \,.$$

Here for a site x in $\mathbb{T}_N^d$, $\square_N(x)$ stands for an hypercube of length N^{-1} centered at x:

$$\square_N(x) \;=\; \{y \in \mathbb{R}^d,\ x_i/N - 1/2N \le y_i < x_i/N + 1/2N \quad \text{for} \quad 1 \le i \le d\}\,.$$

By assumption, the first integral converges to 0 as $N \uparrow \infty$. Since for each $\varepsilon > 0$ ρ_ε is Lipschitz continuous, the third term converges to 0 as $N \uparrow \infty$. At last, by construction of the initial profiles $\rho_\varepsilon(u)$, the second term converges to 0 as ε decreases to 0. In conclusion, we proved that

$$\limsup_{N\to\infty} Q^N\Big[\Big| < \pi_t, H > - \int H(u)\rho(t,u)du\Big| > \delta\Big] \;=\; 0$$

for every continuous function H and every $\delta > 0$. This concludes the proof of Theorem 0.3. □

4. Comments and References

The first proof of the hydrodynamic behavior of asymmetric interacting particle systems was given by Rost (1981). He proved the conservation of local equilibrium for a one-dimensional nearest neighbor totally asymmetric (particles jumps only to the right) simple exclusion process starting from the configuration with all sites at the left of the origin occupied and all others empty. This result was extended by Andjel and Kipnis (1984) for a one-dimensional nearest neighbor totally asymmetric zero range process with rate $g(k) = \mathbf{1}\{k \ge 1\}$ starting from a product measure associated to initial profiles of type $\rho_- \mathbf{1}\{u < 0\} + \rho_+ \mathbf{1}\{u \ge 0\}$. They also proved conservation of local equilibrium for general decreasing profiles in the case where the drift is toward the right (in this case the entropy solution does not present shocks). Benassi and Fouque (1987) proved conservation of local equilibrium for the one-dimensional nearest neighbor asymmetric simple exclusion process starting from a product measure with profile $\rho\mathbf{1}\{u > 0\}$ and extended it in Benassi and Fouque (1988) to one-dimensional attractive zero range processes starting from the same type of initial profile. By the same time and independently, Andjel and Vares (1987) prove the same result for zero range processes with initial profiles of type $\rho_- \mathbf{1}\{u < 0\} + \rho_+ \mathbf{1}\{u \ge 0\}$ and general decreasing profiles if the drift is to the right. Landim (1991a,b) extended the conservation of local equilibrium for attractive zero range processes in any dimension with initial profile given by $\alpha\mathbf{1}\{C\} + \beta\mathbf{1}\{C^c\}$, where C is a cone. The arguments presented in this chapter follow Rezakhanlou (1991) who proved a law of large numbers for the empirical measure and the weak conservation of local equilibrium, as stated in Chapter 3, for attractive particle systems in any dimension starting from bounded initial profiles. The spaces $L^\infty([0,T])$, $L^\infty([0,T], \mathcal{M}_+(\mathbb{T}^d \times \mathbb{R}_+))$ were introduced by Bahadoran (1996a).

Extensions.

(a) *Infinite volume.* The techniques presented in this section permit to prove the hydrodynamic limit for systems evolving on the infinite lattice $\mathbb{Z}^d$. In fact the original proof of Rezakhanlou (1991) is in infinite volume.

(b) *Conservation of local equilibrium.* Benassi, Fouque, Saada and Vares (1991) proved the conservation of local equilibrium for one-dimensional attractive processes starting from general monotone initial profiles. Fouque and Saada (1994) considered one-dimensional exclusion processes starting from product initial measures associated to bumps: $\rho_0 = \alpha + \beta \mathbf{1}\{[u_0, u_1]\}$. Landim (1993) deduced the conservation of local equilibrium from the weak local equilibrium for attractive systems. Together with Rezakhanlou (1991), this proves conservation of local equilibrium for initial measures associated to bounded initial profiles. This argument is explained in details in Chapter 9.

(c) *Non product initial states.* We already pointed out in Chapter 6 that the relative entropy method introduced by Yau (1991) permits to prove the hydrodynamic limit of asymmetric processes up to the appearance of the first shock. With coupling arguments, Venkatraman (1994) proved the law of large numbers for the empirical measure for one-dimensional totally asymmetric nearest neighbor exclusion processes and zero range processes with jump rate $g(k) = \mathbf{1}\{k \geq 1\}$ starting from deterministic initial configurations associated to bounded profiles. Independently, Seppäläinen (1996b) proved the same result for totally asymmetric processes, through the explicit Lax–Oleinik formula for the entropy solution for one-dimensional hyperbolic equations. Bahadoran (1997) extended the previous results to one-dimensional nearest neighbor misanthrope processes.

(d) *Spatially inhomogeneous processes.* Landim (1996) examined the hydrodynamic behavior of a one-dimensional totally asymmetric zero range process with bounded jump rates where the rates at a finite number of sites are slowed down. This result is further discussed below under large deviations. Bahadoran (1996a,b) and Covert and Rezakhanlou (1996) considered independently the case of asymmetric processes where the jump rate varies smoothly in the macroscopic scale. Bahadoran (1996a) considered an attractive zero range process where a particle at site x jumps to site $x+y$ at rate $g(x/N, \eta(x))p(y)$ for a irreducible transition probability $p(\cdot)$ and a rate $g(\cdot,\cdot)$ such that $g(u,\cdot)$ is an nondecreasing function for every u in $\mathbb{R}$ and $g(\cdot, n)$ is a twice continuously differentiable function for every $n \geq 1$. With further mild technical assumptions on the jump rate, the hydrodynamic equation is shown to be $\partial_t \rho + \gamma \cdot \nabla_u \hat{g}(u, \rho(t,u)) = 0$, where $\hat{g}$ is a smooth function depending on the jump rate and the invariant measures of the process. Bahadoran (1996b) extended the result to inhomogeneous exclusion processes where a particle at x jumps to $x+y$ at rate $\alpha(x/N)p(y)\eta(x)[1-\eta(x+y)]$ for a irreducible transition probability $p(\cdot)$ and a twice continuously differentiable function $\alpha(\cdot)$. The same result was obtained independently and at the same time by Covert and Rezakhanlou (1997) for misanthrope processes. In both cases the hydrodynamic equation is of type $\partial_t \rho + \gamma \cdot \nabla_u \{\alpha(u)\hat{h}(\rho(t,u))\} = 0$. Moreover, in these last two models, the invariant measures are not known explicitly and the method presented in this chapter can not be applied in a straightforward manner.

(e) *Continuous spin systems.* Aldous and Diaconis (1995) and Rost (private communication) considered a version of a one-dimensional continuous spin attractive zero range process, called the Hammersley or stick process. The state space is $\mathbb{R}_+^{\mathbb{Z}}$ and the dynamics may be described as follows. At rate $\eta(x)$ a random piece $U\eta(x)$ is taken out from site x and transferred to site $x+1$, where U is a random variable with uniform distribution in the interval $[0,1]$. A simple computation shows that the product measure with marginals distributed according to exponential variables are invariant. Aldous and Diaconis (1995) and Rost derived the hydrodynamic behavior of this model for initial profiles of type $\rho \mathbf{1}\{u<0\}$. Seppäläinen (1996a) extended this result to processes starting from general initial states that include deterministic ones.

(f) *Random rates.* Benjamini, Ferrari and Landim (1996) proved the hydrodynamic limit of two types of asymmetric zero range processes with random rates. In the first model jumps are speeded up by random variables: let $\{\beta_x, x \in \mathbb{Z}^d\}$ be i.i.d. random variables taking finitely many positive values. For a fixed realization, particles at site x jump to site $x+y$ at rate $\beta_x g(\eta(x))p(y)$ for some nondecreasing bounded jump rate g. The second model is one-dimensional. Fix $1/2 < c < 1$ and consider i.i.d. random variables $\{p_x, x \in \mathbb{Z}\}$ taking finitely many values in the interval $[c,1]$. For a fixed realization, at rate $p_x \mathbf{1}\{\eta(x) \geq 1\}$ (resp. $[1-p_{x-1}]\mathbf{1}\{\eta(x) \geq 1\}$) a particle at x jumps to the right (resp. left).

Long time behavior of weakly asymmetric processes. The one dimensional nearest neighbor weakly asymmetric simple exclusion process is the Markov process on $\{0,1\}^{\mathbb{Z}^d}$ whose generator L_N writes as $L_s + N^{-1}L_a$, where L_s (resp. L_a) is the generator of the one-dimensional nearest neighbor symmetric (resp. totally asymmetric) exclusion process. To fix ideas assume that L_a permits only jumps to the right. De Masi, Presutti and Scacciatelli (1989) proved that the hydrodynamic behavior of this process is described by the solution of the Burgers equation with viscosity:

$$\partial_t \rho = (1/2)\Delta\rho - \partial_u\{\rho(1-\rho)\} \tag{4.1}$$

and deduced the nonequilibrium fluctuations. Gärtner (1988) presented an alternative proof of the hydrodynamic behavior of this process and Dittrich and Gärtner (1991) obtained the nonequilibrium fluctuations of this model. Fritz and Maes (1988) derived a similar equation with a non linear second order term as equation of motion for a continuous spin Ginzburg–Landau model in presence of a small external field. Due to the factor N^{-1} in front of the asymmetric part, it is easy to show that N^2 is the correct time renormalization one needs to perform in order to investigate the hydrodynamic behavior of the process.

Dittrich (1990) studies the evolution of this process in time scales of order N^3. He proves that in this scale the behavior is entirely determined by the initial configuration. From this result Dittrich (1992) shows that the hydrodynamic behavior of this process starting from an initial configuration ζ is described in macroscopic times of order N (thus microscopic times of order N^3) by the solution of the discrete version of the Burgers equation with viscosity (4.1) and initial data ζ. More precisely, he shows that for every continuous function with compact support $H: \mathbb{R} \to \mathbb{R}$ and every sequence $\{u_N, n \geq 1\}$

$$\frac{1}{N} \sum_{x \in \mathbb{Z}^d} H(x/N - u_N)\{\eta_{tN^3}(x) - \rho^N(tN, x/N)\}$$

converges to 0 in probability as $N \uparrow \infty$, provided ρ^N is the solution of the discrete partial differential equation

$$\begin{cases} \partial_t \rho^N = (1/2)\Delta_N \rho^N - \nabla_N^- [\rho^N(t, (x+1)/N)\{1 - \rho^N(t, x/N)\}] \, , \\ \rho^N(0, x/N) = \eta(x) \, . \end{cases}$$

In this formula, $\nabla_N^\pm$ and Δ_N stand respectively for the discrete space derivative and the discrete Laplacian: $(\nabla_N^\pm f)(x/N) = N[f((x \pm 1)/N) - f(x/N)]$, $(\Delta_N f)(x/N) = N^2\{f((x+1)/N) + f((x-1)/N) - 2f(x/N)\}$.

Dittrich (1992) proves also that on the space scale N^2 and time scale N^3 the process evolves according to the Burgers equation without viscosity. Fix a piecewise continuous profile $\rho_0: \mathbb{R} \to [0, 1]$ and denote by ν^N the Bernoulli product measure associated to ρ_0 on the scale N^2: $E_{\nu^N}[\eta(x)] = \rho_0(x/N^2)$. Then, for every continuous function H with compact support,

$$\frac{1}{N^2} \sum_{x \in \mathbb{Z}^d} H(x/N^2)\eta_{tN^3}(x)$$

converges in probability, as $N \uparrow \infty$, to $\int_{\mathbb{R}} H(u)\rho(t, u)du$, where ρ is now the entropy solution of the inviscid Burgers equation

$$\partial_t \rho + \partial_u \{\rho(1 - \rho)\} = 0 \, .$$

Tracer particles. In contrast with the investigation of the hydrodynamic limit, where we are interested in the behavior of the whole system, we now tag a single particle and examine its evolution. To fix ideas, consider a zero range process η_t with jump rate $g(\cdot)$ and transition probability $p(\cdot)$ satisfying the assumptions of this chapter. Denote by A_t the position at time t of the tagged particle and assume without loss of generality that $A_0 = 0$. Let ξ_t stand for the process as seen from the tagged particle: $\xi_t = \tau_{A_t}\eta_t$. By Harris (1967) and Port and Stone (1973), for each $\alpha \ge 0$, the Palm measure ν_α^t, which is product with marginals given by

$$\nu_\alpha^t\{\xi,\ \xi(x) = k\} = \begin{cases} \nu_\alpha\{\xi,\ \xi(x) = k\} & \text{if } x \neq 0 \text{ and } k \geq 0\,, \\ \frac{k}{\alpha}\frac{1}{Z(\Phi(\alpha))}\frac{\Phi(\alpha)^k}{g(k)!} & \text{for } x = 0 \text{ and } k \geq 1 \end{cases}$$

is invariant for the process ξ_t.

The mean displacement of the tagged particle is $\mathcal{L}A = \{g(\eta(A))/\eta(A)\}\gamma$, provided $\mathcal{L}$ stand for the generator of the process (A_t, η_t) and γ for the mean drift of the particles: $\gamma = \sum_z zp(z)$. Assume the local equilibrium assumption and that the density around the tagged particle is ρ. In this case the expectation of the mean displacement is $E_{\nu_\rho^t}[g(\xi(0))/\xi(0)]\gamma = h(\rho)\gamma$, where $h(\alpha) = \Phi(\alpha)/\alpha$ if $\alpha > 0$ and $h(0) = \Phi'(0)$. Denote by $a_N(t)$ the macroscopic position of the tagged particle at the macroscopic time t: $a_N(t) = N^{-1}A(tN)$. If at the macroscopic time t the tagged particle is at the macroscopic point u, where, by the hydrodynamic limit, the density is $\rho(t,u)$, we expect $da_N(t)/dt$ to be close to $h(\rho(t,u))\gamma$. Therefore, in the limit as $N \uparrow \infty$, $a(t)$ should be the solution of

$$\begin{cases} \dfrac{da(t)}{dt} = \gamma H(t, a(t)) \\ a(0) = 0\,, \end{cases} \tag{4.2}$$

where $H(t,u) = h(\rho(t,u))$. Since the entropy solution $\rho(t,u)$ might be discontinuous the same lack of smoothness is inherited by $H(t,u)$ and, following Rezakhanlou (1994a), we shall interpret equation (4.2) in the Filippov (1960) sense: a Lipschitz continuous function $a: \mathbb{R}_+ \to \mathbb{R}^d$ is a solution of (4.2) in the Filippov sense if for almost all t, da/dt belongs to the interval

$$\Big[\text{ess lim inf } H(t,a),\ \text{ess lim sup } H(t,a)\Big]\ \gamma\,.$$

Rezakhanlou (1994a) proved the existence and the uniqueness of a solution in the Filippov sense of equation (4.2).

For one-dimensional nearest neighbor simple exclusion processes, the tagged particle may not jump over the other particles. In particular, the mass at the left of the tagged particle is constant in time. Therefore, starting from a product measure with slowly varying parameter associated to an integrable profile, a law of large numbers for the tagged particle follows from the hydrodynamic limit provided the tagged particle never hits a hole (in which case its position may not be well defined). More generally we have

Theorem 4.1 *(Rezakhanlou (1994a)) Consider a one-dimensional nearest neighbor zero range process with bounded jump rate $g(\cdot)$. Let μ^N be a sequence of probability measures satisfying assumptions* **(M1)***–***(M3)** *associated to an integrable profile ρ_0 such that*

$$\int_{\delta_1}^{\delta_2} \rho_0(u)\, du > 0$$

for all $\delta_1 \le 0 \le \delta_2$, $\delta_2 - \delta_1 > 0$. *Add a particle at the origin and denote by* $a_N(t)$ *its macroscopic position. Then, for each fixed* $t \ge 0$, $a_N(t)$ *converges in* L^1 *to* $a(t)$, *the solution of (4.2).*

Rezakhanlou proved this theorem for misanthrope processes, a class that includes at the same time zero range processes and simple exclusion processes (cf. Cocozza (1985)). The behavior of a single tagged particle starting from a sequence of product measures associated to a general profile in higher dimension or in dimension 1 with general finite range transition probability $p(\cdot)$ is still open. There is a partial answer due to Rezakhanlou (1994a) that shows that in the average, particles follow equation (4.2). This result, called propagation of chaos, is discussed below.

The question of a law of large numbers for a tagged particle was already present in Spitzer (1970). Kipnis (1986) proved a strong law of large numbers in the case of a nearest neighbor one-dimensional exclusion process starting from an equilibrium measure ν_α conditioned on the presence of a particle at the origin The asymptotic velocity of the particle was shown to be $(1-\alpha)(p-q)$. This result was extended by Saada (1987a,b) to asymmetric exclusion processes in any dimension and asymmetric zero range processes with jump rate $g(k) = \mathbf{1}\{k \ge 1\}$. Seppäläinen (1996b) gives an alternative proof of Theorem 4.1 in the case of totally asymmetric one-dimensional exclusion processes and zero range processes with jump rate $g(k) = \mathbf{1}\{k \ge 1\}$. We refer the reader to Kipnis (1985) and Ferrari (1996) for clear reviews of the subject.

Seppäläinen (1997a) proves a law of large numbers for a tagged particle in the one-dimensional totally asymmetric simple exclusion process in a scale different from the Euler scale. Bramson et al. (1986) investigate a one-dimensional nearest neighbor symmetric exclusion process speeded up by N^2 in which particles create at rate $1/2$ new particles at the neighboring sites. They prove the existence of a stationary measure for the process as seen from the rightmost particle and compute the asymptotic velocity, as $N \uparrow \infty$, of the rightmost particle. The velocity is related to the velocity of the traveling wave solution of the reaction–diffusion equation with a heaviside function as initial data.

Central limit theorem for a tagged particle. Ferrari and Fontes (1996) obtained a sharp estimate for the position of a tagged particle. Fix $\alpha > 0$ and consider a one-dimensional nearest neighbor simple exclusion process jumping with probability p to the right, $q = 1 - p$ to the left and starting from the equilibrium measure ν_α. To fix ideas set $p > q$. Add, if necessary, a tagged particle at the origin or simply tag the particle already there. Denote by A_t the position of the tagged particle at time t. Ferrari and Fontes (1994a) proved the existence of a Poisson point process of rate $(p-q)(1-\alpha)$ and a stationary process S_t such that $X_t = N_t - S_t + S_0$ and

$$\sup_t E\left[e^{\theta S_t}\right] < \infty$$

for some $\theta > 0$. It follows from this result that

$$\lim_{N\to\infty} \frac{X_{tN} - (p-q)(1-\alpha)tN}{\sqrt{N}} = \sqrt{(p-q)(1-\alpha)}W_t$$

in distribution, where W is a Brownian motion. Moreover, Ferrari (1992) proved that the fluctuations are due to fluctuations of the initial measure showing that

$$\lim_{t\to\infty} \frac{1}{t} \mathbb{E}_{\nu_\alpha} \Big[\Big(X_t - \frac{1}{\alpha} M_0(\eta_0, (p-q)\alpha t) \Big)^2 \Big] = 0 \tag{4.3}$$

if $M_0(\eta, k)$ stands for the number of holes in the interval $\{0, \dots, k\}$ for the configuration η: $M_0(\eta, k) = \sum_{0\le x\le k}\{1 - \eta(x)\}$.

The central limit theorem for the tagged particle just stated is due to Kipnis (1986). He obtained a lower bound for the diffusion coefficient. De Masi and Ferrari (1985) showed that the diffusion coefficient is $(p-q)(1-\alpha)$. The fact that the fluctuations of the tagged particle are due to fluctuations of the initial configuration was first observed by Gärtner and Presutti (1990).

In nonequilibrium, Wick (1985) proved a central limit theorem for the tagged particle in a totally asymmetric simple exclusion process starting from a product measure $\nu_{0,\alpha}$ with density α at the right of the origin and 0 at the left and showed that the diffusion coefficient is $(1-\alpha)$. This result was extended to nearest neighbor one-dimensional exclusion processes by De Masi, Kipnis, Presutti and Saada (1989). In the same context, Ferrari (1992) proved (4.3), giving an alternative proof of the central limit theorem in the case where the initial measure is $\nu_{0,\alpha}$ since $M_0(\eta, k)$ is a sum of i.i.d. random variables.

Propagation of chaos. To fix ideas, consider a zero range process with jump rate $g(\cdot)$ and transition probability $p(\cdot)$ satisfying the assumptions of this chapter. Instead of proving that the asymptotic behavior of a single particle is described by the solutions of equation (4.2), Rezakhanlou (1994a) proved that in the average particles behave according to (4.2).

We fix our attention on initial configurations with a finite number of particles and label all particles. The state space is thus $\mathcal{E} = \cup_{L\ge 0}(\mathbb{Z}^d)^L$, where L indicates the total number of particles. Configurations are denoted by $\mathbf{A} = (A_1, \dots, A_L)$. The microscopic dynamics is clear. Each particle at site y waits a rate $p(z)g(\eta(\mathbf{A}, y))/\eta(\mathbf{A}, y)$ exponential random time at the end of which it jumps to site $y+z$. Here, $\eta(\mathbf{A}, y)$ stands for the total number of particles at site y: $\eta(\mathbf{A}, y) = \sum_i \mathbf{1}\{A_i = y\}$. Denote by $A_i(t)$ the position of the i-th particle at time t. To write the generator of this Markov process, for each positive integer i and each site z, denote by $T_{i,z}\colon \mathcal{E} \to \mathcal{E}$ the transformation that moves the i-th particle by z:

$$(T_{i,z}\mathbf{A})_j = \begin{cases} A_j & if j \neq i\,, \\ A_j + z & \text{if } j = i\,. \end{cases}$$

$\mathbf{A}(t)$ is a Markov process with generator given by

$$(\mathcal{L}f)(\mathbf{A}) = \sum_{\substack{y,z\in\mathbb{Z}^d \\ i\ge 1}} \frac{g(\eta(\mathbf{A}, y))}{\eta(\mathbf{A}, y)} p(z) \mathbf{1}\{A_i = y\} \Big(f(T_{i,z}\mathbf{A}) - f(\mathbf{A}) \Big)\,.$$

Fix a bounded integrable profile $\rho_0: \mathbb{R}^d \to \mathbb{R}_+$ of bounded variation. Denote by μ^N a sequence of product measures on $\mathcal{E}$ such that

(i) (symmetry) $\mu^N\{\mathbf{A}, \mathbf{A} = (z_1, \dots, z_L)\} = \mu^N\{\mathbf{A}, \mathbf{A} = (z_{\sigma_1}, \dots, z_{\sigma_L})\}$ for every $L \ge 1$ and every permutation σ,

(ii) it is associated to the profile ρ_0 in the sense of assumption (**M3**): the sequence $\rho_0^N: \mathbb{R}^d \to \mathbb{R}_+$ that on $\square_N(x)$ is equal to $E_{\mu^N}[\eta(\mathbf{A}, x)]$ converges to ρ_0 in $L^1(\mathbb{R}^d)$.

For each v in $\mathbb{R}^d$, denote by $a^v(t)$ the solution of equation (4.2) with initial condition v instead of 0. Denote by $\mathbb{R}_{\mu^N}$ the probability measure on $D(\mathbb{R}_+, \mathcal{E})$ induced by μ^N and the process $\mathbf{a}^N(t) = (a_1^N(t), \dots, a_L^N(t)) = N^{-1}\mathbf{A}(tN)$.

Theorem 4.2 *(Rezakhanlou (1994a)) For every continuous function $J: D(\mathbb{R}_+, \mathbb{R}^d) \to \mathbb{R}$,*

$$\lim_{N\to\infty} E_{\mathbb{R}_{\mu^N}}\Big[\Big| L^{-1}\sum_{k=1}^{L} J(a_k) - \int J\, dR \Big| \Big] \;=\; 0\,,$$

where R is the probability measure on $D(\mathbb{R}_+, \mathbb{R}^d)$ concentrated on the solutions of (4.2) and such that

$$R\{a,\, a(0) \in \Gamma\} \;=\; \frac{\int_\Gamma \rho_0(u)du}{\int_{\mathbb{R}^d} \rho_0(u)du}\,.$$

It follows from this theorem that for every $n \ge 1$ and every family of continuous functions $J_i: D(\mathbb{R}_+, \mathbb{R}^d) \to \mathbb{R}$, $1 \le i \le n$,

$$\lim_{N\to\infty} E_{\mathbb{R}_{\mu^N}}\Big[\prod_{k=1}^{n} J_k(a_k)\Big] \;=\; \prod_{k=1}^{n} E_R[J_k(a)]\,.$$

This last identity is called propagation of chaos.

Large deviations. This is one of the main open questions in the theory of hydrodynamic limits. There are only two results in this direction. Kipnis and Léonard (1995) proved a large deviation principle for the empirical measure in the non interacting case (particles move according to asymmetric independent random walks). In this case the probability of observing a large deviation decays as $\exp\{-CN^{d+1}\}$, instead of $\exp\{-CN^d\}$ as in the diffusive case considered in Chapter 10. In the interacting case the picture is expected to be completely different. Landim (1996) proved the hydrodynamic behavior of a one-dimensional totally asymmetric attractive zero range process with bounded jump rate $g(\cdot)$, where the jumps from the origin or the jumps of a tagged particle are slowed down by a fixed factor. In the first case, the hydrodynamic behavior is given by the entropy solution of an hyperbolic equation with a boundary condition at the origin. This boundary condition allows the appearance of a Dirac mass at the origin. In the second case the hydrodynamic limit is described by a non entropy solution of the hyperbolic equation. From these results he deduced that in dimension 1, the probability of large deviations are of order bounded below by $\exp\{-CN\}$, in contrast with the non interacting case. This suggests, as pointed out by Varadhan, that the large deviations for the asymmetric exclusion process should be of order $\exp\{-CN\}$ and given by non entropy solutions of the Burger's equation.

Fluctuations of the empirical measure. This is another mainly open question, even in equilibrium. Ferrari and Fontes (1994b) proved the convergence of the finite dimensional distribution of the fluctuation density field in the case of a one-dimensional nearest neighbor exclusion process starting from the product measure $\nu_{\alpha,\beta}$ with density α at the left of the origin and β at the right. In the totally asymmetric case with $\alpha = 0$, the convergence away from the shock was obtained by Benassi and Fouque (1991).

9. Conservation of Local Equilibrium for Attractive Systems

In Chapter 1 we introduced the concept of local equilibrium and proved the conservation of local equilibrium for a superposition of independent random walks. Then, from Chapter 4 to Chapter 8, we proved a weaker version of local equilibrium for a large class of interacting particle systems: we showed that the empirical measure π_t^N converges in probability to an absolutely continuous measure whose density is the solution of some partial differential equation. The purpose of this chapter is to to show that in the case of attractive processes, the conservation of local equilibrium may be deduced from a law of large numbers for local fields, i.e., from the convergence in probability of the averages

$$N^{-d}\sum_{x} H(x/N)\tau_x\Psi(\eta_t) \quad \text{to} \quad \int_{\mathbb{T}^d} H(u)\tilde{\Psi}(\rho(t,u))\, du$$

for every $t \geq 0$, every continuous function H and every bounded cylinder function Ψ. Here $\rho(t,u)$ is the solution of the hydrodynamic equation. This statement is slightly stronger than the convergence of the empirical measures since it involves all local fields.

To fix ideas, we consider in this chapter a zero range process on $\mathbb{N}^{\mathbb{T}_N^d}$ with generator given by

$$(L_N f)(\eta) = \sum_{x\in\mathbb{T}_N^d}\sum_{y\in\mathbb{Z}^d} g(\eta(x))p(y)[f(\eta^{x,x+y}) - f(\eta)] \, , \tag{0.1}$$

where $p(\cdot)$ is a finite range irreducible transition probability: there exists R_0 such that $p(x) = 0$ for all x not in Λ_{R_0}, $\sum_x p(x) = 1$ and for every x, y in $\mathbb{Z}^d$, there exists $M \geq 1$ and $x = x_0, \dots, x_M = y$ such that $p^s(x_{i+1} - x_i) > 0$ for $0 \leq i \leq M-1$, where $p^s(y) = p(y) + p(-y)$. Denote by S_t^N the semigroup associated to the generator L_N defined in (0.1). Notice that in this chapter S_t^N is the semigroup associated to the generator L_N, which has not been speeded up. We consider two cases: asymmetric processes where the mean displacement of each elementary particle does not vanish

$$m = (m_1, \dots, m_d) := \sum_x x p(x) \neq 0 \tag{0.2}$$

and mean-zero asymmetric processes, where the mean displacement vanishes. In the second case, we denote by σ the covariance matrix defined by

$$\sigma_{i,j} = \sum_{x \in \mathbb{Z}^d} x_i x_j p(x) \quad \text{for } 1 \le i, j \le d\,. \tag{0.3}$$

$\sigma = \{\sigma_{i,j},\ 1 \le i, j \le d\}$ is a symmetric non–negative definite matrix that we shall assume to be positive definite to avoid degeneracy of the hydrodynamic equation: there exist $a > 0$ such that,

$$\sum_{i,j} v_i \sigma_{i,j} v_j \ \ge\ a \sum_i v_i^2$$

for every v in $\mathbb{R}^d$.

For a continuous function $\rho_0 : \mathbb{T}^d \to \mathbb{R}_+$, denote by $\nu^N_{\rho_0(\cdot)}$ the product measure with slowly varying parameter associated to ρ_0, this is the product measure on $\mathbb{N}^{\mathbb{T}^d_N}$ with marginals given by

$$\nu^N_{\rho_0(\cdot)}\{\eta,\ \eta(x) = k\} \ = \ \frac{1}{Z(\Phi(\rho_0(x/N)))} \frac{\Phi(\rho_0(x/N))^k}{g(k)!} \tag{0.4}$$

for $k \ge 0$ so that $E_{\nu^N_{\rho_0(\cdot)}}[\eta(x)] = \rho_0(x/N)$ for all x in $\mathbb{T}^d_N$.

Theorem 0.1 *Assume that $m = 0$. For every $t \ge 0$,*

$$\lim_{N\to\infty} S^N_{tN^2} \tau_{[uN]} \nu^N_{\rho_0(\cdot)} \ = \ \nu_{\rho(t,u)}\ ,$$

where $\rho(t,u)$ is the unique weak solution of the nonlinear heat equation

$$\begin{cases} \partial_t \rho = \displaystyle\sum_{1\le i,j\le d} \sigma_{i,j} \partial^2_{u_i,u_j} \Phi(\rho) \\ \rho(0,\cdot) = \rho_0(\cdot) \end{cases} \tag{0.5}$$

and σ is the covariance matrix defined in (0.3).

Theorem 0.2 *Assume that the mean displacement m defined in (0.2) does not vanish and that Φ is strictly concave or convex in the range of $\rho_0(\cdot)$. For every $t \ge 0$ and for every continuity point u of $\rho(t,\cdot)$,*

$$\lim_{N\to\infty} S^N_{tN} \tau_{[uN]} \nu^N_{\rho_0(\cdot)} \ = \ \nu_{\rho(t,u)}\ ,$$

where $\rho(t,u)$ is the unique entropy weak solution of

$$\begin{cases} \partial_t \rho + \displaystyle\sum_{j=1}^d m_j \partial_{u_j} \Phi(\rho) = 0 \\ \rho(0,\cdot) = \rho_0(\cdot) \end{cases} \tag{0.6}$$

and $m = (m_1, \cdots, m_d)$ is the mean drift of each elementary particle defined in (0.2).

1. Replacement Lemma for Attractive Processes

We consider in this section a mean-zero asymmetric zero range process with generator given by (0.1). We have seen in Chapter 5 that the main step in the proof of the hydrodynamic behavior of these processes is the replacement lemma that permits to replace average of cylinder functions by a function of the empirical measure. This theorem was proved under the assumption that the partition function is finite on $\mathbb{R}_+$ excluding, for instance, the case where the rate function $g(\cdot)$ is bounded. We prove in this section a replacement lemma in the attractive case for sequences of measures bounded above by a product invariant measure. The approach illustrates how coupling arguments may replace entropy estimates.

For a fixed time $T > 0$ and for a probability measure μ^N on $\mathbb{N}^{\mathbb{T}_N^d}$, denote by $\mathbb{P}_{\mu^N}^N = \mathbb{P}_{\mu^N}$ the probability measure on the path space $D([0,T], \mathbb{N}^{\mathbb{T}_N^d})$ corresponding to the mean-zero asymmetric zero range process with generator L_N, defined in (0.1), speeded up by N^2 and starting from the initial measure μ^N, and by $\mathbb{E}_{\mu^N}$ expectation with respect to $\mathbb{P}_{\mu^N}$.

Theorem 1.1 *Consider a sequence of probability measures $\{\mu^N, N \geq 1\}$ bounded by some invariant measure $\nu_{\alpha_0}^N$. Assume that the entropy of μ^N with respect to $\nu_{\alpha_0}^N$ is bounded by $K_0 N^d$ for some universal constant K_0: $H(\mu^N \mid \nu_{\alpha_0}^N) \leq K_0 N^d$. For every cylinder Lipschitz function Ψ,*

$$\limsup_{\varepsilon \to 0} \limsup_{N \to \infty}$$
$$\mathbb{E}_{\mu^N}\Big[\int_0^T dt \, \frac{1}{N^d} \sum_x \Big| \frac{1}{(2\varepsilon N + 1)^d} \sum_{|y-x| \leq \varepsilon N} \tau_y \Psi(\eta_t) - \tilde{\Psi}(\eta_t^{\varepsilon N}(x))\Big|\Big] = 0 .$$

Remark 1.2 In the previous theorem we assumed for simplicity that the sequence μ^N is bounded above and has finite entropy with respect to the same invariant measure $\nu_{\alpha_0}^N$. The fact that in both assumptions appears the same reference measure $\nu_{\alpha_0}^N$ is not restrictive: by Remark 5.1.2 a sequence μ^N which has relative entropy with respect to some invariant measure $\nu_{\alpha_2}^N$ bounded by $K_0 N^d$, has entropy with respect to $\nu_{\alpha_1}^N$ bounded by $K_1 N^d$ for some $K_1 = K_1(K_0, \alpha_1, \alpha_2)$.

Proof. Denote by f_t^N the Radon–Nikodym derivative of $S_{tN^2}^N \mu^N$ with respect to $\nu_{\alpha_0}^N$. Set $\bar{f}_t^N = T^{-1} \int_0^T f_t^N dt$. By attractiveness, the probability measures $f_t^N d\nu_{\alpha_0}^N$ are bounded by $\nu_{\alpha_0}^N$ and so is $\bar{f}_t^N d\nu_{\alpha_0}^N$. Moreover, by the entropy computation performed in section 5.2, the Dirichlet form of $\bar{f}_T^N$ is bounded by $C(K_0, T) N^{d-2}$.

With the notation just introduced, we may write the expectation appearing in the statement of the theorem as

$$T \int \bar{f}_T^N(\eta) \frac{1}{N^d} \sum_x \Big| \frac{1}{(2\varepsilon N + 1)^d} \sum_{|y-x| \leq \varepsilon N} \tau_y \Psi(\eta) - \tilde{\Psi}(\eta^{\varepsilon N}(x))\Big| \nu_{\alpha_0}^N(d\eta) .$$

By Corollary 2.3.7, $\tilde{\Psi}$ is a uniformly Lipschitz continuous function. In particular, the arguments presented in the beginning of section 5.3 show that the proof of the replacement lemma may be reduced to the proof of the one and two blocks estimates:

Lemma 1.3 (One block estimate) *Under the hypotheses of Theorem 1.1,*

$$\limsup_{\ell\to\infty}\limsup_{N\to\infty} \int \frac{1}{N^d}\sum_x \Big|\frac{1}{(2\ell+1)^d}\sum_{|y-x|\le\ell}\tau_y\Psi(\eta) - \tilde{\Psi}(\eta^\ell(x))\Big| \bar{f}_T^N(\eta)\,\nu_{\alpha_0}^N(d\eta) \;=\; 0\,.$$

Lemma 1.4 (Two blocks estimate) *Under the hypotheses of Theorem 1.1,*

$$\limsup_{\ell\to\infty}\limsup_{\varepsilon\to 0}\limsup_{N\to\infty} \sup_{|y|\le\varepsilon N}\int N^{-d}\sum_x \big|\eta^\ell(x+y)-\eta^{N\varepsilon}(x)\big|\, \bar{f}_T^N(\eta)\nu_{\alpha_0}^N(d\eta) \;=\; 0\,.$$

We already pointed out in Remark 6.1.14 that in the attractive case either one of the assumptions (**FEM**) or (**SLG**) is fulfilled. The proof of the one block estimate presented in Chapter 5 applies therefore to the present context. Nevertheless, to illustrate how coupling arguments may replace entropy estimates, we present below a proof of the one block estimate in the attractive set–up.

Proof of Lemma 1.3. In the proof of the one block estimate, the assumptions (**FEM**) or (**SLG**) were only used to introduce the indicator $\mathbf{1}\{\eta^\ell(x)\le A\}$ inside the expectation to avoid large densities. In the case of attractive processes, this indicator function can be introduced by a simple coupling argument: since Ψ is a Lipschitz function, $|(2\ell+1)^{-d}\sum_{|y|\le\ell}\tau_y\Psi(\eta)-\tilde{\Psi}(\eta^\ell(0))|$ is bounded above by $C(\Psi)\eta^{\ell+s_\Psi}(0)$, where s_Ψ stands for the linear size of the support of Ψ. In particular,

$$\int \frac{1}{N^d}\sum_x \Big|\frac{1}{(2\ell+1)^d}\sum_{|y-x|\le\ell}\tau_y\Psi(\eta) - \tilde{\Psi}(\eta^\ell(x))\Big| \mathbf{1}\{\eta^\ell(x)\ge A\}\bar{f}_T^N(\eta)\nu_{\alpha_0}^N(d\eta)$$
$$\le\; C(\Psi)\int \frac{1}{N^d}\sum_x \eta^{\ell+s_\Psi}(x)\mathbf{1}\{\eta^\ell(x)\ge A\}\bar{f}_T^N(\eta)\,\nu_{\alpha_0}^N(d\eta)\,.$$

Since $\eta^{\ell+s_\Psi}(0)\mathbf{1}\{\eta^\ell(x)>A\}$ is an increasing function and $\bar{f}_T^N(\eta)\,\nu_{\alpha_0}^N(d\eta)$ is bounded by the product measure $\nu_{\alpha_0}^N$, this expression is less than or equal to

$$C(\Psi)\int \eta^{\ell+s_\Psi}(0)\mathbf{1}\{\eta^\ell(0)\ge A\}\,\nu_{\alpha_0}^N(d\eta) \;\le\; \frac{C(\Psi)}{A}\int \eta(0)^2\nu_{\alpha_0}^N(d\eta)\,.$$

that converges to 0 as $A\uparrow\infty$.

Therefore, in order to prove the one block estimate, it suffices to show that

$$\limsup_{\ell\to\infty}\limsup_{N\to\infty}\int \frac{1}{N^d}\sum_x (\tau_x V_{\Psi,\ell})(\eta)\mathbf{1}\{\eta^\ell(x)\le A\}\bar f_T^N(\eta)\,\nu_{\alpha_0}^N(d\eta) \;=\; 0$$

for every $A > 0$, where $V_{\Psi,\ell}(\eta) = |(2\ell+1)^{-d}\sum_{|y|\le\ell}\tau_y\Psi(\eta) - \tilde\Psi(\eta^\ell(0))|$. This follows from the proof of the one block estimate presented after formula (5.4.1) and the estimate on the Dirichlet form of $\bar f_T^N$ obtained in the beginning of the proof of Theorem 1.1. □

Proof of Lemma 1.4. In the same way, in the proof of the two blocks estimate given in section 5.3, the assumption **(FEM)** was invoked only at formula (5.5.1) to justify the introduction of the indicator function $\mathbf{1}\{\eta^\ell(x)\vee\eta^\ell(x+y)\le A\}$. The arguments presented in the proof of Lemma 1.3 above show that in the attractive case such indicator function may also be introduced. □

Since the proof of the hydrodynamic behavior of mean-zero asymmetric zero range processes presented in Chapter 5 relies almost exclusively on the replacement lemma and we just proved this result in the context of attractive systems, we have

Theorem 1.5 *Consider a mean-zero attractive zero range process with generator given by (0.1) speeded up by N^2 and starting from a measure μ^N satisfying the following three assumptions: μ^N is bounded above by an invariant measure $\nu_{\alpha_0}^N$, it has relative entropy with respect to $\nu_{\alpha_0}^N$ bounded by K_0N^d and*

$$\lim_{N\to\infty} E_{\mu^N}\Big[\Big|N^{-d}\sum_x H(x/N)\eta(x) - \int H(u)\rho_0(u)\,du\Big|\Big] \;=\; 0$$

for every continuous function H and some continuous initial profile $\rho_0\colon\mathbb{T}^d\to\mathbb{R}_+$. Then, for every $t > 0$, the empirical measure π_t^N defined in (4.0.2) converges in probability to the absolutely continuous measure $\rho(t,u)du$ whose density $\rho(t,u)$ is the solution of the nonlinear heat equation (0.5).

2. One Block Estimate Without Time Average

The first step toward the proof that for attractive processes the conservation of local equilibrium follows from a law of large numbers for the empirical measure is a one block estimate without time average. The proof of this estimate relies on the fact that under some assumptions on the sequence of initial measures μ^N all limit points of the sequence

$$N^{-d}\sum_{x\in\mathbb{T}_N^d}\tau_x S^N_{t\theta(N)}\mu^N$$

are invariant. Here $\theta(N)$ stands for the hydrodynamic time renormalization: $\theta(N) = N$ in the asymmetric case and $\theta(N) = N^2$ is the mean-zero case. The condition on the initial measure is simple to state and relies on the behavior of the microscopic density field: we shall assume that for all uniformly Lipschitz continuous function $F : \mathbb{R}_+ \to \mathbb{R}$ (there exists a constant $C(F)$ such that $|F(u) - F(v)| \le C(F)|u - v|$ for all u, v in $\mathbb{R}_+$)

$$\textbf{(MF)} \qquad \lim_{\ell\to\infty}\lim_{N\to\infty} \int_0^T ds\, E_{S^N_{s\theta(N)}\mu^N}\Big[N^{-d}\sum_x F(\eta^\ell(x))\Big] \;=\; \int_0^T A(F,s)\,ds\,.$$

for some continuous function $A(F,\cdot)$.

Before stating the main theorem of this section, we shall discuss assumption (**MF**). In the asymmetric case, consider a sequence $\mu^N = \nu^N_{\rho_0(\cdot)}$ of product measures with slowly varying parameter associated to a continuous profile $\rho_0 \colon \mathbb{T}^d \to \mathbb{R}_+$. It follows from the law of large numbers for the Young measure proved in Chapter 8 that the left hand side of (**MF**) is equal to

$$\int_0^T ds \int_{\mathbb{T}^d} du\, F(\rho(s,u))\,,$$

where $\rho(s,u)$ is the entropy solution of the hyperbolic equation (0.6). Since by Theorem A2.5.11 the entropy solution is $L^1(\mathbb{T}^d)$ continuous in the sense that $\rho(t,\cdot)$ converges to $\rho(s,\cdot)$ in $L^1(\mathbb{T}^d)$ as t approaches s, for every uniformly Lipschitz continuous function F, $\int du\, F(\rho(s,u))$ is time continuous, what proves assumption (**MF**) for attractive asymmetric zero range processes.

We consider now the case of mean-zero attractive asymmetric zero range processes. Fix a sequence of measures μ^N satisfying the assumptions of Theorem 1.5. It follows from the two blocks estimate proved in section 1 that assumption (**MF**) is equivalent to the statement that for every uniformly Lipschitz continuous function F

$$\lim_{\varepsilon\to 0}\lim_{N\to\infty} \mathbb{E}_{\mu^N}\Big[\int_0^T ds\, N^{-d}\sum_x F(\eta_s^{\varepsilon N}(x))\Big] \;=\; \int_0^T A(F,s)\,ds$$

for some continuous function $A(F,\cdot)$. By the hydrodynamic behavior of attractive mean-zero asymmetric zero range processes stated in Theorem 1.5 above, the left hand side is equal to

$$\int_0^T ds \int_{\mathbb{T}^d} du\, F(\rho(s,u))\,.$$

Assumption (**MF**) is therefore satisfied for attractive mean-zero asymmetric zero range processes because by Theorem A2.4.3 the solution $\rho(t,u)$ is uniformly Hölder continuous on each compact set of $(0,\infty)\times\mathbb{T}^d$.

Theorem 2.1 *Let μ^N be a sequence of probability measures bounded by $\nu^N_{\alpha_0}$ and satisfying assumption (**MF**). For every bounded cylinder function Ψ and for every $t > 0$,*

$$\limsup_{\ell\to\infty}\limsup_{N\to\infty}$$

$$E_{S^N_{t\theta(N)}\mu^N}\left[N^{-d}\sum_{x\in\mathbb{T}^d}\left|\frac{1}{(2\ell+1)^d}\sum_{|y-x|\le\ell}\tau_y\Psi(\eta)-\tilde{\Psi}(\eta^\ell(x))\right|\right] = 0\,.$$

Proof. Fix $t > 0$. Let $\mu^N(t)$ be the spatial average of $S^N_{t\theta(N)}\mu^N$:

$$\mu^N(t) = \frac{1}{N^d}\sum_{x\in\mathbb{T}^d_N}\tau_x S^N_{t\theta(N)}\mu^N\,.$$

With this notation, we may rewrite last expectation as

$$E_{\mu^N(t)}\left[\left|\frac{1}{(2\ell+1)^d}\sum_{|x|\le\ell}\tau_x\Psi(\eta)-\tilde{\Psi}(\eta^\ell(0))\right|\right].$$

By attractiveness, $\mu^N(t)$ is bounded above by $\nu^N_{\alpha_0}$, for $\nu^N_{\alpha_0}$ is invariant and translation invariant. This bound on $\mu^N(t)$ in turn permits to show that the sequence $\{\mu^N(t);\ N\ge 1\}$ is weakly relatively compact (cf. Lemma 2.3.9 for a similar statement). Denote by $\mathcal{A}$ the set of limit points of this sequence. Since Ψ is a bounded cylinder function, in order to prove the theorem, it is enough to show that

$$\limsup_{\ell\to\infty}\sup_{\mu\in\mathcal{A}} E_\mu\left[\left|\frac{1}{(2\ell+1)^d}\sum_{|x|\le\ell}\tau_x\Psi(\eta)-\tilde{\Psi}(\eta^\ell(0))\right|\right] = 0\,.$$

Due to the presence of the space average, all measures in $\mathcal{A}$ are translation invariant. Moreover, all elements of $\mathcal{A}$ are bounded above by ν_{α_0} because by Lemma 2.3.9 the set of probability measures $\{\mu\in\mathcal{M}_1(\mathbb{N}^{\mathbb{Z}^d});\ \mu\le\nu_{\alpha_0}\}$ is weakly relatively compact.

Let us assume for the moment that all limit points of the sequence $\{\mu^N(t);\ N\ge 1\}$ are invariant: $\mathcal{A}\subset\mathcal{I}$. This is the content of Proposition 2.2 below. $\mathcal{A}$ is therefore contained in the set of invariant and translation invariant measures bounded by ν_{α_0}. By Theorem 2.6.2, the convex hull of the compact convex set $\{\mu\in\mathcal{I}\cap\mathcal{S};\ \mu\le\nu_{\alpha_0}\}$ is equal to the set of product translation invariant measures ν_α, $0\le\alpha\le\alpha_0$. In particular,

$$\begin{aligned}
&\sup_{\mu\in\mathcal{A}} E_\mu\left[\left|\frac{1}{(2\ell+1)^d}\sum_{|x|\le\ell}\tau_x\Psi(\eta)-\tilde{\Psi}(\eta^\ell(0))\right|\right]\\
&\quad\le \sup_{\substack{\mu\in\mathcal{I}\cap\mathcal{S}\\ \mu\le\nu_{\alpha_0}}} E_\mu\left[\left|\frac{1}{(2\ell+1)^d}\sum_{|x|\le\ell}\tau_x\Psi(\eta)-\tilde{\Psi}(\eta^\ell(0))\right|\right]\\
&\quad= \sup_{\alpha\le\alpha_0} E_{\nu_\alpha}\left[\left|\frac{1}{(2\ell+1)^d}\sum_{|x|\le\ell}\tau_x\Psi(\eta)-\tilde{\Psi}(\eta^\ell(0))\right|\right].
\end{aligned}$$

Since Ψ is a bounded cylinder function,

$$\limsup_{\ell\to\infty}\sup_{\alpha\le\alpha_0} E_{\nu_\alpha}\Big[\Big|\frac{1}{(2\ell+1)^d}\sum_{|x|\le\ell}\tau_x\Psi - E_{\nu_\alpha}[\Psi]\Big|\Big] = 0\,.$$

On the other hand, for a fixed density $\alpha_1>\alpha_0$,

$$E_{\nu_\alpha}\Big[\Big|\tilde\Psi(\eta^\ell(0))-\tilde\Psi(\alpha)\Big|\Big] \le 2\|\Psi\|_\infty E_{\nu_\alpha}\Big[\mathbf{1}\{\eta^\ell(0)\ge\alpha_1\}\Big] + \Big(\sup_{u\le\alpha_1}|\tilde\Psi'(u)|\Big)E_{\nu_\alpha}\Big[|\eta^\ell(0)-\alpha|\Big]$$

and the right hand side converges to 0 as $\ell\uparrow\infty$ uniformly in $[0,\alpha_0]$. □

An elementary coupling argument permits to extend Theorem 2.1 to cylinder Lipschitz functions Ψ.

We now complete the proof of the previous theorem showing that all limit points of the sequence $\{\mu^N(t),\ N\ge1\}$ are invariant measures.

Proposition 2.2 *Under the assumptions of Theorem 2.1, $\mathcal{A}\subset\mathcal{I}$.*

Proof. The proof of this result is divided in several lemmas. Consider a limit point μ^* and assume without loss of generality that the sequence $\mu^N(t)$ converges to μ^*. By Theorem 2.6.3, the translation invariant measure μ^* is invariant if for each density $\alpha\ge0$ there exists a translation invariant measure $\bar\mu_\alpha$ on the product space $\mathbb{N}^{\mathbb{Z}^d}\times\mathbb{N}^{\mathbb{Z}^d}$ with first marginal equal to μ^*, second marginal equal to ν_α and concentrated on ordered configurations:

(i) the first marginal of $\bar\mu_\alpha$ is μ^*,

(ii) the second marginal of $\bar\mu_\alpha$ is ν_α and

(iii) $\bar\mu_\alpha\{(\eta,\xi);\ \eta\le\xi \text{ or } \xi\le\eta\}=1$.

Fix $\alpha\ge0$. To obtain a measure $\bar\mu_\alpha$ satisfying assumptions (i)–(iii), we shall consider a coupled process (η_t,ξ_t) evolving according to the generator $\bar L_N$ defined in (2.5.1) starting from $\bar\mu_\alpha^N=\mu^N\times\nu_\alpha^N$.

For a measure $\bar\mu$ on the product space $\mathbb{N}^{\mathbb{Z}^d}\times\mathbb{N}^{\mathbb{Z}^d}$ and $j=1,2$, let $\pi_j\bar\mu$ stand for the j-th marginal of $\bar\mu$. Denote by $\bar S_t^N$ the semigroup associated to the generator $\bar L_N$. Since each marginal evolves according to the original dynamics and ν_α^N is an invariant measure, at each time the first marginal of the coupled process is $S_t^N\mu^N$ and the second is ν_α^N:

$$\pi_1\bar S_t^N\bar\mu_\alpha^N = S_t^N\mu^N\,,\qquad \pi_2\bar S_t^N\bar\mu_\alpha^N = \nu_\alpha^N\,.$$

Fix $\alpha_1>\alpha\vee\alpha_0$. Denote by $\bar\mu_\alpha^N(t)$ the space average of the measure $\bar\mu_\alpha^N\bar S_{t\theta(N)}^N$:

$$\bar\mu_\alpha^N(t) = \frac{1}{N^d}\sum_{x\in\mathbb{T}_N^d}\tau_x\bar S_{t\theta(N)}^N\bar\mu_\alpha^N$$

and by $\bar\nu^N_{\alpha_1}$ the measure on the product space concentrated on the diagonal and with both marginals equal to $\nu^N_{\alpha_1}$: $\bar\nu^N_{\alpha_1}\{(\eta,\xi),\ \eta=\xi\}=1$, $\pi_j\bar\nu^N_{\alpha_1}=\nu^N_{\alpha_1}$ for $j=1,2$. Since the measures μ^N and ν^N_α are bounded above by $\nu^N_{\alpha_1}$, since the translation invariant measure $\bar\nu^N_{\alpha_1}$ is invariant with respect to the coupled process and since the coupled dynamics is itself attractive, a simple argument shows that $\bar S^N_{t\theta(N)}\bar\mu^N_\alpha$ (and therefore $\bar\mu^N_\alpha(t)$) is bounded above by $\bar\nu^N_{\alpha_1}$. In particular, the sequence $\{\bar\mu^N_\alpha(t);\ N\ge 1\}$ is weakly relatively compact. Denote by $\bar\mu_\alpha$ a limit point and assume, without loss of generality, that the sequence $\bar\mu^N_\alpha(t)$ converges to $\bar\mu_\alpha$. It is clear that $\bar\mu_\alpha$ satisfies requirements (i) and (ii) of Theorem 2.6.3. The third property follows from the next four lemmas and and is closely related to the proof of the entropy inequality at the microscopic level for the Young measures presented in section 8.2 (cf. proof of Lemma 8.2.2). □

Lemma 2.3 *Fix $t_0>0$. For every x such that $p(x)+p(-x)>0$,*

$$\bar\mu_\alpha(t_0)\Big\{(\eta,\xi);\ 0=\eta(0)<\xi(0),\ \xi(x)<\eta(x)\Big\}\ =\ 0\,,$$
$$\bar\mu_\alpha(t_0)\{(\eta,\xi);\ 0=\xi(0)<\eta(0),\ \eta(x)<\xi(x)\}\ =\ 0\,.$$

Proof. Denote by $d_N(t)$ the density of uncoupled particles at time t:

$$d_N(t)\ =\ E_{\bar S^N_{t\theta(N)}\bar\mu^N_\alpha}\left[N^{-d}\sum_z\Big|\eta(z)-\xi(z)\Big|\right]\ =\ E_{\bar\mu^N_\alpha(t)}\Big[\Big|\eta(0)-\xi(0)\Big|\Big]\,.$$

For the dynamics defined by the generator $\bar L_N$, the number of uncoupled particles may only decrease in time. Indeed, a simple computation taking advantage of the translation invariance of the measures $\bar\mu^N_\alpha(t)$ shows that the time derivative of d_N is given by

$$d'_N(t)\ =\ -2\theta(N)E_{\bar\mu^N_\alpha(t)}\Big[\sum_x p(-x)G_{0,x}(\eta,\xi)\big|g(\eta(x))-g(\xi(x))\big|\Big]\,,$$

where, for two distinct sites x, y, $G_{x,y}(\eta,\xi)$ is the cylinder functions equal to 1 if the configurations η and ξ are not ordered at sites x, y and 0 otherwise:

$$G_{x,y}(\eta,\xi)\ =\ \mathbf{1}\{\eta(x)<\xi(x),\eta(y)>\xi(y)\}\ +\ \mathbf{1}\{\eta(x)>\xi(x),\eta(y)<\xi(y)\}\,.$$

Denote by $-\theta(N)f_N(t)$ the time derivative of $d_N(t)$. In Lemma 2.6 below, we prove that the sequence of continuous monotone functions $\{d_N(\cdot),\ N\ge 1\}$ converges uniformly on each compact set of $\mathbb{R}_+$. It is in the proof of this assertion that assumption **(MF)** is required.

From the definition of the sequence f_N, we have that

$$d_N\Big(t+\frac{\delta}{\theta(N)}\Big)\ -\ d_N(t)\ =\ -\theta(N)\int_t^{t+\delta\theta(N)^{-1}}f_N(s)\,ds$$

for every $\delta>0$. By Taylor expansion,

$$\theta(N) \int_t^{t+\delta\theta(N)^{-1}} f_N(s)\, ds \;=\; \delta f_N(t) \;+\; \frac{\delta^2}{2\theta(N)} f_N'(s_N(t)),$$

for some $s_N(t)$ in the interval $[t, t+\delta\theta(N)^{-1}]$. Therefore,

$$f_N(t) \;=\; -\frac{d_N(t+\delta\theta(N)^{-1}) - d_N(t)}{\delta} \;-\; \frac{\delta}{2\theta(N)} f_N'(s_N(t))\,.$$

A simple but rather long computation shows that the derivative of f_N is of order $\theta(N)$. Since by Lemma 2.6 below the sequence d_N converges uniformly on each compact set, for every $t_1 > 0$,

$$\lim_{N\to\infty} \sup_{t\le t_1} f_N(t) \;=\; 0\,. \tag{2.1}$$

Recall the definition of $f_N(t)$. Since by assumption the sequence $\bar\mu_\alpha^N(t_0)$ converges weakly to $\bar\mu_\alpha(t_0)$, it follows from (2.1) that

$$E_{\bar\mu_\alpha(t_0)}\Big[\sum_z p(-z) G_{0,z}(\eta,\xi)\big|g(\eta(z)) - g(\xi(z))\big|\Big] \;=\; 0\,.$$

Since the jump rate is nondecreasing and vanishes at 0, the left hand side is bounded below by

$$g(1)\sum_z p(-z) E_{\bar\mu_\alpha(t_0)}\Big[\mathbf{1}\{\eta(0) < \xi(0), 0 = \xi(z) < \eta(z)\}$$
$$+\; \mathbf{1}\{\xi(0) < \eta(0), 0 = \eta(z) < \xi(z)\}\Big]\,.$$

This proves the lemma because the measure $\bar\mu_\alpha(t_0)$ is translation invariant. □

Lemma 2.4 *Fix $t_0 > 0$. For all x in $\mathbb{Z}^d$ such that $p(x) + p(-x) > 0$,*

$$\bar\mu_\alpha(t_0)\{(\eta,\xi);\ \eta(0) < \xi(0),\ \xi(x) < \eta(x)\} \;=\; 0\,,$$
$$\bar\mu_\alpha(t_0)\{(\eta,\xi);\ \xi(0) < \eta(0),\ \eta(x) < \xi(x)\} \;=\; 0\,.$$

Proof. We prove the first identity and leave the second one to the reader. Fix x in $\mathbb{Z}^d$ so that $p(x)+p(-x) > 0$. For a positive integer m, denote by I_m the indicator function of the set

$$\{(\eta,\xi);\ m = \eta(0) < \xi(0), \xi(x) < \eta(x)\}$$

and by $f_N^m(t)$ its expectation with respect to the measure $\bar\mu_\alpha^N(t)$:

$$f_N^m(t) \;=\; E_{\bar\mu_\alpha^N(t)}[I_m]\,.$$

We shall prove that for all $m \ge 0$, the sequence $\{f_N^m,\ N \ge 1\}$ converges to 0 uniformly over all compact set of $\mathbb{R}_+$. In the previous lemma we proved this statement for $m = 0$ because

$$f_N^0(t) \le 2p(x)g(1)f_N(t)\,.$$

We shall proceed by induction. Fix $m > 0$ and assume that f_N^n converges uniformly to 0 on all compact subsets of $\mathbb{R}_+$ for all $0 \le n \le m$. The time derivative of $f_N^m(t)$, denoted by $\theta(N)h_N^m(t)$, is given by

$$(f_N^m)'(t) = \theta(N)E_{\bar\mu_\alpha^N(t)}[\bar L_N I_m] =: \theta(N)h_N^m(t)\,.$$

Computing $\bar L_N I_m$ we obtain positive terms that correspond to jumps of $I_m(\eta_t,\xi_t)$ from 0 to 1 and negative terms that correspond to jumps of $I_m(\eta_t,\xi_t)$ from 1 to 0. Since I_m takes only the values 0 and 1, the positive terms, whose expectation with respect to $\bar\mu_\alpha^N(t)$ is denoted by $h_N^{m,+}(t)$, are given by $(1-I_m)\bar L_N I_m$:

$$h_N^{m,+}(t) = E_{\bar\mu_\alpha^N(t)}[(1-I_m)\bar L_N I_m]\,.$$

In the same way, the negative terms of $\bar L_N I_m$ are given by $I_m\bar L_N I_m$ and their expectation with respect to $\bar\mu_\alpha^N(t)$ is denoted by $h_N^{m,-}(t)$:

$$h_N^{m,-}(t) = E_{\bar\mu_\alpha^N(t)}[I_m\bar L_N I_m]\,.$$

Among the positive terms there is one which corresponds to a jump of two coupled particles from site 0 to some site x when site 0 is occupied by $m+1$ η-particles and more that $m+2$ ξ-particles and the site x is occupied by more η-particles than ξ-particles. Since the rate of such jump is $g(m+1)$ this term is equal to

$$\begin{aligned} p(x)g(m+1)\mathbf{1}\{(\eta,\xi);\ m+1 = \eta(0) < \xi(0), \xi(x) < \eta(x)\} \\ = p(x)g(m+1)I_{m+1}\,. \end{aligned}$$

In particular,

$$h_N^{m,+}(t) \ge p(x)g(m+1)E_{\bar\mu_\alpha^N(t)}[I_{m+1}] = p(x)g(m+1)f_N^{m+1}(t)$$

because all other terms of $(1-I_m)\bar L_N I_m$ are positives. Therefore, to show that the sequence $\{f_N^{m+1},\, N\ge 1\}$ converges to 0 uniformly over all compact sets of $\mathbb{R}_+$, it is enough to prove the same result for $h_N^{m,+}(\cdot)$. By definition of $h_N^{m,+}$ and $h_N^{m,-}$ we have that

$$f_N^m(t+\delta\theta(N)^{-1}) - f_N^m(t) = \theta(N)\int_t^{t+\delta\theta(N)^{-1}} [h_N^{m,+}(r) + h_N^{m,-}(r)]\,dr\,.$$

By Taylor expansion,

$$\theta(N)\int_t^{t+\delta\theta(N)^{-1}} h_N^{m,+}(r)\,dr = \delta h_N^{m,+}(t) + \frac{\delta^2}{2\theta(N)}(h_N^{m,+})'(s_N(t))$$

for some $s_N(t)$ in $[t, t+\delta\theta(N)^{-1}]$. In particular, since $h_N^{m,+}$ is positive,

$$
\begin{aligned}
0 \le h_N^{m,+}(t) &= \frac{f_N^m(t+\delta\theta(N)^{-1}) - f_N^m(t)}{\delta} - \frac{\delta}{2\theta(N)}(h_N^{m,+})'(s_N(t)) \\
&\quad - \frac{\theta(N)}{\delta}\int_t^{t+\delta\theta(N)^{-1}} h_N^{m,-}(r)\, dr\,.
\end{aligned}
\tag{2.2}
$$

By the induction assumption the sequence $(f_N^m)_{N\ge 1}$ converges to 0 uniformly over all compact sets of $\mathbb{R}_+$. On the other hand, an elementary computation gives that the time derivative of $h_M^{m,+}$ is of order $\theta(N)$. Therefore the first two terms on the right hand side of the equality converge uniformly to 0 as $N\uparrow\infty$ and $\delta\downarrow 0$. It remains to show that the third term converges to 0. It is not difficult to bound the absolute value of $h_N^{m,-}(t)$ by

$$
(g(m)+2g^*)E_{\bar\mu_\alpha^N(t)}[I_m] \;+\; g^*\sum_z p(-z)E_{\bar\mu_\alpha^N(t)}\Big[\{\eta(z)+\xi(z+x)\}I_m\Big]\,,
$$

where $g^* = \sup_{n\ge 0}\{g(n+1)-g(n)\}$. In this last computation we used the trivial bound $g(n)\le g^* n$. Since both marginals of $\bar\mu_\alpha^N(t)$ are bounded above by $\nu_{\alpha_1}^N$ and $E_{\nu_{\alpha_1}}[\eta(0)^2]$ is finite, by Schwarz inequality, there exists a constant $C_0(\alpha_1, g^*, m)$ such that

$$
\big|h_N^{m,-}(t)\big| \;\le\; C_0\sqrt{E_{\bar\mu_\alpha^N(t)}[I_m]} \;=\; C_0 f_N^m(t)^{1/2}\,.
$$

In particular, by the induction assumption, the third term of the right hand side of (2.2) converges to 0 uniformly over compact sets of $\mathbb{R}_+$.

To conclude the proof of the lemma, it remains to notice that

$$
\begin{aligned}
&\bar\mu_\alpha(t_0)\{(\eta,\xi);\ \eta(0)<\xi(0),\ \xi(x)<\eta(x)\} \\
&\quad\le \sum_{m=0}^n \bar\mu_\alpha(t_0)\{(\eta,\xi);\ m=\eta(0)<\xi(0),\ \xi(x)<\eta(x)\} \;+\; \bar\mu_\alpha(t_0)\{\eta(0)>n\} \\
&\quad\le \sum_{m=0}^n \lim_{N\to\infty} f_N^m(t_0) \;+\; \frac{\alpha_1}{n}
\end{aligned}
$$

because by assumption $\bar\mu_\alpha^N(t_0)$ converges to $\bar\mu_\alpha(t_0)$ and both marginals are bounded by $\nu_{\alpha_1}^N$. We just proved that the first term on the right hand side vanishes for all m. The second vanishes as $n\uparrow\infty$. □

Lemma 2.5 *For every $t_0>0$,*

$$
\bar\mu_\alpha(t_0)\{(\eta,\xi);\ \eta\le\xi \text{ or } \xi\le\eta\} \;=\; 1\,.
$$

Proof. Since $\bar\mu_\alpha(t_0)$ is translation invariant, we just need to prove that

$$
\bar\mu_\alpha(t_0)\{(\eta,\xi);\ \eta(0)<\xi(0),\ \xi(x)<\eta(x)\} \;=\; 0
$$

for all x in $\mathbb{Z}^d$ and a similar identity with η and ξ interchanged.

Fix x in $\mathbb{Z}^d$ and denote by $a_N^x(t)$ the function

$$a_N^x(t) = E_{\bar{\mu}_\alpha^N(t)}\Big[\mathbf{1}\{\eta(0) < \xi(0),\ \xi(x) < \eta(x)\}\Big] .$$

We shall prove that for each x in $\mathbb{Z}^d$ a_N^x converges to 0 uniformly on each compact interval of $\mathbb{R}_+$, as $N \uparrow \infty$:

$$\lim_{N\to\infty} a_N^x = 0 \quad \text{locally uniformly} . \tag{2.3}$$

Fix z in $\mathbb{Z}^d$. Since the transition probability is irreducible, there exists a positive integer m and a path $0 = x_0, x_1, \dots, x_m = z$ such that $p(x_i - x_{i+1}) - p(x_{i+1} - x_i) > 0$. Denote by $M(z)$ the length of the smallest path. We shall prove (2.3) by induction on $M(z)$. In Lemma 2.4 above we proved (2.3) for all z such that $M(z) = 1$. Indeed, with the notation introduced in the previous lemma,

$$a_N^x(t) \le \sum_{m=0}^{n} f_N^m(t) + \frac{\alpha_1}{n}$$

for every $n \ge 1$ and f_N^m converges to 0 uniformly on each compact subset of $\mathbb{R}_+$ for each $m \ge 0$.

Fix $m \ge 1$ and assume that (2.3) has been proved for each z such that $M(z) \le m$. Fix z in $\mathbb{Z}^d$ such that $M(z) = m + 1$ and a path $0 = x_0, x_1, \dots, x_{m+1} = z$. To fix ideas, assume that $p(x_m - z) > 0$. If $p(x_m - z) = 0$, in the proof below, we just need to consider the path $x_m - z, 0, x_1, \dots, x_m$ and let $x_m - z$, 0 and x_m play the roles of z, x_m and 0, respectively.

Denote by I^x the indicator function of the set $\{\eta(0) < \xi(0),\ \xi(x) < \eta(x)\}$. We have that

$$a_N^{x_m}(t) = \theta(N) \int_0^t \{b_N^+(s) + b_N^-(s)\}\, ds ,$$

where $b_N^+ = (1 - I^{x_m})\bar{L}_N I^{x_m}$, $b_N^- = I^{x_m}\bar{L}_N I^{x_m}$. Notice that while b_N^+ is positive, b_N^- is negative.

By Taylor expansion, for each $t \ge 0$, $\delta > 0$, we have

$$\begin{aligned} 0 \le b_N^+(t) &= \frac{a_N^{x_m}(t + \delta\theta(N)^{-1}) - a_N^{x_m}(t)}{\delta} \\ &\quad - \theta(N) \int_t^{t+\delta\theta(N)^{-1}} b_N^-(s)\, ds - \frac{\delta}{2\theta(N)} (b_N^+)'(s_N(t)) , \end{aligned}$$

where $s_N(t)$ belongs to the interval $[t, t+\delta\theta(N)^{-1}]$. By the induction assumption, the first term on the right hand side converges to 0 uniformly on each compact subset of $\mathbb{R}_+$. A simple computation shows that $(b_N^+)'$ is of order $\theta(N)$ so that the third term is bounded by δ. Finally, as in the proof of the previous lemma, $b_N^-(t)$ is bounded by $C\sqrt{a_N^{x_m}(t)}$ for some constant $C = C(\alpha_1, g^*)$. In particular, the second term is bounded by δ. This proves that b_N^+ converges to 0 uniformly on each compact subset of $\mathbb{R}_+$.

Since $p(x_m - z) > 0$ among the positive terms of $b_N^+(t)$ there is one that corresponds to a jump of a η-particle from z to x_m when $\eta(0) < \xi(0)$, $\eta(x_m) =$

$\xi(x_m)$ and $\eta(z) > \xi(z)$. This jump happens with rate $p(x_m - z)[g(\eta(z)) - g(\xi(z))]$. Therefore, since $g(k) \ge g(1) > 0$,

$$\begin{aligned} 0 \;&\le\; E_{\bar\mu_\alpha^N(t)}\Big[\mathbf{1}\{\eta(0) < \xi(0),\, \eta(x_m) = \xi(x_m),\, \eta(z) > \xi(z) = 0\}\Big] \\ &\le\; \{p(x_m - z) g(1)\}^{-1} b_N^+(t)\; . \end{aligned} \tag{2.4}$$

The indicator function of the previous equation can be rewritten as

$$\begin{aligned} &1\{\eta(0) < \xi(0),\, \eta(z) > \xi(z) = 0\} \\ &\quad - \; 1\{\eta(0) < \xi(0),\, \eta(x_m) > \xi(x_m),\, \eta(z) > \xi(z) = 0\} \\ &\qquad - \; 1\{\eta(0) < \xi(0),\, \eta(x_m) < \xi(x_m),\, \eta(z) > \xi(z) = 0\}\; . \end{aligned}$$

Since by the induction assumption $E_{\bar\mu_\alpha^N(t)}[\mathbf{1}\{\eta(0) < \xi(0),\, \eta(x_m) > \xi(x_m)\}]$ and

$$\begin{aligned} &E_{\bar\mu_\alpha^N(t)}[\mathbf{1}\{\eta(x_m) < \xi(x_m),\, \eta(z) > \xi(z)\}] \\ &\quad = \; E_{\bar\mu_\alpha^N(t)}[\mathbf{1}\{\eta(0) < \xi(0),\, \eta(z - x_m) > \xi(z - x_m)\}] \end{aligned}$$

converge to 0 uniformly on each compact subset of $\mathbb{R}_+$, the same holds for $E_{\bar\mu_\alpha^N(t)}[\mathbf{1}\{\eta(0) < \xi(0),\, \eta(z) > \xi(z) = 0\}]$ in virtue of (2.4).

It remains now to repeat the proof of Lemma 2.4 to remove the restriction $\xi(z) = 0$. This concludes the proof of the lemma. □

We turn now to the proof that the sequence d_N converges uniformly to 0 on each compact set of $\mathbb{R}_+$.

Lemma 2.6 *The sequence d_N converges uniformly on each compact set of $\mathbb{R}_+$.*

Proof. Fix an interval $[0, u]$. To prove that a sequence of continuous functions converges uniformly on the compact set $[0, u]$, we may proceed as follows: prove first that the limit is unique and then show that all sequences admit a converging subsequence.

To show that the limit is unique, consider a subsequence d_{N_j} that converges uniformly on $[0, u]$ to d_∞. To keep notation simple assume that the sequence d_N converges. The sequence is bounded since

$$\begin{aligned} d_N(t) \;&=\; E_{\bar S^N_{t\theta(N)} \bar\mu^N_\alpha}\Big[N^{-d} \sum_x \big|\eta(x) - \xi(x)\big|\Big] \\ &\le\; E_{S^N_{t\theta(N)}\mu^N}\Big[N^{-d} \sum_x \eta(x)\Big] \;+\; E_{S^N_{t\theta(N)}\nu^N_\alpha}\Big[N^{-d} \sum_x \xi(x)\Big] \;\le\; \alpha_0 + \alpha\; . \end{aligned}$$

By the dominated convergence theorem, for every $0 \le t \le u$,

$$\lim_{N\to\infty} \int_0^t d_N(s)\, ds \;=\; \int_0^t d_\infty(s)\, ds\; .$$

On the other hand, recalling the definition of $d_N(t)$,

$$\int_0^t d_N(s)\,ds = \int_0^t ds\, E_{\bar{S}^N_{s\theta(N)}\bar{\mu}^N_\alpha}\Big[N^{-d}\sum_x \big|\eta(x)-\xi(x)\big|\Big]$$

$$= \int_0^t ds\, E_{S^N_{s\theta(N)}\mu^N}\Big[N^{-d}\sum_x \big|\eta^\ell(x)-\alpha\big|\Big]$$

$$+ \int_0^t ds\, E_{\bar{S}^N_{s\theta(N)}\bar{\mu}^N_\alpha}\Big[N^{-d}\sum_x \tau_x V_\ell(\eta,\xi)\Big]\,,$$

where,

$$V_\ell(\eta,\xi) = \frac{1}{(2\ell+1)^d}\sum_{|y|\le\ell}\big|\eta(y)-\xi(y)\big| - \big|\eta^\ell(0)-\alpha\big|\,.$$

By assumption (**MF**), the first term on the right hand side of the last identity converges to

$$\int_0^t A(F_\alpha,s)\,ds$$

where, for a real positive α, the function F_α is equal to $F_\alpha(u)=|u-\alpha|$. The second term is bounded above by

$$\int_0^t ds E_{\bar{S}^N_{s\theta(N)}\bar{\mu}^N_\alpha}\Big[N^{-d}\sum_x\Big|\frac{1}{(2\ell+1)^d}\sum_{|y-x|\le\ell}\big|\eta(y)-\xi(y)\big| - \big|\eta^\ell(x)-\xi^\ell(x)\big|\Big|\Big]$$

$$+ \int_0^t ds\, E_{S^N_{s\theta(N)}\nu^N_\alpha}\Big[N^{-d}\sum_x\big|\xi^\ell_s(x)-\alpha\big|\Big]\,.$$

By Lemma 8.2.3, the first term vanishes as $N\uparrow\infty$ and then $\ell\uparrow\infty$ (this result was proved for asymmetric systems but the arguments apply to mean-zero processes speeded up by N^2). On the other hand, since the measure ν^N_α is invariant, the second expectation is equal to

$$t\,E_{\nu^N_\alpha}\Big[\big|\xi^\ell(0)-\alpha\big|\Big] \le \frac{t}{(2\ell+1)^{d/2}}E_{\nu^N_\alpha}\Big[\big(\xi(0)-\alpha\big)^2\Big]^{1/2}$$

that vanishes as $\ell\uparrow\infty$. In conclusion, we proved that the time integral of any converging subsequence is equal to the time integral of $A(F_\alpha,\cdot)$. In particular, the unique possible limit point of the sequence d_N is the continuous function $A(F_\alpha,\cdot)$.

To prove that every sequence admits a converging subsequence, fix a subsequence N_j, that we denote by N to keep notation simple and consider a countable and dense subset D of $[0,u]$. Since the functions d_N are uniformly bounded, by Cantor diagonal procedure, we may obtain a subsequence N_j so that d_{N_j} converges at each point of D. Denote by d_∞ the function defined on D thus obtained. It is a nonincreasing function since, for each N, d_N is a nonincreasing function. We may therefore extend the definition of d_∞ to $[0,u]$ taking right limits: $\tilde{d}_\infty(r)=\sup\{d_\infty(s),\ s>r,\ s\in D\}$. Notice that on points r of D, $\tilde{d}_\infty(r)\le d_\infty(r)$ and there might be strict inequality. The function thus obtained, denoted from now on by d_∞ to keep notation simple, is right continuous, nonincreasing and has therefore at most a countable number of discontinuities. Moreover, it is not difficult to see that the subsequence d_{N_j} converges pointwisely to d_∞ at all continuity points of d_∞. In particular, d_{N_j} converges almost surely to d_∞. Since the sequence is uniformly bounded, by the dominated convergence theorem,

$$\lim_{j\to\infty} \int_0^t d_{N_j}(s)\, ds \;=\; \int_0^t d_\infty(s)\, ds$$

for all $t \le u$. We have seen in the first part of the proof that

$$\lim_{j\to\infty} \int_0^t d_{N_j}(s)\, ds \;=\; \int_0^t A(F_\alpha, s)\, ds\ .$$

Therefore, d_∞ is almost surely equal to the continuous function $A(F_\alpha,\cdot)$. Since d_∞ is nonincreasing, the two functions are equal. We have thus a sequence of nonincreasing continuous functions d_{N_j} that converges pointwisely to a continuous function $A(F_\alpha,\cdot)$. This forces the sequence to converge uniformly on $[0,u]$. □

3. Conservation of Local Equilibrium

Recall from Appendix 2 the terminology of weak solutions of the nonlinear partial differential equation

$$\partial_t \rho \;=\; \sum_{1\le i,j\le d} \sigma_{i,j}\partial^2_{u_i,u_j}\Phi(\rho) \tag{3.1}$$

and Theorem A2.4.3 and Theorem A2.4.5 therein that shall be invoked throughout this section.

We start with a weak version of local equilibrium for attractive systems. This question was already investigated in Chapter 6 in the case of mean-zero processes and in Chapter 8 in the case of asymmetric processes. The next result follows from Corollary 6.1.3 and some elementary coupling arguments.

Theorem 3.1 *Fix a continuous profile $\rho_0 : \mathbb{T}^d \to \mathbb{R}_+$ and recall that we denote by $\nu^N_{\rho_0(\cdot)}$ the product measure with slowly varying parameter associated to ρ_0. Consider an attractive mean-zero asymmetric zero range process with generator given by (0.1). For every $t > 0$, every bounded cylinder function Ψ and every continuous function $H : \mathbb{T}^d \to \mathbb{R}$,*

$$\lim_{N\to\infty} E_{S^N_{tN^2}\nu^N_{\rho_0(\cdot)}}\Big[N^{-d}\sum_x H(x/N)(\tau_x\Psi)(\eta)\Big] \;=\; \int du\, H(u)\tilde\Psi(\rho(t,u))\ ,$$

where $\rho(t,u)$ is the unique weak solution of (3.1) with initial data ρ_0.

Proof. This result was proved in Corollary 6.1.3 for smooth enough strictly positive profiles ρ_0. The strategy is thus to approximate ρ_0 by a sequence ρ_0^ε

of such smooth profiles and couple two copies of the process starting from an initial measure with first marginal equal to $\nu^N_{\rho_0(\cdot)}$ and second marginal equal to $\nu^N_{\rho_0^\varepsilon(\cdot)}$. To use then coupling arguments to show that $E_{S^N_{tN^2}\nu^N_{\rho_0(\cdot)}}[\tau_x\Psi]$ is close to $E_{S^N_{tN^2}\nu^N_{\rho_0^\varepsilon(\cdot)}}[\tau_x\Psi] \sim \tilde\Psi(\rho^\varepsilon(t,x/N))$ for ε small and the continuous dependence on the initial data of solutions of the equation (3.1) to prove that $\tilde\Psi(\rho^\varepsilon(t,x/N))$ is close to $\tilde\Psi(\rho(t,x/N))$.

Fix a continuous function $\rho_0\colon \mathbb{T}^d \to \mathbb{R}_+$ and consider a sequence $\rho_0^\varepsilon\colon \mathbb{T}^d \to \mathbb{R}_+$ of strictly positive profiles of class $C^\infty(\mathbb{T}^d)$ converging to ρ_0 uniformly on $\mathbb{T}^d$ and bounded below by ρ_0: $\rho_0 \le \rho_0^\varepsilon$. For each $\varepsilon > 0$, denote by $\rho^\varepsilon(t,u)$ (resp. $\rho(t,u)$) the unique weak solution of equation (3.1) with initial data ρ_0^ε (resp. ρ_0). By Theorem A2.4.5, for each $t \ge 0$, $\rho^\varepsilon(t,\cdot)$ converges uniformly to $\rho(t,\cdot)$.

Assume for a while that

$$
\begin{aligned}
&\lim_{N\to\infty} E_{S^N_{tN^2}\nu^N_{\rho_0(\cdot)}}\Big[N^{-d}\sum_x H(x/N)(\tau_x\Psi)(\eta)\Big] \\
&\quad = \lim_{\varepsilon\to 0}\lim_{N\to\infty} E_{S^N_{tN^2}\nu^N_{\rho_0^\varepsilon(\cdot)}}\Big[N^{-d}\sum_x H(x/N)(\tau_x\Psi)(\eta)\Big]\,.
\end{aligned}
\tag{3.2}
$$

In this case, by Corollary 6.1.3 and Remark 6.1.14, the right hand side is equal to

$$
\lim_{\varepsilon\to 0}\int du\, H(u)\tilde\Psi(\rho^\varepsilon(t,u)) \;=\; \int du\, H(u)\tilde\Psi(\rho(t,u))
$$

because $\rho^\varepsilon(t,\cdot)$ converges uniformly to $\rho(t,\cdot)$ and $\tilde\Psi(\cdot)$ is a continuous bounded function. Therefore, to prove the theorem we just need to justify identity (3.2)

For each $\varepsilon > 0$, denote by $\bar\mu^N_\varepsilon$ the probability measure on the product space $\mathbb{N}^{\mathbb{Z}^d} \times \mathbb{N}^{\mathbb{Z}^d}$ whose first marginal is equal to $\nu^N_{\rho_0(\cdot)}$, second marginal is equal to $\nu^N_{\rho_0^\varepsilon(\cdot)}$ and concentrated on configurations (η,ξ) such that $\eta \le \xi$. This is possible because $\nu^N_{\rho_0(\cdot)}$ and $\nu^N_{\rho_0^\varepsilon(\cdot)}$ are product measures and $\rho_0 \le \rho_0^\varepsilon$.

Recall that we denote by $\bar S^N_t$ the semigroup measure associated to coupling generator $\bar L_N$ defined in (2.5.1). With this notation we may write

$$
\begin{aligned}
&E_{S^N_{tN^2}\nu^N_{\rho_0^\varepsilon(\cdot)}}\Big[N^{-d}\sum_x H(x/N)(\tau_x\Psi)(\eta)\Big] \\
&\quad - E_{S^N_{tN^2}\nu^N_{\rho_0(\cdot)}}\Big[N^{-d}\sum_x H(x/N)(\tau_x\Psi)(\eta)\Big] \\
&\quad = E_{\bar S^N_{tN^2}\bar\mu^N_\varepsilon}\Big[N^{-d}\sum_x H(x/N)\tau_x\{\Psi(\xi)-\Psi(\eta)\}\Big]\,.
\end{aligned}
$$

Since Ψ is a bounded cylinder function, there exists a constant $C = C(\Psi)$ such that $|\Psi(\xi) - \Psi(\eta)| \le C\sum_{|x|\le s_\Psi}|\xi(x) - \eta(x)|$ for any two configurations η, ξ. In particular, the absolute value of the right hand side of the previous equality is bounded above by

$$C(\Psi, H) E_{\tilde{S}^N_{tN^2} \bar{\mu}^N_\varepsilon} \Big[N^{-d} \sum_x |\xi(x) - \eta(x)| \Big] .$$

Since at time 0 the configurations are ordered ($\eta_0 \le \xi_0$) and since the dynamics preserves the order, $\eta_t \le \xi_t$ for all $t \ge 0$. In particular, in the previous formula, $|\xi(x) - \eta(x)| = \xi(x) - \eta(x)$. Since, on the other hand, the total number of particles is conserved, last expectation is equal to

$$C(\Psi, H) \Big\{ E_{\nu^N_{\rho_0^\varepsilon(\cdot)}} \Big[N^{-d} \sum_x \xi(x) \Big] \; - \; E_{\nu^N_{\rho_0(\cdot)}} \Big[N^{-d} \sum_x \eta(x) \Big] \Big\}$$

that converges to

$$C(\Psi, H) \Big\{ \int du\, \rho_0^\varepsilon(u) \; - \; \int du\, \rho_0(u) \Big\}$$

as $N \uparrow \infty$. This expression vanish as $\varepsilon \downarrow 0$ because ρ_0^ε converges to ρ_0 uniformly on $\mathbb{T}^d$, what proves (3.2). □

In the asymmetric context the same statement follows from the law of large numbers for the Young measure proved in Chapter 8.

Theorem 3.2 *Fix a continuous profile $\rho_0 : \mathbb{T}^d \to \mathbb{R}_+$ and recall that we denote by $\nu^N_{\rho_0(\cdot)}$ the product measures with slowly varying parameter associated to ρ_0. Consider an attractive asymmetric zero range process with generator given by (0.1). For every $t > 0$, every bounded cylinder function Ψ and every continuous function $H : \mathbb{T}^d \to \mathbb{R}$,*

$$\lim_{N\to\infty} E_{S^N_{tN} \nu^N_{\rho_0(\cdot)}} \Big[N^{-d} \sum_x H(x/N)(\tau_x \Psi)(\eta) \Big] \; = \; \int du\, H(u) \tilde{\Psi}(\rho(t,u)) \, ,$$

where $\rho(t,u)$ is the unique entropy solution of the differential equation (0.6).

Proof. Due to Theorem 2.1, we just need to show that

$$\lim_{\ell\to\infty} \lim_{N\to\infty} E_{S^N_{tN} \nu^N_{\rho_0(\cdot)}} \Big[N^{-d} \sum_x H(x/N) \tilde{\Psi}(\eta_t^\ell(0)) \Big] \; = \; \int du\, H(u) \tilde{\Psi}(\rho(t,u)) \tag{3.3}$$

for every $t > 0$, every bounded cylinder function Ψ and every continuous function H. Recall the definition of the Young measure $\pi_t^{N,\ell}(dq, du)$ introduced in Chapter 8. With this notation, we may rewrite last expectation as

$$E_{S^N_{tN} \nu^N_{\rho_0(\cdot)}} \Big[\int_{\mathbb{T}^d} \int_{\mathbb{R}_+} H(u) \tilde{\Psi}(q) \pi^{N,\ell}(dq, du) \Big]$$

that converges, by Theorem 8.1.1, to $\int du\, H(u) \tilde{\Psi}(\rho(t,u))$ as $N \uparrow \infty$ and then $\ell \uparrow \infty$ provided ρ_0 is Lipschitz continuous. A simple coupling argument, very similar to the one presented in the proof of Theorem 3.1, permits to extend (3.3) to continuous profiles. □

We have now all tools to prove the conservation of local equilibrium.

Proof of Theorem 0.1. Fix a continuous function $\rho_0 : \mathbb{T}^d \to \mathbb{R}_+$ and denote by $\rho(t,u)$ the solution of (3.1) with initial condition ρ_0.

For each $\varepsilon > 0$, define $\rho_0^{\varepsilon,\pm} : \mathbb{T}^d \to \mathbb{R}_+$ by

$$\rho_0^{\varepsilon,+}(u) = \sup_{v\in B(u,\varepsilon)} \rho_0(v) , \qquad \rho_0^{\varepsilon,-}(u) = \inf_{v\in B(u,\varepsilon)} \rho_0(v) ,$$

where $B(u,\varepsilon)$ is a ball centered at u of radius ε for the max norm: $B(u,\varepsilon) = \{v, \max_{1\le i\le d}|v_i - u_i| \le \varepsilon\}$. Notice that for each $\varepsilon > 0$ $\rho_0^{\varepsilon,\pm}$ is a continuous function and that

$$\rho_0^{\varepsilon,-}(v_1) \le \rho_0(u) \le \rho^{\varepsilon,+}(v_2) \tag{3.4}$$

for each v_1, v_2 in $B(u,\varepsilon)$.

Denote by $\rho^{\varepsilon,\pm}(t,u)$ the solution of (3.1) with initial condition $\rho_0^{\varepsilon,\pm}$. By Theorem A2.4.5,

$$\lim_{\varepsilon\to 0} \rho^{\varepsilon,\pm}(t,u) = \rho(t,u) \tag{3.5}$$

for every $t \ge 0$ and u in $\mathbb{T}^d$. By property (3.4), for all u_1, u_2 in $B(0,\varepsilon)$,

$$\tau_{[u_1N]}\nu^N_{\rho_0^{\varepsilon,-}(\cdot)} \le \nu^N_{\rho_0(\cdot)} \le \tau_{[u_2N]}\nu^N_{\rho_0^{\varepsilon,+}(\cdot)} .$$

Therefore,

$$S^N_{tN^2}\tau_{[v_1N]}\nu^N_{\rho_0^{\varepsilon,-}(\cdot)} \le S^N_{tN^2}\tau_{[uN]}\nu^N_{\rho_0(\cdot)} \le S^N_{tN^2}\tau_{[v_2N]}\nu^N_{\rho_0^{\varepsilon,+}(\cdot)} \tag{3.6}$$

for each v_1, v_2 in $B(u,\varepsilon)$ because the order is preserved by attractive processes.

Fix $t > 0$, u in $\mathbb{T}^d$, $\varepsilon > 0$ and consider a bounded monotone cylinder function Ψ. We claim that

$$\limsup_{N\to\infty} E_{S^N_{tN^2}\tau_{[uN]}\nu^N_{\rho_0(\cdot)}}[\Psi] \le E_{\nu_{\rho(t,u)}}[\Psi] = \tilde{\Psi}(\rho(t,u)) . \tag{3.7}$$

Indeed, for $0 < \lambda < \varepsilon$, consider a sequence of continuous approximations $H_{\lambda,k} : \mathbb{R}^d \to \mathbb{R}_+$ of the function $H_\lambda = \lambda^{-d}\mathbf{1}\{B(0,\lambda/2)\}$ with support contained in $[-\lambda,\lambda]^d$:

$$\begin{aligned} H_\lambda = \lambda^{-d}\mathbf{1}\{B(0,\lambda/2)\} &\le H_{\lambda,k} \le \lambda^{-d}\mathbf{1}\{B(0,\lambda)\} \\ \text{and} \quad \lim_{k\to\infty} H_{\lambda,k} &= H_\lambda \end{aligned}$$

pointwisely. By inequality (3.6) and by definition of $H_{\lambda,k}$ we have that

$$E_{S^N_{tN^2}\tau_{[uN]}\nu^N_{\rho_0(\cdot)}}[\Psi] \le E_{S^N_{tN^2}\nu^N_{\rho_0^{\varepsilon,+}(\cdot)}}\Big[N^{-d}\sum_x H_{\lambda,k}(x/N)\tau_{[uN]+x}\Psi(\eta)\Big]$$

because $\lambda < \varepsilon$. By Theorem 3.1, as $N \uparrow \infty$, the right hand side converges to

$$\int dv\, H_{\lambda,k}(v)\tilde{\Psi}(\rho^{\varepsilon,+}(t,u+v)) .$$

Since $\tilde{\Psi}$ is a bounded function and the sequence $H_{\lambda,k}$ converges pointwisely to the function H_λ, as $k \uparrow \infty$, this expression converges to

$$\int dv\, H_\lambda(v)\tilde{\Psi}(\rho^{\varepsilon,+}(t, u+v)) \, . \tag{3.8}$$

Since H_λ is an approximation of the identity, as λ decreases to 0, the integral converges to

$$\tilde{\Psi}(\rho^{\varepsilon,+}(t,u))$$

because the solution of the parabolic equation (3.1) is continuous and $\tilde{\Psi}(\cdot)$ is a continuous bounded function. Finally, by (3.5), letting $\varepsilon \downarrow 0$ we obtain (3.7).

The same argument with an approximation from below of the function $\lambda^{-d}\mathbf{1}\{B(0,\lambda/2)\}$ and with $\rho^{\varepsilon,-}$ replacing $\rho^{\varepsilon,+}$ shows that

$$\liminf_{N\to\infty} E_{S^N_{tN^2}\tau_{[uN]}\nu^N_{\rho_0(\cdot)}}[\Psi] \;\geq\; \tilde{\Psi}(\rho(t,u)) \, .$$

This concludes the proof of the theorem since, by Lemma 2.1.5, to prove that a sequence of probability measures μ^N converges weakly to a probability measure μ, it is enough to show that the expected value of all bounded monotone cylinder functions converges. □

The proof of the conservation of local equilibrium for asymmetric processes is exactly the same and relies on the following result on the continuous dependence on initial data of entropy solutions of equation (0.6). Recall from the proof of Theorem 0.1 the definition of the profiles $\rho_0^{\varepsilon,\pm}$. For each $\varepsilon > 0$, denote by $\rho^{\varepsilon,\pm}(t,u)$ the entropy solution of equation (0.6) with initial condition $\rho_0^{\varepsilon,\pm}$.

Lemma 3.3 *Fix $t > 0$ and recall the notation introduced in the proof of Theorem 0.1. For every continuity point u of $\rho(t,\cdot)$,*

$$\lim_{\varepsilon\to 0}\sup_{|v-u|\leq\varepsilon} |\rho^{\varepsilon,\pm}(t,v) - \rho(t,u)| \;=\; 0 \, .$$

Proof. The proof of this lemma relies on a result of P. Lax (1957), stated here as Theorem A2.5.10, that asserts that the entropy solution depends continuously on the initial data for-one dimensional equations.

A change of coordinates permits to rewrite equation (0.6) as

$$\partial_t \rho \;+\; \partial_{u_1}\Phi(\rho) \;=\; 0 \, .$$

It is therefore enough to prove the lemma for this equation, which is a one-dimensional equation and for which the variables $u_2, \dots, u_d$ may be interpreted as parameters.

Fix $t > 0$ and a continuity point u of $\rho(t,\cdot)$. In order to prove the lemma, it is enough to show that for every sequence $\{a_\varepsilon,\, \varepsilon > 0\}$ such that $|a_\varepsilon| \leq \varepsilon$, $\rho^{\varepsilon,\pm}(t, u+a_\varepsilon)$ converges to $\rho(t,u)$. Fix such sequence $\{a_\varepsilon,\, \varepsilon > 0\}$. By definition of

$\rho_0^{\varepsilon,\pm}$, $\rho_0^{\varepsilon,\pm}(\cdot + a_\varepsilon)$ converges uniformly to $\rho_0(\cdot)$ because ρ_0 is uniformly continuous. Since u is a continuity point of $\rho(t,\cdot)$, u_1 is a continuity point of $\rho(t,\cdot,u_2,\dots,u_d)$. Therefore, by Theorem A2.5.10, $\rho(t,u+a_\varepsilon)$ converges to $\rho(t,u)$ as $\varepsilon \downarrow 0$, what concludes the proof of the lemma. □

Proof of Theorem 0.2. In the proof of Theorem 0.1, we did not used any property of the solutions of equation (0.5) until formula (3.8). In the hyperbolic setup, we may bound the integral (3.8) by

$$\tilde{\Psi}(\rho(t,u)) + \sup_{a \le \|\rho_0\|_\infty} |\tilde{\Psi}'(a)| \sup_{|v-u|\le\varepsilon} |\rho^{\varepsilon,+}(t,v) - \rho(t,u)|$$

because $\lambda \le \varepsilon$. In the previous lemma we showed that the second term vanishes as $\varepsilon \downarrow 0$. This concludes the proof of the conservation of local equilibrium for asymmetric processes. □

4. Comments and References

The proof of the one block estimate without time average is due to Rezakhanlou (1991). It relies partially on Proposition 2.2 whose proof is very similar to the one of Proposition 5.1 of Andjel (1982) and Theorem VIII.3.9 (a) of Liggett (1985). The idea to deduce the conservation of local equilibrium from a law of large numbers for the local fields is taken from Landim (1991b), Landim (1993).

We review here some aspects of the macroscopic motion of asymmetric processes.

Microscopic structure of the shock. We have seen in this chapter that in the asymmetric case the empirical measure converges in probability to the entropy solution of a first order hyperbolic equation. These equations have the peculiarity that the solutions develop shocks even when the initial data is smooth. An interesting question at the physical level concerns the microscopic structure of the shock. It consists in determining whether there exists a finer scale than the hydrodynamic scale where the density profile becomes smooth or whether such an intermediary scale does not exists and the shock is sharp.

To examine this question consider a nearest neighbor asymmetric simple exclusion process on $\mathbb{Z}$ with probability p to jump to the right and probability $q = 1 - p < p$ to jump to the left. Fix $\alpha < \beta$ and let $\nu_{\alpha,\beta}$ be the inhomogeneous Bernoulli measure with density α at the left of the origin and β at the right. In this case the entropy solution of the hydrodynamic equation is a traveling wave: $\rho(t,u) = \alpha \mathbf{1}\{u < vt\} + \beta \mathbf{1}\{u > vt\}$, where the velocity v of the shock is equal to $(p-q)(1-\alpha-\beta)$.

Add a particle at the origin if this site is empty, tag this particle and let it evolve as a second class particle. This means that at rate p (resp. q) the particle will attempt to jump to the right (resp. to the left). If the site chosen is empty the particle jumps, otherwise nothing happens. In addition, if another particle tries to jump to the site occupied by the tagged particle they exchange position. In particular, if the site on the left (resp. right) of the tagged particle is occupied, at rate p (resp. q) the tagged particle jumps to the left (resp. right). The tagged particle is called a second class particle because the other particles have priority over it and jump to a site even when it is occupied by the tagged particle.

Denote by Z_t the position of the tagged particle at time t. Ferrari (1992) proved that the process as seen from the tagged particle (i.e. $\tau_{Z_t}\eta_t$) converges weakly to an invariant measure $\mu_{\alpha,\beta}$ with asymptotic distribution ν_α and ν_β: $\lim_{x\to\infty} \tau_x\mu_{\alpha,\beta} = \nu_\beta$, $\lim_{x\to-\infty} \tau_x\mu_{\alpha,\beta} = \nu_\alpha$. He proved moreover that for each fixed time t the distribution of $\tau_{Z_t}\eta_t$ has the same asymptotics. This proves the sharpness of the shock since there is a random position from which microscopically to the left (resp. right) we see the invariant measure with density α (resp. β).

The question of the microscopic structure of the shock was examined by several authors. Ferrari (1986) investigated the invariant measures of exclusion processes as seen from a first class tagged particle. Wick (1985) considered the structure of the shock for the totally asymmetric case ($p = 1$) with no particles at the left of the origin ($\alpha = 0$). De Masi, Kipnis, Presutti and Saada (1989) extended the result to the case $p < 1$. In both previous situations the position of the shock is determined by the leftmost particle and the invariant measure for the process as seen from this particle is explicitly known and its asymptotics converge exponentially fast to ν_β. In the case $\alpha > 0$, however, there is no such natural choice. Nevertheless, Ferrari, Kipnis and Saada (1991) proved in this case the existence of a random position X_t from which to the left (resp. right) we see the invariant measure with density α (resp. β). Ferrari (1992) showed that the position of a second class particle has this property.

In higher dimension nothing is known. Alexander et al. (1992) present simulations of the shock in a two-dimensional asymmetric exclusion process.

Closely related to the problem of the microscopic structure of the shock is the question of the position of a second class particle in asymmetric processes.

Asymptotics of a second class particle. Consider the nearest neighbor asymmetric simple exclusion process on $\mathbb{Z}$ with probability p to jump to the right and $q = 1 - p < p$ to jump to the left. Fix $\alpha < \beta$ and recall from the previous subsection the definition of the product measure $\nu_{\alpha,\beta}$ and the definition of the evolution of a second class particle. Denote by Z_t the position of a second class particle initially at the origin. Ferrari and Fontes (1994b) proved that

$$\lim_{N\to\infty} \frac{Z_{tN} - vtN}{\sqrt{N}} = W_t ,$$

weakly in the sense of finite dimensional distributions, provided W_t stands for a Brownian motion with diffusion coefficient

$$D = (p-q)\frac{\alpha(1-\alpha) + \beta(1-\beta)}{\beta - \alpha} \tag{4.1}$$

and v stands for the velocity of the shock which is equal to $(p-q)(1-\alpha-\beta)$.

Wick (1985) proved this central limit theorem in the context of totally asymmetric simple exclusion processes with $\alpha = 0$. De Masi, Kipnis, Presutti and Saada (1989) extended this result to the case $1/2 < p < 1$. Gärtner and Presutti (1990) proved in the case $p = 1$, $\alpha = 0$ that the fluctuations of Z_t arise from the fluctuations of the initial state. Ferrari (1992) extended this result to the case $p < 1$. Ferrari and Fontes (1994b) proved the result that we just described and extended Gärtner and Presutti result to the case $0 < \alpha < \beta$, $1/2 < p < 1$. Ferrari and Fontes (1996) proved a central limit theorem for the current over macroscopic regions for one-dimensional nearest neighbor asymmetric simple exclusion processes starting from an invariant state ν_α.

From the central limit theorem for the second class particle and the microscopic structure of the shock we may deduce the behavior of the state of the process at the discontinuity points of the solution of the hydrodynamic equation. This question is further discussed in the next subsection.

Rezakhanlou (1995) considered the behavior of a second class particle in one-dimensional asymmetric misanthrope processes starting from product initial measures. To fix ideas we shall present the results in the context of exclusion processes with jump rate $p(\cdot)$. Fix a profile $\rho_0 \colon \mathbb{R}_+ \to [0,1]$ in $L^1(\mathbb{R})$ and denote by μ^N a product measure associated to ρ_0 in the sense that

$$\lim_{N\to\infty} \int_{|u|\le A} du \left| E_{\mu^N}[\eta([uN])] - \rho_0(u) \right| = 0$$

for every $A > 0$. Fix a_0 in $\mathbb{R}$ and denote by Z_t^N the position at time t of a second class particle initially at $[a_0 N]$.

Let $\rho(t,u)$ stand for the unique entropy solution of the hyperbolic equation $\partial_t \rho + \gamma \partial_u(\rho[1-\rho]) = 0$, where $\gamma = \sum_x x p(x)$. Recall from section 8.4 the definition of a Filippov solution of the equation

$$\begin{cases} \dfrac{da(t)}{dt} = \gamma\{1 - 2\rho(t, a(t))\}\ , \\ a(0) = a_0\ . \end{cases}$$

Rezakhanlou (1995) proved that there exists at most one solution in the Filippov sense of the previous equation provided

$$\liminf_{\varepsilon\downarrow 0} \frac{1}{\varepsilon} \int_{a_0}^{a_0+\varepsilon} \rho_0(u)\, du \;\le\; \limsup_{\varepsilon\downarrow 0} \frac{1}{\varepsilon} \int_{a_0-\varepsilon}^{a_0} \rho_0(u)\, du\ .$$

Furthermore, he proved that Z_{tN}^N/N converges in probability to $a(t)$ provided

$$\int_{a_0}^{a_0+\delta} [1-\rho_0(u)]\, du \int_{a_0-\delta}^{a_0} \rho_0(u)\, du \ne 0$$

for every positive δ.

All the results cited in this subsection show that the second class particles in asymmetric processes either follow the characteristics or the shocks of the hydrodynamic equation. Ferrari and Kipnis (1995) consider the case of a rarefaction fan, where more than one characteristic starts from the origin. They considered the nearest neighbor asymmetric simple exclusion process with probability p to jump to the right and $q = 1 - p < p$ to jump to the left starting from a product measure $\nu_{\alpha,\beta}$ with density α at the left of the origin and $\beta < \alpha$ at the right. They proved that the second class particle chooses with uniform distribution one of the characteristics and then sticks to it. More precisely, denote by Z_t the position at time t of a second class particle sitting at the origin at time 0. They proved that, as $N \uparrow \infty$, $N^{-1} Z_{tN}$ converges to U_t in distribution, where U_t is a random variable with uniform distribution over the interval $[(1 - 2\alpha)t, (1 - 2\beta)t]$. Furthermore, they showed that for $0 < s < t$, $Z_{tN}/tN - Z_{sN}/sN$ converges to 0 in probability as $N \uparrow \infty$.

Behavior at discontinuity points of the profile or dynamical phase transition. We proved in this section that the sequence $\mu^N S_{t\theta(N)} \tau_{[uN]}$ converges to $\nu_{\rho(t,u)}$ at all continuity points u of $\rho(t, \cdot)$. Nothing is said at the discontinuity points. For diffusive systems this remark is irrelevant since the solutions are Hölder continuous by Nash's theorem. However, for asymmetric processes, where the entropy solution may develop shocks even if the initial profile is smooth, the behavior at the shocks must be examined by different means.

Since the article of Liggett (1975) it is conjectured that the sequence $\mu^N S_{t\theta(N)} \tau_{[uN]}$ converges at a shock to a mixture of extremal invariant measures. The first result in this direction was proved by Wick (1985) who considered the one-dimensional totally asymmetric zero range process on $\mathbb{Z}$ moving to the right with rate $g(k) = \mathbf{1}\{k \geq 1\}$ and starting from a product measure $\nu_{0,\beta}$ with density 0 at the left of the origin and density β at the right. In this case, if $\rho_0(u) = \mathbf{1}\{u > 0\}$ stands for the initial profile, the entropy solution $\rho(t, u)$ is a traveling wave equal to $\rho_0(u - vt)$, where $v = v(\beta) = (1 + \beta)^{-1}$ is the speed of the shock. Wick (1985) proved that for any sequence T_N that increases to ∞ as $N \uparrow \infty$,

$$\lim_{N\to\infty} \frac{1}{T_N} \int_{tN}^{tN+T_N} dr \mu^N S_r \tau_{[v(\beta)tN + u\sqrt{N}]} = \{1 - m(t, u)\}\nu_\beta + m(t, u)\nu_0 ,$$

where m is the solution of the parabolic equation

$$\begin{cases} \partial_t m = (1/2) D(\beta) \Delta m \\ m(0, u) = \mathbf{1}\{u < 0\} \end{cases}$$

and $D(\beta) = (1 + \beta)^{-1}$. Wick called this phenomena a dynamical phase transition. It is closely related to the microscopic structure of the shock and to the asymptotic behavior of a second class particle.

This result was successively improved for one-dimensional nearest neighbor asymmetric exclusion processes by Andjel (1986), Andjel, Bramson and Liggett (1988), De Masi, Kipnis, Presutti and Saada (1989) and Ferrari and Fontes (1994b).

Bramson (1988) presents a short review of some of the previous results. We now describe Ferrari and Fontes (1994b). Consider an asymmetric simple exclusion process that attempts to jump with rate p to the right and rate q to the left. For α, $\beta > 0$, denote by $\nu_{\alpha,\beta}$ the product measure on $\mathbb{N}^{\mathbb{Z}}$ that has density α at the left of the origin and β at the right and by ρ_0 the associates profile: $\rho_0 = \alpha\mathbf{1}\{u < 0\} + \beta\mathbf{1}\{u \geq 0\}$. The entropy solution $\rho(t, u)$ is a traveling wave given by $\rho(t, u) = \alpha\mathbf{1}\{u < vt\} + \beta\mathbf{1}\{u > vt\}$, where the speed of the shock $v = v(\alpha, \beta)$ is equal to $(p - q)(1 - \alpha - \beta)$. Recall the definition of the diffusion coefficient D given in (4.1) and denote by $w(t, u)$ the probability of a Brownian motion with diffusion coefficient D to be less than u at time t: $w(t, u) = (2\pi tD)^{-1/2} \int_{-\infty}^{u} da \exp\{-a^2/2tD\}$. Ferrari and Fontes (1994b) proved that

$$\lim_{N\to\infty} \nu_{\alpha,\beta} S_{tN} \tau_{[vtN+a\sqrt{N}]} = (1 - w(t, a))\nu_\alpha + w(t, a)\nu_\beta .$$

By the time this note was written, Ferrari, Fontes and Vares (private communication) obtained the behavior at the shock for an increasing piecewise constant initial profile and with a finite number of discontinuities.

Stationary measures of asymmetric systems. Derrida, Domany and Mukamel (1992) consider a one-dimensional simple exclusion process on $\{0, \dots, N\}$, where particles jump only to the right. Particles are create with intensity $\alpha > 0$ at 0 and are destroyed at N with intensity $\beta > 0$. The generator of this process is therefore

$$(L_N f)(\eta) = \sum_{x=0}^{N-1} \eta(x)[1 - \eta(x+1)][f(\eta^{x,x+1}) - f(\eta)] + (L_- f)(\eta) + (L_+ f)(\eta) ,$$

where L_-, L_+ are the boundary generators given by

$$\begin{aligned} (L_+ f)(\eta) &= \beta\eta(N-1)[f(\eta - \mathfrak{d}_N) - f(\eta)] , \\ (L_- f)(\eta) &= \alpha[1 - \eta(0)][f(\eta + \mathfrak{d}_0) - f(\eta)] . \end{aligned}$$

Derrida, Domany and Mukamel (1992) present a general method to derive an explicit formula for the stationary measure and implement it in the case $\alpha = \beta = 1$. The computations were extended by Schütz and Domany (1993) to the general case $\alpha, \beta > 0$. Derrida, Evans and Mallick (1995) compute the fluctuations of the current for this model.

The method introduced by Derrida, Domany and Mukamel (1992) was extended by Derrida et al. (1993) for asymmetric exclusion processes with first and second class particles. Derrida et al. (1993) compute the stationary measure of a totally asymmetric simple exclusion process evolving on $\mathbb{T}_N$ with first and second–class particles. They deduce from this result the profile of the first class particles as seen from a second class particle in the stationary regime. This profile determines the microscopic shape of the shock linking two different densities. Speer (1994) extend this result to the infinite volume case. Ferrari, Fontes and Kohayakawa (1994) propose an alternative method to describe the invariant measure of the exclusion process in infinite volume with first and second–class particles.

Foster and Godrèche (1994) and Evans et al. (1995) examine the stationary measures of exclusion processes with two types of species. Janowsky and Lebowitz (1994) investigate the stationary measures of an inhomogeneous totally asymmetric exclusion process. Here at all sites but one particles jump to the right at rate one. In the remaining site, particles jump with rate $0 < r < 1$. Schütz (1993) obtains the stationary measure of a deterministic model with a stochastic defect.

10. Large Deviations from the Hydrodynamic Limit

In Chapters 4 and 5 we proved a law of large numbers for the empirical density of reversible interacting particle systems. A natural development of the theory is to investigate the large deviations from the hydrodynamic limit.

To avoid technical problems related to the lack of regularity of the rate function, we concentrate on symmetric simple exclusion processes. Moreover, for historical reasons, we decided to consider the process starting from an equilibrium product state: ν_α^N, for some density $0 < \alpha < 1$. In fact, the same approach applies to a process starting from any sequence η^N of deterministic configurations associated to a profile $\rho : \mathbb{T}^d \to [0,1]$ (cf. Remark 1.2).

In the case where the process starts from ν_α^N, two distinct types of large deviations of the same order arise. The first one corresponds to large deviations from the initial state. It is very simple since it reduces to large deviations of i.i.d. random variables. The second one comes from the stochastic character of the evolution. Since we are mainly interested in the latter, we ignore in this introduction the static large deviations.

We claim that in order to prove an upper bound large deviations, we just need to find a family of mean-one positive martingales that can be expressed as function of the empirical measure. Indeed, denote by $\exp\{C_N J_\beta(\pi_\cdot^N)\}$ such a martingale indexed by β in $\mathcal{A}$ and fix a compact set $\mathcal{K}$ on the path space $D([0,T], \mathcal{M}_+)$. We have

$$
\begin{aligned}
Q^N[\pi^N \in \mathcal{K}] &= E_{Q^N}\Big[e^{-C_N J_\beta(\pi_\cdot^N)} e^{C_N J_\beta(\pi_\cdot^N)} \mathbf{1}\{\pi^N \in \mathcal{K}\}\Big] \\
&\le \exp\Big\{ - C_N \inf_{\pi \in \mathcal{K}} J_\beta(\pi_\cdot)\Big\} E_{Q^N}\Big[e^{C_N J_\beta(\pi_\cdot^N)} \mathbf{1}\{\pi^N \in \mathcal{K}\}\Big] \\
&\le \exp\Big\{ - C_N \inf_{\pi \in \mathcal{K}} J_\beta(\pi_\cdot)\Big\} .
\end{aligned}
$$

The last inequality follows from the fact that $\exp\{C_N J_\beta(\pi_\cdot)\}$ is a mean-one positive martingale. Therefore, minimizing over β in $\mathcal{A}$, we have that

$$
\limsup_{N\to\infty} \frac{1}{C_N} \log Q^N[\pi^N \in \mathcal{K}] \le - \sup_{\beta \in \mathcal{A}} \inf_{\pi \in \mathcal{K}} J_\beta(\pi_\cdot) . \tag{0.1}
$$

To conclude the proof of the upper bound, it remains to justify the exchange between the supremum and the infimum. This is done through a minimax theorem relying on some regularity of J_β. The upper bound rate function obtained in this way is equal to $\sup_{\beta \in \mathcal{A}} J_\beta(\pi_\cdot)$. Of course the upper bound may be bad if we considered to few positive martingales or not the relevant ones.

This argument shows that we have to build positive martingales. Following Donsker and Varadhan (1975a,b), (1976), in the context of Markov processes, the relevant positive martingales are obtained as small Markovian perturbations of the original process.

To clarify this general philosophy we return to the case of symmetric simple exclusion processes. For each H in $C^{1,2}([0,T] \times \mathbb{T}^d)$ consider the time inhomogeneous Markov process with generator at time t given by

$$(L_{N,t}^H f)(\eta) = \\ (1/2)N^2 \sum_{|x-y|=1} \eta(x)[1-\eta(y)] e^{H(t,y/N)-H(t,x/N)} [f(\eta^{x,y}) - f(\eta)] \; .$$

This is a small perturbation of the original process in the following sense. At time t instead of jumping from x to $x \pm e_i$ with rate $1/2$, a particle jumps with rate $(1/2)\{1 \pm N^{-1}(\partial_{u_i} H)(t, x/N)\}$. We introduced therefore a small (of order N^{-1}) space and time dependent asymmetry in the jump rate.

For each H in $C^{1,2}([0,T] \times \mathbb{T}^d)$, denote by P_H^N the probability measure on $D([0,T], \{0,1\}^{\mathbb{T}_N^d})$ corresponding to the inhomogeneous Markov process η_t with generator L_N^H. When $H = 0$, we denote P_H^N simply by P^N.

Denote by $(dP_H^N / dP^N)(t)$ the Radon–Nikodym derivative of P_H^N with respect to P^N restricted to the σ-algebra generated by $\{\eta_s,\ 0 \le s \le t\}$. Of course, $(dP_H^N / dP^N)(t)$ is a mean-one positive martingale. The explicit formula for the Radon–Nikodym derivative of a Markov process with respect to another one (cf. Proposition A1.7.3) and a simple computation (cf. section 2) shows that $(dP_H^N / dP^N)(t)$ is equal to

$$\exp N^d \Big\{ < \pi_t^N, H_t > - < \pi_0^N, H_0 > - \int_0^t < \pi_s^N, \partial_s H_s + (1/2)\Delta H_s > ds \\ - (1/2) \sum_{i=1}^d N^{-d} \sum_{x \in \mathbb{T}_N^d} \int_0^t (\partial_{u_i} H(s, x/N))^2 \eta_s(x)(1 - \eta_s(x+e_i))\, ds \\ + O_H(N^{-1}) \Big\}$$

where $O_H(N^{-1})$ is a constant bounded in absolute value by $C(H)N^{-1}$ for some finite constant $C(H)$.

Unfortunately, $(dP_H^N / dP^N)(t)$ is not a function of the empirical measure due to the second integral term. This is the main difficulty in the proof of a large deviations principle: we have to show that the integral term can be rewritten as a function of the empirical measure.

In view of the replacement lemma proved in Chapter 5, the idea is clear. Denote by $F(\alpha)$ the polynomial $\alpha(1-\alpha)$. We would like to show that

$$N^{-d} \sum_{x \in \mathbb{T}_N^d} \int_0^s G(s, x/N)) \Big\{ \eta_s(x)(1 - \eta_s(x + e_i)) - F(\eta_s^{\varepsilon N}(x)) \Big\} \, ds$$

is small as $N \uparrow \infty$ and than $\varepsilon \downarrow 0$. However, this time, since we are interested in large deviations events with probability of order $\exp\{-CN^d\}$, we need to show that this difference is superexponentially small, i.e., that for any $\delta > 0$

$$\limsup_{\varepsilon \to 0} \limsup_{N \to \infty} \frac{1}{N^d} \log P^N \left[\left| \int_0^T V_{N,\varepsilon}(s, \eta_s) \, ds \right| > \delta \right] = -\infty .$$

where

$$V_{N,\varepsilon}(t, \eta) = N^{-d} \sum_{x \in \mathbb{T}_N^d} G(t, x/N) \Big\{ \eta(x)(1 - \eta(x + e_i)) - F(\eta^{\varepsilon N}(x)) \Big\} .$$

This is the content of section 3. In section 4 we prove the upper bound following the strategy presented above. In order to prove that we may exchange the infimum with the maximum on the right hand side of (0.1), we rely on Lemma A2.3.3 that allows such replacement provided $\mathcal{K}$ is compact and each J_β is lower semicontinuous. In possession of the upper bound for compact sets, the passage to closed sets is standard and presented with all details at the end of section 4.

The strategy of the proof of the lower bound is also easy to understand. We start proving a law of large numbers for the empirical measure evolving according to the perturbations considered in the proof of the upper bound. More precisely, denote by Q_H^N the probability on the path space $D([0,T], \mathcal{M}_+)$ corresponding to the inhomogeneous Markov process π_t^N with generator L_N^H. We show that for each H in $C^{1,2}([0,T] \times \mathbb{T}^d)$, Q_H^N converges weakly to the measure Q_H concentrated on an absolutely continuous deterministic path $\pi^H(t, du)$ whose density is the solution of a differential equation involving H (cf. Proposition 5.1).

Denote by I the large deviations rate function obtained in the proof of the upper bound. The second step consists in proving that the entropy of P_H^N with respect to P^N divided by N^d converges to $I(\pi^H)$:

$$\lim_{N \to \infty} N^{-d} H\Big(P_H^N \,|\, P^N \Big) = I(\pi^H) . \tag{0.2}$$

At this point it is not difficult to obtain a lower bound large deviations. Consider an open set $\mathcal{O}$. Fix H in $C^{1,2}([0,T] \times \mathbb{T}^d)$ and recall that we denote by $\pi^H(t, du)$ the hydrodynamic limit of the empirical measure evolving according to the generator L_N^H. For each H such that $\pi^H(t, du)$ belongs to $\mathcal{O}$, we have

$$N^{-d} \log Q^N[\mathcal{O}] = N^{-d} \log E_{P_H^N} \left[\frac{dP^N}{dP_H^N} \mathbf{1}\{\pi \in \mathcal{O}\} \right] .$$

Since $\mathcal{O}$ contains $\pi^H(t, du)$, under P_H^N the probability of the event $\{\pi \in \mathcal{O}\}$ is close to 1. We may therefore remove the restriction $\{\pi \in \mathcal{O}\}$ in the last expectation. Moreover, by Jensen inequality, the right hand side is bounded below by

$$E_{P_H^N}\left[N^{-d}\log\frac{dP^N}{dP_H^N}\right] = -N^d H\Big(P_H^N \mid P^N\Big)\,.$$

By (0.2) the right hand side converges to $I(\pi^H)$. We have thus proved that

$$\liminf_{N\to\infty} N^{-d}\log Q^N[\mathcal{O}] \;\ge\; -\inf_{H\in C^{1,2}([0,T]\times\mathbb{T}^d);\ \pi^H\in\mathcal{O}} I(\pi^H)\,.$$

To conclude the proof of the lower bound we are left to show that

$$\inf_{H\in C^{1,2}([0,T]\times\mathbb{T}^d);\ \pi^H\in\mathcal{O}} I(\pi^H) \;=\; \inf_{\pi\in\mathcal{O}} I(\pi)$$

or, equivalently, that each path π with finite rate function ($I(\pi)<\infty$) can be approximated by a sequence π^{H_n}, H_n in $C^{1,2}([0,T]\times\mathbb{T}^d)$, such that $\lim_{n\to\infty} I(\pi^{H_n}) = I(\pi)$.

The arguments presented above for the proof of the upper bound explain in part why it is sometimes easier to prove a large deviation principle at a higher level. It may happen that the natural positive martingales to consider in the problem cannot be expressed as functions of the process under investigation but only as functions of higher level processes.

For instance, one might be interested in studying equilibrium occupation time large deviations, i.e., large deviations for the functional $\int_0^t ds\,\eta_s(0)$. Unfortunately, no natural martingale is function of $\int_0^t ds\,\eta_s(0)$ only. There is therefore no direct way to prove an upper bound large deviations for this functional. If instead we investigate higher level processes like δ_{η_s} or π_s^N, a large deviations principle can be proved through the method presented above. To get from these results a large deviations principle for the original functional $\int_0^t ds\,\eta_s(0)$, we just apply a contraction principle (cf. Benois (1994) for one-dimensional superposition of independent random walks and Landim (1992) for symmetric simple exclusion processes in dimension $d\neq 2$.)

1. The Rate Function

Fix once for all a density α in $(0,1)$. We state in this section the large deviations principle for the empirical measure. We start introducing all apparatus required to define the rate function.

Recall from Chapter 4 that we denote by $\mathcal{M}_+ = \mathcal{M}_+(\mathbb{T}^d)$ the space of all positive measures on $\mathbb{T}^d$ with finite total mass. Denote by ω the elements of $\mathcal{M}_+$ and by $\mathcal{M}_{+,1} = \mathcal{M}_{+,1}(\mathbb{T}^d)$ the closed subset of $\mathcal{M}_+$ of all positive measures on $\mathbb{T}^d$ with total mass bounded by 1. Hereafter, absolutely continuous measures with respect to the Lebesgue measure are called absolutely continuous measures. In the case where $\omega\in\mathcal{M}_+$ or $\pi\in D([0,T],\mathcal{M}_+)$ are absolutely continuous, we denote respectively by θ and ρ their density: $\omega(du)=\theta(u)du$, $\pi(t,du)=\rho(t,u)du$.

We mentioned in the introduction that there are two distinct types of large deviations arising in this problem. Static large deviations from the initial product measure and dynamics large deviations due to the stochastic character of the evolution. This translates in the decomposition of the rate function in two pieces. We define now the first one which is associated to the static large deviations. This one is very simple because it reduces to large deviations of independent Bernoulli random variables.

For each continuous function $\gamma : \mathbb{T}^d \to (0,1)$, define $h_\gamma : \mathcal{M}_+ \to \mathbb{R}$ and $h : \mathcal{M}_+ \to \bar{\mathbb{R}}_+$ by

$$\begin{aligned} h_\gamma(\omega) &= < \omega, \log \frac{\gamma(1-\alpha)}{(1-\gamma)\alpha} > + < \lambda, \log \frac{1-\gamma}{1-\alpha} > , \\ h(\omega) &= \sup_\gamma h_\gamma(\omega) \cdot \end{aligned} \tag{1.1}$$

In this formula and below λ stands for the Lebesgue measure on $\mathbb{T}^d$ and $< \omega, f >$ for the integral of f with respect to ω. Since for each γ, $h_\gamma(\cdot)$ is linear, h is convex and lower semicontinuous. h is the piece of the rate function associated to large deviations from the initial measure.

Recall that in simple exclusion processes there is at most one particle per site. In particular, in the hydrodynamic limit, as the scale parameter $N \uparrow \infty$, all trajectories are absolutely continuous with density bounded by 1. This space plays therefore a particular role and deserves a special notation.

Denote by $\mathcal{M}^o_{+,1}$ the subset of $\mathcal{M}_{+,1}$ of all absolutely continuous measures with density bounded by 1:

$$\mathcal{M}^o_{+,1} = \Big\{ \omega \in \mathcal{M}_{+,1};\ \omega(du) = \theta(u)du \quad \text{and} \quad 0 \le \theta(u) \le 1 \quad \text{a.e.} \Big\} .$$

$\mathcal{M}^o_{+,1}$ is a closed subset of $\mathcal{M}_+$ endowed with the weak topology. This property is inherited by $D([0,T], \mathcal{M}^o_{+,1})$: $D([0,T], \mathcal{M}^o_{+,1})$ is a closed subset of $D([0,T], \mathcal{M}_+)$ for the Skorohod topology.

In section 5 we obtain an explicit formula for h and we show that h is infinite outside $\mathcal{M}^o_{+,1}$. We now turn to the piece of the rate function associated to the dynamics.

For each smooth function H in $C^{1,2}([0,T] \times \mathbb{T}^d)$, define the functionals J_H, $\ell_H : D([0,T], \mathcal{M}^o_{+,1}) \to \mathbb{R}$ by

$$\begin{aligned} J_H(\pi) &= \ell_H(\pi) - (1/2) \int_0^T dt \int_{\mathbb{T}^d} du\, \|(\nabla H)(t,u)\|^2 F(\rho(t,u)) , \\ \ell_H(\pi) &= < \pi_T, H_T > - < \pi_0, H_0 > \\ &\quad - \int_0^T dt < \pi_t, \partial_t H_t + (1/2)\Delta H_t > . \end{aligned} \tag{1.2}$$

In this formula ∇H stands for the gradient of H: $\nabla H = (\partial_{u_1} H, \ldots, \partial_{u_d} H)$, $\|\nabla H\|$ for the Euclidean norm of the gradient: $\|\nabla H\|^2 = \sum_{1 \le i \le d} (\partial_{u_i} H)^2$ and $F : [0,1] \to \mathbb{R}_+$ for the polynomial $F(\alpha) = \alpha(1-\alpha)$.

Since F is concave and ℓ_H linear, J_H is convex for each H in $C^{1,2}([0,T]\times\mathbb{T}^d)$. Moreover, we claim that J_H is lower semicontinuous. Since ℓ_H is linear, we just need to show that the integral part is upper semicontinuous.

Let π_n be a sequence in $D([0,T],\mathcal{M}^o_{+,1})$ converging to some π. In particular, $\pi_n(t,\cdot)$ converges to $\pi(t,\cdot)$ for almost all $0 \le t \le T$. For each such t, since π belongs to $D([0,T],\mathcal{M}^o_{+,1})$, $\pi_n(t,[a,b])$ converges to $\pi(t,[a,b])$ for all closed hypercubes $[a,b]$. Recall from (5.1.8) that ι_ε stands for an approximation of the identity. Denote by $\rho * \iota_\varepsilon$ the convolution of the density ρ with ι_ε and keep in mind that $(\rho * \iota_\varepsilon)(t,u) = (2\varepsilon)^{-d}\pi(t,[u-\varepsilon,u+\varepsilon])$. Since $\rho * \iota_\varepsilon$ converges, as $\varepsilon \downarrow 0$, to ρ in $L^1([0,T]\times\mathbb{T}^d)$,

$$\begin{aligned}&\int_0^T dt \int_{\mathbb{T}^d} du\, \|(\nabla H)(t,u)\|^2 F(\rho(t,u))\\ &\quad = \lim_{\varepsilon\to 0}\int_0^T dt \int_{\mathbb{T}^d} du\, \|(\nabla H)(t,u)\|^2 F((\rho * \iota_\varepsilon)(t,u))\,.\end{aligned}$$

Since $\pi_n(t,[u-\varepsilon,u+\varepsilon])$ converges to $\pi(t,[u-\varepsilon,u+\varepsilon])$ for almost all (t,u) and since F is concave, the right hand side is equal to

$$\begin{aligned}&\lim_{\varepsilon\to 0}\lim_{n\to\infty}\int_0^T dt \int_{\mathbb{T}^d} du\, \|(\nabla H)(t,u)\|^2 F\Big((2\varepsilon)^{-d}\int_{[u-\varepsilon,u+\varepsilon]^d}\rho_n(t,v)\,dv\Big)\\ &\ge \lim_{\varepsilon\to 0}\lim_{n\to\infty}\int_0^T dt \int_{\mathbb{T}^d} du\, \|(\nabla H)(t,u)\|^2 (2\varepsilon)^{-d}\int_{[u-\varepsilon,u+\varepsilon]^d} dv F(\rho_n(t,v))\,.\end{aligned}$$

Now, since ρ_n is uniformly bounded by 1 and since H is smooth, integrating by parts the space variable it is easy to see that we may interchange limits. This shows that the last expression is equal to

$$\lim_{n\to\infty}\int_0^T dt \int_{\mathbb{T}^d} du\, \|(\nabla H)(t,u)\|^2 F(\rho_n(t,u))\,.$$

Therefore the integral term is upper semicontinuous.

Since in the proof of the hydrodynamic behavior we considered the empirical measure π^N as evolving on $D([0,T],\mathcal{M}_+)$, we need to extend the definition of J_H to this larger space. The natural way is to set J_H to be ∞ outside $D([0,T],\mathcal{M}^o_{+,1})$:

$$J_H(\pi) \;=\; \infty \quad \text{if } \pi \notin D([0,T],\mathcal{M}^o_{+,1})\,. \tag{1.3}$$

Notice that since $D([0,T],\mathcal{M}^o_{+,1})$ is closed, J_H is still a convex lower semicontinuous function on $D([0,T],\mathcal{M}_+)$.

We are now ready to define the large deviations rate function. Let I, I_0 : $D([0,T],\mathcal{M}_+) \to \bar{\mathbb{R}}_+$ be defined as

$$I_0(\pi) \;=\; \sup_{H\in C^{1,2}([0,T]\times\mathbb{T}^d)} J_H(\pi)\,, \qquad I(\pi) \;=\; I_0(\pi) \,+\, h(\pi_0)\,. \tag{1.4}$$

In this formula π_0 stands for trajectory π at time 0. Of course I is identically equal to ∞ outside $D([0,T],\mathcal{M}^o_{+,1})$:

$$I(\pi) \;=\; \infty \qquad \text{if} \qquad \pi \notin D([0,T], \mathcal{M}_{+,1}^{o})\,. \tag{1.5}$$

In section 5 we present an explicit formula for rate function I_0.

Recall that we consider for simplicity nearest neighbor symmetric simple exclusion processes, i.e., a simple exclusion process with transition probability p such that $p(x) = 0$ if $|x| > 1$ and $p(e_j) = p(-e_j) = 1/2$ for $1 \le j \le d$, where e_j stands for the jth vector of the canonical basis of $\mathbb{R}^d$. Denote by Q^N (resp. P^N) the probability on the path space $D([0,T], \mathcal{M}_+)$ (resp. $D([0,T], \{0,1\}^{\mathbb{T}_N^d})$) corresponding to the Markov process π_t^N (resp. η_t) with generator L_N introduced in Definition 2.2.1 accelerated by N^2 and starting from the invariant state ν_α^N. Expectation with respect to P^N is denote by E^N.

We have now all elements to state the large deviations principle for the empirical measure.

Theorem 1.1 *For each closed set $\mathcal{C}$ and each open set $\mathcal{O}$ of $D([0,T], \mathcal{M}_+)$,*

$$\limsup_{N\to\infty} N^{-d} \log Q^N[\mathcal{C}] \;\le\; -\inf_{\pi\in\mathcal{C}} I(\pi)\,,$$

$$\liminf_{N\to\infty} N^{-d} \log Q^N[\mathcal{O}] \;\ge\; -\inf_{\pi\in\mathcal{O}} I(\pi)\,.$$

Remark 1.2 We prove in this chapter large deviations starting from the equilibrium measure ν_α^N for historical reasons. It is straightforward to extend this proof to processes starting from a product measure with slowly varying parameter or to a process starting from a sequence of deterministic configurations η^N in $\{0,1\}^{\mathbb{T}_N^d}$ associated to some profile $\gamma : \mathbb{T}^d \to [0,1]$ in the following sense:

$$\lim_{N\to\infty} N^{-d} \sum_{x\in\mathbb{T}_N^d} H(x/N)\eta^N(x) \;=\; \int H(u)\rho(u)\, du$$

for all continuous functions H.

2. Weakly Asymmetric Simple Exclusion Processes

The investigation of large deviations from the initial measure is quite simple since the occupation variables $\{\eta(x),\ x \in \mathbb{T}_N^d\}$ under ν_α^N are independent Bernoulli random variables. We are reduced therefore to study large deviations of i.i.d. random variables: for each continuous function $\gamma : \mathbb{T}^d \to (0,1)$, let $\nu_{\gamma(\cdot)}^N$ denote the product measure on $\{0,1\}^{\mathbb{T}_N^d}$ with marginals given by:

$$\nu_{\gamma(\cdot)}^N \Big\{\eta;\ \eta(x) = 1\Big\} \;=\; \gamma(x/N) \tag{2.1}$$

for all x in $\mathbb{T}_N^d$. An elementary computation shows that $d\nu_{\gamma(\cdot)}^N / d\nu_\alpha^N$ is equal to

$$\frac{d\nu^N_{\gamma(\cdot)}}{d\nu^N_\alpha} = \prod_{x\in\mathbb{T}^d_N} \Big(\frac{\gamma(x/N)}{\alpha}\Big)^{\eta(x)} \Big(\frac{1-\gamma(x/N)}{1-\alpha}\Big)^{1-\eta(x)}$$
$$= \exp \sum_{x\in\mathbb{T}^d_N} \Big\{\eta(x) \log \frac{\gamma(x/N)}{\alpha} + (1-\eta(x)) \log \frac{1-\gamma(x/N)}{1-\alpha}\Big\} . \tag{2.2}$$

In order to keep notation simple, for a continuous function $\gamma : \mathbb{T}^d \to (0, 1)$, define $h^N_\gamma : \mathcal{M}_+ \to \mathbb{R}$ as

$$h^N_\gamma(\omega) = < \omega, \log \frac{\gamma(1-\alpha)}{(1-\gamma)\alpha} > + < \lambda_N, \log \frac{1-\gamma}{1-\alpha} > . \tag{2.3}$$

In this formula λ_N stands for the discrete approximation of the Lebesgue measure and is defined by $\lambda_N = N^{-d}\sum_x \delta_{x/N}$. With this notation we may write the Radon–Nikodym derivative $d\nu^N_{\gamma(\cdot)}/d\nu^N_\alpha$ as

$$\frac{d\nu^N_{\gamma(\cdot)}}{d\nu^N_\alpha} = \exp N^d \Big\{h^N_\gamma(\pi_0)\Big\} \cdot$$

Notice that $d\nu^N_{\gamma(\cdot)}/d\nu^N_\alpha$ is a continuous function of the empirical measure. Moreover, for each continuous $\gamma : \mathbb{T}^d \to (0, 1)$, h^N_γ converges uniformly to h_γ in $\mathcal{M}_+$.

We turn now to the large deviations coming from the dynamics. We have seen in the beginning of this chapter that in order to prove a large deviations principle, we need to construct mean-one positive martingales. For Markov processes, this is done by considering small perturbations of the original dynamics and taking Radon–Nikodym derivatives. In the context of symmetric simple exclusion processes, the relevant perturbations to introduce are small time and space dependent asymmetries in the jump rate. More precisely, for each function H in $C^{1,2}([0, T] \times \mathbb{T}^d)$, consider the time inhomogeneous Markov process whose generator at time t is given by

$$(L^H_{N,t} f)(\eta) = (1/2) \sum_{|x-y|=1} \eta(x)[1-\eta(y)] e^{H(t,y/N)-H(t,x/N)} [f(\eta^{x,y}) - f(\eta)] . \tag{2.4}$$

The interpretation is simple, at time t and site x, instead of a symmetric jump rate equal to $1/2$ to each neighbor, the jump rate from x to $x \pm e_i$ is equal to $(1/2)\{1 \pm N^{-1}(\partial_{u_i} H)(t, x/N)\}$.

For each continuous function $\gamma : \mathbb{T}^d \to (0, 1)$ and H in $C^{1,2}([0, T] \times \mathbb{T}^d)$, denote by $P^N_{\gamma,H}$ (resp. $Q^N_{\gamma,H}$) the probability measure on the path space $D([0, T], \{0, 1\}^{\mathbb{T}^d_N})$ (resp. $D([0, T], \mathcal{M}_+)$) corresponding to the inhomogeneous Markov process η_t (resp. π^N_t) with generator L^H_N defined in (2.4) accelerated by N^2 and starting from $\nu^N_{\gamma(\cdot)}$. In the case where γ is constant equal to α, we denote $P^N_{\gamma,H}$ simply by P^N_H.

In section A1.7, we compute the Radon–Nikodym derivative dP_H^N/dP^N. It is equal to

$$\exp N^d \Big\{ < \pi_T^N , H_T > \; - \; < \pi_0^N , H_0 >$$
$$- N^{-d} \int_0^T dt \, \exp\Big\{ - N^d < \pi_t^N , H_t > \Big\} (\partial_t + L) \exp\Big\{ N^d < \pi_t^N , H_t > \Big\} \Big\} .$$

Taylor's expansion up to the second order and the elementary inequality $|e^u - 1 - u - (1/2)u^2| \le (1/6)|u|^3 e^{|u|}$ permit to rewrite the Radon–Nikodym derivative dP_H^N/dP^N as

$$\exp N^d \Big\{ < \pi_T^N , H_T > - < \pi_0^N , H_0 > - \int_0^T < \pi_s^N , \partial_s H_s + (1/2)\Delta H_s > ds$$
$$- (1/2) \sum_{i=1}^d N^{-d} \sum_{x \in \mathbb{T}_N^d} \int_0^T (\partial_{u_i} H(t, x/N))^2 \eta_t(x)(1 - \eta_t(x + e_i)) \, dt$$
$$+ \; O_H(N^{-1}) \Big\} ,$$

where $O_H(N^{-1})$ is a constant bounded in absolute value by $C(H)N^{-1}$.

In order to write this Radon–Nikodym derivative in a simple form, we introduce some notation. For each smooth function H in $C^{1,2}([0,T] \times \mathbb{T}^d)$, recall the linear functional ℓ_H introduced in (1.2). We consider from now on ℓ_H as defined on $D([0,T], \mathcal{M}_+)$. Thus $\ell_H : D([0,T], \mathcal{M}_+) \to \mathbb{R}$ is given by

$$\ell_H(\pi) \; = < \pi_T , H_T > \; - \; < \pi_0 , H_0 > \; - \; \int_0^T dt \; < \pi_t , \partial_t H_t + (1/2)\Delta H_t > \; . \tag{2.5}$$

With this notation we may write the Radon–Nikodym derivative $dP_{\gamma,H}^N/dP^N$, which is equal to $d\nu_{\gamma(\cdot)}^N/d\nu_\alpha^N \times dP_H^N/dP^N$, as

$$\frac{dP_{\gamma,H}^N}{dP^N} \; = \; \exp N^d \Big\{ \ell_H(\pi^N) \; + \; h_\gamma^N(\pi_0^N) \; + \; O_H(N^{-1})$$
$$- (1/2) \sum_{i=1}^d N^{-d} \sum_{x \in \mathbb{T}_N^d} \int_0^T (\partial_{u_i} H(t, x/N))^2 \eta_t(x)(1 - \eta_t(x + e_i)) \, dt \Big\} . \tag{2.6}$$

We have seen in the introduction that prove upper bound large deviations we need the mean-one positive martingales to be function of the process. Unfortunately, $dP_{\gamma,H}^N/dP^N$ is not a function of the empirical measure. The first step in the proof of a large deviations principle is therefore to show that $dP_{\gamma,H}^N/dP^N$ is superexponentially close to a function of the empirical measure. Here superexponentially means that the difference between the Radon–Nikodym derivative $dP_{\gamma,H}^N/dP^N$ and a function of the empirical measure has expectation of order smaller than

$\exp\{-CN^d\}$ for all $C > 0$. We need such a small order because we are interested in large deviations events that have probability of order $\exp\{-CN^d\}$.

In view of the replacement lemma of Chapter 5, there is at least one possible approach. We proved there that for each continuous function G and cylinder function Ψ,

$$\limsup_{\varepsilon\to 0}\limsup_{N\to\infty} P^N\Big[\Big|\int_0^T dt\, N^{-d}\sum_{x\in\mathbb{T}_N^d} G(t,x/N)\Big[\tau_x\Psi(\eta) - \tilde{\Psi}(\eta^{\varepsilon N}(x))\Big]\Big| > \delta\Big] = 0$$

for all $\delta > 0$. In this formula $\tilde{\Psi}(\alpha) = E_{\nu_\alpha}[\Psi]$ and that $\eta^{\varepsilon N}(\cdot)$ is a function of the empirical measure. One could hope to prove that this probability is in fact superexponentially small and obtain in consequence that the Radon–Nikodym derivative $dP^N_{\gamma,H}/dP^N$ is superexponentially close to a function of the empirical measure. This is the content of the main theorem of next section.

3. A Superexponential Estimate

We prove in this section the main ingredient required in the investigation of large deviations of the empirical measure: a superexponential estimate that allows the replacement of cylinder functions by functions of the density field. Such replacement lemma was already proved in Chapter 5 to close the differential equation satisfied by the empirical measure. We shall prove below that the probability involved is superexponentially small, that is, of order smaller that $\exp\{-CN^d\}$ for all $C > 0$.

Recall from Chapter 5 that for any cylinder function Ψ we defined $\tilde{\Psi} : [0,1] \to \mathbb{R}$ as $\tilde{\Psi}(\alpha) = E_{\nu_\alpha}[\Psi]$.

Theorem 3.1 *For each $G \in C([0,T]\times\mathbb{T}^d)$, each cylinder function Ψ and each $\varepsilon > 0$, let*

$$V_{N,\varepsilon}^{G,\Psi}(t,\eta) = V_{N,\varepsilon}(t,\eta) = N^{-d}\sum_{x\in\mathbb{T}_N^d} G(t,x/N)\Big[\tau_x\Psi(\eta) - \tilde{\Psi}(\eta^{\varepsilon N}(x))\Big].$$

Then, for any $\delta > 0$

$$\limsup_{\varepsilon\to 0}\limsup_{N\to\infty}\frac{1}{N^d}\log P^N\Big[\Big|\int_0^T V_{N,\varepsilon}(t,\eta_t)\,dt\Big| > \delta\Big] = -\infty. \tag{3.1}$$

Proof. Since for any positive sequences a_N and b_N

$$\limsup_{N\to\infty} N^{-d}\log(a_N+b_N) \le \max\left\{\limsup N^{-d}\log a_N\,,\ \limsup N^{-d}\log b_N\right\}, \tag{3.2}$$

it is enough to prove (3.1) without the absolute value for $V_{N,\varepsilon}^{G,\Psi}$ and $V_{N,\varepsilon}^{-G,\Psi}$.

For each positive a, by Chebychev exponential inequality, the probability on the left hand side of (3.1) without the absolute value is bounded above by

$$\exp\{-a\delta N^d\}E^N\left[\exp\left\{\int_0^T aN^d V_{N,\varepsilon}(t,\eta_t)\,dt\right\}\right].$$

To conclude the proof of the theorem it is therefore enough to show that

$$\limsup_{\varepsilon\to 0}\limsup_{N\to\infty}\frac{1}{N^d}\log E^N\left[\exp\left\{\int_0^T aN^d V_{N,\varepsilon}(t,\eta_t)\,dt\right\}\right] \le 0 \tag{3.3}$$

for every positive a, because in this case we would have proved that the left hand side of (3.1) is bounded above by $-a\delta$ for every positive a and it would remain to let a increase to ∞.

Denote by $S_t^{a,V}$ the semigroup associated to the inhomogeneous Markov process with generator $N^2L_N + aN^dV_{N,\varepsilon}(t,\eta)$ and notice that this generator is symmetric with respect to the product measure ν_α^N. By Feynman–Kac formula, the expectation in (3.3) is equal to $< S_T^{a,V}\mathbf{1},\mathbf{1} >_\alpha$ (cf. section A1.7) provided $< \cdot,\cdot >_\alpha$ stands for the inner product in $L^2(\nu_\alpha^N)$ and $\mathbf{1}$ for the function on the configuration space $\{0,1\}^{\mathbb{T}_N^d}$ identically equal to 1.

Denote by $\lambda_{a,V}(t)$ the largest eigenvalue of the symmetric operator $N^2L_N + aN^dV_{N,\varepsilon}(t,\eta)$ on $L^2(\nu_\alpha^N)$. By the spectral representation (Lemma A1.7.2), $< S_T^{a,V}\mathbf{1},\mathbf{1} >_\alpha$ is less than or equal to $\exp\{\int_0^T \lambda_{a,V}(t)\,dt\}$. Moreover, by (A3.1.1) the largest eigenvalue $\lambda_{a,V}(t)$ is equal to the variational formula

$$\sup_f\left\{\int aN^dV_{N,\varepsilon}(t,\eta)f(\eta)\nu_\alpha^N(d\eta) \;-\; N^2D_N(f)\right\}.$$

In this formula, the supremum is carried over all densities with respect to ν_α^N (that is all integrable positive functions f with $\int f d\nu_\alpha^N = 1$) and D_N is the convex and lower semicontinuous functional defined in Corollary A1.10.3 and given by $D_N(f) = < -L_N\sqrt{f},\sqrt{f} >$.

Up to this point we proved that the expression in (3.3) is bounded above by

$$\int_0^T dt\,\sup_f\left\{\int aV_{N,\varepsilon}(t,\eta)f(\eta)\nu_\alpha^N(d\eta) \;-\; N^{2-d}D_N(f)\right\}.$$

Since Ψ is a cylinder function and G is continuous, both are bounded. In particular, the supremum is bounded by a finite constant $C(a,G,\Psi)$. Therefore, by the dominated convergence theorem, in order to prove (3.3) we have to show that

$$\limsup_{\varepsilon\to 0}\limsup_{N\to\infty}\sup_f\Big\{\int aV_{N,\varepsilon}(t,\eta)f(\eta)\nu_\alpha^N(d\eta) \;-\; N^{2-d}D_N(f)\Big\} \;\le\; 0$$

for every positive a and $0\le t\le T$.

Recall that $V_{N,\varepsilon}(t,\eta)$ is bounded in absolute value by a constant $C(\Psi,H)$ depending only on Ψ and H. In particular, in last formula the expression inside braces is negative whenever the density f is such that $D_N(f)$ is greater than $aC(H,\Psi)N^{d-2}$. We may therefore restrict the supremum over all densities with functional $D_N(f)$ bounded by $aC(H,\Psi)N^{d-2}$. To conclude the proof it remains therefore to show that

$$\limsup_{\varepsilon\to 0}\limsup_{N\to\infty}\sup_{f;\,D_N(f)\le CN^{d-2}}\int aV_{N,\varepsilon}(t,\eta)f(\eta)\nu_\alpha^N(d\eta) \;\le\; 0$$

for every $0\le t\le T$, $a>0$ and positive constant C. This is exactly the content of the replacement lemma stated in Chapter 5 (cf. Lemma 5.5.7). □

Corollary 3.2 *The statement of Theorem 3.1 remains in force if the probability $P^N_{\gamma,H}$ replaces P^N.*

Proof. By the explicit formula (2.6) for the Radon–Nikodym derivative and since for simple exclusion processes there is at most one particle per site,

$$\|dP^N_{\gamma,H}/dP^N\|_\infty \le \exp\{C(\gamma,H,T)N^d\}\,. \tag{3.4}$$

On the other hand, we have

$$\begin{aligned}&E^N_{\gamma,H}\left[\mathbf{1}\Big\{\Big|\int_0^T V_{N,\varepsilon}(t,\eta_t)\,dt\Big|>\delta\Big\}\right]\\ &\quad= E^N\left[\frac{dP^N_{\gamma,H}}{dP^N}\mathbf{1}\Big\{\Big|\int_0^T V_{N,\varepsilon}(t,\eta_t)\,dt\Big|>\delta\Big\}\right]\,.\end{aligned}$$

It remains to bound $dP^N_{\gamma,H}/dP^N$ by its L^∞ norm and to apply Theorem 3.1. □

4. Large Deviations Upper Bound

In the previous two sections we essentially proved that the martingales $dP^N_{\gamma,H}/dP^N$ are functions of the empirical measure. We are therefore ready to carry out the strategy presented in the introduction of this chapter to prove the upper bound large deviations.

We shall prove first an upper bound for compact sets. Recall the definition of the linear functional ℓ_H given in (2.5). For a function H in $C^{1,2}([0,T]\times\mathbb{T}^d)$ and $\delta,\,\varepsilon>0$, let $B_{H,\delta,\varepsilon}$ denote the set of trajectories $(\eta_t)_{0\le t\le T}$ defined by

$$B_{H,\delta,\varepsilon} = \left\{ \eta \in D([0,T], \{0,1\}^{\mathbb{T}_N^d}); \sum_{i=1}^{d} \left| \int_0^T dt\, V_{N,\varepsilon}^{H,i}(t,\eta_t) \right| \le \delta \right\}, \tag{4.1}$$

where

$$V_{N,\varepsilon}^{H,i}(t,\eta) = N^{-d} \sum_{x \in \mathbb{T}_N^d} (\partial_{u_i} H(t,x/N))^2 \Big\{ \eta(x)(1-\eta(x+e_i)) - F(\eta^{\varepsilon N}(x)) \Big\} .$$

For each positive integer N and $\varepsilon > 0$, denote by $\iota_{\varepsilon,N}$ the approximation of the identity defined by

$$\iota_{\varepsilon,N}(u) = \frac{1}{(2\varepsilon + N^{-1})^d} \mathbf{1}\{[-\varepsilon,\varepsilon]\}(u) .$$

With this notation $\eta_t^{\varepsilon N}(x) = (\pi_t^N * \iota_{\varepsilon,N})(x/N)$. In particular,

$$\begin{aligned} & N^{-d} \sum_{x \in \mathbb{T}_N^d} (\partial_{u_i} H(t,x/N))^2 F(\eta_t^{\varepsilon N}(x)) \\ & \quad = N^{-d} \sum_{x \in \mathbb{T}_N^d} (\partial_{u_i} H(t,x/N))^2 F((\pi_t^N * \iota_{\varepsilon,N})(x/N)) . \end{aligned}$$

Notice moreover that $\pi_t^N * \iota_{\varepsilon,N}$ belongs to $\mathcal{M}_{+,1}^o$ because there is at most one particle per site. In view of this identity and (2.6), on $B_{H,\delta,\varepsilon}$ the Radon–Nikodym derivative $dP_{\gamma,H}^N/dP^N$ can be written as a function of the empirical measure modulo some small errors.

On the other hand, by Theorem 3.1, the set $B_{H,\delta,\varepsilon}$ has probability superexponentially close to 1: for each $\delta > 0$ and H in $C^{1,2}([0,T] \times \mathbb{T}^d)$,

$$\limsup_{\varepsilon \to 0} \limsup_{N \to \infty} \frac{1}{N^d} \log P^N[B_{H,\delta,\varepsilon}^c] = -\infty . \tag{4.2}$$

These two observations will provide the large deviations upper bound for compact sets.

Let $\mathcal{O}$ denote an open set of $D([0,T], \mathcal{M}_+)$ and fix $\delta > 0$, H in $C^{1,2}([0,T] \times \mathbb{T}^d)$ and a continuous function $\gamma : \mathbb{T}^d \to (0,1)$. By (3.2), for each $\varepsilon > 0$,

$$\begin{aligned} & \limsup_{N \to \infty} \frac{1}{N^d} \log Q^N[\mathcal{O}] \\ & \quad \le \max \left\{ \limsup_{N \to \infty} \frac{1}{N^d} \log P^N[\pi \in \mathcal{O}, B_{H,\delta,\varepsilon}], U(H,\delta,\varepsilon) \right\}, \end{aligned} \tag{4.3}$$

where

$$U(H,\delta,\varepsilon) = \limsup_{N \to \infty} \frac{1}{N^d} \log P^N[B_{H,\delta,\varepsilon}^c] .$$

Keep in mind that by (4.2), $\lim_{\varepsilon \to 0} U(H,\delta,\varepsilon) = -\infty$ for each fixed H and δ.

We may rewrite the probability on the right hand side of (4.3) as

$$E^N_{\gamma,H}\left[\frac{dP^N}{dP^N_{\gamma,H}}\,\mathbf{1}\Big\{\pi\in\mathcal{O}\,,\;B_{H,\delta,\varepsilon}\Big\}\right].$$

As observed above, on $B_{H,\delta,\varepsilon}$ the derivative $dP^N_{\gamma,H}/dP^N$ can be written as a function of the empirical density modulo small errors. More precisely, on $B_{H,\delta,\varepsilon}$, $dP^N_{\gamma,H}/dP^N$ is equal to

$$\exp N^d\Big\{\ell_H(\pi^N)\;+\;h^N_\gamma(\pi^N_0)\;+\;O(\delta)\;+\;O_H(N^{-1})\Big\}$$
$$\times\;\exp N^d\Big\{\;-\;(1/2)\int_0^T <\lambda^N,\|\nabla H_t\|^2F((\pi^N_t*\iota_{\varepsilon,N})(\cdot))>\,dt\Big\}\,. \tag{4.4}$$

Because $\ell_H(\cdot)$ is linear, H belongs to $C^{1,2}([0,T]\times\mathbb{T}^d)$ and there is at most one particle per site, $\ell_H(\pi^N)=\ell_H(\pi^N*\iota_{\varepsilon,N})+o_H(\varepsilon)$, where $o_H(\varepsilon)$ is a constant depending on H that vanishes in the limit as $\varepsilon\downarrow 0$. On the other hand, due to the definition of $\iota_{\varepsilon,N}$ and the exclusion rule, $\pi^N*\iota_{\varepsilon,N}$ belongs to $D([0,T],\mathcal{M}^o_{+,1})$. Furthermore,

$$<\lambda^N,\|\nabla H_t\|^2F((\pi^N_t*\iota_{\varepsilon,N})(\cdot))>$$
$$=\;\int_{\mathbb{T}^d}du\,\|(\nabla H)(t,u)\|^2F((\pi^N_t*\iota_{\varepsilon,N})(u))\;+\;O_H(N^{-1})\,,$$

where $O_H(N^{-1})$ is bounded by a function of H that vanishes as $N\uparrow\infty$. In particular, from (1.2) and (1.3)

$$\ell_H(\pi^N)\;-\;(1/2)\int_0^T<\lambda^N,\|\nabla H_t\|^2F((\pi^N_t*\iota_{\varepsilon,N})(\cdot))>\,dt$$
$$=\;J_H(\pi^N*\iota_{\varepsilon,N})\;+\;O_H(N^{-1})\;+\;o_H(\varepsilon)\,.$$

For similar reasons $h^N_\gamma(\pi^N_0)=h_\gamma(\pi^N_0)+O_\gamma(N^{-1})$ where $O_\gamma(N^{-1})$ is bounded by a function of γ that vanishes as $N\uparrow\infty$. In conclusion, on $B_{H,\delta,\varepsilon}$, $dP^N_{\gamma,H}/dP^N$ is equal to

$$\exp N^d\Big\{J_H(\pi^N*\iota_{\varepsilon,N})+h_\gamma(\pi^N_0)+O(\delta)+O_H(N^{-1})+O_\gamma(N^{-1})+o_H(\varepsilon)\Big\}\,.$$

In particular, replacing in last expectation the Radon–Nikodym derivative by the previous expression, taking logarithms and dividing by N^d, we obtain that

$$\frac{1}{N^d}\log P^N[\pi\in\mathcal{O}\,,\;B_{H,\delta,\varepsilon}]\;\le\;O(\delta)\;+\;O_H(N^{-1})\;+\;O_\gamma(N^{-1})\;+\;o_H(\varepsilon)$$
$$+\;\sup_{\pi\in\mathcal{O}}\Big\{\;-\;J_H(\pi*\iota_{\varepsilon,N})\;-\;h_\gamma(\pi_0)\Big\}\,.$$

Letting $N\uparrow\infty$, recalling (4.3) and minimizing over H, γ, δ and ε,

$$\limsup_{N\to\infty} \frac{1}{N^d} \log Q^N[\mathcal{O}] \;\le\; \inf_{H,\gamma,\delta,\varepsilon} \;\sup_{\pi\in\mathcal{O}} J_{H,\gamma,\delta,\varepsilon}(\pi)\;,$$

where

$$J_{H,\gamma,\delta,\varepsilon}(\pi) \;=\; \max\Big\{ -\, J_H(\pi * \iota_\varepsilon) \;-\; h_\gamma(\pi_0) + O(\delta) + o_H(\varepsilon)\,,\; U(H,\delta,\varepsilon)\Big\}$$

and ι_ε is the approximation of the identity defined in (5.1.8). We have thus proved an upper bound large deviations for every open set $\mathcal{O}$ of $D([0,T],\mathcal{M}_+)$. Since for each H in $C^{1,2}([0,T]\times\mathbb{T}^d)$, continuous function $\gamma:\mathbb{T}^d\to(0,1)$, $\delta>0$ and $\varepsilon>0$, $J_{H,\gamma,\delta,\varepsilon} : D([0,T],\mathcal{M}_+)\to\bar{\mathbb{R}}$ is upper semicontinuous, by Lemma A2.3.3, for every compact set $\mathcal{K}$,

$$\limsup_{N\to\infty} \frac{1}{N^d} \log Q^N[\mathcal{K}] \;\le\; \sup_{\pi\in\mathcal{K}} \;\inf_{H,\gamma,\delta,\varepsilon} J_{H,\gamma,\delta,\varepsilon}(\pi)\;.$$

By (4.2) and the definition of $J_{H,\gamma,\delta,\varepsilon}$, for each fixed γ, H, δ and π

$$\lim_{\varepsilon\to 0} J_{H,\gamma,\delta,\varepsilon}(\pi) \;\le\; -J_H(\pi) \;-\; h_\gamma(\pi_0) \;+\; O(\delta)$$

because J_H is lower semicontinuous and $\pi * \iota_\varepsilon$ converges to π as $\varepsilon\downarrow 0$. Letting $\delta\downarrow 0$, by definition of the rate function I given in (1.4), we get

$$\limsup_{N\to\infty} \frac{1}{N^d} \log Q^N[\mathcal{K}] \;\le\; \sup_{\pi\in\mathcal{K}} \;\inf_{H,\gamma} -\Big\{J_H(\pi) \;+\; h_\gamma(\pi_0)\Big\} \;=\; -\inf_{\pi\in\mathcal{K}} I(\pi)$$

for every compact set $\mathcal{K}$ of $D([0,T],\mathcal{M}_+)$.

We conclude this section extending the large deviations upper bound to closed sets. By remark (3.2), it is enough to find a sequence of compact sets $\mathcal{K}_n$ such that

$$\limsup_{N\to\infty} \frac{1}{N^d} \log Q^N[\mathcal{K}_n^c] \;\le\; -n\;. \tag{4.5}$$

This is the so called exponential tightness of the sequence Q^N. To construct such sequence, we proceed in two steps. We first show that for every continuous function $H:\mathbb{T}^d\to\mathbb{R}$ and $\varepsilon>0$,

$$\lim_{\delta\to 0}\limsup_{N\to\infty} \frac{1}{N^d} \log Q^N\Big[\sup_{|t-s|\le\delta}\Big| <\pi_t, H> - <\pi_s, H>\Big| \;>\; \varepsilon\Big] \;=\; -\infty\;. \tag{4.6}$$

From this estimate we prove the exponential tightness.

First of all, for N sufficiently large, we have that

$$\Big\{\sup_{|t-s|\le\delta}\Big| <\pi^N_t, H> - <\pi^N_s, H>\Big| > \varepsilon\Big\}$$

$$\subset \bigcup_{k=0}^{[T\delta^{-1}]} \Big\{\sup_{k\delta\le t<(k+1)\delta}\Big| <\pi^N_t, H> - <\pi^N_{k\delta}, H>\Big| > \varepsilon/4\Big\}\;.$$

We have here $\varepsilon/4$ instead of $\varepsilon/3$ due to the presence of jumps. Since we start from equilibrium, by remark (3.2), in order to prove (4.6), it is enough to show that

$$\lim_{\delta\to 0}\limsup_{N\to\infty}\frac{1}{N^d}\log Q^N\Big[\sup_{0\le t<\delta}\Big|<\pi_t,H>-<\pi_0,H>\Big|>\varepsilon\Big] = -\infty \tag{4.7}$$

for every $\varepsilon>0$ and H in $C^{1,2}([0,T]\times\mathbb{T}^d)$.

Fix a constant a that will increase to ∞ after $N\uparrow\infty$ and $\delta\downarrow 0$. Denote by $A_t^{a,H}$ the integral $(1/2)N^2\sum_{|x-y|=1}\int_0^t(\exp\{a[H(y/N)-H(x/N)]\}-1)\eta_s(x)[1-\eta_s(y)]ds$. We introduced $A_t^{a,H}$ because $M_t^{a,H}=\exp\{N^d a<\pi_t^N,H>-N^d a<\pi_0^N,H>-A_t^{a,H}\}$ is a positive martingale equal to 1 at time 0.

In order to prove (4.7), it is enough to prove the same statement (with $a\varepsilon$ instead of ε) for $N^{-d}\log M_t^{a,H}$ and for $N^{-d}A_t^{a,H}$. On the one hand, a simple computation, similar to the one performed to express the Radon–Nikodym derivative dP_H^N/dP^N as a function of the empirical measure, shows that $N^{-d}A_t^{a,H}$ is bounded by $C(a,H)t$ because there is at most one particle per site. In particular, because $t\le\delta$, for δ small enough the probability in (4.7) with $N^{-d}A_t^{a,H}$ vanishes.

On the other hand, in order to prove (4.7) for $N^{-d}\log M_t^{a,H}$, we first observe that we can neglect the absolute value. Without the absolute value, by Chebychev exponential inequality, the probability is bounded above by $\exp\{-a\varepsilon N^d\}$ because $M_t^{a,H}$ is a positive martingale equal to 1 at time 0. This concludes the proof of (4.6).

It is now a simple game to obtain a sequence of compact sets satisfying (4.5). Consider a sequence H_ℓ of $C^2(\mathbb{T}^d)$ functions dense in $C(\mathbb{T}^d)$ for the uniform topology. For each $\delta>0$ and $\varepsilon>0$, denote by $\mathcal{C}_{\ell,\delta,\varepsilon}$ the set of all paths π_t such that

$$\mathcal{C}_{\ell,\delta,\varepsilon} = \Big\{\pi_\cdot\in D([0,T],\mathcal{M}_+);\ \sup_{|t-s|\le\delta}\Big|<\pi_t,H_\ell>-<\pi_s,H_\ell>\Big|\le\varepsilon\Big\}.$$

We just proved that

$$\lim_{\delta\to 0}\limsup_{N\to\infty}\frac{1}{N^d}\log Q^N\Big[\pi\notin\mathcal{C}_{\ell,\delta,\varepsilon}\Big] = -\infty$$

for each $\ell\ge 1$ and $\varepsilon>0$. In particular, for each positive integers ℓ, m and n, there exists $\delta=\delta(\ell,m,n)$ such that

$$Q^N\Big[\pi\notin\mathcal{C}_{\ell,\delta,1/m}\Big] \le \exp\{-N^d nm\ell\}$$

for all N large enough. We may extend this inequality to all positive integers N by modifying δ if necessary. Consider the set $\mathcal{K}_n^o$ defined by

$$\mathcal{K}_n^o = \bigcap_{\ell\ge 1,\ m\ge 1}\mathcal{C}_{\ell,\delta(\ell,m,n),1/m}\,.$$

It is quite simple to check that $\mathcal{K}_n = \mathcal{K}_n^o \cap D([0,T], \mathcal{M}_{+,1})$ is a compact set for each $n \geq 1$. On the other hand, since there is at most one particle per site, $Q^N[\mathcal{K}_n] = Q^N[\mathcal{K}_n^o]$. Furthermore, by construction,

$$Q^N\Big[\pi \notin \mathcal{K}_n^o\Big] \;\leq\; \sum_{\ell\geq 1,\, m\geq 1} \exp\{-N^d nm\ell\} \;\leq\; C\exp\{-N^d n\}$$

for some universal constant C. In particular,

$$\limsup_{N\to\infty} \frac{1}{N^d} \log Q^N\Big[\pi \notin \mathcal{K}_n^o\Big] \;\leq\; -n\,.$$

This concludes the proof of the large deviations upper bound for closed sets.

5. Large Deviations Lower Bound

We start this section proving the hydrodynamic behavior of the inhomogeneous Markov process η_t with generator $L^H_{N,t}$ defined in (2.4).

Proposition 5.1 *Fix a continuous function $\gamma : \mathbb{T}^d \to (0,1)$ and H in $C^{1,2}([0,T]\times\mathbb{T}^d)$. The sequence of probabilities $Q^N_{\gamma,H}$ converges in distribution to the probability measure concentrated on the absolutely continuous path $\pi_t(du) = \rho(t,u)du$ whose density $\rho(t,u)$ is the unique weak solution of the partial differential equation*

$$\begin{cases} \partial_t\rho \;=\; (1/2)\Delta\rho \;-\; \displaystyle\sum_{i=1}^d \partial_{u_i}\Big(\rho(1-\rho)\partial_{u_i}H\Big)\,, \\ \rho(0,\cdot) \;=\; \gamma(\cdot)\,. \end{cases} \tag{5.1}$$

The proof presented in Chapter 5 for symmetric zero range processes can easily be adapted to the case of inhomogeneous weakly asymmetric simple exclusion processes described above.

We now obtain an explicit form for the large deviations rate function I.

Lemma 5.2 *Recall the definition of h and h_γ introduced in (1.1). Then*

$$h(\omega) \;=\; \int du \left\{\theta(u)\log\frac{\theta(u)}{\alpha} \;+\; [1-\theta(u)]\log\frac{1-\theta(u)}{1-\alpha}\right\}$$

if $\omega(du) = \theta(u)du$ for some $0 \leq \theta(u) \leq 1$. $h(\omega) = \infty$ otherwise. Moreover, if $\omega(du) = \theta(u)du$ for some continuous function $\theta : \mathbb{T}^d \to (0,1)$,

$$h(\omega) \;=\; h_\theta(\omega)\,. \tag{5.2}$$

The proof is elementary and left to the reader.

To obtain an explicit formula for I_0, for each absolutely continuous path π in $D([0,T], \mathcal{M}^o_{+,1})$ consider the inner product $< \cdot , \cdot >_\pi$ on $C^{1,2}([0,T] \times \mathbb{T}^d)$ defined by

$$< G, H >_\pi = \int_0^T dt \int_{\mathbb{T}^d} du \Big\{ (\nabla H)(t,u) \cdot (\nabla G)(t,u) \Big\} F(\rho(t,u)) .$$

Denote by $\mathcal{N}(\pi)$ the kernel of this inner product and by $\mathcal{H}_1(\pi)$ the Hilbert space obtained by completing $C^{1,2}([0,T] \times \mathbb{T}^d)\Big|_{\mathcal{N}(\pi)}$.

Lemma 5.3 *Assume that $I_0(\pi) < \infty$ and denote by ρ the density of π: $\pi(t,du) = \rho(t,u)du$. There exists H in $\mathcal{H}_1(\pi)$ such that*

$$I_0(\pi) \;=\; \frac{1}{2} \int_0^T dt \int_{\mathbb{T}^d} du \, \|(\nabla H)(t,u)\|^2 F(\rho(t,u)) .$$

Proof. Assume that $I_0(\pi) < \infty$ and recall the definition of the linear functional ℓ_G given in (2.5). By definition of $I_0(\pi)$, for every G in $C^{1,2}([0,T] \times \mathbb{T}^d)$,

$$\ell_G(\pi) \;-\; (1/2) \int_0^T dt \int_{\mathbb{T}^d} du \, \|(\nabla G)(t,u)\|^2 F(\rho(t,u)) \;\le\; I_0(\pi) .$$

Fix a real a. Considering aG instead of G in the previous formula and minimizing over a, we obtain that

$$\big| \ell_G(\pi) \big| \;\le\; \sqrt{2 I_0(\pi)} \Big\{ \int_0^T dt \int_{\mathbb{T}^d} du \, \|(\nabla G)(t,u)\|^2 F(\rho(t,u)) \Big\}^{1/2} .$$

Notice that $\ell_\cdot(\pi)$ is a linear functional on $C^{1,2}([0,T] \times \mathbb{T}^d)$. We just proved that it is bounded in $\mathcal{H}_1(\pi)$. We may therefore extend $\ell_\cdot(\pi)$ to $\mathcal{H}_1(\pi)$. In this case, by Riesz' representation theorem, there exists H in $\mathcal{H}_1(\pi)$ such that

$$\ell_G(\pi) \;=\; \int_0^T dt \int_{\mathbb{T}^d} du \, \big(\nabla G \cdot \nabla H \big)(t,u) F(\rho(t,u))$$

for each G in $\mathcal{H}_1(\pi)$.

It is now easy to derive the explicit formula for the rate function I_0. By definition of I_0 and by the above representation of the linear functional $\ell_\cdot(\pi)$,

$$\begin{aligned} I_0(\pi) \;=\; & \\ & \sup_{G \in C^{1,2}([0,T] \times \mathbb{T}^d)} \Big\{ \ell_G(\pi) \;-\; (1/2) \int_0^T dt \int_{\mathbb{T}^d} du \, \|(\nabla G)(t,u)\|^2 F(\rho(t,u)) \Big\} \\ =\; & \sup_{G \in C^{1,2}([0,T] \times \mathbb{T}^d)} \Big\{ \int_0^T dt \int_{\mathbb{T}^d} du \, \big(\nabla G \cdot \nabla H \big)(t,u) F(\rho(t,u)) \\ & \qquad - (1/2) \int_0^T dt \int_{\mathbb{T}^d} du \, \|(\nabla G)(t,u)\|^2 F(\rho(t,u)) \Big\} \end{aligned}$$

Adding and subtracting $(1/2)\int_0^T dt \int du \,\|(\nabla H)(t,u)\|^2 F(\rho(t,u))$, we rewrite last supremum as

$$
\begin{aligned}
&(1/2)\int_0^T dt \int_{\mathbb{T}^d} du \,\|(\nabla H)(t,u)\|^2 F(\rho(t,u)) \\
&\quad - (1/2) \sup_{G\in C^{1,2}([0,T]\times\mathbb{T}^d)} \Big\{ \int_0^T dt \int_{\mathbb{T}^d} du \,\|\nabla(G-H)(t,u)\|^2 F(\rho(t,u)) \Big\}
\end{aligned}
$$

Since $C^{1,2}([0,T]\times\mathbb{T}^d)$ is dense in $\mathcal{H}_1(\pi)$, this last expression is equal to

$$
(1/2)\int_0^T dt \int_{\mathbb{T}^d} du \,\|(\nabla H)(t,u)\|^2 F(\rho(t,u))
$$

and the proof is concluded. □

For each continuous $\gamma : \mathbb{T}^d \to (0,1)$ and H in $C^{1,2}([0,T]\times\mathbb{T}^d)$, denote by $\rho^{\gamma,H}$ the weak solution of (5.1). Let $\pi^{\gamma,H}$ stand for the path on $D([0,T],\mathcal{M}_+)$ with density $\rho^{\gamma,H}$: $\pi^{\gamma,H}(t,du) = \rho^{\gamma,H}(t,u)du$. Multiplying both sides of equation (5.1) by H and integrating by parts we obtain that

$$
\ell_H(\pi^{\gamma,H}) \;=\; \int_0^T dt \int_{\mathbb{T}^d} du \,\|(\nabla H)(t,u)\|^2 F(\rho_t^{\gamma,H}) \, .
$$

In particular, from last lemma,

$$
\begin{aligned}
I_0(\pi^{\gamma,H}) \;&=\; (1/2)\int_0^T dt \int_{\mathbb{T}^d} du \,\|(\nabla H)(t,u)\|^2 F(\rho^{\gamma,H}(t,u)) \\
&=\; \ell_H(\pi^{\gamma,H}) \;-\; (1/2)\int_0^T dt \int_{\mathbb{T}^d} du \,\|(\nabla H)(t,u)\|^2 F(\rho^{\gamma,H}(t,u)) \\
&=\; J_H(\pi^{\gamma,H}) \, .
\end{aligned}
$$

This last equality and (5.2) show that

$$
I(\pi^{\gamma,H}) \;=\; h_\gamma(\pi_0^{\gamma,H}) \;+\; J_H(\pi^{\gamma,H}) \, . \tag{5.3}
$$

The previous two lemmas permit to interpret the large deviations rate function as an entropy. This statement is clarified in next lemma.

Lemma 5.4 *Recall from section A1.8 the definition of the entropy. For each continuous function $\gamma : \mathbb{T}^d \to (0,1)$ and smooth function H in $C^{1,2}([0,T]\times\mathbb{T}^d)$,*

$$
\lim_{N\to\infty} N^{-d} H\Big(P^N_{\gamma,H} \,\big|\, P^N\Big) \;=\; I(\pi^{\gamma,H}) \, .
$$

Proof. Recall the definition of the set $B_{H,\delta,\varepsilon}$ given in (4.1). By Corollary 3.2, the probability of $B^c_{H,\delta,\varepsilon}$ with respect to $P^N_{\gamma,H}$ is superexponentially small. In other words, statement (4.2) remains correct if we replace P^N by $P^N_{\gamma,H}$.

By the explicit formula for the entropy,

$$N^{-d} H\Big(P^N_{\gamma,H} \,|\, P^N\Big) \;=\; N^{-d} E^N_{\gamma,H}\Big[\log \frac{dP^N_{\gamma,H}}{dP^N}\Big]\,.$$

Since by (3.4) $N^{-d}\log(dP^N_{\gamma,H}/dP^N)$ is bounded by $C(\gamma, H, T)$ and since by (4.2) (with $P^N_{\gamma,H}$ instead of P^N) the probability of $B^c_{H,\delta,\varepsilon}$ is superexponentially small, the right hand side of last identity is equal to

$$N^{-d} E^N_{\gamma,H}\Big[\log \frac{dP^N_{\gamma,H}}{dP^N} \mathbf{1}\{B_{H,\delta,\varepsilon}\}\Big] \;+\; o_N(1)$$

for all $\delta > 0$ and each ε small enough ($\varepsilon < \varepsilon(\delta)$). By (4.4), on $B_{H,\delta,\varepsilon}$, $N^{-d}\log(dP^N_{\gamma,H}/dP^N)$ is equal to

$$\begin{aligned}&\ell_H(\pi^N) \;+\; h^N_\gamma(\pi^N_0) \;-\; (1/2)\int_0^T <\lambda^N, \|\nabla H_t\|^2 F((\pi^N_t * \iota_{\varepsilon,N})(\cdot))>\, dt\\ &\quad + O(\delta) \;+\; O_H(N^{-1})\,.\end{aligned}$$

Since this expression is bounded and the probability of $B^c_{H,\delta,\varepsilon}$ with respect to $P^N_{\gamma,H}$ vanishes as $N\uparrow\infty$, last expectation (and therefore the entropy) is equal to

$$\begin{aligned}&E^N_{\gamma,H}\Big[\ell_H(\pi) \;+\; h_\gamma(\pi_0) \;-\; (1/2)\int_0^T <\lambda^N, \|\nabla H_t\|^2 F((\pi_t * \iota_{\varepsilon,N})(\cdot))>\, dt\Big]\\ &\quad + O(\delta) \;+\; o_N(1)\end{aligned}$$

for all δ positive and all ε small enough. Notice also that we replaced h^N_γ by h_γ because h^N_γ converges uniformly to h_γ. We may also replace the discrete Lebesgue measure λ^N by λ. The details were given in the proof of the upper bound.

All expression inside the expectation are continuous with respect to the Skorohod topology. By Proposition 5.1 the sequence $Q^N_{\gamma,H}$ converges weakly to the probability concentrated on the weak solution of (5.1). In particular, as $N\uparrow\infty$, the previous expectation converges to

$$\begin{aligned}&\ell_H(\pi^{\gamma,H}) \;+\; h_\gamma(\pi^{\gamma,H}_0)\\ &\quad -\; (1/2)\int_0^T dt \int_{\mathbb{T}^d} du\, \|\nabla H(t,u)\|^2 F((\pi^{\gamma,H}_t * \iota_\varepsilon)(u)) \;+\; O(\delta)\,.\end{aligned}$$

It remains to let $\varepsilon\downarrow 0$, then $\delta\downarrow 0$ and recall identity (5.3). □

Denote by $D^o([0,T],\mathcal{M}_+)$ the subset of $D([0,T],\mathcal{M}_+)$ consisting of all paths $\pi^{\gamma,H}$ associated to some continuous function $\gamma : \mathbb{T}^d \to (0,1)$ and some H in $C^{1,2}([0,T]\times\mathbb{T}^d)$. That is, the set of all trajectories π of $D([0,T],\mathcal{M}_+)$ that are absolutely continuous ($\pi(t,du) = \rho(t,u)du$) and for which there exists a continuous $\gamma : \mathbb{T}^d \to (0,1)$ and H in $C^{1,2}([0,T]\times\mathbb{T}^d)$ so that ρ is solution of (5.1).

We are now ready to prove the lower bound large deviations.

Proof of the lower bound. Let $\mathcal{O}$ be an open set of $D([0,T],\mathcal{M}_+)$. We shall first prove that

$$\liminf_{N\to\infty} \frac{1}{N^d} \log Q^N[\mathcal{O}] \;\geq\; -I(\pi) \tag{5.4}$$

for all paths π in $\mathcal{O} \cap D^o([0,T],\mathcal{M}_+)$. Since π belongs to $D^o([0,T],\mathcal{M}_+)$, there exists a continuous $\gamma : \mathbb{T}^d \to (0,1)$ and H in $C^{1,2}([0,T]\times\mathbb{T}^d)$ such that $\pi = \pi^{\gamma,H}$.

Denote by $P^N_{\gamma,H,\mathcal{O}}$ the probability on $D([0,T],\{0,1\}^{\mathbb{T}^d_N})$ given by

$$P^N_{\gamma,H,\mathcal{O}}[A] \;=\; \frac{1}{P^N_{\gamma,H}[\pi^N_\cdot \in \mathcal{O}]} P^N_{\gamma,H}[A, \pi^N_\cdot \in \mathcal{O}]$$

for all measurable set A of $D([0,T],\{0,1\}^{\mathbb{T}^d_N})$. By Jensen's inequality,

$$N^{-d}\log Q^N[\mathcal{O}] \;\geq\; E^N_{\gamma,H,\mathcal{O}}\Big[N^{-d}\log\frac{dP^N}{dP^N_{\gamma,H}}\Big] \;+\; N^{-d}\log Q^N_{\gamma,H}[\mathcal{O}]\,.$$

By Proposition 5.1, since $\mathcal{O}$ is a neighborhood that contains $\pi^{\gamma,H}$, the second expression on the right hand side converges to 0 as $N\uparrow\infty$. The first one is equal to

$$\frac{1}{Q^N_{\gamma,H}[\mathcal{O}]}\Big\{-N^{-d}H\Big(P^N_{\gamma,H}\,\big|\,P^N\Big) \;-\; E^N_{\gamma,H}\Big[N^{-d}\log\frac{dP^N}{dP^N_{\gamma,H}}\mathbf{1}\{\pi^N_\cdot \in \mathcal{O}^c\}\Big]\Big\}\,.$$

Once again, by Proposition 5.1, $Q^N_{\gamma,H}[\mathcal{O}]$ converges to 1 as $N\uparrow\infty$. Since by (3.4) the expression $N^{-d}\log(dP^N/dP^N_{\gamma,H})$ is bounded, the second term inside braces vanishes as $N\uparrow\infty$. Therefore,

$$\liminf_{N\to\infty} Q^N[\mathcal{O}] \;\geq\; \lim_{N\to\infty} -N^{-d}H\Big(P^N_{\gamma,H}\,\big|\,P^N\Big)\,.$$

Recall the statement of Lemma 5.4 to conclude the proof of (5.4) for paths in $\mathcal{O}\cap D^o([0,T],\mathcal{M}_+)$.

To conclude the proof of the lower bound it remains to show that all paths π with finite rate function ($I(\pi)<\infty$) can be approximated by a sequence π_n in $D^o([0,T],\mathcal{M}_+)$ such that $\lim_{n\to\infty} I(\pi_n) = I(\pi)$. This is the content of next lemma. □

Lemma 5.5 *For each π with finite rate function ($I(\pi)<\infty$), there exist a sequence π_n in $D^o([0,T],\mathcal{M}_+)$ converging to π in $D([0,T],\mathcal{M}_+)$ and such that*

$$\lim_{n\to\infty} I(\pi_n) \;=\; I(\pi)$$

Proof. In this proof we adopt the following terminology. A sequence of paths π_n is said to approximate π if π_n converges to π in $D([0,T],\mathcal{M}_+)$ and $\lim_{n\to\infty} I(\pi_n) = I(\pi)$.

The proof is divided in three steps. We first show that we can approximate each path with finite rate function by absolutely continuous paths with density bounded below by a strictly positive constant and bounded above by a constant strictly smaller than 1. In this case, we say that the density is bounded away from 0 and 1.

Let π in $D([0,T],\mathcal{M}_+)$ with $I(\pi)<\infty$. In particular, by (1.5), π is absolutely continuous: $\pi(t,du)=\rho(t,u)du$ with density ρ bounded below by 0 and above by 1. Denote by $\widetilde{\pi}^1$ (resp. $\widetilde{\pi}^0$) the constant path with density equal to 1 (0): $\widetilde{\pi}^1(t,du)=du$ for $0\le t\le T$ (resp. $\widetilde{\pi}^0(t,du)=0$ for $0\le t\le T$). Notice that both $\widetilde{\pi}^0$ and $\widetilde{\pi}^1$ belong to $D^o([0,T],\mathcal{M}_+)$ and $I(\widetilde{\pi}^1)=-\log\alpha<\infty$ ($I(\widetilde{\pi}^0)=-\log(1-\alpha)<\infty$).

For $0\le\varepsilon<1$, define π^ε by $\pi^\varepsilon=(1-\varepsilon)\pi+(\varepsilon/2)\widetilde{\pi}^1+(\varepsilon/2)\widetilde{\pi}^0$. Denote by ρ^ε the density of π^ε: $\pi^\varepsilon(t,du)=\rho^\varepsilon(t,u)du$. By construction,

$$\varepsilon/2 \;\le\; \rho^\varepsilon(t,u) \;\le\; 1-(\varepsilon/2)\,. \tag{5.5}$$

Moreover, it is clear that π^ε converges to π as $\varepsilon\downarrow 0$. In particular, by lower semicontinuity of the rate function, $I(\pi)\le\liminf_{\varepsilon\to 0}I(\pi^\varepsilon)$. On the other hand, by convexity, $I(\pi^\varepsilon)\le(1-\varepsilon)I(\pi)+(\varepsilon/2)I(\widetilde{\pi}^0)+(\varepsilon/2)I(\widetilde{\pi}^1)$. Therefore, $\limsup_{\varepsilon\to 0}I(\pi^\varepsilon)\le I(\pi)$.

In conclusion, for each path with finite rate function we constructed an approximating sequence π^ε of absolutely continuous trajectories with density bounded away from 0 and 1.

It remains to show that we can approximate an absolutely continuous path $\pi(t,du)=\rho(t,u)du$ with density bounded away from 0 and 1 by a sequence π_n in $D^o([0,T],\mathcal{M}_+)$.

Consider such a path π in $D([0,T],\mathcal{M}^o_{+,1})$ For each $\varepsilon>0$, denote by ι_ε a smooth approximation of the identity:

$$\iota_\varepsilon\ge 0\,,\qquad \text{supp}\;\iota_\varepsilon\subset[-\varepsilon,\varepsilon]^d\,,\qquad \int_{\mathbb{R}^d}\iota_\varepsilon(u)\,du \;=\; 1\,.$$

Let ρ^ε denote the spatial convolution of ρ with ι_ε: $\rho^\varepsilon(t,u)=\int\rho(t,u-v)\iota_\varepsilon(v)dv$ and $\pi^\varepsilon(t,du)=\rho^\varepsilon(t,u)du$. It is easy to check that π^ε converges to π in $D([0,T],\mathcal{M}_+)$. In particular, by lower semicontinuity of the rate function, $I(\pi)\le\liminf_{\varepsilon\to 0}I(\pi^\varepsilon)$. On the other hand, by convexity and translation invariance of I,

$$I(\pi^\varepsilon)\;\le\;\int I(T_u\pi)\iota_\varepsilon(u)\,du \;=\; I(\pi)\,.$$

In this formula $T_u\pi$ stands for the translation of π by u: $(T_u\pi)(t,dv)=\rho(t,v-u)dv$. Therefore $\limsup_{\varepsilon\to 0}I(\pi^\varepsilon)\le I(\pi)$. Notice that the density ρ^ε is bounded below and above by the same values that bound ρ.

To conclude the proof of the lemma we have to show that we can approximate every absolutely continuous path with density bounded away from 0 and 1 and which is smooth in the space variable by a sequence in $D^o([0,T],\mathcal{M}_+)$.

Fix such a path $\pi(t,du)=\rho(t,u)du$. For each $0\le t\le 1$, extend the definition of ρ to $[T,T+1]$ by setting $\rho(T+t,u)=\widetilde{\rho}(t,u)$, where $\widetilde{\rho}(t,u)$ is the solution of the heat equation with initial condition $\rho(T,\cdot)$:

$$\begin{cases} \partial_t \widetilde{\rho} = (1/2)\Delta \widetilde{\rho} \\ \widetilde{\rho}(0, \cdot) = \rho(T, \cdot)\,. \end{cases}$$

For $0 \le t < 1$, denote by $\sigma_t \rho$ the time translation of ρ: $(\sigma_t \rho)(s, u) = \rho(t+s, u)$ for (s, u) in $[0, T] \times \mathbb{T}^d$. By extension we define $(\sigma_t \pi)(s, du) = (\sigma_t \rho)(s, u)du$. We leave to the reader, as an exercise, to check that $I_0(\sigma_t \pi) \le I_0(\pi)$. This result follows from the variational formula for the rate function I_0 and Lemma 5.3. It is a quite natural result since we extended ρ following the hydrodynamic equation.

For each $0 < \varepsilon < 1$, denote by β_ε a smooth one-dimensional approximation of the identity: $\beta_\varepsilon \ge 0$, $\int \beta_\varepsilon(s)ds = 1$, supp $\beta_\varepsilon \subset [0, \varepsilon]$. Notice that the support of β_ε is contained in $\mathbb{R}_+$.

Let $\rho^\varepsilon(t, u) = \int \beta_\varepsilon(s)(\sigma_s \rho)\, ds$ and $\pi^\varepsilon(t, du) = \rho^\varepsilon(t, u)du$. Clearly, π^ε converges to π as $\varepsilon \downarrow 0$. In particular, by lower semicontinuity,

$$I(\pi) \le \liminf_{\varepsilon \to 0} I(\pi^\varepsilon)\,.$$

In contrast, by convexity, $I(\pi^\varepsilon) \le \int \beta_\varepsilon(s) I(\sigma_s \pi)\, ds$. We left to the reader to show that $I_0(\sigma_s \pi) \le I_0(\pi)$. On the other hand, since $h((\sigma_s \pi)_0) = h(\pi_s)$ and since, by the explicit formula for h, $\lim_{s \to 0} h(\pi_s) = h(\pi_0)$, we have $\lim_{s \to 0} h((\sigma_s \pi)_0) = h(\pi_0)$. In conclusion, $\limsup_{\varepsilon \to 0} I(\pi^\varepsilon) \le I(\pi)$.

Finally, we claim that π^ε belongs to $D^o([0, T], \mathcal{M}_+)$ for every $\varepsilon > 0$. Since π^ε has a smooth density bounded away from 0 and 1, we may solve equation (5.1) in H and obtain that H is of class $C^{1,2}([0, T] \times \mathbb{T}^d)$. Since $\gamma = \rho(0, \cdot)$, γ is a continuous function bounded away from 0 and 1, which proves that π^ε belongs to $D^o([0, T], \mathcal{M}_+)$. □

6. Comments and References

We presented in this chapter the ideas of Kipnis, Olla and Varadhan (1989) and Donsker and Varadhan (1989) who were the first to investigate the large deviations of the empirical measure from the hydrodynamic limit.

Extensions.

(a) *Infinite volume.* A large deviations principle for the empirical measure for symmetric simple exclusion processes and mean-zero asymmetric zero range processes on the lattice $\mathbb{Z}^d$ starting from an equilibrium state were obtained by Landim (1992) and Benois, Kipnis and Landim (1995). Based on the investigation of the time evolution of the $\mathcal{H}_{-1}$ norm and on estimates derived in Yau (1994) this result was extended by Landim and Yau (1995) to one-dimensional Ginzburg–Landau processes in infinite volume starting from a large class of non equilibrium states, including deterministic initial configurations. They also proved a similar result for attractive zero range processes through coupling techniques.

(b) *Nonconservative systems.* Large deviations from the hydrodynamic limit for reaction–diffusion models were considered by Jona–Lasinio, Landim and Vares (1993) and Landim (1991c). In contrast with conservative systems described by parabolic equations, the rate function involves exponential terms and cannot be interpreted as due to a simple stochastic perturbation of the hydrodynamic equation. Jona–Lasinio (1991), (1992) discusses further this issue.

(c) *Nongradient systems.* Quastel (1995a) proved a large deviations principle for the empirical measure in the case of a one-dimensional nongradient Ginzburg–Landau model. His proof applies in any dimension provided the diffusion coefficient is Lipschitz continuous. Quastel and Yau (1997) derived weak solutions of the incompressible Navier Stokes equations from an interacting particle system where particles have velocity. The generator has two part: The first one corresponds to displacements of particles and the second to collisions. Both density and momenta are conserved by the dynamics. They deduce furthermore the large deviations from the hydrodynamic limit. Two different large deviations rates arise in this model: The probability to violate the divergence free condition decays at rate at least $\exp\{-CN^{1-d}\}$, while the probability to violate the momentum conservation decays at rate $\exp\{-CN^{2-d}\}$.

Onsager–Machlup time–reversal relation. Consider, to fix ideas, the symmetric simple exclusion process and recall the notation introduced in this section. Fix a density α. For a profile $\gamma\colon \mathbb{T}^d \to [0,1]$, the specific entropy $N^{-d}H(\nu^N_{\gamma(\cdot)}|\nu^N_\alpha)$ converges, as $N \uparrow \infty$, to the entropy functional $S_\alpha(\gamma)$ that may be written as the integral of a density $s_\alpha(\gamma)$:

$$S_\alpha(\gamma) \;=\; \int_{\mathbb{T}^d} du\, s_\alpha(\gamma(u)) \, ,$$

where $s_\alpha(a) = a\log\{a/\alpha\} + (1-a)\log\{(1-a)/(1-\alpha)\}$.

Denote by $\rho_\alpha\colon \mathbb{T}^d \to [0,1]$ the constant profile with density α: $\rho_\alpha(u) = \alpha$ for u in $\mathbb{T}^d$. ρ_α is an equilibrium state of the heat equation (4.2.1) and its basin of attraction consists of all profiles with total mass equal to α: for every profile $\gamma\colon \mathbb{T}^d \to [0,1]$ with density α ($\int_{\mathbb{T}^d} du\,\gamma(u) = \alpha$), the solution of the hydrodynamic equation with initial profile γ relaxes, as $t \uparrow \infty$, to ρ_α.

Denote by $L^\infty_{+,1,\alpha}(\mathbb{T}^d)$ the space of positive, measurable functions on $\mathbb{T}^d$ bounded by 1 and with density α. For $T_1 < T_2$, denote by $I_{[T_1,T_2]}(\cdot)$ the large deviations rate functional for the empirical measure on the time interval $[T_1, T_2]$. Define the quasi–potential $V_\alpha\colon L^\infty_{+,1,\alpha}(\mathbb{T}^d) \to \mathbb{R}_+$ by

$$V_\alpha(\gamma) \;=\; \inf I_{(-\infty,0]}(\rho) \, ,$$

where the infimum is taken over all trajectories $\rho\colon (-\infty,0] \times \mathbb{T}^d \to [0,1]$ such that $\rho(0,\cdot) = \gamma$, $\lim_{t\to-\infty} \rho(t,\cdot) = \rho_\alpha$. The quasi–potential $V_\alpha(\gamma)$ represents the price to create a fluctuation from ρ_α to γ.

The Onsager–Machlup principle (Onsager and Machlup (1953)) states that for reversible systems the infimum is attained by the reversed trajectory: $V_\alpha(\gamma) =$

$I_{(-\infty,0]}(\rho^*)$, where ρ^* is the time reversed trajectory: $\rho^*(t,u) = \rho_{\gamma(\cdot)}(-t,u)$ if $\rho_{\gamma(\cdot)}$ stands for the solution of the heat equation with initial data γ. Furthermore, by the Boltzmann–Einstein relation, V_α and S_α coincide up to an additive constant.

The Onsager–Machlup time–reversal relation has been conjectured for a wide class of systems by Eyink (1990). It has been proved by Gabrielli, Jona–Lasinio, Landim (1996) for a particular class of non reversible, conservative interacting particle systems showing that reversibility is not a necessary condition for the Onsager–Machlup principle. Gabrielli et al. (1997) proved the Onsager–Machlup principle for non reversible and non conservative systems obtained superposing speeded up symmetric simple exclusion processes with Glauber dynamics. The evolution of these later processes at the macroscopic level are described by reaction–diffusion equations (De Masi, Ferrari and Lebowitz (1986)).

The Onsager–Machlup time–reversal relation is further discussed in Eyink, Lebowitz, Spohn (1996) in a general context. The question remains open for a general non reversible system associated to a reaction–diffusion equation.

Metastability. Consider a reaction–diffusion model obtained by the superposition of a speeded up symmetric simple exclusion process with a Glauber dynamics. This is a Markov process on $\{0,1\}^{\mathbb{T}_N^d}$ whose generator is

$$
\begin{aligned}
(L_N f)(\eta) \;&=\; N^2 \sum_{|x-y|=1} \eta(x)[1-\eta(y)][f(\eta^{x,y}) - f(\eta)] \\
&+ \sum_{x\in\mathbb{T}_N^d} (\tau_x c)(\eta)[f(\sigma_x\eta) - f(\eta)]\;.
\end{aligned}
\tag{6.1}
$$

In this formula $c(\eta)$ is a cylinder function and $\sigma_x\eta$ stands for the configuration η with the spin at x flipped:

$$
(\sigma_x\eta)(y) \;=\; \begin{cases} 1-\eta(x) & \text{if } y = x\,, \\ \eta(y) & \text{otherwise}\,. \end{cases}
$$

The hydrodynamic behavior (De Masi, Ferrari and Lebowitz (1986)) of this system is described by a reaction–diffusion equation:

$$
\partial_t \rho \;=\; (1/2)\Delta\rho \;+\; F(\rho)\,, \tag{6.2}
$$

where $F(\cdot)$ is a polynomial given by $F(\alpha) = E_{\nu_\alpha}[\{1-2\eta(0)\}c(\eta)]$, provided ν_α stands for the Bernoulli product measure with density α. To fix ideas, assume that $F = -W'$ is a double well potential and denote by α_- (resp. α_+) the local (resp. global) minima of the potential W. Assume that the constant profiles ρ_{α_-}, ρ_{α_+} are stable equilibrium points of the differential equation (6.2). Denote by B_{α_-}, B_{α_+} (resp. ∂B_{α_-}, ∂B_{α_+}) the basins of attraction (resp. the boundary of the basins of attraction for the weak topology) of these equilibrium points.

Fix a profile γ_0 in B_{α_-} and consider a system starting from $\nu^N_{\gamma_0(\cdot)}$. Since ρ_{α_-} is a stable equilibrium point of the hydrodynamic equation, the empirical measure converges in probability, as $N\uparrow\infty$ and $t\uparrow\infty$ to $\rho_{\alpha_-}(u)du$. The typical behavior of the empirical measure is therefore to relax towards the equilibrium ρ_{α_-} and to fluctuate around it due to the stochastic character of the dynamics. Since ρ_{α_-} is a stable equilibrium point of the hydrodynamic equation, we expect the system to remains a long time in a stationary situation around $\rho_{\alpha_-}(u)du$. In particular, two systems starting from two different profiles γ_1 and γ_2 in the basin of attraction of ρ_{α_-} will rapidly be very close one to the other. In this sense the process loses memory of its starting point. Eventually one large fluctuation will drive the empirical measure out of the basin of attraction B_{α_-} to the basin of attraction of ρ_{α_+}, where it stays for a much larger time because ρ_{α_+} is the global minimum of the potential W. Denote by T_N the first time the process leaves B_{α_-}: $T_N = \inf\{t \geq 0;\ \pi_t^N \notin B_{\alpha_-}\}$. Due to the loss of memory of the system, it is natural to conjecture that T_N correctly renormalized converges in some sense to an exponential distribution:

$$\frac{T_N}{\mathbb{E}_{\nu^N_{\gamma_0}}[T_N]} \to \exp(1)\ . \tag{6.3}$$

Cassandro, Galves, Olivieri and Vares (1984) proposed a mathematical formulation to describe this phenomenon. For fixed N in $\mathbb{N}$ and $R > 0$, denote by $A_R^N(s)$ the time average

$$A_R^N(s) \;=\; \frac{1}{R}\int_s^{s+R} \delta_{\pi_t^N}\, dt\ ,$$

where δ_μ stands for the Dirac mass at μ. Cassandro, Galves, Olivieri and Vares suggested to characterize the meta–stable behavior of the process by proving the existence of a sequence $R_N \uparrow \infty$ such that $A_{R_N}^N(sE_{\nu^N_{\gamma_0(\cdot)}}[T_N])$ converges in distribution to a jump process $A(s)$ given by

$$A(s) \;=\; \begin{cases} \delta_{\rho_{\alpha_-}(u)du} & \text{for } s < T\,, \\ \delta_{\rho_{\alpha_+}(u)du} & \text{for } s \geq T\,, \end{cases}$$

where T is a mean-one exponential random variable. They proved this result as well as (6.3) for the total magnetization in a Curie–Weiss model.

The meta–stable behavior of interacting particle systems has been widely studied. We refer to Penrose and Lebowitz (1987) and to the recent book by Olivieri and Vares (1997) for clear and detailed presentations of the subject.

Exit points from a basin of attraction. We keep here the notation introduced in the previous subsection. Recall that V_{α_-} stands for the quasi–potential corresponding to the stable equilibrium ρ_{α_-}. Denote by A_{α_-} the subset of the boundary ∂B_{α_-} where the quasi–potential assumes its minimum:

$$A_{\alpha_-} \;=\; \Big\{\gamma \in \partial B_{\alpha_-};\ V_{\alpha_-}(\gamma) = \inf_{\gamma' \in \partial B_{\alpha_-}} V_{\alpha_-}(\gamma')\Big\}\ .$$

Since the quasi–potential represents the price needed to create a fluctuation from the equilibrium point ρ_{α_-}, it is natural to expect that the system leaves B_{α_-} through A_{α_-}: for every $\delta > 0$,

$$\lim_{N\to\infty} \mathbb{P}_{\nu^N_{\gamma_0(\cdot)}} \Big[d(\pi_{T_N}, A_{\alpha_-}) > \delta \Big] = 0 . \tag{6.4}$$

This statement has been proved by Comets (1987) for a mean field, non conservative Glauber dynamics under some technical assumptions on the quasi–potential. For local dynamics this question and the meta–stable behavior of the empirical measure remain open problems.

Escape from unstable equilibrium points. To examine the escape from an unstable equilibrium point, consider the same reaction–diffusion process with generator given by (6.1). Recall that the hydrodynamic behavior is described by the reaction–diffusion equation (6.2), where $F(\alpha) = E_{\nu_\alpha}[\{1-2\eta(0)\}c(\eta)]$. Denote by $W(\cdot)$ the potential associated to the reaction part: $W'(\alpha) = -F(\alpha)$ and fix an unstable equilibrium point α^*: $W'(\alpha^*) = 0$, $W''(\alpha^*) < 0$. Consider a system starting from the Bernoulli product measure with density α^*. By the conservation of local equilibrium, proved for these systems by De Masi, Ferrari, Lebowitz (1986), for a fixed macroscopic time t, the system remains close to α^*: $\lim_{N\to\infty} S^N_t \tau_{[uN]} \nu^N_{\alpha^*} = \nu_{\alpha^*}$ for every $t > 0$, u in $\mathbb{T}^d$, where $\{S^N_t,\, t \ge 0\}$ stands for the semigroup of the Markov process with generator L_N defined in (6.1). However, since α^* is an unstable equilibrium point, one might believe that in a longer time scale the process escapes from the unstable equilibrium point and converges to some stable equilibrium point.

Since the behavior depends drastically on the shape of the potential W close to the saddle point, we introduce two distinct examples. Following the review of Vares (1991), let the jump rate $c(\eta)$ in (6.1) be equal to

$$\text{(a)}\quad c(\eta) = 1 - \gamma\chi(0)[\chi(1)+\chi(-1)] + \gamma^2\chi(1)\chi(-1) ,$$

$$\text{(b)}\quad c(\eta) = \Big\{1 - \frac{\chi(0)\chi(1)}{2}\Big\}\Big\{1 - \frac{\chi(0)\chi(-1)}{2}\Big\}\big\{1 - c\,\chi(0)\chi(2)\chi(3)\chi(4)\big\} .$$

In this formula, $\chi(x)$ stands for $2\eta(x) - 1$ and $1/2 < \gamma \le 1$, $1/4 < c \le 1$. The upper bounds are required to ensure that the jump rates are positive and the lower bounds to guarantee that $1/2$ is an unstable equilibrium point.

In example (a) the potential is equal to $W(\alpha) = 4^{-1}(1-2\gamma)(2\alpha-1)^2 + 8^{-1}\gamma^2(2\alpha-1)^4$ and is quadratic in a neighborhood of the unstable equilibrium point $\alpha^* = 1/2$. In example (b) the potential is $W(\alpha) = 8^{-1}(4^{-1}-c)(2\alpha-1)^4 + 2^{-4}c(2\alpha-1)^6$ and is quartic close to the unstable equilibrium point $\alpha^* = 1/2$.

De Masi, Presutti, Vares (1986), De Masi, Pellegrinotti, Presutti and Vares (1994) proved for example (a) that

$$\lim_{N\to\infty} S^N_{t\log N}\, \nu^N_{\alpha^*} = \begin{cases} \nu_{1/2} & \text{if } t < (2A)^{-1} , \\ (1/2)\{\nu_{\alpha_-} + \nu_{\alpha_+}\} & \text{if } t > (2A)^{-1} \end{cases} \tag{6.5}$$

and that

$$\lim_{N\to\infty} S^N_{(1/2A)\log N + t}\, \nu^N_{\alpha^*} = \int_0^1 \nu_\alpha\, \lambda_t(d\alpha) ,$$

where $A = 2(2\gamma - 1)$, convergence is meant by weak convergence, λ_t is a Lebesgue absolutely continuous probability measure that converges, as $t \uparrow \infty$, to $(1/2)\{\delta_{\alpha_-} + \delta_{\alpha_+}\}$ and α_-, α_+ are the stable equilibrium points of the potential W.

Notice in this example that, although being produced by a stochastic fluctuation, the escape from the unstable equilibrium point occurs at a deterministic time in the macroscopic scale $\log N$.

For example (b) Calderoni, Pellegrinotti, Presutti, Vares (1989) proved that

$$\lim_{N\to\infty} S^N_{t\sqrt{N}} \nu^N_{\alpha^*} = a(t)\nu_{\alpha^*} + [1 - a(t)](1/2)\{\nu_{\alpha_-} + \nu_{\alpha_+}\} ,$$

where $a(t) = P[S > t]$ and S is the explosion time of the diffusion $dZ_t = AZ_t^3 dt + dW_t$ with $A = 2[c - (1/4)]$, $Z_0 = 0$ and W_t a standard Brownian motion. Therefore, in the case where the potential is quartic around the saddle point, the escape time is a random variable. This is the expected behavior for every higher degree of degeneracy.

De Masi and Presutti (1991) present a proof of (6.5). An extension to infinite volume is contained in De Masi, Pellegrinotti, Presutti, Vares (1994). Giacomin (1994), (1995) investigates the problem in higher dimension.

Rare events. Related to the question of metastability is the investigation of occurrence times of rare events. To fix ideas, consider an interacting particle system ξ_t evolving on the lattice $\mathbb{Z}^d$ with invariant measure ν, not necessarily unique. Let $\{A_N, N \geq 1\}$ be a sequence of events such that $\lim_{N\to\infty} \nu(A_N) = 0$ and define the sequence of hitting times $T_N = \inf\{t \geq 0; \xi_t \in A_N\}$. There are two natural questions regarding the asymptotic behavior of T_N: its magnitude and the limit distribution of T_N correctly renormalized.

This type of question has been analyzed for Markov processes with good recurrence properties, by Bellman and Harris (1951) and Harris (1952) for recurrent, discrete time Markov chains on a countable set; by Aldous (1982, 1989) and by Aldous and Brown (1992, 1993) for finite state Markov chains and by Korolyuk and Sil'vestrov (1984) and Cogburn (1985) in the case of Harris recurrent chains.

In the context of interacting particle systems, the hitting times of rare events were first examined by Lebowitz and Schonmann (1987). They considered the first time the density in a cube $\Lambda_N = \{-N, \dots, N\}^d$ performs a large fluctuation in the case of attractive stochastic spin systems starting from equilibrium. Assuming rapid convergence to equilibrium they deduced estimates on the order of magnitude of the hitting times T_N and they proved that T_N/β_N, for some suitably chosen sequence β_N, converges in distribution to a mean-one exponential random variable. These results were extended to the one-dimensional supercritical contact process by Galves, Martinelli and Olivieri (1989).

As in the question of metastability, the main difficulty in the investigation of hitting times of rare events is to show that the process loses memory in the scale where the phenomenon occurs. Typically, nonconservative systems lose memory much faster than conservative ones. There are, however, results for conservative dynamics. For the one-dimensional, nearest neighbor, totally asymmetric zero range process with jump rate $g(k) = \mathbf{1}\{k \geq 1\}$, Ferrari, Galves and Landim (1994) proved the estimate

$$\sup_{t\geq 0} \Big| \mathbb{P}_{\bar{\nu}_\varphi} \Big[\varphi(1-\varphi)^N T_N > t \Big] - e^{-t} \Big| \leq (N+3)(1-\varphi)^{N/2}$$

if T_N stands for the first time the sites $\{1, \dots, N\}$ become empty and $\bar{\nu}_\varphi$ for the product, translation invariant, stationary measure with marginals given by $\bar{\nu}_\varphi\{\eta, \eta(x) = k\} = (1-\varphi)\varphi^k$. Ferrari, Galves and Liggett (1995) studied the same problem for the one-dimensional nearest neighbor symmetric simple exclusion process. They proved that

$$\sup_{t\geq 0} \Big| \mathbb{P}_{\nu_\alpha} \Big[\beta_N \alpha^N T_N > t \Big] - e^{-t} \Big| \leq C_0 \, \alpha^{C_1 N}$$

for some sequence $0 < \beta \leq \beta_N \leq \beta' < 1$. In this formula ν_α stands for the Bernoulli product measure with density α and C_0, C_1 for two finite constants. Asselah and Dai Pra (1997) extends this result to higher dimension for the first time the empirical density in the cube $\{1, \dots, N\}^d$ reaches a level $\alpha' > \alpha$.

Large deviations of asymmetric models. This issue, which remains one of the main open questions in the theory of hydrodynamic behavior of interacting particle systems, is discussed in Chapter 8.

11. Equilibrium Fluctuations of Reversible Dynamics

In Chapters 4 to 7 we examined the hydrodynamic behavior of several mean-zero interacting particle systems and proved a law of large numbers under diffusive rescaling for the empirical measure. We now investigate the fluctuations of the empirical measure around the hydrodynamic limit starting from an equilibrium state. To fix ideas, we consider the nearest neighbor symmetric zero range process. The reader shall notice, however, that the approach presented below applies to a large class of reversible models including nongradient systems. The generator of this process is

$$(L_N f)(\eta) = \sum_{x,y\in\mathbb{T}_N^d} p(y)g(\eta(x))[f(\eta^{x,x+y}) - f(\eta)]\,, \tag{0.1}$$

where $p(y) = 1/2$ if $|y| = 1$ and 0 otherwise and g is a rate function satisfying the assumptions of Definition 2.3.1.

We proved in Chapters 5 and 6 that for a class of zero range processes starting from a sequence of probability measures $\{\mu^N,\, N \geq 1\}$ associated to a profile $\rho_0\colon \mathbb{T}^d \to \mathbb{R}_+$, the empirical measure π^N converges in probability to an absolutely continuous measure whose density is the weak solution of the non linear heat equation

$$\begin{cases} \partial_t \rho = (1/2)\sum_{j=1}^{d} \partial_{u_j}\Big(\Phi'(\rho)\partial_{u_j}\rho\Big) \\ \rho(0,\cdot) = \rho_0(\cdot)\,. \end{cases} \tag{0.2}$$

To investigate the equilibrium fluctuations of π^N, we fix once for all a density $\alpha > 0$ and we denote by $Y_\cdot^N$ the density fluctuation field that acts on smooth functions H as

$$Y_t^N(H) = N^{-d/2} \sum_{x\in\mathbb{T}_N^d} H(x/N)(\eta_{tN^2}(x) - \alpha)\,. \tag{0.3}$$

Notice the diffusive rescaling of time on the right hand side of this identity. The aim of this chapter is to prove that $Y_\cdot^N$ converges to a stationary Gaussian process with a given space–time correlations.

To state the main theorem of this section we need to introduce some notation. Consider the lattice $\mathbb{Z}^d$ endowed with the lexicographical order. Let $h_0 \equiv 1$ and for each $z > 0$ (resp. $z < 0$), define $h_z : \mathbb{T}^d \to \mathbb{R}$ by $h_z(u) = \sqrt{2}\cos(2\pi z \cdot u)$ (resp. $h_z(u) = \sqrt{2}\sin(2\pi z \cdot u)$). Here $\cdot$ denotes the inner product of $\mathbb{R}^d$. It is well known that the set $\{h_z, z \in \mathbb{Z}^d\}$ is an orthonormal basis of $L^2(\mathbb{T}^d)$: each function f in $L^2(\mathbb{T}^d)$ can be written as

$$f = \sum_{z\in\mathbb{Z}^d} < f, h_z > h_z .$$

In this formula and below $< \cdot, \cdot >$ stands for the inner product of $L^2(\mathbb{T}^d)$.

Consider on $L^2(\mathbb{T}^d)$ the positive, symmetric linear operator $\mathcal{L} = (1 - \Delta)$. A simple computation shows that the functions h_z are eigenvectors:

$$\mathcal{L} h_z = \gamma_z h_z ,$$

where $\gamma_z = 1 + 4\pi^2\|z\|^2$. For a positive integer k, denote by $\mathcal{H}_k$ the Hilbert space obtained as the completion of $C^\infty(\mathbb{T}^d)$ endowed with the inner product $< \cdot, \cdot >_k$ defined by

$$< f, g >_k = < f, \mathcal{L}^k g > .$$

It is easy to check that $\mathcal{H}_k$ is the subspace of $L^2(\mathbb{T}^d)$ consisting of all functions f such that

$$\sum_{z\in\mathbb{Z}^d} < f, h_z >^2 \gamma_z^k < \infty .$$

In particular, if we denote $L^2(\mathbb{T}^d)$ by $\mathcal{H}_0$,

$$\mathcal{H}_0 \supset \mathcal{H}_1 \supset \mathcal{H}_2 \supset \cdots \tag{0.4}$$

Moreover, on $\mathcal{H}_k$ the inner product $< \cdot, \cdot >_k$ can be expressed as

$$< f, g >_k = \sum_{z\in\mathbb{Z}^d} < f, h_z >< g, h_z > \gamma_z^k .$$

For each positive integer k, denote by $\mathcal{H}_{-k}$ the dual of $\mathcal{H}_k$ relatively to the inner product $< \cdot, \cdot >$. $\mathcal{H}_{-k}$ can be obtained as the completion of $L^2(\mathbb{T}^d)$ with respect to the inner product obtained from the quadratic form $< f, f >_{-k}$ defined by

$$< f, f >_{-k} = \sup_{g\in\mathcal{H}_k} \Big\{ 2 < f, g > - < g, g >_k \Big\} .$$

It is again easy to check that $\mathcal{H}_{-k}$ consists of all sequences $\{< f, h_z >, z \in \mathbb{Z}^d\}$ such that

$$\sum_{z\in\mathbb{Z}^d} < f, h_z >^2 \gamma_z^{-k} < \infty$$

and that the inner product $< f, g >_{-k}$ of two functions f, g in $\mathcal{H}_{-k}$ can be written as

$$< f, g >_{-k} = \sum_{z\in\mathbb{Z}^d} < f, h_z >< g, h_z > \gamma_z^{-k} .$$

It follows also from the explicit characterization of $\mathcal{H}_{-k}$ and from (0.4) that

$$\cdots \subset \mathcal{H}_2 \subset \mathcal{H}_1 \subset \mathcal{H}_0 \subset \mathcal{H}_{-1} \subset \mathcal{H}_{-2} \subset \cdots$$

We shall consider the density fluctuation field Y_t^N as taking values in the Sobolev space $\mathcal{H}_{-k}$ for some large enough k. Fix a time $T > 0$, a positive integer k_0 and denote by $D([0,T], \mathcal{H}_{-k_0})$ (resp. $C([0,T], \mathcal{H}_{-k_0})$) the space of $\mathcal{H}_{-k_0}$ valued functions, that are right continuous with left limits (resp. continuous), endowed with the uniform weak topology: a sequence $\{Y_j,\, j \geq 1\}$ converges to a path Y if $Y_j(t)$ converges weakly to $Y(t)$ uniformly in time, i.e., if for all f in $\mathcal{H}_{k_0}$,

$$\lim_{j\to\infty} \sup_{0\leq t\leq T} \Big| < Y_j(t)\,,\, f > - < Y(t)\,,\, f > \Big| = 0\,.$$

Denote by Q_N the probability measure on $D([0,T], \mathcal{H}_{-k_0})$ induced by the density fluctuation field Y^N introduced in (0.3) and the product measure ν_α^N, by $\mathbb{P}_N$ the probability measure on $D([0,T], \mathbb{N}^{\mathbb{T}_N^d})$ induced by the probability measure ν_α^N and the Markov process η_t speeded up by N^2 and denote by $\mathbb{E}_N$ expectation with respect to $\mathbb{P}_N$.

Theorem 0.1 *Fix a positive integer $k_0 > 2+(d/2)$. Let Q be the probability measure concentrated on $C([0,T], \mathcal{H}_{-k_0})$ corresponding to the stationary generalized Ornstein–Uhlenbeck process with mean 0 and covariance*

$$E_Q\Big[Y_t(H) Y_s(G)\Big] = \frac{\chi(\alpha)}{(2\pi(t-s)\Phi'(\alpha))^{d/2}} \int_{\mathbb{R}^d} du \int_{\mathbb{R}^d} dv\, \bar{H}(u) \exp\Big\{ - \frac{(u-v)^2}{2(t-s)\Phi'(\alpha)} \Big\} \bar{G}(v) \tag{0.5}$$

for every $0 \leq s \leq t$ and H, G in $\mathcal{H}_{k_0}$. Here $\chi(\alpha)$ stands for the static compressibility given by $\chi(\alpha) = \mathbf{Var}(\nu_\alpha, \eta(0))$ and $\bar{H},\ \bar{G}\colon \mathbb{R}^d \to \mathbb{R}$ are periodic functions with period $\mathbb{T}^d$ and equal to H, G on $\mathbb{T}^d$. Then, the sequence Q_N converges weakly to the probability measure Q.

Theorem 0.1 relates the covariance of the equilibrium density fluctuation to the diffusion coefficient of the hydrodynamic equation (0.2), a parameter determined by the non equilibrium evolution. In the mathematical physics literature this result is called a fluctuation–dissipation theorem since it connects the non equilibrium dissipative feature of the system to its equilibrium fluctuations.

The proof of Theorem 0.1 relies on Holley and Stroock's theory of generalized Ornstein–Uhlenbeck processes that we now explain. Denote by $\mathfrak{A}$ the nonnegative self adjoint operator $(1/2)\Phi'(\alpha)\Delta$ defined on a domain of $L^2(\mathbb{T}^d)$, by $\{T_t,\, t \geq 0\}$ the semigroup associated to $\mathfrak{A}$ and by $\mathfrak{B}$ the linear operator $\Phi(\alpha)\nabla$. For $t \geq 0$, let $\mathcal{F}_t$ be the σ–algebra on $D([0,T], \mathcal{H}_{-k_0})$ generated by $Y_s(H)$ for $s \leq t$ and H in $C^\infty(\mathbb{T}^d)$ and set $\mathcal{F} = \sigma(\cup_{t\geq 0}\mathcal{F}_t)$.

Theorem 0.2 *Fix a positive integer $k_1 \geq 2$. Let Q be a probability measure on the space $\{C([0,T], \mathcal{H}_{-k_1}), \mathcal{F}\}$. Assume that for each H in $C^\infty(\mathbb{T}^d)$,*

$$M_t^{\mathfrak{A},H} = Y_t(H) - Y_0(H) - \int_0^t Y_s(\mathfrak{A}H)\,ds \qquad (0.6)$$
$$\text{and} \quad (M_t^{\mathfrak{A},H})^2 - \|\mathfrak{B}H\|_2^2\, t$$

are $L^1(Q)$ $\mathcal{F}_t$–martingales. Then, for all $0 \le s < t$, H in $C^\infty(\mathbb{T}^d)$ and subsets A of $\mathbb{R}^d$,

$$Q\Big[Y_t(H) \in A \,\big|\, \mathcal{F}_s\Big] =$$
$$\int_A \frac{1}{\sqrt{2\pi \int_0^{t-s} \|\mathfrak{B}T_r H\|_2^2\,dr}} \exp\Big\{\frac{-(y - Y_s(T_{t-s}H))^2}{2\int_0^{t-s} \|\mathfrak{B}T_r H\|_2^2\,dr}\Big\}\,dy \quad Q \text{ a.s.}\,. \qquad (0.7)$$

In particular, condition (0.6) and the knowledge of the restriction of Q to $\mathcal{F}_0$ uniquely determines Q on $\{C([0,T],\mathcal{H}_{-k_1}),\mathcal{F}\}$.

The proof of this theorem is postponed to section 4. It states that for each distribution q on $\mathcal{H}_{-k_1}$, there exists a unique probability measure Q on $C([0,T],\mathcal{H}_{-k_1})$ that solves the martingale problem (0.6) and such that Q restricted to $\mathcal{F}_0$ is equal to q. In our setting, for any fixed time t_0, the limit distribution of $Y_{t_0}^N$ is easy to deduce: we shall prove at the beginning of section 2 that $Y_{t_0}^N$ converges in law to a mean-zero Gaussian field with covariance given by

$$E_Q[Y(H)Y(G)] = \chi(\alpha) < H, G > \qquad (0.8)$$

for each smooth function G, H in $\mathcal{H}_{k_0}$. In particular, Theorem 0.2 reduces the proof of Theorem 0.1 to the verification that the sequence Q_N converges to a probability measure Q that solves the martingale problem (0.6).

Relation (0.6) and the equal time covariances $E_Q[Y_t(G)Y_t(H)]$ given by (0.8) permit to deduce the space time covariances $E_Q[Y_s(G)Y_t(H)]$: an expansion argument gives that for $0 \le s < t$

$$E_Q\Big[Y_s(G)Y_t(H)\Big] = \chi(\alpha) < T_{t-s}G, H > ,$$

which is precisely the right hand side of (0.5). Moreover, by (0.6), for each H in $\mathcal{H}_{k_1}$, $W_t^H = \|\mathfrak{B}H\|_2^{-1} M_t^{\mathfrak{A},H}$ is a martingale with quadratic variation equal to t. Therefore, by the martingale characterization of Brownian motion due to Levy, W_t^H is a Brownian motion and we may rewrite (0.6) as

$$Y_t(H) = Y_0(H) + \int_0^t Y_s(\mathfrak{A}H)\,ds + \|\mathfrak{B}H\|_2 W_t^H\,, \qquad (0.9)$$

where W_t is a generalized Brownian motions with covariance

$$E_Q\Big[W_s^G W_t^H\Big] = (s \wedge t) \int_{\mathbb{T}^d} \frac{\nabla G(u)}{\|\nabla G\|_2} \cdot \frac{\nabla H(u)}{\|\nabla H\|_2}\,du\,.$$

To deduce this last relation we used the identity $\Phi'(\alpha) = \Phi(\alpha)/\chi(\alpha)$ that follows from the equality $\Phi'(\alpha_0) = [R'(\Phi(\alpha_0))]^{-1}$ and a straightforward computation of $R'(\Phi)$. Equation (0.9) suggests the following formal stochastic differential equation for Y_t:

$$dY_t = (1/2)\Phi'(\alpha)\Delta Y_t dt + \sqrt{\Phi(\alpha)}\nabla dW_t .$$

We conclude this section sketching the strategy of the proof of Theorem 0.1. Theorem 0.2 reduces the proof of Theorem 0.1 to the verification of three properties: (a) that the sequence of probability measures is tight, (b) that the restriction to $\mathcal{F}_0$ of all limit points Q of the sequence Q_N are Gaussian fields with covariance given by (0.8) and (c) that all limit points Q solve the martingale problem (0.6). The first two properties are straightforward. To check that all limit points solve the martingale problem, we consider a collection of martingales associated to the empirical measure. For each smooth function $G\colon \mathbb{T}^d \to \mathbb{R}$, denote by M_t^G and by N_t^G the martingales defined by

$$M_t^G = Y_t^N(G) - Y_0^N(G) - \int_0^t N^2 L_N N^{-d/2} \sum_{x\in\mathbb{T}_N^d} G(x/N)[\eta_s(x) - \alpha]\, ds ,$$

$$N_t^G = (M_t^G)^2 - \int_0^t \Big\{N^2 L_N \Big(Y_s^N(G)\Big)^2 - 2Y_s^N(G) N^2 L_N Y_s^N(G)\Big\}\, ds .$$

A simple computation permits to rewrite these martingales as

$$\begin{aligned} M_t^G &= Y_t^N(G) - Y_0^N(G) \\ &\quad - \int_0^t (1/2)N^{-d/2} \sum_{x\in\mathbb{T}_N^d} (\Delta_N G)(x/N)[g(\eta_s(x)) - \Phi(\alpha)]\, ds , \\ N_t^G &= (M_t^G)^2 \\ &\quad - \int_0^t (1/2)N^{-d} \sum_{j=1}^d \sum_{x\in\mathbb{T}_N^d} [g(\eta_s(x+e_j)) + g(\eta_s(x))][(\partial_{u_j}^N G)(x/N)]^2\, ds , \end{aligned} \tag{0.10}$$

where Δ_N stands for the discrete Laplacian and $(\partial_{u_j}^N G)(x/N)$ is equal to $N\{G((x+e_j)/N) - G(x/N)\}$ for $1 \le j \le d$. We took advantage here from the fact that $\sum_x (\Delta_N G)(x/N) = 0$ to add the expression $\Phi(\alpha)\sum_x(\Delta_N G)(x/N)$ to the martingale M_t^G in order to obtain the mean-zero cylinder function $g(\eta(x)) - \Phi(\alpha)$.

To prove that all limit points of the sequence Q^N solve the martingale problem (0.6) it remains to close the equations in terms of the fluctuation field Y_t^N. This is easy for the martingale N_t^G: an elementary computation shows that for every continuous function H, the $L^2(\mathbb{P}_N)$ norm of

$$\int_0^t N^{-d} \sum_{x\in\mathbb{T}_N^d} H(x/N)[g(\eta_s(x)) - \Phi(\alpha)]\, ds$$

is bounded above by $t^2 N^{-d} < H, H > \mathbf{Var}(\nu_\alpha, g)$ because ν_α^N is a product invariant measure. In particular, by the definition of the linear operator $\mathfrak{B}$, in the limit $N \uparrow \infty$, $(M_t^G)^2 - \|\mathfrak{B}H\|_2^2 t$ is a martingale.

To close the equation for the martingale M_t^H, we follow an approach proposed by Rost (1983). For each Lipschitz cylinder function Ψ, denote by $Y_t^{N,\Psi}$ the Ψ–fluctuation field defined by

$$Y_t^{N,\Psi}(H) = N^{-d/2} \sum_{x \in \mathbb{T}_N^d} H(x/N)[\tau_x \Psi(\eta_t) - \tilde{\Psi}(\alpha)] .$$

Notice that the integral part of the martingale M_t^G is equal to

$$\int_0^t Y_s^{N,g}((1/2)\Delta_N H) ds .$$

Since non conserved quantities fluctuates in a much faster scale than conserved quantities, in the time scale where the density changes, the non conserved quantities should average out and only their projection on the density fluctuation field should persist in the limit. In substance, there should exists a constant $C_\alpha(\Psi)$ such that

$$\int_0^t ds \left\{ Y_s^{N,\Psi}(H) - C_\alpha(\Psi) Y_s^N(H) \right\}$$

vanishes as $N \uparrow \infty$ for every smooth function H. This is the content of the Boltzmann–Gibbs principle stated in the next section, where we prove convergence to 0 in $L^2(\mathbb{P}_N)$ of the above integral term with $C_\alpha(\Psi) = \tilde{\Psi}'(\alpha)$. This convergence and some elementary estimates ensure that in the limit $Y_t(G) - Y_0(G) - \int_0^t Y_s(\mathfrak{A}H) ds$ is a martingale, concluding the proof of the convergence of the density fluctuation fields to the stationary generalized Ornstein–Uhlenbeck process satisfying (0.6).

1. The Boltzmann–Gibbs Principle

We show in this section that the martingales M_t^G introduced just before (0.10) can be expressed in terms of the fluctuation fields Y_t. This replacement of the cylinder function $g(\eta(0)) - \Phi(\alpha)$ by $\Phi'(\alpha)[\eta(0) - \alpha]$ constitutes the main step toward the proof of the equilibrium fluctuations.

Theorem 1.1 (Boltzmann–Gibbs principle) *For every cylinder Lipschitz function Ψ, every continuous function G on $\mathbb{T}^d$ and every $t > 0$,*

$$\lim_{N \to \infty} \mathbb{E}_N \left[\left(\int_0^t ds N^{-d/2} \sum_{x \in \mathbb{T}_N^d} G(x/N) \tau_x V_\Psi(\eta_s) \right)^2 \right] = 0 ,$$

where

$$V_\Psi(\eta) = \Psi(\eta) - \tilde{\Psi}(\alpha) - \tilde{\Psi}'(\alpha)[\eta(0) - \alpha] .$$

Proof. We first localize the problem. Fix a positive integer K that shall increase to ∞ after N. For each N, we subdivide $\mathbb{T}_N^d$ in non overlapping cubes of length K: let $M = [N/K]$, where $[r]$ stands for the integer part of r, and denote by $\{B_j,\, 1 \le j \le M^d\}$ non overlapping cubes of linear size K: for each j

$$B_j = y_j + \{1, \dots, K\}^d \quad \text{for some } y_j \text{ in } \mathbb{T}_N^d \quad \text{and} \quad B_i \cap B_j = \phi \quad \text{if} \quad i \neq j .$$

Denote by B_0 the set of sites not included in one of the cubes B_i. By construction the cardinality of B_0 is bounded by dKN^{d-1}.

Recall from section 7.2 that Λ_{s_Ψ} is the smallest cube centered at the origin that contains the support of Ψ. Denote by B_i^o the interior of the cube B_i, i.e., the sites x in B_i that are at a distance at least s_Ψ from the boundary:

$$B_i^o = \{x \in B_i,\, d(x, \mathbb{T}_N^d - B_i) > s_\Psi\} .$$

We defined the interior B_i^o so that $\tau_x \Psi$ is measurable with respect to $\sigma(\eta(z),\, z \in B_i)$ for all x in B_i^o. In particular, under ν_α^N, $\tau_x\Psi$ and $\tau_y\Psi$ are independent for x and y in the interior of distinct cubes. Let B^o stand for the set of all interior points of $\mathbb{T}_N^d$ and B^1 for its complement:

$$B^o = \bigcup_{i=1}^{M^d} B_i^o , \quad B^1 = \mathbb{T}_N^d - B^o .$$

The cardinality of B^1 is bounded by $dN^d\{C(\Psi)K^{-1} + KN^{-1}\}$ for some finite constant $C(\Psi)$ depending only on Ψ.

With the notation just introduced, we have that

$$\begin{aligned} N^{-d/2} \sum_{x \in \mathbb{T}_N^d} G(x/N)\tau_x V_\Psi(\eta) &= N^{-d/2} \sum_{x \in B^1} G(x/N)\tau_x V_\Psi(\eta) \\ &+ N^{-d/2} \sum_{i=1}^{M^d} \sum_{x \in B_i^o} [G(x/N) - G(y_i/N)]\tau_x V_\Psi(\eta) \\ &+ N^{-d/2} \sum_{i=1}^{M^d} G(y_i/N) \sum_{x \in B_i^o} \tau_x V_\Psi(\eta) . \end{aligned} \tag{1.1}$$

We claim that the expected value of the L^2 norm of the time integral of the first two expressions on the right hand side vanishes as $N \uparrow \infty$ and then $K \uparrow \infty$.

To show that the first expression vanishes in the limit, apply Schwarz inequality to bound the expected value by

$$t^2 E_{\nu_\alpha^N}\Big[\Big(N^{-d/2} \sum_{x \in B^1} G(x/N)\tau_x V_\Psi(\eta)\Big)^2\Big]$$

because ν_α^N is invariant. Since the cylinder function V_Ψ has mean zero with respect to the product measure ν_α^N, the last expression reduces to

$$t^2 N^{-d} \sum_{\substack{x,y\in B^1 \\ |x-y|\le 2s_\Psi}} G(x/N)G(y/N)E_{\nu_\alpha^N}\Big[\tau_x V_\Psi(\eta)\tau_y V_\Psi(\eta)\Big]$$

that vanishes in the limit as $N\uparrow\infty$ and then $K\uparrow\infty$ because the cardinality of B^1 is bounded above by $dN^d\{C(\Psi)K^{-1}+KN^{-1}\}$ and V_Ψ belongs to $L^2(\nu_\alpha^N)$ (since Ψ is Lipschitz).

For similar reasons and because G is assumed to be continuous, the expectation of the square of the time integral of the second expression on the right hand side of (1.1) vanishes in the limit as $N\uparrow\infty$.

For each $1\le i\le M^d$, denote by ξ_i the configuration $\{\eta(x), x\in B_i\}$ and by L_{B_i} the restriction of the generator L_N to the cube B_i:

$$(L_{B_i}f)(\eta) \;=\; (1/2)\sum_{\substack{x,y\in B_i \\ |x-y|=1}} g(\eta(x))[f(\eta^{x,y})-f(\eta)]\,.$$

Consider a $L^2(\nu_\alpha)$ cylinder function f measurable with respect to $\sigma(\eta(x), x\in B_1)$ and denote by f_i the translation of f that makes it measurable with respect to $\sigma(\eta(x), x\in B_i)$. By definition of the generator L_{B_i}, $L_{B_i}f_i$ is also measurable with respect to the σ-algebra $\sigma(\eta(x), x\in B_i)$.

By Proposition A1.6.1, for every $t>0$,

$$\begin{aligned}\mathbb{E}_N\Big[\Big(\int_0^t ds\, N^{-d/2}\sum_{i=1}^{M^d} G(y_i/N)L_{B_i}f_i(\xi_i(s))\Big)^2\Big] \\ \le\; 20t\Big\langle V_{G,f}^N, (-N^2L_N)^{-1}V_{G,f}^N\Big\rangle_\alpha\,,\end{aligned} \tag{1.2}$$

where

$$V_{G,f}^N(\eta) \;=\; N^{-d/2}\sum_{i=1}^{M^d} G(y_i/N)L_{B_i}f_i(\xi_i)\,.$$

In this formula and below $<\cdot,\cdot>_\alpha$ stands for the inner product in $L^2(\nu_\alpha^N)$. By the variational formula for the $\mathcal{H}_{-1}$ norm, the right hand side of the last expression is equal to

$$20t\,\sup_h\Big\{2\int V_{G,f}^N(\eta)h(\eta)\nu_\alpha^N(d\eta) - N^2 < h, (-L_N)h >_\alpha\Big\}\,,$$

where the supremum is taken over all functions h in $L^2(\nu_\alpha^N)$. The linear term in the previous supremum is equal to

$$2N^{-d/2}\sum_{i=1}^{M^d} G(y_i/N)\int L_{B_i}f_i(\xi_i)h(\eta)\nu_\alpha^N(d\eta)\,.$$

Integrating by parts the expression $\int L_{B_i} f_i(\xi_i) h(\eta) d\nu_\alpha^N$ and applying Schwarz inequality, we obtain that it is bounded above by

$$\frac{1}{2\gamma} < (-L_{B_i}) f_i, f_i >_\alpha + \frac{\gamma}{2} < (-L_{B_i}) h, h >_\alpha$$

for every $\gamma > 0$. The summation over i of the second term is less than or equal to $(\gamma/2) < h, (-L_N) h >_\alpha$ because L_{B_i} is the restriction of the generator L_N to B_i. Therefore, taking $\gamma = N^{2+(d/2)} |G(y_i/N)|^{-1}$, we obtain that the right hand side of (1.2) is bounded above by

$$20tN^{-d-2} \sum_{i=1}^{M^d} G(y_i/N)^2 < (-L_{B_i}) f_i, f_i >_\alpha \le \frac{t\|G\|_\infty^2}{K^d N^2} < (-L_{B_1}) f_1, f_1 >_\alpha$$

because the dynamics is translation invariant and f_i is defined as the translation of f_1. The last expression vanishes as $N \uparrow \infty$.

Up to this point we reduced the proof of the theorem to the proof that

$$\lim_{K\to\infty} \inf_f \lim_{N\to\infty}$$
$$\mathbb{E}_N \Big[\Big(\int_0^t ds\, N^{-d/2} \sum_{i=1}^{M^d} G(y_i/N) \Big\{ \sum_{x \in B_i^o} \tau_x V_\Psi(\eta_s) - L_{B_i} f_i(\xi_i(s)) \Big\} \Big)^2 \Big] = 0\,,$$

where the infimum is taken over all $L^2(\nu_\alpha)$ functions f measurable with respect to $\sigma(\eta(x), x \in B_1)$ and f_i stands for the translation of f that makes it measurable with respect to $\sigma(\eta(x), x \in B_i)$.

By Schwarz inequality, the expectation appearing in the previous expression is bounded above by

$$t^2 N^{-d} \sum_{i=1}^{M^d} G(y_i/N)^2 E_{\nu_\alpha^N} \Big[\Big\{ \sum_{x \in B_1^o} \tau_x V_\Psi(\eta) - L_{B_1} f_1(\xi_1) \Big\}^2 \Big]$$

because the product measure ν_α^N is invariant and translation invariant and because the support of $\tau_x V_\Psi - L_{B_i} f_i$, $\tau_y V_\Psi - L_j f_j$ are disjoints for x in B_i^o, y in B_j^o, $i \neq j$. As N increases to infinity, this expression converges to

$$t^2 K^{-d} \|G\|_2^2\, E_{\nu_\alpha^{B_1}} \Big[\Big\{ \sum_{x \in B_1^o} \tau_x V_\Psi - L_{B_1} f_1 \Big\}^2 \Big]\,. \tag{1.3}$$

Recall that L_{B_1} stands for the restriction to B_1 of the generator L_N. Denote by $R(L_{B_1})$ the range of the generator L_{B_1} in $L^2(\nu_\alpha^{B_1})$, i.e., the space generated by $L_{B_1} f$, f in $L^2(\nu_\alpha^{B_1})$ and by $R^\perp(L_{B_1})$ the space orthogonal to $R(L_{B_1})$. Fix a $\sigma(\eta(x), x \in B_1)$ measurable function h in $L^2(\nu_\alpha^{B_1})$. The formula

$$\inf_{f \in L^2(\nu_\alpha^{B_1})} E_{\nu_\alpha^{B_1}} \Big[\{h - L_{B_1} f\}^2 \Big]$$

corresponds to the projection of h on $R^{\perp}(L_{B_1})$. Denote by $\Sigma_{B_1,L}$ the space of all configurations of $\mathbb{N}^{B_1}$ with total number of particles equal to L, by $\nu_{B_1,L}$ the restriction of $\nu_\alpha^{B_1}$ to $\Sigma_{B_1,L}$: $\nu_{B_1,L}(\cdot) = \nu_\alpha^{B_1}(\cdot \mid \sum_{x\in B_1}\eta(x) = L)$ and by $\Sigma^0_{B_1,L}$ the space of $L^2(\nu_{B_1,L})$-mean zero functions. $\Sigma^0_{B_1,L}$ has codimension 1 and $R(L_{B_1})$ is a subset of $\Sigma^0_{B_1,L}$ because $\nu_{B_1,L}$ is invariant for the dynamics generated by L_{B_1}. On the other hand, the kernel of L_{B_1} in $L^2(\nu_{B_1,L})$ reduces to the constant functions since $L_{B_1}f = 0$ implies that $< f, L_{B_1}f >= 0$ that in turn forces f to be constant. In particular, the dimension of Ker L_{B_1} (and thus the codimension of $R(L_{B_1})$) is equal to 1. Therefore $R(L_{B_1}) = \Sigma^0_{B_1,L}$ because $R(L_{B_1}) \subset \Sigma^0_{B_1,L}$ and the codimension of the latter space is equal to 1. This shows that on $L^2(\nu_{B_1,L})$ $R^{\perp}(L_{B_1})$ is the one-dimensional space of constant functions. Thus $R^{\perp}(L_{B_1})$ consists of all functions that depend on the configuration η only through its total number of particles. In particular, the infimum over all f in $L^2(\nu_\alpha^{B_1})$ of the expression (1.3) is equal to

$$t^2 K^{-d}\|G\|_2^2\, E_{\nu_\alpha^{B_1}}\Big[\Big\{E_{\nu_\alpha^{B_1}}\Big[\sum_{x\in B_1^o}\tau_x V_\Psi \,\big|\, \eta^{B_1}(y_1)\Big]\Big\}^2\Big]. \tag{1.4}$$

In this formula $\eta^{B_1}(y_1)$ stands for the average number of particles of the configuration η on the cube B_1: $\eta^{B_1}(y_1) = K^{-d}\sum_{x\in B_1}\eta(x)$. For x in B_1^o, denote by $\tilde\Psi_K(\eta^{B_1}(y_1))$ the conditional expectation of $\tau_x\Psi$ with respect to $\eta^{B_1}(y_1)$:

$$\tilde\Psi_K(\eta^{B_1}(y_1)) \;=\; E_{\nu_\alpha^{B_1}}\Big[\tau_x\Psi \,\big|\, \eta^{B_1}(y_1)\Big]$$

and notice that this expression does not depend on x because ν_α is homogeneous. With this notation, we may rewrite (1.4) as

$$t^2\frac{|B_1^o|^2}{K^d}\|G\|_2^2\, E_{\nu_\alpha^{B_1}}\Big[\Big\{\tilde\Psi_K(\eta^{B_1}(y_1)) - \tilde\Psi(\alpha) - \tilde\Psi'(\alpha)[\eta^{B_1}(y_1) - \alpha]\Big\}^2\Big].$$

In Corollary A2.1.7 we prove that the absolute value of the difference between $\tilde\Psi_K(\alpha)$ and $\tilde\Psi(\alpha)$ is bounded above by $C(\Psi)K^{-d}$ on all compact sets of $\mathbb{R}_+$. We may therefore estimate the previous expectation by

$$\begin{aligned}&2t^2K^d\|G\|_2^2\, E_{\nu_\alpha^{B_1}}\Big[\Big\{\tilde\Psi_K(\eta^{B_1}(y_1)) - \tilde\Psi(\eta^{B_1}(y_1))\Big\}^2\Big]\\ &\quad + 2t^2K^d\|G\|_2^2\, E_{\nu_\alpha^{B_1}}\Big[\Big\{\tilde\Psi(\eta^{B_1}(y_1)) - \tilde\Psi(\alpha) - \tilde\Psi'(\alpha)[\eta^{B_1}(y_1) - \alpha]\Big\}^2\Big]\Big\}.\end{aligned}$$

Since Ψ is Lipschitz, $|\tilde\Psi_K(\alpha)|$ and $|\tilde\Psi(\alpha)|$ are bounded by $C(\Psi)(1+\alpha)$. We may thus introduce the indicator function $\mathbf{1}\{\eta^{B_1}(y_1) \le A\}$ inside both expectations. By Corollary A2.1.7, the first one is bounded above by $C(A,\alpha,\Psi)K^{-2d}$. On the other hand, by Taylor's expansion up to the second order and since ν_α is a product measure, the second expectation is also bounded by $C(A,\alpha,\Psi)K^{-2d}$. This concludes the proof of the Boltzmann–Gibbs principle. □

2. The Martingale Problem

In section 3 we prove that the sequence of probability measures Q_N is tight and that all limit points are concentrated on continuous paths. In view of Theorem 0.2, to conclude the proof of the equilibrium fluctuations, it remains to show that all limit points Q of the sequence Q_N solve the martingale problem (0.6) and to characterize their restriction to $\mathcal{F}_0$. We start with the latter question which is easier. Fix a limit point Q^* and assume without loss of generality that Q^N converges to Q^*.

Lemma 2.1 *For every continuous function $H: \mathbb{T}^d \to \mathbb{R}$ and every $t > 0$,*

$$\lim_{N\to\infty} \log \mathbb{E}_N \Big[\exp\{iY_t(H)\} \Big] \;=\; -\frac{\chi(\alpha)}{2} < H, H > \; .$$

Proof. Since ν_α^N is a product invariant measure,

$$\log \mathbb{E}_N \Big[\exp\{iY_t(H)\} \Big] \;=\; \sum_{x\in\mathbb{T}_N^d} \log E_{\nu_\alpha^N} \Big[\exp\big\{ iN^{-d/2} H(x/N)[\eta(0) - \alpha] \big\} \Big] \, . \tag{2.1}$$

Since by assumption (2.3.2), $\eta(0)$ has a finite exponential moment, by Taylor expansion, last expectation is equal to

$$1 \;-\; \frac{\chi(\alpha)}{2N^d} H(x/N)^2 \;+\; O(N^{-3d/2})\, .$$

In this formula $\chi(\alpha)$ stands for the static compressibility $\mathbf{Var}(\nu_\alpha, \eta(0))$. Therefore, the right hand side of (2.1) is equal to

$$-\;\frac{\chi(\alpha)}{2} N^{-d} \sum_{x\in\mathbb{T}_N^d} H(x/N)^2 \;+\; O(N^{-d/2})\, .$$

As $N \uparrow \infty$ this expression converges to $-(1/2)\chi(\alpha) < H, H >$, what concludes the proof of the lemma. □

Corollary 2.2 *Restricted to $\mathcal{F}_0$, Q^* is a Gaussian field with covariance given by*

$$E_{Q^*}[Y_0(G)Y_0(H)] \;=\; \chi(\alpha) < H, G > \; . \tag{2.2}$$

Proof. Fix a positive integer n, θ in $\mathbb{R}^n$ and $H_1, \dots, H_n$ in $\mathcal{H}_{k_0}$. Since Y_0 is linear and since, by assumption, Q_N converges weakly to Q^*, by the previous lemma,

$$\log E_{Q^*} \Big[\exp\big\{ i \sum_{j=1}^n \theta_j Y_0(H_j) \big\} \Big] \;=\; \lim_{N\to\infty} \mathbb{E}_N \Big[\exp\big\{ iY_0\big(\sum_{j=1}^n \theta_j H_j\big) \big\} \Big]$$

$$=\; -\;\frac{\chi(\alpha)}{2} < \sum_{j=1}^n \theta_j H_j, \sum_{j=1}^n \theta_j H_j > \; \cdot$$

The Q^* joint distribution of $(Y_0(H_1), \dots, Y_0(H_n))$ is thus Gaussian with covariance given by (2.2), what concludes the proof of the lemma. □

We turn now to the dynamic part of the problem.

Proposition 2.3 *Q^* solves the martingale problem (0.6).*

Proof. Fix H in $C^2(\mathbb{T}^d)$ and denote by $M_t^{\mathfrak{A},H}$, $N_t^{\mathfrak{A},H}$ the random processes defined by

$$M_t^{\mathfrak{A},H} = Y_t(H) - Y_0(H) - \int_0^t ds\, Y_s(\mathfrak{A}H)$$

$$\text{and} \quad N_t^{\mathfrak{A},H} = (M_t^{\mathfrak{A},H})^2 - \|\mathfrak{B}H\|_2^2 t\,.$$

By definition, $M_t^{\mathfrak{A},H}$ is $\mathcal{F}_t$–measurable. Thus, in order to prove that $M_t^{\mathfrak{A},H}$ is a martingale, we just need to check that

$$E_{Q^*}\Big[M_t^{\mathfrak{A},H} U\Big] = E_{Q^*}\Big[M_s^{\mathfrak{A},H} U\Big]$$

for all $0 \le s \le t \le T$ and U of the form $U = \mathbf{1}\{Y_{s_i}(H_i) \in A_i,\ 1 \le i \le n\}$, where n is a positive integer, $0 \le s_1 \le \cdots \le s_n \le s$, H_i are in $C^2(\mathbb{T}^d)$ and A_i are measurable subsets of $\mathbb{R}^d$ for $1 \le i \le n$.

Recall from the introduction that for each H in $C^2(\mathbb{T}^d)$, the process M_t^H defined by

$$M_t^H = Y_t^N(H) - Y_0^N(H) - \int_0^t N^2 L_N N^{-d/2} \sum_{x\in\mathbb{T}_N^d} H(x/N)[\eta_s(x) - \alpha]\, ds$$

is a martingale so that $\mathbb{E}_N[M_t^H U]$ is equal to $\mathbb{E}_N[M_s^H U]$. To conclude the proof of the first statement of the proposition it remains to show that these two expectations converge respectively to $E_{Q^*}[M_t^{\mathfrak{A},H} U]$ and $E_{Q^*}[M_s^{\mathfrak{A},H} U]$.

By the Boltzmann–Gibbs principle, since U is bounded,

$$\lim_{N\to\infty} \mathbb{E}_N\Big[M_t^H U\Big] = \lim_{N\to\infty} \mathbb{E}_N\Big[\Big\{Y_t(H) - Y_0(H) - (1/2)\Phi'(\alpha) \int_0^t ds\, Y_s(\Delta_N H)\Big\} U\Big]\,.$$

A simple computation shows that the square of the expression inside braces on the right hand side of the equality has uniformly (in N) bounded expectation. In particular, since U is bounded and Q_N converges weakly to Q^*, which is concentrated on continuous paths, $\mathbb{E}_N[M_t^H U]$ converges to $E_{Q^*}[M_t^{\mathfrak{A},H} U]$. Since the same argument applies to the expectation $\mathbb{E}_N[M_s^H U]$, the first statement of the proposition is proved.

The argument that shows that the process $N_t^{\mathfrak{A},H}$ is a martingale is similar to the one presented above to prove that $M_t^{\mathfrak{A},H}$ is a martingale. It relies on the martingale N_t^H introduced in (0.10), on the Boltzmann–Gibbs principle and on the fact that the martingale M_t^H has uniformly bounded fourth moments. We leave the details to the reader. $\square$

3. Tightness

We prove in this section that the sequence of probability measures Q_N is tight and that all limit points are concentrated on continuous paths. We first review some aspects of the uniform weak topology on $D([0,T],\mathcal{H}_{-k})$ introduced in the beginning of the chapter. Throughout this section k stands for a positive integer larger than $2+(d/2)$.

For $\delta > 0$ and a path Y in $D([0,T],\mathcal{H}_{-k})$, define the uniform modulus of continuity $w_\delta(Y)$ by

$$w_\delta(Y) = \sup_{\substack{|s-t|\le\delta \\ 0\le s,t\le T}} \|Y_t - Y_s\|_{-k} .$$

The first result provides sufficient conditions for a subset to be weakly relatively compact.

Lemma 3.1 *A subset A of $D([0,T],\mathcal{H}_{-k})$ is relatively compact for the uniform weak topology if*

(i) $\sup_{Y\in A}\sup_{0\le t\le T} \|Y_t\|_{-k} < \infty$

(ii) $\lim_{\delta\to 0}\sup_{Y\in A} w_\delta(Y) = 0$.

The proof of this result is standard and left to the reader. From this lemma we deduce a criterion for tightness of a sequence of probability measures P_N defined on $D([0,T],\mathcal{H}_{-k})$.

Lemma 3.2 *A sequence $\{P_N, N \ge 1\}$ of probability measures defined on $D([0,T],\mathcal{H}_{-k})$ is tight provided for every $0\le t\le T$,*

$$\lim_{A\to\infty}\limsup_{N\to\infty} P_N\Big[\sup_{0\le t\le T}\|Y_t\|_{-k} > A\Big] = 0$$

and

$$\lim_{\delta\to 0}\limsup_{N\to\infty} P_N\Big[w_\delta(Y)\ge\varepsilon\Big] = 0$$

for every $\varepsilon > 0$.

We have now all elements to prove the tightness of the sequence Q_N introduced in the beginning of the chapter.

Proposition 3.3 *The sequence of probability measures Q_N is tight. Moreover, all limit points are concentrated on continuous paths.*

The proof of this proposition is divided in several lemmas. Following Holley and Stroock (1978), we start with a key estimate. For each z in $\mathbb{Z}^d$, denote by M_t^z and N_t^z the martingales $M_t^{h_z}$ and $N_t^{h_z}$ introduced just before (0.10). To keep notation simple let

$$\Gamma_1^z(s) = (1/2)N^{-d/2} \sum_{x\in\mathbb{T}_N^d} (\Delta_N h_z)(x/N)[g(\eta_s(x)) - \Phi(\alpha)]$$

$$\Gamma_2^z(s) = (1/2)N^{-d} \sum_{j=1}^{d} \sum_{x\in\mathbb{T}_N^d} [(\partial_{u_j}^N h_z)(x/N)]^2 [g(\eta_s(x)) + g(\eta_s(x+e_j))]$$

so that $M_t^z = Y_t^N(h_z) - Y_0^N(h_z) - \int_0^t \Gamma_1^z(s)ds$ and $N_t^z = (M_t^z)^2 - \int_0^t \Gamma_2^z(s)ds$.

Lemma 3.4 *There exists a finite constant $C(\alpha, T)$ depending only on α and T such that for every z in $\mathbb{Z}^d$,*

$$\limsup_{N\to\infty} \mathbb{E}_N\Big[\sup_{0\le t\le T} | < Y_t\, ,\, h_z > |^2 \Big] \le C(\alpha, T)\Big\{ < h_z, h_z > + < \Delta h_z, \Delta h_z > \Big\} .$$

In this formula and below $< Y_t, h_z >$ stands for the inner product of $Y_t \in \mathcal{H}_{-k}$ and $h_z \in \mathcal{H}_k$.

Proof. Rewrite $< Y_t^N, h_z >$ as $M_t^z + Y_0^N(h_z) + \int_0^t \Gamma_1^z(s)ds$. A straightforward computation shows that the limit, as N increases to ∞, of $\mathbb{E}_N[| < Y_0\, ,\, h_z > |^2]$ is equal to $\chi(\alpha) < h_z, h_z >$ because ν_α^N is a product measure.

On the other hand, since $M_\cdot^z$ is a martingale, by Doob inequality,

$$\mathbb{E}_N\Big[\sup_{0\le t\le T} |M_t^z|^2 \Big] \le 4\,\mathbb{E}_N\Big[|M_T^z|^2 \Big] .$$

By definition of the martingale $N_\cdot^z$, the right hand side is equal to

$$4\,\mathbb{E}_N\Big[\int_0^T ds\, (1/2)N^{-d} \sum_{1\le i\le d} \sum_{x\in\mathbb{T}_N^d} [g(\eta_s(x)) + g(\eta_s(x+e_i))](\partial_{u_i}^N h_z(x/N))^2 \Big] ,$$

where $\partial_{u_i}^N h_z(x/N)$ stands for $N[h_z((x+e_i)/N) - h_z(x/N)]$. As $N \uparrow \infty$, this expression converges to $4T\Phi(\alpha) \int du \|\nabla h_z(u)\|_2^2$, where $\|\nabla h_z(u)\|_2$ is the $L^2(\mathbb{T}^d)$ norm of the gradient of h_z.

Finally, by definition of h_z and Schwarz inequality,

$$\mathbb{E}_N\Big[\sup_{0\le t\le T}\Big(\int_0^t ds\,\Gamma_1^z(s)\Big)^2\Big]$$
$$\le T\,\mathbb{E}_N\Big[\int_0^T ds\,\Big((1/2)N^{-d/2}\sum_{x\in\mathbb{T}_N^d}(\Delta_N h_z)(x/N)[g(\eta_s(x))-\Phi(\alpha)]\Big)^2\Big]\,.$$

Since the initial measure ν_α^N is product and invariant, as $N\uparrow\infty$, this expression converges to $(1/4)T^2 < \Delta h_z, \Delta h_z >$ $\mathbf{Var}(\nu_\alpha, g)$, what concludes the proof of the lemma. □

Corollary 3.5 *For $k > 2 + (d/2)$,*

(a) $\displaystyle \limsup_{N\to\infty}\mathbb{E}_N\Big[\sup_{0\le t\le T}\|Y_t\|_{-k}^2\Big] < \infty$

(b) $\displaystyle \lim_{n\to\infty}\limsup_{N\to\infty}\mathbb{E}_N\Big[\sup_{0\le t\le T}\sum_{|z|\ge n}(< Y_t, h_z >)^2\gamma_z^{-k}\Big] = 0\,.$

Proof. The first expression is bounded above by

$$\sum_{z\in\mathbb{Z}^d}\gamma_z^{-k}\,\mathbb{E}_N\Big[\sup_{0\le t\le T}(< Y_t, h_z >)^2\Big]\,.$$

By the previous lemma, the limsup as $N\uparrow\infty$ of this sum is less than or equal to

$$C(\alpha,T)\sum_{z\in\mathbb{Z}^d}\gamma_z^{-k}\Big\{1+ < \Delta h_z, \Delta h_z >\Big\}$$

for some finite constant $C(\alpha,T)$ depending only on α and T because h_z has $L^2(\mathbb{T}^d)$ norm equal to 1. By definition of h_z and γ_z, this expression is equal to

$$C(\alpha,T)\sum_{z\in\mathbb{Z}^d}\frac{1+(2\pi\|z\|)^4}{(1+(2\pi\|z\|)^2)^k} \le C(\alpha,T)\sum_{z\in\mathbb{Z}^d}\frac{1}{(1+(2\pi\|z\|)^2)^{k-2}}\,.$$

This estimate proves the first statement. The second claim follows by the same argument. □

It follows from Lemma 3.2 and Corollary 3.5 that in order to prove that the sequence Q_N is tight, we only have to show that for every $\varepsilon > 0$,

$$\lim_{\delta\to0}\limsup_{N\to\infty}\mathbb{P}_N\Big[w_\delta(Y) > \varepsilon\Big] = 0\,.$$

In view of part (b) of the previous corollary, this result follows from the following lemma:

Lemma 3.6 *For every positive integer n and every $\varepsilon > 0$,*

$$\lim_{\delta\to 0}\limsup_{N\to\infty}\mathbb{P}_N\Big[\sup_{\substack{0\le|s-t|\le\delta\\0\le s,t\le T}}\sum_{|z|\le n}(<Y_t-Y_s,h_z>)^2\gamma_z^{-k}>\varepsilon\Big] = 0\,.$$

Proof. To prove this lemma we just have to show that

$$\lim_{\delta\to 0}\limsup_{N\to\infty}\mathbb{P}_N\Big[\sup_{\substack{0\le|s-t|\le\delta\\0\le s,t\le T}}(<Y_t-Y_s,h_z>)^2>\varepsilon\Big] = 0$$

for every z in $\mathbb{Z}^d$ and $\varepsilon>0$. Fix z in $\mathbb{Z}^d$ and recall the definition of the martingale M^z. Since $<Y_t,h_z>=<Y_0,h_z>+M_t^z+\int_0^t\Gamma_1^z(s)ds$, the lemma follows from the next two results. □

Lemma 3.7 *Fix a function G in $C^2(\mathbb{T}^d)$. For every $\varepsilon>0$,*

$$\lim_{\delta\to 0}\limsup_{N\to\infty}\mathbb{P}_N\Big[\sup_{\substack{0\le s,t\le T\\|t-s|\le\delta}}|M_t^G-M_s^G|>\varepsilon\Big] = 0\,.$$

Proof. Denote by $w'_\delta(M^G)$ the modified modulus of continuity defined as

$$w'_\delta(M^G) = \inf_{\{t_i\}}\max_{0\le i<r}\sup_{t_i\le s<t<t_{i+1}}|M_t^G-M_s^G|\,,$$

where the first infimum is taken over all partitions of $[0,T]$ such that

$$\begin{cases}0=t_0<t_1<\cdots<t_r=T\\ t_{i+1}-t_i>\delta\quad 0\le i<r\,.\end{cases}$$

Since $\sup_t|M_t^G-M_{t-}^G| = \sup_t|<Y_t,G>-<Y_{t-},G>|$ is bounded by $C(G)N^{-(1+(d/2))}$ and

$$w_\delta(M^G)\;\le\;2w'_\delta(M^G)+\sup_t|M_t^G-M_{t-}^G|\,,$$

in order to prove the lemma we just need to show that

$$\lim_{\delta\to 0}\limsup_{N\to\infty}\mathbb{P}_N\Big[w'_\delta(M^G)>\varepsilon\Big] = 0 \tag{3.1}$$

for every $\varepsilon>0$.

By Proposition 4.1.6, to prove (3.1) it is enough to check that for every $\varepsilon>0$,

$$\lim_{\delta\to 0}\limsup_{N\to\infty}\sup_{\substack{\tau\in\mathfrak{T}_T\\0\le\theta\le\delta}}\mathbb{P}_N\Big[|M_{\tau+\theta}^G-M_\tau^G|>\varepsilon\Big] = 0\,,$$

where $\mathfrak{T}_T$ stands for all stopping times bounded by T. By Chebychev inequality, the last probability is less than or equal to

$$\frac{1}{\varepsilon^2}\mathbb{E}_N\Big[(M^G_{\tau+\theta} - M^G_\tau)^2\Big] = \frac{1}{\varepsilon^2}\mathbb{E}_N\Big[(M^G_{\tau+\theta})^2 - (M^G_\tau)^2\Big]$$

because M^G_t is a martingale and τ a bounded stopping time. By (0.10) this expression is bounded above by

$$\frac{1}{2\varepsilon^2}\mathbb{E}_N\Big[\int_0^\delta ds\, N^{-d}\sum_{j=1}^d \sum_{x\in\mathbb{T}^d_N}[g(\eta_s(x+e_j)) + g(\eta_s(x))] \times \\ \times\ [N\{G((x+e_j)/N) - G(x/N)\}]^2\Big]$$

because ν^N_α is invariant, τ a stopping time and θ is bounded above by δ. The limit as $N\uparrow\infty$ of this last expression is less than or equal to $\delta\varepsilon^{-2}\|\nabla G\|_2^2\Phi(\alpha)$, what concludes the proof of the lemma. □

We now turn to the additive functional $\int_0^t N^2 L_N < Y^N_s, G > ds$. From (0.10), this expression is equal to

$$\frac{1}{2}\int_0^t ds\, N^{-d/2}\sum_{x\in\mathbb{T}^d_N}(\Delta_N G)(x/N)[g(\eta_s(x)) - \Phi(\alpha)]$$

Lemma 3.8 *Fix a function G in $C^2(\mathbb{T}^d)$. For every $\varepsilon > 0$,*

$$\lim_{\delta\to 0}\limsup_{N\to\infty} \\ \mathbb{P}_N\Big[\sup_{\substack{0\le s,t\le T\\ |t-s|\le\delta}}\Big|\int_s^t dr\, N^{-d/2}\sum_{x\in\mathbb{T}^d_N}(\Delta_N G)(x/N)[g(\eta_r(x)) - \Phi(\alpha)]\Big| > \varepsilon\Big] = 0\,.$$

Proof. By Chebychev and Schwarz inequality, the probability is bounded above by

$$\frac{\delta}{\varepsilon^2}\mathbb{E}_N\Big[\int_0^T dr\,\Big(N^{-d/2}\sum_{x\in\mathbb{T}^d_N}(\Delta_N G)(x/N)[g(\eta_r(x)) - \Phi(\alpha)]\Big)^2\Big]$$

because $|t-s|\le\delta$ and $s,\, t\le T$. Since ν^N_α is an invariant product measure and $g(\eta(x)) - \Phi(\alpha)$ has mean zero, this expression is equal to

$$\frac{\delta T}{\varepsilon^2}N^{-d}\sum_{x\in\mathbb{T}^d_N}[\Delta_N G(x/N)]^2 E_{\nu^N_\alpha}\Big[\{g(\eta(x)) - \Phi(\alpha)\}^2\Big]$$

that vanishes as $N\uparrow\infty$ and then $\delta\downarrow 0$. □

This concludes the proof of the tightness of the sequence Q_N.

We conclude this section showing that tightness can be proved in a stronger norm in the case where the jump rate $g(\cdot)$ is bounded. The argument applies to reversible nongradient systems and may be skipped without prejudice by those who are satisfied with convergence in $D([0,T],\mathcal{H}_{-k})$ for $k > 2 + (d/2)$.

We shall prove that there exists a finite constant $C(\alpha, T)$ depending only on α and T such that for every z in $\mathbb{Z}^d$,

$$\limsup_{N\to\infty} \mathbb{E}_N \Big[\sup_{0\le t\le T} | < Y_t \, , \, h_z > |^2 \Big] \le C(\alpha, T) \Big\{ < h_z, h_z > + < (-\Delta) h_z, h_z > \Big\} . \tag{3.2}$$

The reader should compare this estimate with Lemma 3.4 and check that Corollary 3.4 with $k > 1 + (d/2)$ instead of $k > 2 + (d/2)$ follows from estimate (3.2). In particular, by Lemma 3.6, the sequence Q_N introduced in the beginning of the chapter is tight (and therefore, by section 2, converges) in $D([0,T], \mathcal{H}_{-k})$ for $k > 1 + (d/2)$.

In view of the proof of Lemma 3.4, in order to deduce estimate (3.2), we just need to show that

$$\limsup_{N\to\infty} \mathbb{E}_N \Big[\sup_{0\le t\le T} \Big| \int_0^t \Gamma_1^z(s)\, ds \Big|^2 \Big] \le C(\alpha, T) < h_z, (-\Delta) h_z > . \tag{3.3}$$

For x in $\mathbb{Z}^d$ and $1 \le i \le d$, denote by $W_{x,x+e_i}$ the current over the bond $\{x, x+e_i\}$. In the case of the symmetric nearest neighbor zero range process, the current is $W_{x,x+e_i} = (1/2)[g(\eta(x)) - g(\eta(x+e_i))]$. Notice that $\Gamma_1^z(s)$ may be expressed in terms of the current:

$$\Gamma_1^z(s) = N^{-d/2} \sum_{x,j} [\partial_{u_j}^N h_z(x/N)] N W_{x,x+e_j}(s) . \tag{3.4}$$

The proof of (3.3) is divided in two steps. We first obtain an exponential estimate of the current fluctuation fields and then apply the Garsia–Rodemich–Rumsey inequality to deduce (3.3).

Lemma 3.9 *For every $a > 0$, continuous functions G_i, $i = 1, \dots, d$ and $0 \le s < t$,*

$$\mathbb{E}_N \Big[\exp \Big\{ a \Big| \int_s^t dr\, N^{-d/2} \sum_{x,i} N W_{x,x+e_i}(r) G_i(x/N) \Big| \Big\} \Big] \le 2 \exp \Big\{ \|g\|_\infty (t-s) a^2 N^{-d} \sum_{x,i} G_i(x/N)^2 \Big\}$$

for all N large enough.

The proof of this lemma follows closely the one of Lemma 7.6.2. In sake of completeness, we prove it again in the gradient context of zero range processes with bounded rates.

Proof. Since $e^{|x|} \le e^x + e^{-x}$, it is enough to show that

$$\mathbb{E}_N\Big[\exp\Big\{a\int_s^t dr\, N^{-d/2}\sum_{x,i} NW_{x,x+e_i}(r)G_i(x/N)\Big\}\Big]$$
$$\le \exp\Big\{\|g\|_\infty (t-s)a^2 N^{-d}\sum_{x,i} G_i(x/N)^2\Big\}$$

for all N large enough and all sets of continuous functions G_i, $i = 1, \dots, d$. Fix such functions. By Feynman–Kac formula and by stationarity of ν_α,

$$\mathbb{E}_N\Big[\exp\Big\{a\int_s^t N^{-d/2}\sum_{x,i} G_i(x/N)NW_{x,x+e_i}(r)dr\Big\}\Big]$$
$$= \mathbb{E}_N\Big[\exp\Big\{a\int_0^{t-s} N^{-d/2}\sum_{x,i} G_i(x/N)NW_{x,x+e_i}(r)dr\Big\}\Big] \le e^{(t-s)\lambda_N(G)},$$

where $\lambda_N(G)$ is the largest eigenvalue of the symmetric operator $N^2L_N + aN^{-d/2}\sum_{x,i} G_i(x/N)NW_{x,x+e_i}$. By inequality (A3.1.1),

$$\lambda_N(G) \le \sup_f \Big\{aN^{-d/2}\sum_{x,i} G_i(x/N) < NW_{x,x+e_i} f >_\alpha - N^2 D_N(f)\Big\},$$

where the supremum is carried over all densities f and where $< \cdot >_\alpha$ stands for expectation with respect to ν_α^N.

Recall that for symmetric nearest neighbor zero range processes, the current takes the form $W_{x,x+e_i} = (1/2)[g(\eta(x)) - g(\eta(x+e_i))]$. A change of variables permits to rewrite $< W_{x,x+e_i} f >_\alpha$ as

$$(1/2)\Phi(\alpha)\int \Big[f(\eta + \mathfrak{d}_x) - f(\eta + \mathfrak{d}_{x+e_i})\Big]\nu_\alpha(d\eta).$$

Since $a - b = (\sqrt{a} - \sqrt{b})(\sqrt{a} + \sqrt{b})$, by Schwarz inequality, last expression is bounded above by

$$\frac{\Phi(\alpha)}{4\gamma}\int \Big(\sqrt{f(\eta + \mathfrak{d}_x)} - \sqrt{f(\eta + \mathfrak{d}_{x+e_i})}\Big)^2 \nu_\alpha(d\eta)$$
$$+ \frac{\gamma\Phi(\alpha)}{4}\int \Big(\sqrt{f(\eta + \mathfrak{d}_x)} + \sqrt{f(\eta + \mathfrak{d}_{x+e_i})}\Big)^2 \nu_\alpha(d\eta)$$

for every $\gamma > 0$. The first term is just a multiple of the piece of the Dirichlet form corresponding to jumps over the bond $\{x, x+e_i\}$ and is equal to $(2\gamma)^{-1}I_{x,x+e_i}(f)$. The second one, by a new change of variables, is bounded above by $\gamma\|g\|_\infty$ because the jump rate $g(\cdot)$ is bounded and f is a density with respect to ν_α. From this estimate it follows that $aN^{-d/2}\sum_{x,i} G_i(x/N) < NW_{x,x+e_i} f >_\alpha$ is bounded above by

$$(2\gamma)^{-1}\sum_{x,i} I_{x,x+e_i}(f) + \gamma\|g\|_\infty a^2 N^{2-d}\sum_{x,i} G_i(x/N)^2$$

for every density f. Choosing $\gamma = N^{-2}$, we conclude that the largest eigenvalue $\lambda_N(G)$ defined above is bounded by

$$\|g\|_\infty a^2 N^{-d} \sum_{x,i} G_i(x/N)^2 ,$$

what proves the lemma. □

Lemma 3.10 *There exists a finite constant $C(\alpha, T)$ depending only on α and T such that for each z in $\mathbb{Z}^d$,*

$$\limsup_{N\to\infty} \mathbb{E}_N \Big[\sup_{0\le t\le T} \Big| \int_0^t \Gamma_1^z(s)ds \Big|^2 \Big] \;\le\; C(\alpha,T) < \|\nabla h_z\|_2^2 > \cdot$$

Proof. Recall from section 7.6 the statement of the Garsia–Rodemich–Rumsey inequality. Fix $a_0 > 0$ and set

$$g(t) = \int_0^t \Gamma_1^z(s)ds\,, \quad \psi(u) = \exp\{a_0 N^d u\} - 1\,.$$

By the just mentioned inequality with $p(u) = \sqrt{u}$, we have that

$$\sup_{0\le t\le T} |g(t)| \;\le\; 4 \int_0^T \psi^{-1}\Big(\frac{4B}{u^2}\Big) \frac{1}{\sqrt{u}}\, du\,,$$

where

$$B \;=\; \int_0^T ds \int_0^T dt\, \psi\Big(\frac{|g(t)-g(s)|}{\sqrt{|t-s|}}\Big)\,.$$

With our choice of ψ, $\psi^{-1}(u) = (a_0 N^d)^{-1} \log(1+u)$. Integrating by parts we get that

$$\int_0^T \log\Big\{1 + \frac{4B}{u^2}\Big\} \frac{1}{\sqrt{u}}\, du \;\le\; 2\sqrt{T}\Big\{ \log\Big(1 + \frac{4B}{T^2}\Big) \;+\; 4\Big\}$$

because $B/(4B+u^2) \le 1/4$. Therefore, by the two previous estimates,

$$\mathbb{E}_N\Big[\exp\Big\{\frac{a_0 N^d}{8\sqrt{T}} \sup_{0\le t\le T} |g(t)|\Big\}\Big] \;\le\; \mathbb{E}_N\Big[\exp\Big\{\log\Big(1+\frac{4B}{T^2}\Big) \;+\; 4\Big\}\Big]$$

$$= e^4 \Big\{ 1 + \frac{4}{T^2} \mathbb{E}_N[B] \Big\}\,.$$

Recall the definition of B, the explicit formula (3.4) for $\Gamma_1^z(r)$ and apply Lemma 3.9 with $a = a_0 N^d / \sqrt{|t-s|}$ and $G_j = \partial_{u_j}^N h_z$, to obtain that the previous expression is less than or equal to

$$e^4\Big\{1-4+\frac{8}{T^2}\int_0^T ds\int_0^T dt\,\exp\Big\{a_0^2\|g\|_\infty N^d\sum_{x,j}[(\partial_{u_j}^N h_z)(x/N)]^2\Big\}$$

$$\le\; 8e^4\exp\Big\{a_0^2\|g\|_\infty N^d\sum_{x,j}[(\partial_{u_j}^N h_z)(x/N)]^2\Big\}\;.$$

In conclusion, we have proved that

$$\mathbb{E}_N\Big[\exp\Big\{\frac{a_0N^d}{8\sqrt{T}}\sup_{0\le t\le T}|g(t)|-a_0^2\|g\|_\infty N^d\sum_{x,j}[(\partial_{u_j}^N h_z)(x/N)]^2\Big\}\Big]\;\le\;8e^4\;.$$

Maximizing over a_0, we get that

$$\mathbb{E}_N\Big[\exp\Big\{\frac{\Big(\sup_{0\le t\le T}|g(t)|\Big)^2}{2^8\|g\|_\infty TN^{-d}\sum_{x,j}[(\partial_{u_j}^N h_z)(x/N)]^2}\Big\}\Big]\;\le\;8e^4\;.$$

In particular, by Jensen inequality,

$$\mathbb{E}_N\Big[\Big(\sup_{0\le t\le T}|g(t)|\Big)^2\Big]\;\le\;2^8(4+\log 8)\|g\|_\infty TN^{-d}\sum_{x,j}[(\partial_{u_j}^N h_z)(x/N)]^2\;,$$

what concludes the proof of the lemma. □

4. Generalized Ornstein–Uhlenbeck Processes

We conclude this chapter proving Theorem 0.2, a particular case of a general result due to Holley and Stroock (1978). The reader should notice that the proof below does not rely on the special form of the operators $\mathfrak{A}$ and $\mathfrak{B}$.

By Ito's formula and (0.6), for each fixed $s\ge 0$ and H in $C^\infty(\mathbb{T}^d)$, $\{X_t^s(H),\,t\ge 0\}$ defined by

$$X_t^s(H)\;=\;\exp\Big\{i\Big(Y_{t\vee s}(H)-Y_s(H)-\int_s^{s\vee t}Y_r(\mathfrak{A}H)dr\Big)+(1/2)\|\mathfrak{B}H\|_2^2(t-s)^+\Big\}$$

is a martingale (in fact, as already mentioned in the beginning of the chapter, $\|\mathfrak{B}H\|_2^{-1}M_t^{\mathfrak{A},H}$ is a Brownian motion for each H in $C^\infty(\mathbb{T}^d)$).

We now claim that for each $S>0$ and smooth function H in $C^\infty(\mathbb{T}^d)$, $\{Z_t^{H,S},\,t\ge 0\}$ defined by

$$Z_t^{H,S}\;=\;\exp\Big\{iY_{t\wedge S}(H_{(S-t)^+})+(1/2)\int_0^{t\wedge S}\|\mathfrak{B}H_{S-r}\|_2^2dr\Big\}$$

is a martingale. Here, for $t>0$, H_t stands for T_tH.

To prove this claim, fix $0\le t_1<t_2\le S$ and set $s_{n,j}=t_1+(j/n)(t_2-t_1)$ for $0\le j\le n$. It is easy to check that the function defined on $\mathbb{R}_+^2$ that associates

to each (s,t) the value $Y_s(H_t)$ is continuous because $\{Y_s,\ 0 \le s \le T\}$ is weakly compact on $\mathcal{H}_{-k}$ and $T_t H$ is uniformly continuous. It follows from this continuity and from the expansion $H_{t+\varepsilon} = H_t + \varepsilon\mathfrak{A}H_t + o(\varepsilon)$ that

$$\prod_{j=0}^{n-1} X^{s_{n,j}}_{s_{n,j+1}}(H_{S-s_{n,j}})$$

converges a.s. and in $L^1(Q)$ to $Z^{H,S}_{t_2}/Z^{H,S}_{t_1}$ as $n \uparrow \infty$.

Let G be a bounded $\mathcal{F}_{t_1}$–measurable function. Since the convergence takes place in $L^1(Q)$,

$$E_Q\Big[\frac{Z^{H,S}_{t_2}}{Z^{H,S}_{t_1}} G\Big] = \lim_{n\to\infty} E_Q\Big[\prod_{j=0}^{n-1} X^{s_{n,j}}_{s_{n,j+1}}(H_{S-s_{n,j}})\, G\Big]\,.$$

Taking conditional expectation with respect to $\mathcal{F}_{s_{n,n-1}}$, we reduce the range of the product in last expectation to $0 \le j \le n-2$ because $X^s_\cdot(H)$ is a martingale for each $s \ge 0$ and smooth H. repeating this argument, we obtain that

$$E_Q\Big[\frac{Z^{H,S}_{t_2}}{Z^{H,S}_{t_1}} G\Big] = E_Q[G]$$

what proves that $\{Z^{H,S}_t,\ t \ge 0\}$ is a $L^1(Q)$ martingale.

In particular,

$$E_Q\Big[e^{iY_t(H)}\,\big|\,\mathcal{F}_s\Big] = \exp\Big\{-(1/2)\int_0^t \|\mathfrak{B}H_{t-r}\|_2^2\,dr\Big\} E_Q\Big[Z^{H,t}_t\,\big|\,\mathcal{F}_s\Big]\,.$$

Since $Z^{H,t}_\cdot$ is martingale, the conditional expectation on the right hand side of last formula is equal to $Z^{H,t}_s$. Therefore, a change of variables gives that

$$E_Q\Big[e^{iY_t(H)}\,\big|\,\mathcal{F}_s\Big] = \exp\Big\{iY_s(H_{t-s}) - (1/2)\int_0^{t-s} \|\mathfrak{B}H_r\|_2^2\,dr\Big\}\,.$$

This equation states that conditionally to $\mathcal{F}_s$, $Y_t(H)$ has a Gaussian distribution of mean $Y_s(H_{t-s})$ and variance $\int_0^{t-s} \|\mathfrak{B}H_r\|_2^2 dr$. This is precisely (0.7). A standard Markov argument guarantees then the uniqueness of finite dimension distributions, which in turn gives the uniqueness of Q.

5. Comments and References

The first rigorous result in equilibrium fluctuations was obtained by Martin–Löf (1976) for a superposition of independent Markov processes on $\mathbb{R}^d$. Due to the absence of interaction between particles, the hydrodynamic equation is linear and the martingales introduced in (0.10) are functions of the density field. With the present techniques, the equilibrium fluctuations follows therefore from the Holley–Stroock theory of generalized Ornstein–Uhlenbeck processes and some compactness arguments.

To prove the equilibrium fluctuations of interacting systems, Rost (1983) introduced the Boltzmann–Gibbs principle described in section 1. Brox and Rost (1984) proved the validity of the principle for attractive zero range processes: they showed that for a fixed density α, in the Hilbert space generated by the ν_α–mean zero cylinder functions and the inner product defined by $\ll g, h \gg_\alpha = \sum_{x \in \mathbb{Z}^d} < \tau_x g, h >_\alpha$, the semigroup S_t, as $t \uparrow \infty$, acts as a projection on the space generated by the cylinder function associated to the conserved quantity:

$$\lim_{t \to \infty} \ll g, S_t h \gg_\alpha = \frac{1}{\chi(\alpha)} \ll g, \eta(0) - \alpha \gg_\alpha \ll h, \eta(0) - \alpha \gg_\alpha \cdot$$

From this result they deduced the Boltzmann–Gibbs principle and, in Rost (1985), the equilibrium fluctuations as stated in this chapter.

The Boltzmann–Gibbs principle was extended by De Masi, Presutti, Spohn and Wick (1986) for exclusion processes with speed change (from which they deduced the equilibrium fluctuations for gradient lattice gases), by Spohn (1985, 1986, 1987) for interacting Brownian motions and by Zhu (1990) for one-dimensional Ginzburg–Landau lattice models.

Landim and Vares (1994) proposed an alternative proof of the Boltzmann–Gibbs principle in dimension 1 based on a superexponential replacement lemma, at the fluctuations level, for blocks of size $\varepsilon\sqrt{N}$. The proof we present here of the Boltzmann–Gibbs principle is due to Chang (1994) and Chang and Yau (1992). It was extended to nongradient Ginzburg–Landau lattice models by Lu (1994) and to nongradient generalized exclusion processes by Chang (1995) and Sellami (1998). Gielis, Koukkous and Landim (1997) proved the equilibrium fluctuations for symmetric zero range processes in random environment.

Bertini et al. (1994) prove, for one-dimensional stochastic Ising dynamics with a Kac potential at the critical temperature, that the fluctuation field correctly renormalized converges in distribution to the solution of a stochastic partial differential equation obtained by adding a white noise to a Ginzburg–Landau equation. This analysis is extended in Fritz and Rüdiger (1995) to infinite volume for a wider class of initial states and temperatures close to the critical one.

The nonequilibrium fluctuations are a much less understood question and constitutes one of the main open problems in the theory of hydrodynamic limit of interacting particle systems. Until now only partial results for gradient models are known. Comets and Eisele (1988) prove the hydrodynamic limit and the nonequilibrium large deviations for a non conservative mean field stochastic Ising model. Ravishankar (1992a) proves the nonequilibrium fluctuations for symmetric simple exclusion process in any dimension. De Masi, Presutti and Scacciatelli (1989), Dittrich and Gärtner (1991) prove nonequilibrium fluctuations for the one-dimensional nearest neighbor weakly asymmetric simple exclusion process. Ravishankar (1992b) extends this result to 2-dimensional weakly asymmetric simple exclusion processes. Ferrari, Presutti and Vares (1988) proved a nonequilibrium version of the Boltzmann–Gibbs principle for symmetric simple exclusion processes in dimension 1 and extended the result for the one-dimensional nearest neighbor symmetric zero range process with jump rate given by $g(k) = \mathbf{1}\{k \geq 1\}$ starting from a local equilibrium. These ideas were applied to a superposition of Kawasaki and Glauber dynamics by De Masi, Ferrari and Lebowitz (1986) and to particles systems with unbounded spins associated to reaction–diffusion equations by Boldrighini, De Masi and Pellegrinotti (1992). Later, Chang and Yau (1992), using a logarithmic Sobolev inequality extended to one-dimensional Ginzburg–Landau lattice models with strictly convex potentials Chang's proof of the Boltzmann–Gibbs principle to the nonequilibrium setting.

A clear presentation of Brox and Rost proof of the Boltzmann–Gibbs principle can be found in De Masi, Ianiro, Pellegrinotti and Presutti (1984) or in Spohn (1991).

The tightness argument and the theory of generalized Ornstein–Uhlenbeck processes presented in this chapter are taken from Holley and Stroock (1978) and Chang (1994). These ideas were successfully applied in Holley and Stroock (1979a,b) to investigate the equilibrium fluctuations of non conservative spin flip dynamics.

Appendix 1. Markov Chains on a Countable Space

We present in this chapter an overview on continuous time Markov chains on countable state spaces. We refer the reader to Kemeny, Snell and Knapp (1966), Breiman (1968), Gikhman et Skorohod (1969) and Ethier and Kurtz (1986) for a detailed and comprehensive exposition of general properties of discrete and continuous time jump Markov chains.

The first two sections are devoted to the construction of a continuous time Markov process on a countable state space and to the investigation of the basic properties of the underlying discrete time skeleton chain. At the end of the second section we compute the Radon–Nikodym derivative between two jump Markov processes. In sections 3 and 4 the basic tools in the theory of Markov processes are introduced: semigroups, generators, adjoint and reversible processes.

In section 5 we introduce a class of martingales in the context of Markov processes. With this collection of martingales we derive a bound in section 6 for the variance of additive functionals of Markov processes and we prove the Feynman–Kac formula in section 7. This Feynman–Kac formula permits to compute explicitly the Radon–Nikodym derivative of a time inhomogeneous Markov process with respect to another, generalizing the formula obtained in section 2.

In section 8 we review the elementary properties of the relative entropy of a probability measure with respect to a reference measure. The explicit formula for the relative entropy permits in section 9 to show that in the context of Markov processes the entropy of the state of the process with respect to an invariant measure does not increase in time. In fact, we show that the time derivative of this entropy is bounded above by the Dirichlet form. This estimate leads us in section 10 to examine the main properties of the Dirichlet form and in section 11 to prove a maximal inequality for reversible Markov processes.

1. Discrete Time Markov Chains

Throughout this chapter E stands for a countable state space. The elements of E are denoted by the last characters of the alphabet.

Let $p: E \times E \to \mathbb{R}_+$ be a transition probability:

$$p(x,y) \geq 0 \qquad \text{and} \qquad \sum_{y \in E} p(x,y) = 1$$

for every $x \in E$. Denote by $\Omega = E^{\mathbb{N}}$ the path space endowed with the Borel σ-algebra $\mathcal{B} = \mathcal{B}(E)$ and by ω the elements of Ω. For $n \ge 0$ let $X_n \colon \Omega \to E$ be the state of the chain at time n:

$$X_n(\omega) = \omega_n$$

and let $\theta_n \colon \Omega \to \Omega$ be the time translation by n units:

$$\Big(\theta_n(\omega)\Big)_j = \omega_{n+j}$$

for all $j \ge 0$.

Proposition 1.1 *For each x in E, there exists a unique probability measure on $(\Omega, \mathcal{B})$, denoted by P_x, such that*

$$P_x[X_0 = x_0\,,\, X_1 = x_1\,, \dots,\, X_n = x_n] = \delta_{x,x_0}\, p(x_0, x_1) \cdots p(x_{n-1}, x_n)$$

for every $n \ge 0$. In this formula $\delta_{x,y}$ stands for the delta of Kronecker. Moreover, if E_x stands for the expectation with respect to P_x, for every bounded, $\mathcal{B}$–measurable function f,

$$E_x\big[f \circ \theta_n \mid X_0, \dots, X_n\big] = E_x\big[f \circ \theta_n \mid X_n\big] = E_{X_n}\big[f\big]\,.$$

Proof. The existence of the probability measure P_x follows from Kolmogorov's theorem (cf. Corollary 2.19 of Breiman (1968)). We just need to check that this pre-probability defined on $\bigcup_{n \ge 0} \sigma(X_0, \dots, X_n)$ may be extended to $\mathcal{B}$. The formula for the conditional expectation follows from an elementary computation. □

The first identity of the second statement of the proposition establishes that the behavior of the Markov chain in the future depends on the past only through the present or, equivalently, that conditioned on the present, the past and the future are independent. This result explains the following definition.

Definition 1.2 Let $p \colon \mathbb{N} \times E \times E \to [0,1]$ be a collection of transition probabilities. A sequence of random variables $\{X_n;\ n \ge 0\}$ defined on a probability space $(\Omega, \mathcal{A}, P)$ and taking values on a countable space E is a Markov chain with transition probability p if for every $n \ge 0$,

$$P\Big[X_{n+1} = y \,\Big|\, X_0 = x_0, \dots, X_n = x_n\Big] = P\Big[X_{n+1} = y \,\Big|\, X_n = x_n\Big] = p(n, x_n, y)$$

for every $(x_0, \dots, x_n, y)$ in E^{n+2}. The Markov chain is said to be homogeneous if the transition probability p does not depend on n, i.e., if there exists a transition probability $p \colon E \times E \to [0,1]$ such that

$$P\Big[X_{n+1} = y \,\Big|\, X_n = x\Big] = p(x, y)$$

for every (x, y) in $E \times E$ and every $n \geq 0$.

The second property imposes the process to be time translation invariant in the following sense. The probability, for a process starting from x at time 0, to be at state y at time n is equal to the probability, for a process that is at x at time m, to be at y at time $m + n$.

If μ is a probability measure on E, we denote by P_μ the probability measure on the path space Ω for a process whose initial position is distributed according to μ:

$$P_\mu[\,\cdot\,] := \sum_{x \in E} \mu(x)\, P_x[\,\cdot\,]\,.$$

Hereafter E_μ stands for expectation with respect to P_μ.

In order to investigate the equilibrium states of a Markov chain, in the space of bounded measurable functions on E, denoted by $C_b(E)$, we introduce the operator P defined by

$$(Pf)(x) = \sum_{y \in E} p(x, y)\, f(y)\,.$$

Notice that for every f in $C_b(E)$ and every probability measure μ on E,

$$E_\mu[f(X_n)] = < \mu, P^n f >\,,$$

where P^n stands for the n-th power of P and $< \mu, g >$ for the integral of a bounded function g with respect to μ.

We endow $C_b(E)$ with its natural topology, the topology of the pointwise bounded convergence: a sequence $(f_j)_{j \geq 1}$ of bounded functions converges boundedly pointwise to f if it converges pointwisely and remains uniformly bounded:

$$\lim_{j \to \infty} f_j(x) = f(x) \qquad \text{for every} \quad x \in E\,,$$

$$\limsup_{j \to \infty} \sup_{x \in E} \Big| f_j(x) \Big| < \infty\,.$$

By duality we may extend the operator P to the space of probability measures on E, denoted by $\mathcal{M}_1(E)$. In this way, for μ in $\mathcal{M}_1(E)$, μP stands for the probability measure defined by

$$(\mu P)(y) = \sum_{x \in E} \mu(x)\, p(x, y)\,.$$

In this context to find a probability measure invariant under the evolution, i.e., a probability measure under which X_0 and X_n have the same distribution for every $n \geq 0$, we need to look for a solution of

$$< \pi P, f > = < \pi, f >$$

for every bounded function f.

Of course, there exists always a solution to this problem in the case where E is finite. It is enough, for instance, to examine the action of P on the compact and convex set $\mathcal{M}_1(E)$ or to take any limit point of the sequence

$$\frac{1}{N}\sum_{i=0}^{N-1}\mu P^i$$

where μ is any probability measure.

In the countable case an invariant probability measure may not exist. For example, in the case of a nearest neighbor symmetric random walk on $\mathbb{Z}$. However, if there is a state x such that the expectation of the first return to x is finite under P_x, then there exists an invariant probability measure (cf. Definition 7.26 and Theorem 7.34, Breiman (1968)). Such state x is said to be positive recurrent.

The next natural question in the investigation of a Markov chain is the problem of the uniqueness of an invariant probability measure. If the Markov chain is indecomposable, i.e., if there are no two sets A and B with $A \cap B = \phi$ and such that starting from any site x in A (resp. B) the chain remains in A (resp. B), there is at most an invariant probability measure (cf. Theorem 7.16, Breiman (1968)).

2. Continuous Time Markov Chains

Let E be a countable space, $\lambda: E \to (0,\infty)$ a bounded function and p a transition probability on E that vanishes on the diagonal: $p(x,x) = 0$ for every x in E. Consider the space $\Omega = (E \times (0,\infty))^{\mathbb{N}}$ endowed with the Borel σ-algebra that makes the variables (ξ_n, τ_n) measurable. For each x in E, let P_x be the probability measure under which

(a) ξ_n is a Markov chain with transition probability p starting from x,

(b) Given the sequence $(\xi_n)_{n\geq 0}$, the random variables τ_n are independent and distributed according to an exponential law of parameter $\lambda(\xi_n)$.

Notice that the conditional distribution of the vector $(\tau_0,\dots,\tau_n)$ given $\{\xi_k;\ k \in \mathbb{N}\}$ depends only on $(\xi_0,\xi_1,\dots,\xi_n)$ and therefore that the conditional distribution of $(\tau_0,\dots,\tau_n)$ given $(\xi_0,\xi_1,\dots,\xi_n)$ is still

$$\prod_{i=0}^{n}\lambda(\xi_i)\, e^{-\lambda(\xi_i)u_i}\, \mathbf{1}\{u_i > 0\}\, du_i\ .$$

The next result follows easily from this remark.

Proposition 2.1 *For $n \geq 0$, let*

$$T_0 = 0\ , \qquad T_{n+1} = T_n + \tau_n\ .$$

For each probability P_x the sequence (ξ_n, T_n) is an inhomogeneous Markov chain on $E \times (0,\infty)$ with transition probability given by

$$P\big[\xi_{n+1} = y \ , \ t \le T_{n+1} \le t + dt \ \big| \ \xi_n = x \ , \ T_n = s\big] \\ = \ p(x,y)\, \lambda(x)\, e^{-\lambda(x)(t-s)}\, \mathbf{1}\{t > s\}\, dt \ .$$

Proof. It is enough to show that for each integer $n \ge 1$ and each pair of measurable bounded functions F, G,

$$\begin{aligned} & E_x\Big[F(\xi_{n+1}, T_{n+1})\, G(\xi_1, T_1, \dots, \xi_n, T_n)\Big] \\ & \quad = \ E_x\Big[E_{(\xi_n, T_n)}\big[F(\xi_1, T_1)\big]\, G(\xi_1, T_1, \dots, \xi_n, T_n)\Big] \ , \end{aligned} \tag{2.1}$$

where the expectation on the right hand side is defined by

$$E_{(x,t)}\big[F(\xi_1, T_1)\big] \ = \ \sum_{y \in E} p(x,y) \int_{\mathbb{R}} ds\, \lambda(x) e^{-\lambda(x)(s-t)}\, \mathbf{1}\{s \ge t\} F(y,s) \ .$$

Identity (2.1) is an elementary consequence of the definition of the probability P_x. Indeed, the left hand side is equal to

$$\sum_{\substack{x_i \in E \\ 1 \le i \le n+1}} E_x\Big[F(x_{n+1}, T_{n+1})\, G(x_1, T_1, \dots, x_n, T_n) \Big| \xi_1 = x_1, \dots, \xi_{n+1} = x_{n+1}\Big] \\ \times\, P_x\Big[\ \xi_1 = x_1, \dots, \xi_{n+1} = x_{n+1}\Big] \ .$$

For each x in E, denote by f_x the density $\lambda(x) e^{-\lambda(x)s} \mathbf{1}\{s > 0\}$. By definition of P_x the previous sum is equal to

$$\sum_{x_1, \dots, x_n \in E} \Bigg[\prod_{0 \le i \le n-1} p(x_i, x_{i+1})\Bigg] \int_0^\infty ds_1 \cdots \int_0^\infty ds_n \Bigg[\prod_{0 \le k \le n-1} f_{x_k}(s_{k+1})\Bigg] \\ \times\, G(x_1, s_1, \dots, x_n, s_1 + \cdots + s_n) \\ \times \Bigg\{ \sum_{y \in E} p(x_n, y) \int_0^\infty ds_n\, f_{x_n}(s) F(y, s_1 + \cdots + s_n + s) \Bigg\}$$

provided x_0 stands for x. The notation introduced in the beginning of the proof permits to rewrite, after a change of variables, the last sum inside braces as

$$E_{(x_n, s_1 + \cdots + s_n)}\big[F(\xi_1, T_1)\big] \ .$$

The next to last expression is thus equal to

$$E_x\Big[E_{(\xi_n, T_n)}\big[F(\xi_1, T_1)\big]\, G(\xi_1, T_1, \cdots, \xi_n, T_n)\Big] \ ,$$

what concludes the proof of the lemma. □

The next result, whose proof relies on the loss of memory of exponential random variables, will be used throughout this chapter.

Corollary 2.2 *For every bounded, $\mathcal{B}(\Omega)$-measurable function H, on the set $\{T_n \le t\}$,*

$$E_x\Big[\mathbf{1}\{T_{n+1} > t\} H(\{(\xi_j, T_j - t);\ j \ge n+1\}) \,\big|\, (\xi_n, T_n)\Big] \\ = e^{-\lambda(\xi_n)(t - T_n)}\, E_{\xi_n}\Big[H\Big(\{(\xi_j, T_j);\ j \ge 1\}\Big)\Big]\,. \tag{2.2}$$

Proof. It is enough to prove the corollary for functions that depend only on a finite number of coordinates. To avoid too long formulas, consider a function $H: (E \times (0,\infty))^2 \to \mathbb{R}$ that depends only on two coordinates. By Proposition 2.1, the left hand side of (2.2) is equal to

$$E_{(\xi_n, T_n)}\Big[\mathbf{1}\{T_1 > t\} H(\xi_1, T_1 - t, \xi_2, T_2 - t)\Big] \\ = E_{(\xi_n, T_n)}\Big[H(\xi_1, T_1 - t, \xi_2, \tau_1 + T_1 - t)\,\big|T_1 > t\Big]\; P_{(\xi_n, T_n)}\Big[T_1 > t\Big]\,. \tag{2.3}$$

Since T_1 is distributed according to an exponential random variable of parameter $\lambda(\xi_n)$, on the set $T_1 > t$, the variable $T_1 - t$ is also distributed according to an exponential variable of parameter $\lambda(\xi_n)$. In particular, the right hand side of the previous identity may be rewritten as

$$e^{-\lambda(\xi_n)(t - T_n)}\, E_{\xi_n}\Big[H(\xi_1, T_1, \xi_2, \tau_1 + T_1)\Big]\,.$$

Identity (2.2) can also be proved recalling the explicit formulas for the transition probabilities. We may indeed rewrite the left hand side of (2.3) as

$$\sum_{x_1, x_2} p(\xi_n, x_1) p(x_1, x_2) \int_{\mathbb{R}} ds_1 \int_{\mathbb{R}} ds_2\, \lambda(\xi_n) e^{-\lambda(\xi_n)(s_1 - T_n)} \lambda(x_1) e^{-\lambda(x_1)(s_2 - s_1)} \\ \times\ \mathbf{1}\{s_1 > T_n \vee t\}\, \mathbf{1}\{s_2 > s_1\}\, H(x_1, s_1 - t, x_2, s_2 - t)\,.$$

On the set $\{T_n \le t\}$ the indicator function of the set $\{s_1 \ge T_n\}$ can be removed. A change of variables permits now to rewrite this sum as

$$e^{-\lambda(\xi_n)(t - T_n)} \sum_{x_1, x_2} p(\xi_n, x_1) p(x_1, x_2) \int_{\mathbb{R}}\int_{\mathbb{R}} ds_1\, ds_2\, \lambda(\xi_n)\, e^{-\lambda(\xi_n) s_1}\, \mathbf{1}\{s_1 > 0\} \\ \times\ \lambda(x_1)\, e^{-\lambda(x_1)(s_2 - s_1)}\, \mathbf{1}\{s_2 > s_1\}\, H(x_1, s_1, x_2, s_2)\,.$$

This double integral is exactly equal to the expectation of H for the Markov chain starting from ξ_n. This last expression is thus equal to

$$e^{-\lambda(\xi_n)(t - T_n)} E_{\xi_n}\left[H(\xi_1, T_1, \xi_2, T_2)\right]\,. \quad \square$$

We now construct a continuous time jump process according to the following prescription. For each $t \in \mathbb{R}_+$ and ω in Ω, let

$$X_t(\omega) = \xi_n(\omega) \qquad \text{if} \quad T_n(\omega) \le t < T_{n+1}(\omega) .$$

This means that the process visits successively the sites occupied by the discrete time Markov chain ξ_n staying at each site an exponential time of parameter $\lambda(\xi_n)$. Notice that $T_n \uparrow \infty$ P_x almost surely and that there is no ambiguity in the definition of X_t because λ was assumed to be bounded and strictly positive.

For each $t \ge 0$, denote by $\theta_t \colon \Omega \to \Omega$ the time translation by t in the path space Ω. To define θ_t rigorously we need to introduce some notation. For each $t \ge 0$, the random variable $n_t \colon \Omega \to \mathbb{N}$ indicates the interval $[T_n, T_{n+1})$ that contains t:

$$n_t(\omega) := \max\{n;\ T_n(\omega) \le t\} .$$

With this notation, $\theta_t\omega$ is the sequence of elements of $E \times (0,\infty)$ whose n-th coordinate, denoted by $(\theta_t\omega)_n$, is equal to

$$(\theta_t\omega)_n = \begin{cases} (\xi_{n_t}, 0) & \text{for } n = 0 , \\ (\xi_{n_t+n}, T_{n_t+n} - t) & \text{for } n \ge 1 . \end{cases}$$

A simple computation shows that

$$X_s(\theta_t\omega) = X_{s+t}(\omega)$$

for every s, $t \ge 0$. Indeed, by definition,

$$\begin{aligned} X_s(\theta_t\omega) &= \sum_{j\ge 0} \xi_j(\theta_t\omega)\mathbf{1}\{T_j(\theta_t\omega) \le s < T_{j+1}(\theta_t\omega)\} \\ &= \sum_{j\ge 0} \xi_{n_t+j}(\omega)\mathbf{1}\big\{(T_{n_t+j}(\omega) - t)^+ \le s < T_{n_t+j+1}(\omega) - t\big\} . \end{aligned}$$

Since $s \ge 0$, we may replace $(T_{n_t+j}(\omega)-t)^+$ by $T_{n_t+j}(\omega)-t$. A change of variables gives that

$$\begin{aligned} X_s(\theta_t\omega) &= \sum_{j\ge n_t} \xi_j(\omega)\mathbf{1}\{T_j(\omega) \le s+t < T_{j+1}(\omega)\} \\ &= \sum_{j\ge 0} \xi_j(\omega)\mathbf{1}\{T_j(\omega) \le s+t < T_{j+1}(\omega)\} \\ &= X_{s+t}(\omega) \end{aligned}$$

because, by definition of n_t, $T_j(\omega) \le t \le t+s$ for $j \le n_t$.

For $t \ge 0$, denote by $\mathcal{F}_t$ the σ-algebra all events prior to time t: $\mathcal{F}_t = \sigma\{X_s;\ 0 \le s \le t\}$ and denote by $\mathcal{F}$ the σ-algebra generated by the random variables $\{X_s;\ s \ge 0\}$.

Proposition 2.3 *For each x in E, under P_x $\{X(t),\ t \ge 0\}$ is Markov process with a time homogeneous transition probability.*

Proof. To prove the Markov property we need to show that for every $\mathcal{F}$–measurable set A and every $\mathcal{F}_t$-measurable set B,

$$P_x\left[\theta_t^{-1}(A)\cap\{X_t=y\}\cap B\right] = P_x[B\cap\{X_t=y\}]\,P_y[A]\,.$$

In fact we shall prove a stronger property: for every $n\geq 0$

$$P_x\Big[\theta_t^{-1}(A)\cap\{X_t=y\}\cap B\cap\{T_n\leq t<T_{n+1}\}\Big]$$
$$= P_x\Big[B\cap\{X_t=y\}\cap\{T_n\leq t<T_{n+1}\}\Big]\,P_y[A]\,.$$

By the π–λ class Theorem (cf. Theorem 1.4.3 in Chow and Teicher (1988)), it is enough to check this identity for sets of the form

$$A = \bigcap_{j=0}^{k_1}\{X_{s_j^1}\in F_j^1\}\,,\qquad 0\leq s_0^1<\cdots<s_{k_1}^1\,,$$
$$B = \bigcap_{j=0}^{k_2}\{X_{s_j^2}\in F_j^2\}\,,\qquad 0\leq s_0^2<\cdots<s_{k_2}^2\leq t\,,$$

where, for $i=1,2$, k_i is a positive integer and $\{F_j^i;\ 0\leq j\leq k_i\}$ is a collection of subsets of E.

On the set $\{T_n\leq t<T_{n+1}\}$, the event B is a function of $\xi_0,\ldots,\xi_n$ and $T_0,\ldots,T_n$, whereas $\theta_t^{-1}(A)$ can be written in terms of $\{\xi_j;\ j\geq n\}$ and of $\{T_j-t;\ j\geq n+1\}$. We may therefore rewrite the left hand side of the previous identity as

$$E_x\Big[F(\xi_0,T_0,\ldots,\xi_{n-1},T_{n-1},y,T_n)\,\mathbf{1}\{\xi_n=y\}\times$$
$$\times\ G(y,\{(\xi_j,T_j-t);\ j\geq n+1\})\,\mathbf{1}\{T_n\leq t<T_{n+1}\}\Big]\,.$$

Taking the conditional expectation with respect to $\{(\xi_j,T_j);\ 0\leq j\leq n\}$ and applying the Markov property for the process (ξ_j,T_j) proved in the previous proposition, this last expression is equal to

$$E_x\Big[F(\xi_0,T_0,\ldots,\xi_{n-1},T_{n-1},y,T_n)\,\mathbf{1}\{\xi_n=y\}\,\mathbf{1}\{T_n\leq t\}\times$$
$$\times\ E_x\Big[\mathbf{1}\{T_{n+1}>t\}\,G(y,\{(\xi_j,T_j-t);\ j\geq n+1\})\,\Big|(\xi_n,T_n)\Big]\Big]\,.$$

By Corollary 2.2, this expectation is equal to

$$E_x\Big[F(\xi_0,T_0,\ldots,\xi_{n-1},T_{n-1},y,T_n)\,\mathbf{1}\{\xi_n=y\}\,\mathbf{1}\{T_n\leq t\}\,e^{-\lambda(\xi_n)(t-T_n)}\Big]\times$$
$$\times\ E_y\Big[G(y,\{(\xi_j,T_j);\ j\geq 1\})\Big]\,.$$

By Corollary 2.2 and a simple computation taking advantage of the explicit form of the set A, this expression may be rewritten as

$$E_x\Big[F(\xi_0, T_0, \ldots, \xi_{n-1}, T_{n-1}, y, T_n)\, \mathbf{1}\{\xi_n = y\}\, \mathbf{1}\{T_n \le t < T_{n+1}\}\Big]\, P_y[A]\,.$$

The first term is exactly

$$P_x\Big[B \cap \{X_t = y\} \cap \{T_n \le t < T_{n+1}\}\Big]\,,$$

what concludes the proof. □

Definition 2.4 A collection of variables $\{X_t;\ t \ge 0\}$ defined on a probability space $(\Omega, \mathcal{A}, P)$ and taking values in a countable space E is a homogeneous, continuous time Markov chain if

(a) (Markov property). For every $s, t \ge 0$

$$P\Big[X_{s+t} = y \,\big|\, \sigma\{X_r;\ r \le t\}\Big] \;=\; P\Big[X_{s+t} = y \,\big|\, X_t\Big]$$

for every site y of E.

(b) (Homogeneity). For every x in E, let P_x be the probability on the path space Ω defined by

$$P_x[\,\cdot\,] \;:=\; P[\,\cdot\,|\, X_0 = x]\,.$$

Then, for every $s, t \ge 0$ and y in E,

$$P\Big[X_{s+t} = y \,\big|\, X_t\Big] \;=\; P_{X_t}\Big[X_s = y\Big]\,.$$

(c) (Jump property). There exists a sequence of strictly increasing stopping times $(T_n)_{n\ge 0}$ such that $T_0 = 0$, X_t is constant on the interval $[T_n, T_{n+1})$ and $X_{T_n-} \ne X_{T_n}$ for every $n \ge 0$.

Proposition 2.5 . (Converse of Proposition 2.3.) *If $(X_t)_{t\ge 0}$ is a homogeneous continuous time Markov chain, then*

(a) The skeleton chain defined by $\xi_n = X_{T_n}$ for $n \ge 0$ is a discrete time Markov chain with transition probability $p(x, y)$ given by

$$p(x, y) \;=\; P\Big[X_{T_1} = y \big| X_0 = x\Big]\,.$$

(b) Recall the definition of the probability P_x introduced in Definition 2.4 (b). Under P_x T_1 has an exponential distribution whose parameter is denoted by $\lambda(x)$. Conditionally to the sequence $(\xi_n)_{n\ge 0}$ the variables $\tau_j = T_{j+1} - T_j$ are independent and have exponential distributions of parameter $\lambda(\xi_j)$.

(c) (Uniqueness in distribution). To each continuous time homogeneous Markov chain we just associated a transition probability $p(\,\cdot\,, \cdot\,)$ and a jump rate $\lambda(\cdot)$.

Two continuous time, homogeneous Markov chains having the same transition probability p and bounded jump rate λ have the same distribution.

The assumption on the boundness of the jump rate λ can be weakened in the proof of the uniqueness in distribution of the process. Nevertheless, an assumption that guarantees the divergence of the sequence of stopping times T_n must be imposed. The reader can find in Doob (1953) (pp. 266 ff.) a proof of this proposition.

The assumption $X_{T_n -} \neq X_{T_n}$ in Definition 2.4 guarantees that the transition probability p of the skeleton chain vanishes on the diagonal.

From now on, we shall abbreviate homogeneous, continuous time Markov chains as Markov chains. Furthermore, we shall always assume the jump rate λ to be positive and bounded and the transition probability p to vanish on the diagonal.

Given two Markov chains on the same countable space E, we may construct on the same path space two probability measures P_x and $\overline{P}_x$ corresponding respectively to the pairs $(\lambda(x), p(x,y))$ and $(\overline{\lambda}(x), \overline{p}(x,y))$. If we consider the paths only up to a certain time t, the probability measures P and $\overline{P}$ are equivalent if the allowed jumps are the same, i.e., if for every x in E, the sets $\{y \in E,\ p(x,y) \neq 0\}$ and $\{y \in E,\ \overline{p}(x,y) \neq 0\}$ are the same.

Proposition 2.6 *The Radon–Nikodym derivative $\frac{dP}{d\overline{P}}$ restricted to $\mathcal{F}_t$ is given by the formula*

$$\frac{dP}{d\overline{P}}\Big|_{\mathcal{F}_t} = \exp\left\{ \int_0^t \left[\lambda(X_s) - \overline{\lambda}(X_s)\right] ds - \sum_{s \leq t} \log \frac{\lambda(X_{s-}) p(X_{s-}, X_s)}{\overline{\lambda}(X_{s-}) \overline{p}(X_{s-}, X_s)} \right\} .$$

Proof. The assumption $p(x,x) = 0$ ensures that the function $p(X_{s-}, X_s)$ vanishes everywhere but at the jumps. In particular, for almost all realization of the process the sum reduces to a finite sum of terms.

Fix $n \geq 1$ and let $F\colon (E \times (0,\infty))^n \to \mathbb{R}$ be a bounded measurable function. The expectation under P_x of $F(\xi_1, T_1, \ldots, \xi_n, T_n)$ is equal to

$$\sum_{x_1,\ldots,x_n \in E} \left[\prod_{0 \leq i \leq n-1} p(x_i, x_{i+1}) \right] \left\{ \int_0^\infty ds_1 \cdots \int_0^\infty ds_n \right.$$
$$\left. \left[\prod_{0 \leq k \leq n-1} \lambda(x_k) e^{-\lambda(x_k) s_{k+1}} \right] F(x_1, s_1, \ldots, x_n, s_1 + \cdots + s_n) \right\} .$$

In this last formula $x_0 = x$. taking the ratio of the densities, we obtain that this sum is equal to

$$\overline{E}_x \left[F \exp\left\{ - \int_0^{T_n} \left[\lambda(X_s) - \overline{\lambda}(X_s)\right] ds + \sum_{s \leq T_n} \log \frac{\lambda(X_{s-}) p(X_{s-}, X_s)}{\overline{\lambda}(X_{s-}) \overline{p}(X_{s-}, X_s)} \right\} \right] .$$

For $k \geq 1$, consider a sequence of times $0 \leq s_1 < \cdots < s_k \leq t$ and a bounded measurable function $F: E^k \to \mathbb{R}$. On the set $\{T_n \leq t < T_{n+1}\}$, $F(X_{s_1}, \ldots, X_{s_k})$ is a function of $(\xi_1, T_1, \ldots, \xi_n, T_n)$. Thus, by Corollary 2.2, we obtain that the expectation of $F(X_{s_1}, \ldots, X_{s_k})$ under P_x is equal to

$$\sum_{n \geq 0} E_x \left[\tilde{F}(\xi_1, T_1, \ldots, \xi_n, T_n) \mathbf{1}\{T_n \leq t\} e^{-\lambda(\xi_n)(t-T_n)}\right]$$

for some function $\tilde{F}$. By the first part of the proof, this sum is equal to

$$\sum_{n \geq 0} \overline{E}_x \Bigg[\tilde{F}(\xi_1, T_1, \ldots, \xi_n, T_n) \mathbf{1}\{T_n \leq t\} e^{-\lambda(\xi_n)(t-T_n)} \times$$
$$\times \exp\left\{ - \int_0^{T_n} \left[\lambda(X_s) - \overline{\lambda}(X_s)\right] ds + \sum_{s \leq T_n} \log \frac{\lambda(X_{s-}) p(X_{s-}, X_s)}{\overline{\lambda}(X_{s-}) \overline{p}(X_{s-}, X_s)} \right\} \Bigg] .$$

By Corollary 2.2, this expression may be rewritten as

$$\sum_{n \geq 0} \overline{E}_x \Bigg[\exp\left\{ - \int_0^{t} \left[\lambda(X_s) - \overline{\lambda}(X_s)\right] ds + \sum_{s \leq t} \log \frac{\lambda(X_{s-}) p(X_{s-}, X_s)}{\overline{\lambda}(X_{s-}) \overline{p}(X_{s-}, X_s)} \right\} \times$$
$$\times \tilde{F}(\xi_1, T_1, \ldots, \xi_n, T_n) \mathbf{1}\{T_n \leq t < T_{n+1}\} \Bigg]$$

and this sum is equal to

$$\overline{E}_x \Bigg[F(X_{s_1}, \ldots, X_{s_k})$$
$$\exp\left\{ - \int_0^{t} \left[\lambda(X_s) - \overline{\lambda}(X_s)\right] ds + \sum_{s \leq t} \log \frac{\lambda(X_{s-}) p(X_{s-}, X_s)}{\overline{\lambda}(X_{s-}) \overline{p}(X_{s-}, X_s)} \right\} \Bigg] .$$

Since every bounded $\mathcal{F}_t$–measurable function may be approximated by functions depending only on a finite number of coordinates, the proposition is proved. □

3. Kolmogorov's Equations, Generators

We now introduce the matrices $P_t(x, y)$ that represent the probability to be at site y at time t for the Markov chain that starts from x:

$$P_t(x, y) := P_x[X_t = y] .$$

It follows from the Markov property of the process that theses matrices form a semigroup:

$$P_{t+s}(x,y) = \sum_{z\in E} P_t(x,z)\, P_s(z,y)$$

for every (x,y) in E^2. In terms of operators this identity becomes

$$P_{t+s} = P_t \circ P_s \, .$$

On the other hand, since λ is assumed to be bounded, it follows from the previous identity that the operators P_t are continuously differentiable in time and satisfy the so called Chapman–Kolmogorov equations:

$$\begin{cases} P_0(x,y) = \delta_{x,y} \\ \partial_t\, P_t(x,y) = \sum_{z\in E} \lambda(x) p(x,z) \left[P_t(z,y) - P_t(x,y)\right] . \end{cases} \tag{3.1}$$

In this formula $\delta_{x,y}$ is equal to 1 if $x = y$ and 0 otherwise. The integral version of equation (3.1) may be obtained computing $P_x[X_t = y]$ by decomposing the event $\{X_t = y\}$ according to the time spent at x before the first jump and the first site visited after x. At the end of this section we give a sketch of an analytical proof of this formula.

Introducing the operators

$$L(x,y) = \begin{cases} \lambda(x)\, p(x,y) & \text{if } y \neq x \, , \\ -\lambda(x) & \text{if } y = x \, , \end{cases}$$

the last formula becomes

$$\partial_t\, P_t = L\, P_t \, .$$

In particular,

$$\lim_{t\to 0} \frac{P_t - I}{t} = L$$

so that for t small

$$P_t(x,y) = \delta_{x,y} + t\, L(x,y) + o(t) \, .$$

In a similar way we can show that the operators $\{P_t,\, t \geq 0\}$ satisfy the second Chapman–Kolmogorov equations:

$$\partial_t\, P_t(x,y) = \sum_{z\in E} P_t(x,z) \lambda(z) p(z,y) - P_t(x,y)\lambda(y) \tag{3.2}$$

so that

$$\partial_t\, P_t = P_t\, L \, .$$

Notice that the matrix L is such that for every x in E,

$$L(x,y) \geq 0 \quad \text{if} \quad y \neq x \qquad L(x,x) < 0 \qquad \text{and} \qquad L\mathbf{1} = 0 \, , \tag{3.3}$$

where $\mathbf{1}$ stands for the function on E constant equal to 1.

Like in the discrete case in the space $C_b(E)$ of bounded measurable functions we introduce the operators $(P_t)_{t\geq 0}$ and L defined by

$$\begin{aligned} (P_t f)(x) &= \sum_{y\in E} P_t(x,y)f(y)\,, \\ (Lf)(x) &= \sum_{y\in E} \lambda(x)p(x,y)\big[f(y)-f(x)\big]\,. \end{aligned} \tag{3.4}$$

Since the jump rate λ is assumed to be bounded, $C_b(E)$ is closed under P_t and L. The next result is proved at the end of this section.

Lemma 3.1 *For every bounded function f, the sequence $\{t^{-1}(P_t f - f),\ t>0\}$ converges boundedly pointwise to Lf as $t \downarrow 0$.*

By duality we can extend the operators P_t and L to the space of probability measures on E:

$$\begin{aligned} (\mu P_t)(x) &= \sum_{y\in E} \mu(y)P_t(y,x)\,, \\ (\mu L)(x) &= \sum_{y\in E} \mu(y)L(y,x)\,. \end{aligned}$$

Notice that $<\mu L, \mathbf{1}> = <\mu, L\mathbf{1}> = 0$ for every probability measure μ. In particular, μL is not a probability measure.

Conversely, to each operator L satisfying properties (3.3) is associated a unique pair of jump rate $\lambda(\cdot)$ and transition probability $p(\cdot,\cdot)$ vanishing on the diagonal. Moreover, solving the differential equation

$$\begin{cases} P_0 = I \\ \partial_t P_t = L P_t\,, \end{cases}$$

where I stands for the identity, we obtain a unique solution $P_t = e^{tL}$.

We call the operator L the generator of the semigroup P_t. The Trotter–Kato formula is easily deduced:

$$P_t = \lim_{n\to\infty} \left(I + \frac{t}{n} L\right)^n\,.$$

It is also worthwhile to examine the effect on the generator of a time scale modification in the semigroup. A simple computation shows that for each $c>0$ the semigroup $\{P_{tc},\ t\geq 0\}$ is associated to the generator cL. Therefore a time scale modification $t \to ct$ translates into a multiplication by c of the jump rate λ.

We conclude this section with a proof of Lemma 3.1 and formulas (3.1), (3.2). They rely on the following estimates that will proved later in this section.

$$\left| \frac{P_t(x,x)-1}{t} + \lambda(x) \right| \leq 2\overline{\lambda}^2 t\,, \tag{3.5}$$

$$\left|\frac{P_t(x,y)}{t} - \lambda(x)p(x,y)\right| \le \overline{\lambda}^2 t\left\{p(x,y) + \sum_{n\ge 0} p_{n+2}(x,y)\frac{(\overline{\lambda}t)^n}{n!}\right\}, \tag{3.6}$$

where $\overline{\lambda}$ stands for the upper bound for the jump rate λ:

$$\overline{\lambda} = \sup_{x\in E} \lambda(x) .$$

Proof of Lemma 3.1. The estimate (3.6) permits to bound the expression

$$\left|\frac{(P_t f)(x) - f(x)}{t} - (Lf)(x)\right| ,$$

which is equal to

$$\left|\sum_{y\in E}\left[t^{-1}P_t(x,y) - \lambda(x)p(x,y)\right]\left[f(y) - f(x)\right]\right| ,$$

by

$$\begin{aligned} 2\|f\|_\infty\overline{\lambda}^2 t \sum_{y\in E}\left\{p(x,y) + \sum_{n\ge 0} p_{n+2}(x,y)\frac{(\overline{\lambda}t)^n}{n!}\right\} \\ \le 2\|f\|_\infty\overline{\lambda}^2 t\left(1 + e^{\overline{\lambda}t}\right) . \end{aligned}$$

This proves that the sequence converges uniformly to Lf as $t \downarrow 0$. □

We turn now to the proof of formula (3.1). By the semigroup property of the operators $\{P_s,\ s\ge 0\}$, we may rewrite $h^{-1}[P_{t+h}(x,y) - P_t(x,y)]$ as

$$\begin{aligned} LP_t(x,y) + \sum_{z\ne x}\left[\frac{P_h(x,z)}{h} - \lambda(x)p(x,z)\right]P_t(z,y) \\ + \left[\frac{P_h(x,x) - 1}{h} + \lambda(x)\right]P_t(x,y) . \end{aligned}$$

Estimates (3.5), (3.6) and similar arguments to the ones presented in the proof of Lemma 3.1 show that the last two terms vanish as $h \downarrow 0$, what proves (3.1)

The proof of formula (3.2) is even simpler. By the semigroup property, we have that the difference $h^{-1}[P_{t+h}(x,y) - P_t(x,y)]$ is equal to

$$\sum_{z\ne x} P_t(x,z)\frac{P_h(z,y)}{h} + P_t(x,y)\frac{P_h(y,y) - 1}{h} .$$

Formula (3.2) follows thus from the dominated convergence theorem and from estimates (3.5), (3.6).

It remains to prove inequalities (3.5), (3.6). We start with (3.5). To compute the probability to be at x at time t for a Markov chain starting from x, we may decompose the event $\{X_t = x\}$ according to the number of jumps before time t. Thus,

$$P_x\Big[X_t = x\Big] = \sum_{n\geq 0} P_x\Big[\xi_n = x,\ T_n \leq t < T_{n+1}\Big] \,.$$

Keep in mind that under P_x T_1 is an exponential random variable of parameter $\lambda(x)$ and that the transition probability vanishes on the diagonal. In particular,

$$\begin{aligned}
&\left|\frac{P_t(x,x)-1}{t} + \lambda(x)\right| \\
&\quad \leq \left|\frac{1}{t}\Big\{1 - P_x\Big[T_1 > t\Big]\Big\} - \lambda(x)\right| + \frac{1}{t}\sum_{n\geq 2} P_x\Big[\xi_n = x,\ T_n \leq t < T_{n+1}\Big] \\
&\quad \leq t\lambda^2(x) + \frac{1}{t}P_x\Big[T_2 \leq t\Big] \,.
\end{aligned}$$

The random variable T_2 is the sum of two independent exponential random variables of parameter $\lambda(x)$ and $\lambda(\xi_1)$. Since the jump rate $\lambda(\cdot)$ is bounded above by $\overline{\lambda}$, if U_1, U_2 stand for two independent, identically distributed, exponential random variables of parameter $\overline{\lambda}$, for every state y of E,

$$P_y\Big[T_2 \leq t\Big] \;\leq\; P\Big[U_1 + U_2 \leq t\Big] \;\leq\; (1/2)\overline{\lambda}^2 t \,,$$

what proves inequality (3.5).

We now turn to the estimate (3.6). With the same decomposition of the event $\{X_t = x\}$ performed in the proof of inequality (3.5), we have that

$$\begin{aligned}
&\left|\frac{P_t(x,y)}{t} - \lambda(x)p(x,y)\right| \\
&\quad \leq \left|t^{-1}P_x\Big[\xi_1 = y,\ T_1 \leq t < T_2\Big] - \lambda(x)p(x,y)\right| \\
&\quad\quad + t^{-1}\sum_{n\geq 2} P_x\Big[\xi_n = y,\ T_n \leq t < T_{n+1}\Big] \,.
\end{aligned}$$

By Corollary 2.2 and the properties of the skeleton chain (ξ_n) stated in Proposition 2.5, this expression is bounded above by

$$\begin{aligned}
&\left|t^{-1}E_x\Big[\mathbf{1}\{\xi_1 = y\}\mathbf{1}\{T_1 \leq t\}\, e^{-\lambda(x)(t-T_1)}\Big] - \lambda(x)p(x,y)\right| \\
&\quad + t^{-1}\sum_{n\geq 2} p_n(x,y)P_x\Big[T_n \leq t \big| \xi_n = y\Big] \,.
\end{aligned}$$

Like in the proof of inequality (3.5), we may bound the probability $P_x\Big[T_n \leq t \big| \xi_n = y\Big]$ by $P_x\Big[U_1 + \cdots + U_n \leq t\Big]$ provided $(U_j)_{j\geq 1}$ stands for a sequence of independent, identically distributed, exponential random variables of parameter $\overline{\lambda}$. A simple computation shows than that the last expression is bounded above by

$$\overline{\lambda}^{-2} t \left\{ p(x,y) + \sum_{n \geq 0} p_{n+2}(x,y) \frac{(\overline{\lambda} t)^n}{(n+2)!} \right\}. \qquad \square$$

4. Invariant Measures, Reversibility and Adjoint Processes

In the same way that we considered invariant probability measures for discrete time Markov chains, we may investigate invariant measures for continuous time processes, i.e, probability measures μ such that $\mu P_t = \mu$ for every $t \geq 0$.

Proposition 4.1 *A probability measure μ is invariant if and only if $\mu L = 0$.*

Proof. Of course, if $\mu P_t = \mu$ for every $t \geq 0$, taking the time derivative at $t = 0$, we obtain that $\mu L = 0$. Conversely, if $\mu L = 0$ it follows from the Trotter–Kato formula that $\mu P_t = \mu$ for every $t \geq 0$. $\square$

Corollary 4.2 *A probability measure μ is invariant for P_t if and only if the probability measure*

$$\nu(x) = \frac{\mu(x)\,\lambda(x)}{\sum_{y \in E} \mu(y)\,\lambda(y)}$$

is invariant for the discrete time Markov chain with transition probability $p(\cdot,\cdot)$.

Proof. The identity $\mu L = 0$ can be rewritten as

$$\sum_{x \in E} \mu(x)\,\lambda(x)\,p(x,y) = \mu(y)\,\lambda(y)$$

for every y in E. $\square$

It follows from the previous result that under the assumption of indecomposability, that is in general easy to verify, we have the existence of a unique invariant probability measure when E is finite. In the countable case it will also be necessary to check conditions that guaranty the positive recurrence. Most of the time, however, it will be important to characterize more or less explicitly the invariant measures. In order to do it we have to solve the $|E|$ linear equations:

$$\sum_{x \in E} \mu(x)\,\lambda(x)\,p(x,y) = \mu(y)\,\lambda(y)$$

for every y in E. It is stronger but easier to try to solve the system with $|E|^2$ equations

$$\mu(x)\,\lambda(x)\,p(x,y) = \mu(y)\,\lambda(y)\,p(y,x)$$

known to the physicists as the "detailed balance" condition.

It is clear that a summable solution μ of these equations is automatically an invariant probability measure if properly renormalized. In particular, it is *the* invariant probability measure if the chain is indecomposable. Furthermore, an invariant measure satisfying the detailed balance conditions possesses certain special properties. To state them we need to introduce some notation.

For an invariant probability measure μ on E, denote by $L^2(\mu)$ the space of the square integrable functions with respect to μ. We extend the operators P_t and L, originally defined on $C_b(E)$, to the space $L^2(\mu)$: for f in $L^2(\mu)$, we denote by $P_t f$, Lf the functions defined by (3.4). An elementary computation, relying on Schwarz inequality and Proposition 4.1, shows that the operators P_t are contractions on $L^2(\mu)$ and that L is a bounded operator:

$$< \mu, \left[Lf\right]^2 > \le 4\overline{\lambda}^2 < \mu, f^2 > .$$

Moreover, it follows from the estimates (3.5), (3.6) that for every function f in $L^2(\mu)$ $t^{-1}[P_t f - f]$ converges to Lf in $L^2(\mu)$ as $t \downarrow 0$:

$$\lim_{t \to 0} < \mu, \left\{ t^{-1}[P_t f - f] - Lf \right\}^2 > = 0 . \tag{4.1}$$

Proposition 4.3 *The probability measure μ satisfies the detailed balance condition if and only if the operators P_t are self–adjoint in $L^2(\mu)$, i.e., if and only if for every $t \ge 0$, f, g in $L^2(\mu)$,*

$$\sum_x \mu(x) f(x) P_t g(x) = \sum_x \mu(x) g(x) P_t f(x)$$

or, briefly,

$$< f, P_t g >_\mu = < P_t f, g >_\mu . \tag{4.2}$$

Here $< \cdot, \cdot >_\mu$ stands for the inner product of $L^2(\mu)$.

Proof. Fix two sites x, y, set $f = \mathbf{1}\{x\}$, $g = \mathbf{1}\{y\}$ and take the time derivative at $t = 0$ in identity (4.2) to obtain that

$$\mu(x) L(x, y) = \mu(y) L(y, x)$$

which is the detailed balance condition. Inversely, it follows from the previous identity that

$$< f, Lg >_\mu = < Lf, g >_\mu$$

for every f, g in $L^2(\mu)$. It remains to recall Trotter–Kato formula to conclude the proof. □

A probability measure satisfying the detailed balance conditions is said to be reversible. The previous result states therefore that a probability measure is reversible if and only if the generator is self adjoint in $L^2(\mu)$. We shall now look for conditions that guarantee that the adjoint of the generator L in $L^2(\mu)$ is also a generator.

Proposition 4.4 *Let μ be a probability measure. The adjoint of L in $L^2(\mu)$, denoted by L^*, is a generator if and only if μ is invariant. In this case $P_t^* = e^{tL^*}$ is also the adjoint of P_t in $L^2(\mu)$ and the semigroup P_t^* is characterized by the pair (λ^*, p^*) given by*

$$\begin{cases} \lambda^*(x) = \lambda(x) , & \textit{for } x \in E \\ p^*(x,y) = \dfrac{\lambda(y)\mu(y)p(y,x)}{\lambda(x)\mu(x)} , & \textit{for } x, y \in E . \end{cases}$$

Moreover,

$$< f, Lg >_\mu = < L^* f, g >_\mu \tag{4.3}$$

for every f, g in $L^2(\mu)$.

Proof. A simple computation shows that the adjoint L^* of L in $L^2(\mu)$ is given by the formula

$$\mu(x)\, L^*(x,y) = \mu(y)\, L(y,x) . \tag{4.4}$$

In particular, L^* satisfies always the first two properties of generators:

$$\begin{cases} L^*(x,x) = -\lambda(x) < 0 , & \text{for } x \in E \\ L^*(x,y) \ge 0 , & \text{for } x \ne y . \end{cases}$$

Therefore, L^* is a generator if and only if

$$\sum_{y \in E} L^*(x,y) = 0$$

for every x in E and hence if and only if μ is invariant because the explicit expression (4.4) permits to rewrite the previous sum as

$$\frac{1}{\mu(x)} \sum_{y\in E} \mu(y) L(y,x) .$$

On the other hand, the Trotter–Kato formula shows that the semigroup P_t^* associated to the generator L^* is the adjoint of the semigroup P_t in $L^2(\mu)$.

Finally, we have already seen in the first part of the proof that the jump rate λ and λ^* coincide. The explicit formula (4.4) permits than to compute $p^*(x,y)$. On the other hand, formula (4.3) follows from the identity $\mu(x)\lambda(x)p(x,y) = \mu(y)\lambda(y)p^*(y,x)$ and a change of variables:

$$< f, Lg >_\mu = \sum_{x,y\in E} \mu(x)\lambda(x)p(x,y)f(x)g(y) \; - \; \sum_{x\in E} \mu(x)\lambda(x)f(x)g(x) .$$

Since $\mu(x)\lambda(x)p(x,y) = \mu(y)\lambda(y)p^*(y,x)$, the first term on the right hand side can be written as

$$\sum_{x,y\in E} \mu(y)\lambda(y)p^*(y,x)f(x)g(y) = \sum_{x,y\in E} \mu(x)\lambda(x)p^*(x,y)f(y)g(x) .$$

From the previous two identities it is easy to conclude the proof of the proposition. □

The process P_t^*, that is defined without ambiguity when the original process admits a unique invariant measure, is called the adjoint process. The adjoint process is closely connected to the process reversed in time. This is the content of the next proposition.

Proposition 4.5 *If the semigroup P_t^* is the adjoint of the semigroup P_t with respect to the invariant probability measure μ, then for every $n > k \geq 0$, every sequence of times $0 \leq t_1 < \cdots < t_n$ and every sequence of bounded functions $\{f_j;\ 1 \leq j \leq n\}$,*

$$
\begin{aligned}
&E_\mu\Big[f_1(X_{t_1})\cdots f_n(X_{t_n})\Big] \\
&\quad = \sum_{x\in E} \mu(x)\, f_k(x)\, E_x\Big[f_{k+1}(X_{t_{k+1}-t_k})\cdots f_n(X_{t_n-t_k})\Big] \times \\
&\quad\quad \times\ E_x^*\Big[f_k(X_{t_k-t_{k-1}})\cdots f_1(X_{t_k-t_1})\Big]\,,
\end{aligned}
$$

where E_x^* stands for the expectation with respect to the Markov chain with transition probability P_t^* starting from x.

Proof. The proof is straightforward. We just have to apply successively the identity

$$\mu(x)\, P_t(x,y) \;=\; \mu(y)\, P_t^*(y,x)\,. \qquad \square$$

We shall sometimes consider the symmetric part, denote by S, of the generator L in $L^2(\mu)$. It is given by

$$S \;=\; 2^{-1}(L + L^*)\,.$$

A simple computation shows that the symmetric part S is itself a generator characterized by the parameters λ^s and p^s given by

$$
\begin{aligned}
\lambda^s(x) \;&=\; \lambda(x)\,, \qquad x \in E \\
p^s(x,y) &= \frac{1}{2}\left[p(x,y) + \frac{\lambda(y)\mu(y)p(y,x)}{\lambda(x)\mu(x)}\right].
\end{aligned}
$$

Furthermore, S satisfies the detailed balance condition:

$$\mu(x)\, S(x,y) \;=\; \mu(y)\, S(y,x)\,.$$

5. Some Martingales in the Context of Markov Processes

The purpose of this section is to introduce a class of martingales in the context of Markov processes. Consider a bounded function $F: \mathbb{R}_+ \times E \to \mathbb{R}$ smooth in the first coordinate uniformly over the second: for each x in E, $F(\cdot, x)$ is twice continuously differentiable and there exists a finite constant C such that

$$\sup_{(s,x)} \Big| (\partial_s^j F)(s,x) \Big| \le C \tag{5.1}$$

for $j = 1, 2$. In this formula $(\partial_s^j F)$ stands for the j-th time derivative of $F(\cdot, x)$.

To each function F satisfying assumption (5.1), define $M^F(t)$ and $N^F(t)$ by

$$M^F(t) = F(t, X_t) - F(0, X_0) - \int_0^t ds\, (\partial_s + L) F(s, X_s) ,$$

$$N^F(t) = (M^F(t))^2 - \int_0^t ds \Big\{ LF(s, X_s)^2 - 2F(s, X_s) LF(s, X_s) \Big\} .$$

Lemma 5.1 *Denote by $\{\mathcal{F}_t, t \ge 0\}$ the filtration induced by the Markov process: $\mathcal{F}_t = \sigma(X_s,\ s \le t)$. The processes $M^F(t)$ and $N^F(t)$ are $\mathcal{F}_t$-martingales.*

Proof. We start showing that $M^F(t)$ is a martingale. Fix $0 \le s < t$. We need to check that $E_x[M^F(t) | \mathcal{F}_s] = M^F(s)$, i. e., that

$$E_x\Big[F(t, X_t) \,\Big|\, \mathcal{F}_s \Big] = F(s, X_s) + \int_s^t E_x\Big[(\partial_r + L) F(r, X_r) \,\Big|\, \mathcal{F}_s \Big]\, dr .$$

For each $r \ge 0$, denote by $F_r: E \to \mathbb{R}$ (resp. by $F_r': E \to \mathbb{R}$) the function that at x takes the value $F(r, x)$ (resp. $(\partial_r F)(r, x)$). By the Markov property and a change of variables in the integral, the previous identity is reduced to

$$(P_{t-s} F_t)(X_s) = F_s(X_s) + \int_0^{t-s} \Big\{ (P_r F_{s+r}')(X_s) + (P_r L F_{s+r})(X_s) \Big\}\, dr .$$

Since for $t = s$ this identity is trivially satisfied, we just need to check that the time derivative of both expressions are equal, i. e., that

$$\partial_t (P_{t-s} F_t)(x) = (P_{t-s} F_t')(x) + (P_{t-s} L F_t)(x)$$

for every x in E and $0 \le s < t$.

To prove this identity we compute the left hand side. Fix $h > 0$ and rewrite the difference $h^{-1}\{(P_{t+h-s} F_{t+h})(x) - (P_{t-s} F_t)(x)\}$ as

$$h^{-1} E_x\Big[F_{t+h}(X_{t-s+h}) - F_t(X_{t-s+h}) \Big] + h^{-1} E_x\Big[F_t(X_{t-s+h}) - F_t(X_{t-s}) \Big] . \tag{5.2}$$

The first expression is equal to

$$\frac{1}{h}\int_t^{t+h} dr\, E_x\Big[F_r'(X_{t-s+h}) - F_t'(X_{t-s+h})\Big]$$
$$+\; E_x\Big[F_t'(X_{t-s+h}) - F_t'(X_{t-s})\Big] \;+\; E_x\Big[F_t'(X_{t-s})\Big]\,.$$

Since by assumption (5.1) $(\partial_r F)(\cdot, x)$ is Lipschitz continuous uniformly on x, the first term vanishes as $h \downarrow 0$. The second term also vanishes as $h \downarrow 0$ because the semigroup P_t is continuous. Therefore, as $h \downarrow 0$, the first term in (5.2) converges to $(P_{t-s}F_t')(x)$.

The second expression in (5.2) is equal to

$$\frac{1}{h}\int_{t-s}^{t-s+h} dr\, E_x\Big[LF_t(X_r)\Big]$$

that converges, as $h \downarrow 0$, to $(P_{t-s}LF_t)(x)$. This proves that $M^F(t)$ is a martingale.

We show now that $N^F(t)$ is a martingale. A simple computation and the first part of the lemma show that $M^F(t)^2$ is equal to

$$F^2(t, X_t) \;-\; 2F(t, X_t)\int_0^t ds\,(\partial_s + L)F(s, X_s) \;+\; \Big(\int_0^t ds\,(\partial_s + L)F(s, X_s)\Big)^2 \tag{5.3}$$

plus a martingale term. Since $F^2(t, X_t) - \int ds(\partial_s + L)F^2(s, X_s)$ is a martingale, $F^2(t, X_t)$ is equal to a martingale added to

$$\int ds\,(\partial_s + L)F^2(s, X_s)\,. \tag{5.4}$$

The second and third expression in (5.3) can be rewritten as

$$-2M_0^F(t)\int_0^t ds\,(\partial_s + L)F(s, X_s) \;-\; \Big(\int_0^t ds\,(\partial_s + L)F(s, X_s)\Big)^2\,, \tag{5.5}$$

where $M_0^F(t) = M^F(t) + F(0, X_0)$. By Ito's formula, the first term in this expression is equal to a martingale added to

$$-\,2\int_0^t ds\, F(s, X_s)(\partial_s + L)F(s, X_s)$$
$$+\; 2\int_0^t ds\,\Big(\int_0^s dr\,(\partial_r + L)F(r, X_r)\Big)(\partial_s + L)F(s, X_s)\,.$$

An integration by parts shows that the second term of this formula cancels with the second term of (5.5). The remaining expression added to (5.4) is just the integral term that we need to subtract in order to turn $(M^F(t))^2$ in a martingale. This concludes the proof of the lemma. □

6. Estimates on the Variance of Additive Functionals of Markov Processes

Consider a Markov process with generator L satisfying the assumptions of section 3 and having an invariant measure denoted by π. Let S be the symmetric part of the generator L: $S = 2^{-1}(L + L^*)$. For a function f in $L^2(\pi)$, denote by $\|f\|_1$ its $\mathcal{H}_1$ norm defined by

$$\|f\|_1^2 = < f, (-S)f >_\pi$$

and notice that $\|f\|_1^2 = - < f, Lf >_\pi$. Recall from section 4 that we denote by $p^s(x, y)$ the transition probability associated to the symmetric part of the generator. With this notation,

$$\|f\|_1^2 = \frac{1}{2} \sum_{x,y \in E} \pi(x)\lambda(x)p^s(x, y)\Big\{f(y) - f(x)\Big\}^2 .$$

In particular, by Schwarz inequality, the $\mathcal{H}_1$ norm $\|f\|_1^2$ of any function f in $L^2(\pi)$ is bounded above by $2\|\lambda\|_\infty \|f\|_0^2$, provided $\|f\|_0$ stands for the $L^2(\pi)$ norm of f.

Let $\mathcal{H}_1$ be the Hilbert space generated by $L^2(\pi)$ and the inner product $< f, g >_1 = < f, (-S)g >_\pi$: $\mathcal{H}_1 = \overline{L^2(\pi)|_{\mathcal{N}}}$, if $\mathcal{N}$ stands for the kernel of the inner product $< \cdot, \cdot >_1$. (In the context of this chapter, $\mathcal{N}$ is the space generated by constant functions because we assumed the skeleton discrete time Markov chain to be indecomposable). Since $\|f\|_1^2 \le 2\|\lambda\|_\infty \|f\|_0^2$, $L^2(\pi) \subset \mathcal{H}_1$.

Denote by $\|\cdot\|_{-1}$ the dual norm of $\mathcal{H}_1$ with respect to $L^2(\pi)$: for f in $L^2(\pi)$, let

$$\|f\|_{-1}^2 = \sup_{g \in L^2(\pi)} \Big\{2 < f, g >_\pi - \|g\|_1^2\Big\} . \tag{6.1}$$

We claim that $\|f\|_{-1}^2 \ge (2\|\lambda\|_\infty)^{-1} \|f\|_0^2$ for each function f in $L^2(\pi)$. Indeed, since $\|g\|_1^2 \le 2\|\lambda\|_\infty \|g\|_0^2$, the supremum on the right hand side of (6.1) is bounded below by

$$\sup_{g \in L^2(\pi)} \Big\{2 < f, g >_\pi - 2\|\lambda\|_\infty \|g\|_0^2\Big\} .$$

By Schwarz inequality this supremum is equal to $(2\|\lambda\|_\infty)^{-1} \|f\|_0^2$.

Denote by $\mathcal{H}_{-1}$ the subset of $L^2(\pi)$ of all functions with finite $\|\cdot\|_{-1}$ norm. It follows from the definition of the $\mathcal{H}_{-1}$ norm that

$$2 < f, g >_\pi \le \frac{1}{A}\|f\|_{-1}^2 + A\|g\|_1^2 \tag{6.2}$$

for every f in $\mathcal{H}_{-1}$, g in $L^2(\pi)$ and $A > 0$. Since $L^2(\pi)$ is dense in $\mathcal{H}_1$, this inequality may be extended to functions g in $\mathcal{H}_1$. Moreover, we claim that $SL^2(\pi) = \{Sf,\ f \in L^2(\pi)\}$ is contained in $\mathcal{H}_{-1}$ and

$$\|Sf\|_{-1}^2 = \|f\|_1^2$$

for all f in $L^2(\pi)$. Indeed, fix f in $L^2(\pi)$. A simple computation shows that

$$2 < g, Sf >_\pi \le < g, (-S)g >_\pi + < f, (-S)f >_\pi = \|g\|_1^2 + \|f\|_1^2$$

for every g in $L^2(\pi)$. In particular, $\|Sf\|_{-1} \le \|f\|_1$. Taking $g = -f$ in the supremum that defines $\|f\|_{-1}^2$ we deduce the reverse inequality.

We are now ready to state the main result of this section.

Proposition 6.1 *For each function g in $\mathcal{H}_{-1}$ and $t > 0$,*

$$E_\pi\Big[\Big(\frac{1}{\sqrt{t}}\int_0^t g(X_s)\, ds\Big)^2\Big] \le 20\,\|g\|_{-1}^2 .$$

Proof. Since L is a generator, for every $\gamma > 0$, $(\gamma - L)^{-1}$ is a bounded operator in $L^2(\pi)$. In particular, since $\mathcal{H}_{-1} \subset L^2(\pi)$, there exists f_γ in $L^2(\pi)$ such that

$$\gamma f_\gamma - L f_\gamma = g .$$

Taking on both sides of this equation the inner product with respect to f_γ and applying Schwarz inequality (6.2), we obtain that

$$\gamma < f_\gamma, f_\gamma >_\pi \le \|g\|_{-1}^2 \quad \text{and} \quad \|f_\gamma\|_1^2 \le \|g\|_{-1}^2 . \tag{6.3}$$

For $\gamma > 0$ consider the process $M_\gamma(t)$ defined by $M_\gamma(t) = f_\gamma(X_t) - f_\gamma(X_0) - \int_0^t Lf_\gamma(X_s)ds$. By Lemma 5.1, $M_\gamma(t)$ is a martingale if f_γ is bounded. Approximating f_γ, that belongs to $L^2(\pi)$, by bounded functions we deduce that $M_\gamma(t)$ is a martingale in $L^2(\pi)$. With this notation we may rewrite the expectation appearing in the statement of the lemma as

$$\frac{1}{t}E_\pi\Big[\Big(M_\gamma(t) - f_\gamma(X_t) + f_\gamma(X_0) + \int_0^t \gamma f_\gamma(X_s)ds\Big)^2\Big] .$$

Since π is an invariant measure for the Markov process, by Schwarz inequality, this expression is bounded above by

$$\frac{4}{t}\Big\{\{2 + (\gamma t)^2\} < f_\gamma, f_\gamma >_\pi + E_\pi\Big[M_\gamma(t)^2\Big]\Big\} .$$

By Lemma 5.1 the quadratic variation of the martingale $M_\gamma(t)$ is equal to the time integral of $Lf_\gamma(X_s)^2 - 2f_\gamma(X_s)Lf_\gamma(X_s)$. In particular, since the probability measure π is invariant, the expectation of $M_\gamma(t)^2$ is equal to $2t\|f_\gamma\|_1^2$ because $- < f, Lf >_\pi = \|f\|_1^2$ for every f in $L^2(\pi)$. Therefore, the expectation appearing on the statement of the lemma is less than or equal to

$$\frac{4}{t}\Big\{\{2 + (\gamma t)^2\} < f_\gamma, f_\gamma >_\pi + 2t\|f_\gamma\|_1^2\Big\}$$
$$\le \Big\{4\{2 + (\gamma t)^2\}(t\gamma)^{-1} + 8\Big\}\|g\|_{-1}^2$$

in virtue of (6.3). To conclude the proof of the proposition, it remains to choose $\gamma = t^{-1}$. □

Notice that we *did not* assume the process to be reversible.

Remark 6.2 Under the assumption of indecomposability, it is not difficult to show that $L^2(\pi) = SL^2(\pi) \oplus \mathbf{1}$, where $\mathbf{1}$ is the subspace of $L^2(\pi)$ generated by the constants and that $\mathcal{H}_{-1} = SL^2(\pi)$. Thus, for each g in $\mathcal{H}_{-1}$, there exists f in $L^2(\pi)$ such that $Sf = g$. This argument remains in force if S is replaced by L.

7. The Feynman–Kac Formula

Consider a bounded function $V: \mathbb{R}_+ \times E \to \mathbb{R}$ satisfying assumption (5.1) and a bounded function $F_0: E \to \mathbb{R}$. Fix $T > 0$ and denote by $F: [0,T] \times E \to \mathbb{R}$ the solution of the differential equation

$$\begin{cases} (\partial_t u)(t,x) \;=\; (Lu)(t,x) + V(T-t,x)u(t,x)\,, \\ u(0,x) \;=\; F_0(x)\,. \end{cases} \tag{7.1}$$

Proposition 7.1 *The solution F has the following stochastic representation:*

$$F(T,x) \;=\; E_x\Big[e^{\int_0^T V(s,X_s)\,ds}\, F_0(X_T)\Big]\,.$$

Proof. Consider the process $\{A_t,\, 0 \le t \le T\}$ given by

$$A_t \;=\; F(T-t, X_t)\exp\Big\{\int_0^t V(s,X_s)\,ds\Big\}\,.$$

By Lemma 5.1, A_t can be rewritten as

$$\Big\{M_0^F(t) + \int_0^t ds\,(\partial_s + L)F(T-s,X_s)\Big\}\exp\Big\{\int_0^t V(s,X_s)\,ds\Big\}\,,$$

where $M_0^F(t)$ is the martingale $F(T-t,X_t) - \int_0^t(\partial_s + L)F(T-s,X_s)ds$. Ito's formula now gives that

$$A_t \;-\; \int_0^t ds\, e^{\int_0^s V(r,X_r)\,dr}\Big\{F(T-s,X_s)V(s,X_s) \;+\; (\partial_s + L)F(T-s,X_s)\Big\}$$

is a martingale. Since F is the solution of (7.1), the integral term vanishes showing that A_t is martingale. In particular, $E_x[A_T] = E_x[A_0]$, which proves the lemma. □

The proof of the previous lemma shows that

$$e^{\int_0^t V(r,X_r)\,dr} F(X_t) \;-\; \int_0^t ds\, e^{\int_0^s V(r,X_r)\,dr}\Big\{(LF)(X_s) + V(s,X_s)F(X_s)\Big\} \tag{7.2}$$

is a martingale for each bounded function $F: E \to \mathbb{R}$.

For $t \geq 0$, denote by $L_t: E \times E \to \mathbb{R}$ the operator defined by $L_t(x,y) = L(x,y) + V(t,y)\delta_{x,y}$, where $\delta_{x,y}$ stands for the delta of Kronecker. Denote furthermore, for $0 \leq s \leq t$, by $P^V_{s,t}: E \times E \to \mathbb{R}_+$ the operator given by

$$P^V_{s,t}(x,y) = E\Big[e^{\int_s^t V(r,X_r)\,dr} \mathbf{1}\{X_t = y\} \,\Big|\, X_s = x\Big] . \tag{7.3}$$

A simple computation, relying on Markov property, permits to rewrite $P^V_{s,t}(x,y)$ as

$$P^V_{s,t}(x,y) = E_x\Big[e^{\int_0^{t-s} V(s+r,X_r)\,dr} \mathbf{1}\{X_{t-s} = y\}\Big] .$$

The collection $\{P^V_{s,t},\ 0 \leq s \leq t\}$ can be extended to act on bounded functions:

$$(P^V_{s,t} f)(x) = E_x\Big[e^{\int_0^{t-s} V(s+r,X_r)\,dr} f(X_{t-s})\Big] .$$

Property (7.2) applied to the bounded function $F(z) = \mathbf{1}\{z = y\}$ and a simple computation shows that $\{P^V_{s,t},\ 0 \leq s \leq t\}$ is a semigroup associated to the operator L_t: $P^V_{s,t} P^V_{t,u} = P^V_{s,u}$ for all $0 \leq s \leq t \leq u$ and

$$\begin{cases} (\partial_t P^V_{s,t})(x,y) = (P^V_{s,t} L_t)(x,y) , \\ P^V_{s,s}(x,y) = \delta_{x,y} \end{cases} \tag{7.4}$$

for $t \geq s \geq 0$.

The arguments presented in section 3 permit to prove also the first Chapman–Kolmogorov equations:

$$\begin{cases} (\partial_s P^V_{s,t})(x,y) = -(L_s P^V_{s,t})(x,y) , \\ P^V_{t,t}(x,y) = \delta_{x,y} \end{cases}$$

for $0 \leq s \leq t$.

Assume now that L is a reversible generator with respect to an invariant state ν. Since V is bounded, $L_t = L + V(t,\cdot)$ is also a symmetric operator in $L^2(\nu)$. Denote by Γ_t the largest eigenvalue of $L + V_t$:

$$\Gamma_t = \sup_{\|f\|_2 = 1} \Big\{ < V_t, f^2 >_\nu + < Lf, f >_\nu \Big\} .$$

By definition of the semigroup $\{P^V_{s,t},\ t \geq s \geq 0\}$,

$$E_\nu\Big[e^{\int_0^t V(r,X_r)\,dr}\Big] = < P^V_{0,t}\mathbf{1}, \mathbf{1} >_\nu .$$

On the other hand, by the first Chapman–Kolmogorov equation, for $0 \leq s \leq t$,

$$\frac{d}{ds} < P^V_{s,t}\mathbf{1}, P^V_{s,t}\mathbf{1} >_\nu = -2 < L_s P^V_{s,t}\mathbf{1}, P^V_{s,t}\mathbf{1} >_\nu$$

because L_s is a symmetric operator. By definition of Γ_s, this expression is bounded below by

$$-2\Gamma_s < P^V_{s,t}\mathbf{1}, P^V_{s,t}\mathbf{1} >_\nu \ .$$

Therefore, by Gronwall inequality and since $P_{t,t}$ is the identity,

$$< P^V_{0,t}\mathbf{1}, P^V_{0,t}\mathbf{1} >_\nu \le \exp\Big\{ \int_0^t \Gamma_s ds \Big\} .$$

Since, by Schwarz inequality,

$$< P^V_{0,t}\mathbf{1}, \mathbf{1} >_\nu \le < P^V_{0,t}\mathbf{1}, P^V_{0,t}\mathbf{1} >_\nu^{1/2} ,$$

we have proved the following lemma:

Lemma 7.2 *Assume that the Markov process is reversible with respect to an invariant probability measure ν. Let $V: \mathbb{R}_+ \times E \to \mathbb{R}$ be a bounded function. For each $t \ge 0$ denote by Γ_t the largest eigenvalue of the operator $L + V(t, \cdot)$. Then,*

$$E_\nu\Big[\exp\Big\{ \int_0^t V(r, X_r)\, dr \Big\}\Big] \le \exp\Big\{ \int_0^t \Gamma_s\, ds \Big\} . \qquad (7.5)$$

In the case where ν is only an invariant measure, the previous argument shows that (7.5) remains in force provided Γ_s is the largest eigenvalue of $S = (1/2)\{L + L^*\}$, the symmetric part of L in $L^2(\nu)$.

The Feynman–Kac formula presented in Proposition 7.1 permits to obtain an explicit formula for the Radon–Nikodym derivative of a time inhomogeneous Markov process with respect to another, generalizing Proposition 2.6 to the inhomogeneous case. Fix a function $F: \mathbb{R}_+ \times E \to \mathbb{R}$ satisfying assumptions (5.1).

Denote by $\mathbb{M}^F(t)$ the process defined by

$$\mathbb{M}^F(t) = \exp\Big\{ F(t, X_t) - F(0, X_0) - \int_0^t ds\, e^{-F(s,X_s)} (\partial_s + L) e^{F(s,X_s)} \Big\} .$$

We claim that $\mathbb{M}^F(t)$ is a mean 1 positive martingale. Indeed, fix x_0 in E and define V, $H: \mathbb{R}_+ \times E \to \mathbb{R}$ by

$$\begin{aligned} V(t,x) &= -\exp\{-F(t,x)\}(\partial_s + L)\exp\{F(t,x)\} , \\ H(t,x) &= \exp\{F(t,x) - F(0,x_0)\} . \end{aligned} \qquad (7.6)$$

It follows from the proof of Proposition 7.1 that

$$\mathbb{M}^F(t) = H(t, X_t)\exp\{ \int_0^t V(s, X_s) ds \}$$

is a martingale. Since this martingale is equal to 1 P_{x_0} almost surely at time 0, it is a mean 1 positive martingale.

Fix a time $T > 0$ and for each x_0 in E, define on $\mathcal{F}_T$ the probability measure $P^F_{x_0}$ by

$$E^F_{x_0}\big[G\big] \;=\; E_{x_0}\big[G\,\mathbb{M}^F(T)\big] \tag{7.7}$$

for all bounded $\mathcal{F}_T$-measurable functions G. A simple computation shows that the conditional expectation is

$$E^F_{x_0}\big[G\,\big|\,\mathcal{F}_s\big] \;=\; \frac{1}{\mathbb{M}^F(s)} E_{x_0}\big[G\,\mathbb{M}^F(T)\,\big|\,\mathcal{F}_s\big]\,.$$

For $0 \le s \le t \le T$, define the functions $Q^F_{s,t}\colon \mathbb{E}\times\mathbb{E} \to \mathbb{R}_+$ by $Q^F_{s,t}(x,y) = P^F_{x_0}[X_t = y | X_s = x]$. Recall the definition of the function V introduced in (7.6) and of the semigroup $\{P^V_{s,t},\, 0 \le s \le t\}$ given in (7.3). It follows from the formula for the conditional expectation that $Q^F_{s,t}(x,y) = P^V_{s,t}(x,y)\exp\{F(t,y) - F(s,x)\}$.

We claim that $\{Q^F_{s,t},\, 0 \le s \le t \le T\}$ is a semigroup of transition probabilities. Indeed, it is clear that $Q^F_{s,t}(x,y) \ge 0$ for all x, y in E. It follows from the explicit formula for the conditional expectation that $\sum_{y\in E} Q^F_{s,t}(x,y) = 1$ for all $0 \le s \le t \le T$ and x in E. Finally, by definition of $Q^F_{s,t}$ and because $\{P^V_{s,t},\, 0 \le s \le t\}$ is a semigroup,

$$\sum_{z\in E} Q^F_{s,t}(x,z) Q^F_{t,u}(z,y) \;=\; Q^F_{s,u}(x,y)$$

for all $s \le t \le u$ and all x, y in E. This proves that $\{Q^F_{s,t},\, 0 \le s \le t \le T\}$ is a semigroup of transition probabilities. We now compute the generator associated to this semigroup. Since $Q^F_{s,t}(x,y) = P^V_{s,t}(x,y)\exp\{F(t,y) - F(s,x)\}$, by the second Chapman–Kolmogorov equations (7.4),

$$\begin{aligned}\partial_t Q^F_{s,t}(x,y) \;&=\; \partial_t\Big\{P^V_{s,t}(x,y)\exp\{F(t,y) - F(s,x)\}\Big\} \\ &=\; \exp\{F(t,y) - F(s,x)\}\Big\{(P^V_{s,t}L_t)(x,y) \;+\; P^V_{s,t}(x,y)(\partial_t F(t,y))\Big\}\,.\end{aligned}$$

The explicit formula for L_t and some elementary computations lead to the formula

$$\partial_t Q^F_{s,t}(x,y) \;=\; (Q^F_{s,t} L^F_t)(x,y)\,,$$

where

$$L^F_t(x,y) \;=\; L(x,y)\exp\{F(t,y) - F(t,x)\} \;-\; \delta_{x,y}\exp\{F(t,y)\}L\exp\{F(t,y)\}\,.$$

In particular, for any bounded function $H\colon E \to \mathbb{R}$,

$$(L^F_t H)(x) \;=\; \sum_{y\in E} \lambda(x)p(x,y)\exp\{F(t,y) - F(t,x)\}[H(y) - H(x)]\,. \tag{7.8}$$

We summarize in the next proposition the result just proved.

Proposition 7.3 *Fix a function $F\colon \mathbb{R}_+ \times E \to \mathbb{R}$ satisfying assumptions (5.1). Fix a time $T > 0$ and for each x_0 in E, define on $\mathcal{F}_T$ the probability measure $P^F_{x_0}$*

by (7.7). Under $P^F_{x_0}$, X_t is a time inhomogeneous Markov process with generator $\{L^F_t, 0 \le t \le T\}$ given by (7.8) and starting from x_0.

8. Relative Entropy

Let π be a reference probability measure on E. For a probability measure μ denote by $H(\mu|\pi)$ the relative entropy of μ with respect to π defined by the variational formula:

$$H(\mu|\pi) \;=\; \sup_f \left\{ < \mu, f > \;-\; \log < \pi, e^f > \right\} .$$

In this formula the supremum is carried over all bounded functions f and $< \mu, f >$ stands for the integral of f with respect to μ. From now on, to keep notation and terminology simple, we denote $H(\mu|\pi)$ by $H(\mu)$ and refer to it as the entropy of μ.

Notice that the addition of a constant to the function f does not change the value of $< \mu, f > - \log < \pi, e^f >$. We may therefore restrict the supremum to bounded positive functions. The next result follows easily from the variational formula for the entropy.

Proposition 8.1 *The entropy is positive, convex and lower semicontinuous.*

To show that the entropy is positive, just observe that $< \mu, f > - \log < \pi, e^f >$ vanishes for any constant function f.

We shall repeatedly use the entropy to estimate the expectation of a function with respect to a probability measure μ in terms of integrals with respect to the reference measure π. Indeed, the entropy inequality gives that

$$< \mu, f > \le \alpha^{-1} \left\{ \log < \pi, e^{\alpha f} > \; + H(\mu) \right\}$$

for every positive constant α. For indicator functions this inequality takes a simple form.

Proposition 8.2 *Let A be a subset of E.*

$$\mu[A] \;\le\; \frac{\log 2 + H(\mu)}{\log\left(1 + \frac{1}{\pi[A]}\right)} \, .$$

Proof. By the entropy inequality, for every $\alpha > 0$,

$$\mu[A] \;\le\; \frac{1}{\alpha} \log\left[1 + (e^\alpha - 1)\pi[A]\right] \;+\; \frac{1}{\alpha} H(\mu) \, .$$

It is enough to choose α in such a way that $(e^\alpha - 1)\pi[A] = 1$, i.e., to take

$$\alpha = \log\left(1 + \frac{1}{\pi[A]}\right). \qquad \square$$

The next result presents an explicit formula for the entropy.

Theorem 8.3 *The entropy $H(\mu)$ is given by the formula*

$$H(\mu) = \sum_{x\in E} \pi(x)\frac{\mu(x)}{\pi(x)} \log \frac{\mu(x)}{\pi(x)}$$

$$= \sum_{x\in E} \mu(x) \log \frac{\mu(x)}{\pi(x)} .$$

if μ is absolutely continuous with respect to π and is equal to ∞ otherwise.

Proof. If the probability measure μ is not absolutely continuous with respect to π, since E is assumed to be countable, a simple choice of bounded functions f shows that the entropy $H(\mu)$ is infinite.

Assume now that μ is absolutely continuous with respect to π and that the set E is finite. The functional $\Phi\colon \mathbb{R}^E \to \mathbb{R}$ defined by

$$\Phi(f) = < \mu, f > - \log < \pi, e^f >$$

is concave and assumes its maximum where its gradient vanishes: for every function f such that

$$< \mu, h > = \frac{< \pi e^f, h >}{< \pi, e^f >}$$

for every h in $\mathbb{R}^E$. The invariance of Φ by the addition of a constant permits to choose among these functions one such that $< \pi, e^f >$ is equal to 1. In particular,

$$f = \log \frac{d\mu}{d\pi}$$

and

$$\sup_f \Phi(f) = \Phi\left(\log \frac{d\mu}{d\pi}\right) = < \mu, \log \frac{d\mu}{d\pi} > \cdot$$

To extend this result to the countable case, it is enough to remark that the variational formula for the entropy can be rewritten as

$$H(\mu) = \lim_{k\to\infty} \sup_{f\in\mathcal{D}(E_k)} \Phi(f) ,$$

where $(E_k)_{k\geq 1}$ stands for an increasing sequence of finite subsets of E whose union is equal to E and $\mathcal{D}(E_k)$ for the set of functions that are constant on the complement of E_k. The arguments presented in the first part of the proof shows that the supremum in last formula is equal to

$$\sum_{x \in E_k} \pi(x) \frac{\mu(x)}{\pi(x)} \log \frac{\mu(x)}{\pi(x)} + \pi(E_k^c) \frac{\mu(E_k^c)}{\pi(E_k^c)} \log \frac{\mu(E_k^c)}{\pi(E_K^c)} \cdot$$

This expression is increasing in k because the function $u \log u$ is convex. Moreover, as $k \uparrow \infty$, it converges, to

$$\sum_{x \in E} \mu(x) \log \frac{\mu(x)}{\pi(x)} \cdot \qquad \square$$

This explicit formula for the relative entropy involving the function $u \log u$ explains the relation between the entropy and the expectation of functions of type e^f in the entropy inequality. Indeed, by Legendre duality, the convex functions $\Phi(u) = (1+u)\log(1+u) - u$ and $\Psi(v) = e^v - 1 - v$ form a pair of Young functions (cf. Neveu (1972) for the terminology and an introduction to the Orlicz spaces associated to pairs of Young functions). In this case we have that

$$uv \le \Phi(u) + \Psi(v)$$

for $u, v \ge 0$. A simple computation shows that this inequality holds for $u \ge -1$ and $v \in \mathbb{R}$. Changing variables, we obtain that

$$uv \le e^v + u \log u - u \tag{8.1}$$

for $u \ge 0$ et $v \in \mathbb{R}$. Taking u as the density of μ with respect to π and v as a function f plus a constant, after an integration with respect to π, we obtain the inequality

$$\int f \, d\mu \le e^{C_0} \int e^f \, d\pi + H(\mu) - 1 - C_0 .$$

Minimizing over the constant C_0, we obtain the entropy inequality.

9. Entropy and Markov Processes

Consider a Markov chain on a countable space E with an invariant measure denoted by π. Let $(P_t)_{t \ge 0}$ be the semigroup associated to the Markov chain. The relative entropy with respect to the invariant measure plays an important role in the investigation of the time evolution of the process. First of all, since $\varphi(u) = u \log u$ is strictly convex and vanish only at $s = 0$ and $s = 1$, the relative entropy of μP_t with respect to π does not increase in time. This is the content of the first proposition.

Proposition 9.1 *For every probability measure μ, we have*

$$H(\mu P_t) \le H(\mu) .$$

Moreover, $H(\mu P_t) = H(\mu) < \infty$ implies that $\mu = \pi$ if the chain is indecomposable.

Proof. The proof relies on the explicit formula obtained in the previous section and on the convexity of the function $\varphi(u) = u \log u$. Assume, without loss of generality, that μ is absolutely continuous with respect to π. A simple computation shows that μP_t shares the same property for all $t \geq 0$. The explicit formula for the entropy permits to write

$$\begin{aligned} H(\mu P_t) &= \sum_x \pi(x) \varphi\Big(\frac{1}{\pi(x)} \sum_y \mu(y) P_t(y,x) \Big) \\ &= \sum_x \pi(x) \varphi\Big(\sum_y \frac{\mu(y)}{\pi(y)} \frac{\pi(y) P_t(y,x)}{\pi(x)} \Big) \\ &\leq \sum_x \pi(x) \sum_y \varphi\Big(\frac{\mu(y)}{\pi(y)} \Big) \frac{\pi(y) P_t(y,x)}{\pi(x)} \\ &= H(\mu) . \end{aligned}$$

Since $\varphi(u)$ is strictly convex, there is equality only if $\mu(y)/\pi(y)$ is constant for every y such that $P_t(x,y) > 0$ for some (and therefore for all) x. □

The entropy furnishes an upper bound for the distance between two probability measures. Denote by $\| \cdot \|$ the total variation distance:

$$\|\mu - \pi\| \; := \; 2 \sup_{A \subset E} \big| \mu[A] - \pi[A] \big| \, .$$

In this formula, the supremum is taken over all subsets of E. The reason for the constant 2 is explained by the explicit formula for the total variation distance. Indeed, a straightforward argument shows that the total variation distance can be written as

$$\|\mu - \pi\| \; = \; \sum_{x \in E} |\mu(x) - \pi(x)| \; = \; \sum_{x \in E} \pi(x) |f(x) - 1|$$

provided f stands for the Radon–Nikodym derivative of μ with respect to π. It follows now from Schwarz inequality and the elementary inequality $3(a-1)^2 \leq (2a+4)(a \log a - a + 1)$ for $a \geq 0$ that

$$\|\mu - \pi\|^2 \; \leq \; 2 H(\mu) \, .$$

On the other hand, if P_t^* stands for the adjoint of P_t in $L^2(\pi)$ and if f (resp. f_t) stands for the density of μ (resp. μP_t) with respect to π,

$$f_t(x) \; = \; (P_t^* f)(x) \, .$$

In particular, the density f_t is solution of

$$\begin{cases} f_0 = f \\ \partial_t f_t = L^* f_t \end{cases} \qquad (9.1)$$

This observation permits to deduce a simple estimate for the time derivative of the entropy of μP_t.

Theorem 9.2 *Let μ be a probability measure with finite entropy: $H(\mu) < \infty$. For every t, $h \geq 0$, we have that*

$$H(\mu P_{t+h}) - H(\mu P_t) = \int_t^{t+h} ds < f_s, L \log f_s >_\pi$$
$$\leq \int_t^{t+h} ds\, 2 < \sqrt{f_s}, L\sqrt{f_s} >_\pi .$$

Moreover,

$$2 < \sqrt{f_s}, L\sqrt{f_s} >_\pi = - \sum_{x,y \in E} \pi(x)\, L(x,y) \left[\sqrt{f_s(y)} - \sqrt{f_s(x)}\right]^2 .$$

In these formulas we used the notation introduced in Proposition 4.3.

Proof. By the explicit formula for the entropy, the difference $H(\mu P_{t+h}) - H(\mu P_t)$ is equal to

$$\sum_{x \in E} \pi(x) \int_t^{t+h} ds \,\{1 + \log f_s(x)\} L^* f_s(x) .$$

Recall that we denoted by $\overline{\lambda}$ the upper bound of the jump rate $\lambda(\cdot)$. It is easy to check that the absolute value of $L^* f_s$ is bounded above by the integrable (with respect to π) positive function $\overline{\lambda}\{f_s(x) + \sum_y p^*(x,y) f_s(y)\}$. In particular, by Fubini lemma

$$\sum_{x \in E} \pi(x) \int_t^{t+h} ds\, L^* f_s(x) = \int_t^{t+h} ds \sum_{x \in E} \pi(x) L^* f_s(x)$$
$$= \int_t^{t+h} ds < \pi, L^* f_s > .$$

Since L is the adjoint of L^* in $L^2(\pi)$, this expression vanishes.

On the other hand, $\log f_s(x) L^* f_s(x)$ can be rewritten as

$$\sum_{\substack{y \in E \\ y \neq x}} L^*(x,y) \log f_s(x) f_s(y) - \lambda(x) f_s(x) \log f_s(x) .$$

Since the entropy deceases in time, $f_s \log f_s$ is integrable for $t \leq s \leq t+h$. By inequality (8.1), the positive part of $\log f_s(x) f_s(y)$ is also integrable. In particular, by Fubini's lemma, we may interchange the time integral with the space integral to obtain that

$$\sum_{x \in E} \pi(x) \int_t^{t+h} ds \log f_s(x) L^* f_s(x) = \int_t^{t+h} ds < \log f_s, L^* f_s >_\pi .$$

This concludes the proof of the first part of the theorem in view of the definition of the adjoint L^*. To prove the inequality, recall that $a[\log b - \log a] \le 2\sqrt{a}[\sqrt{b}-\sqrt{a}]$ for a, $b \ge 0$. From this inequality it follows that the right hand side of the previous identity is bounded above by

$$2\int_t^{t+h} ds < \sqrt{f_s}, L\sqrt{f_s} >_\pi .$$

It remains to show that

$$2 < \sqrt{f_s}, L\sqrt{f_s} >_\pi = - \sum_{x,y\in E} \pi(x)\, L(x,y) \left[\sqrt{f_s(y)} - \sqrt{f_s(x)}\right]^2 . \qquad (9.2)$$

Since L^* is the adjoint of L in $L^2(\pi)$,

$$2 < \sqrt{f_s}, L\sqrt{f_s} >_\pi = < \sqrt{f_s}, L\sqrt{f_s} >_\pi + < \sqrt{f_s}, L^*\sqrt{f_s} >_\pi .$$

The right hand side is equal to

$$\sum_{x,y\in E} \pi(x)\, L(x,y) \sqrt{f_s(x)} \left[\sqrt{f_s(y)} - \sqrt{f_s(x)}\right]$$
$$+ \sum_{x,y\in E} \pi(x)\, L^*(x,y) \sqrt{f_s(x)} \left[\sqrt{f_s(y)} - \sqrt{f_s(x)}\right] .$$

Since L^* is the adjoint of L, $\pi(x)L(x,y) = \pi(y)L^*(y,x)$. A change of variables in the second sum of the last expression permits to conclude. □

10. Dirichlet Form

At the end of the previous section we deduced that the time derivative of the entropy of μS_t is bounded above by a function of the density of μP_t. It is therefore natural to introduce, for every function f in $L^2(\pi)$, the Dirichlet form $\mathfrak{D}(f)$ of f defined by

$$\mathfrak{D}(f) := - < f, Lf >_\pi = - \sum_{x\in E} f(x)\, Lf(x)\, \pi(x) .$$

This sum is well defined because the generator L is a bounded operator in $L^2(\pi)$. The next result follows from the computation performed at the end of last section.

Proposition 10.1 *The Dirichlet form of a function f in $L^2(\pi)$ is positive and equal to*

$$\mathfrak{D}(f) = \frac{1}{2} \sum_{x,y\in E} \pi(x)\, L(x,y)\, [f(y) - f(x)]^2 .$$

Notice that if the Dirichlet form of a function f vanishes, $\mathfrak{D}(f) = 0$, and if the process is indecomposable, then f is constant. Furthermore, if a function $F: \mathbb{R} \to \mathbb{R}$ is such that $|F(a) - F(b)| \le |a - b|$, then

$$\mathfrak{D}(F \circ f) \le \mathfrak{D}(f) . \tag{10.1}$$

This inequality applies to the function $F(x) = x \wedge M$ where M is a fixed real and to $F(x) = |x|$. Furthermore, the Dirichlet form is convex: let p be a probability measure on $\mathbb{N}$. By Schwarz inequality,

$$\mathfrak{D}(\sum_j p_j f_j) \le \sum_j p_j \mathfrak{D}(f_j) . \tag{10.2}$$

In the case, where the probability measure π is reversible, there exists a variational formula for the Dirichlet form $\mathfrak{D}(f)$.

Theorem 10.2 (Variational formula for the Dirichlet form) *For every positive function f of $L^2(\pi)$,*

$$\mathfrak{D}(f) := - \inf_g \sum_{x \in E} \pi(x) \frac{f^2(x)}{g(x)} Lg(x) .$$

In this formula, the infimum is taken over all bounded positive functions bounded below by a strictly positive constant.

Proof. We first show that the supremum is bounded above by the Dirichlet form. Fix a function g that is bounded and is bounded below by strictly positive constant. Set $\alpha = \{g/f\}\mathbf{1}\{f > 0\}$ so that α vanishes where f vanishes. With this definition,

$$< \frac{f^2}{g}, P_t g >_\pi = < \frac{f}{\alpha}, P_t f \alpha >_\pi .$$

Since π is reversible, this last sum is equal to

$$\begin{aligned} &\sum_{x,y} \pi(x) f(x) f(y) P_t(x,y) \frac{\alpha(y)}{\alpha(x)} \\ &\quad = \frac{1}{2} \sum_{x,y} \pi(x) f(x) f(y) P_t(x,y) \left(\frac{\alpha(y)}{\alpha(x)} + \frac{\alpha(x)}{\alpha(y)} \right) . \end{aligned}$$

Since for every positive real a, $a + a^{-1} \ge 2$, this expression is bounded below by

$$\sum_{x,y} \pi(x) f(x) f(y) P_t(x,y) = < f, P_t f >_\pi :$$

Subtracting $< f^2 >_\pi$ to both terms and multiplying by $-t^{-1}$, we deduce that

$$\frac{1}{t} < \frac{f^2}{g}, (g - P_t g) >_\pi \le \frac{1}{t} < f, (f - P_t f) >_\pi .$$

Letting $t \downarrow 0$, we obtain that

$$- < \frac{f^2}{g}, Lg >_\pi \leq \mathfrak{D}(f)$$

for every fixed function g because, by Lemma 3.1, $t^{-1}(g - P_t g)$ converges boundedly pointwise to $-Lg$ and by (4.1) $t^{-1}(f - P_t f)$ converges in $L^2(\mu)$ to $-Lf$.

Formally, to prove the converse inequality, it is enough to take $g = f$. To prove it rigorously, however, we need to approximate f by bounded positive functions bounded below by strictly positive constants.

For each positive integer M, let f_M be the function defined by

$$f_M(x) = M^{-1} + f(x) \wedge M .$$

Notice that f_M is bounded and bounded below by a strictly positive constant and that Lf_M is bounded. Moreover, since π is reversible, we have that

$$- < \frac{f^2}{f_M}, Lf_M > = \frac{1}{2} \sum_{x,y \in E} \pi(x) L(x,y) \Big\{ \frac{f(y)^2}{f_M(y)} - \frac{f(x)^2}{f_M(x)} \Big\} [f_M(y) - f_M(x)] .$$

Since f_M converges pointwisely to f as $M \uparrow \infty$, by Fatou lemma and the explicit formula for the Dirichlet form,

$$\mathfrak{D}(f) \leq \liminf_{M \to \infty} - < \frac{f^2}{f_M}, Lf_M > ,$$

what concludes the proof of the theorem. □

The next result is a simple consequence of this proposition.

Corollary 10.3 *The functional*

$$D(f) = \mathfrak{D}(\sqrt{f})$$

defined for all densities with respect to π is convex and lower semicontinuous.

In principle the domain of the Dirichlet form corresponds to the domain of $\sqrt{-L}$. The explicit form for the Dirichlet form given above in Proposition 10.1 permits, however, to extend the definition of the Dirichlet form to $L^2(\pi)$ if we allow it to assume the value $+\infty$. From now on we will always implicitly assume that $\mathfrak{D}(\cdot)$ is defined on $L^2(\pi)$.

11. A Maximal Inequality for Reversible Markov Processes

We conclude this chapter with a maximal inequality for reversible Markov processes due to Kipnis and Varadhan (1986). We assume throughout this section that X_t is a Markov process reversible with respect to some invariant state π.

Theorem 11.1 *Fix $g: E \to \mathbb{R}$. For each $T > 0$ and $A > 0$, we have that*

$$P_\pi \Big[\sup_{0 \le t \le T} |g(X_t)| \ge A \Big] \;\le\; \frac{e}{A} \sqrt{< g, g >_\pi \; + T \mathfrak{D}(g)}\,. \tag{11.1}$$

Proof. Denote by G the subset $G = \{x,\ |g(x)| \ge A\}$ and by τ the hitting time of G: $\tau = \inf\{t \ge 0,\ X_t \in G\}$. For each $\lambda > 0$, define the function $\varphi_\lambda : E \to \mathbb{R}_+$ by

$$\varphi_\lambda(x) \;=\; \varphi(\lambda, x) \;=\; E_x\Big[e^{-\lambda\tau}\Big]\,.$$

Of course φ_λ is identically equal to 1 on G. On the other hand, for x not in G, decomposing the chain according to the first site visited, the Markov property gives that

$$\varphi(\lambda, x) \;=\; \int_0^t ds\, \lambda(x) e^{-\{\lambda(x)+\lambda\}s} \sum_{y \in E} p(x,y)\varphi(\lambda,y) \;+\; e^{-\{\lambda(x)+\lambda\}t}\varphi(\lambda,x)\,.$$

Dividing by t and letting $t \downarrow 0$, we get that

$$\lambda(x) \sum_{y \in E} p(x,y)[\varphi(\lambda,y) - \varphi(\lambda,x)] \;=\; \lambda\varphi(\lambda,x)\,.$$

for x in G^c. We have thus proved that $L\varphi_\lambda = \lambda\varphi_\lambda$ on G^c.

By definition of the stopping time τ, the left hand side of (11.1) is equal to $P_\pi[\tau \le T]$. Since

$$P_x[\tau \le T] \;\le\; e^{\lambda T}\, \varphi(\lambda, x)\,,$$

by Schwarz inequality we have that

$$\begin{aligned} P_\pi \Big[\sup_{0 \le t \le T} |g(X_t)| \ge A \Big] \;&\le\; e^{\lambda T} \sum_{x \in E} \pi(x)\, \varphi(\lambda, x) \\ &\le\; e^{\lambda T} \sqrt{\sum_{x \in E} \pi(x)\, \varphi(\lambda, x)^2}\,. \end{aligned}$$

The expression inside the square root is clearly bounded above by

$$\sum_{x \in E} \pi(x)\, \varphi_\lambda(x)^2 \;+\; \frac{1}{\lambda}\, \mathfrak{D}(\varphi_\lambda)\,. \tag{11.2}$$

To conclude the proof of the theorem it remains to show that φ_λ is the function that minimizes the functional $J_\lambda(f)$ defined by

$$J_\lambda(f) = \sum_{x\in E} \pi(x) f^2(x) + \frac{1}{\lambda}\mathfrak{D}(f) \tag{11.3}$$

among all functions that are equal to 1 on G. Indeed, since $(|g| \wedge A)A^{-1}$ is equal to 1 on G, (11.2) is bounded above by

$$A^{-2}\Big\{\sum_{x\in E} \pi(x)(|g(x)| \wedge A)^2 + \frac{1}{\lambda}\mathfrak{D}(|g| \wedge A)\Big\}\,.$$

By property (10.1) of the Dirichlet form, this expression is bounded above by

$$A^{-2}\Big\{\sum_{x\in E} \pi(x)\, g(x)^2 + \frac{1}{\lambda}\mathfrak{D}(g)\Big\}\,.$$

It remains to choose $\lambda = T^{-1}$.

To conclude the proof of the theorem we need thus to show that φ_λ is the function that minimizes the functional $J_\lambda(f)$ defined by (11.3) among all functions that are equal to 1 on G.

A function $h: E \to \mathbb{R}$ that is equal to 1 on the set G and that minimizes the functional $J_\lambda(f)$ must satisfy the identity

$$\lambda h(x) = \sum_{y\in E} \lambda(x)\, p(x,y)\,[h(y) - h(x)] = Lh(x)$$

for $x \in G^c$. There is at most one function identically equal to 1 on G and satisfying the previous relation on G^c. Indeed, assume h_1 and h_2 share this property. Then $\bar{h} = h_1 - h_2$ vanishes on G and $L\bar{h} = \lambda\bar{h}$ on G^c. Multiplying this identity by $\bar{h}$ and integrating it with respect to π we obtain that

$$\lambda < \bar{h}, \bar{h} >_\pi = < \bar{h}, L\bar{h} >_\pi$$

because $\bar{h}$ vanishes on G. Since the left hand side is positive and the right hand side is negative in virtue of Proposition 10.1, both expressions vanish, so that $h_1 = h_2$.

Since we proved in the beginning that $L\varphi_\lambda = \lambda\varphi_\lambda$ on G^c, φ_λ is the unique function that minimizes the functional $J_\lambda(f)$ among all functions that are equal to 1 on G. This concludes the proof of the theorem. □

Appendix 2. The Equivalence of Ensembles, Large Deviation Tools and Weak Solutions of Quasi–Linear Differential Equations

In the first two sections of this chapter we prove a uniform local central limit expansion for exponential families of independent and identically distributed random variables. This expansion permits to prove the equivalence of ensembles: the marginals of the canonical measures $\nu_{N,K}$ converge to the marginals of the grand canonical measure ν_α as $N \uparrow \infty$ and $K/N^d \to \alpha$, uniformly on compact sets for the density α.

In section 2, in the context of generalized exclusion processes, we prove a second order expansion for the expectation of a cylinder function with respect to the canonical measure. Such an expansion is needed in the proof of a sharp estimate for the largest negative eigenvalue of the generator of a reversible generalized exclusion process restricted to a finite cube, to be proved in the next appendix.

In section 3 we prove two general results on large deviations that are used in the investigation of the large deviations of the empirical measure around its hydrodynamic limit.

Finally, in sections 4 and 5, we fix the terminology of weak solutions of quasi–linear parabolic and hyperbolic differential equations and state several results without proofs that are quoted throughout the book.

Note: Throughout this section, to keep notation simple, for a positive integer k, we denote by $k_\star$ the number $2k + 1$.

1. Local Central Limit Theorem and Equivalence of Ensembles

Let $p(\cdot)$ be a probability distribution on $\mathbb{N}$ and assume that $0 < p(0) < 1$. To fix ideas one may consider the one site marginal of a zero range distribution: $p(k) = Z(\varphi)^{-1}\{\varphi^k/g(k)!\}$ for some $\varphi > 0$. Define the partition function $Z : \mathbb{R}_+ \to \mathbb{R}_+$ associated to this distribution:

$$Z(\varphi) = \sum_{k \geq 0} \varphi^k p(k)$$

and denote by φ^* the radius of convergence of this series. We shall assume that $Z(\cdot)$ diverges at the boundary of its domain of definition:

$$\lim_{\varphi \to \varphi^*} Z(\varphi) = \infty . \tag{1.1}$$

Define the exponential family of distributions $\{p_\varphi(\cdot),\ \varphi \in [0, \varphi^*)\}$ by

$$p_\varphi(k) = \frac{1}{Z(\varphi)} \varphi^k p(k) \quad \text{for } k \text{ in } \mathbb{N}.$$

Denote by $R(\varphi)$ the mean of the distribution p_φ:

$$R(\varphi) = \sum_{k \geq 0} k p_\varphi(k) = \varphi\, \partial_\varphi \log Z(\varphi) .$$

A simple computation shows that $\log Z(e^\lambda)$ is a strictly convex function in λ and therefore that $R(\cdot)$ is strictly increasing. In particular, if $[0, \alpha^*)$ stands for the range of $R(\cdot)$, R is a bijection from $[0, \varphi^*)$ to $[0, \alpha^*)$.

Notice that α^* is equal to ∞ unless the probability distribution $p(\cdot)$ is finitely supported. Indeed, assume there exists k' such that $p(k) = 0$ for $k \geq k'$. Denote by κ the size of the support of $p(\cdot)$: $\kappa = \max\{k;\ p(k) > 0\}$. It is easy to show that in this case $\alpha^* = \kappa$. Otherwise, assumption (1.1) guarantees that $\alpha^* = \infty$. The proof presented in section 2.3 for zero range distributions applies without modifications to the present general context. Hence, for each α_0 in $[0, \alpha^*)$ there is a unique distribution p_{φ_0} with expectation α_0 and if we denote by $\Phi(\cdot)$ the inverse function of $R(\cdot)$: $\Phi = R^{-1}$,

$$\alpha = \sum_{k \geq 0} k\, p_{\Phi(\alpha)}(k) \quad \text{for each } \alpha \text{ in } [0, \alpha^*).$$

For each $0 \leq \varphi < \varphi^*$ and each finite subset Λ of $\mathbb{Z}^d$, denote by $\bar{\nu}_\varphi$ (resp. $\bar{\nu}_\varphi^\Lambda$) the product measure on $\mathbb{N}^{\mathbb{Z}^d}$ (resp. $\mathbb{N}^\Lambda$) with one site marginal equal to p_φ. Configurations of $\mathbb{N}^{\mathbb{Z}^d}$ or $\mathbb{N}^\Lambda$ are indifferently denoted by Greek letters η, ξ or ζ. For each positive integer K, let $\nu_{\Lambda,K}$ be the measure $\bar{\nu}_\varphi^\Lambda$ conditioned on the hyperplane $\Sigma_{\Lambda,K}^\kappa$ of all configurations of $\mathbb{N}^\Lambda$ with K particles and at most κ particles per site:

$$\begin{aligned} \Sigma_{\Lambda,K}^\kappa &= \Big\{\eta \in \mathbb{N}^\Lambda;\ \sum_{x \in \Lambda} \eta(x) = K \text{ and } \eta(y) \leq \kappa \text{ for } y \in \Lambda\Big\} , \\ \nu_{\Lambda,K}(\cdot) &= \bar{\nu}_\varphi^\Lambda\Big(\cdot \ \Big|\sum_{x \in \Lambda} \eta(x) = K\Big) . \end{aligned} \tag{1.2}$$

We shall often omit the superscript κ in $\Sigma_{\Lambda,K}^\kappa$. Notice that the right hand side of the second line does not depend on the parameter φ. The probability measures $\nu_{\Lambda,K}(\cdot)$ are called the canonical measures and the product measures $\bar{\nu}_\varphi^\Lambda$ are called the grand canonical measures. To keep notation simple, we abbreviate $\Sigma_{\Lambda_N,K}$ and $\nu_{\Lambda_N,K}$ by $\Sigma_{N,K}$ and $\nu_{N,K}$ and we denote the linear size of Λ_N (equal to $2N+1$) by $N_\star$. The purpose of this section is to show that the finite dimensional marginals of the canonical measures $\nu_{N,K}$ converge, as $N \uparrow \infty$ and $K/N_\star^d \to \alpha$, uniformly on compact sets of the parameter φ, to the finite dimensional marginals of the grand canonical measure with density α :

Theorem 1.1 *For every positive integer ℓ, every bounded function $f : \mathbb{N}^{\Lambda_\ell} \to \mathbb{R}$ and every $B > 0$,*

$$\lim_{N\to\infty} \sup_{0\le K\le BN_\star^d} \Big| E_{\nu_{N,K}}[f] \; - \; E_{\bar\nu_{\Phi(K/N_\star^d)}}[f] \Big| \; = \; 0 \, . \tag{1.3}$$

The proof of this result relies on a local central limit theorem for independent random variables with distribution p_φ. To prove such an expansion uniformly over compact sets in the parameter φ, a set of three assumptions is required. To state these hypotheses, for each positive integer k, denote respectively by $\mu_k(\varphi)$, $\omega_k(\varphi)$ and $\gamma_k(\varphi)$ the central moment, the absolute central moment and the cumulant of order k of the distribution p_φ:

$$\mu_k(\varphi) \; = \; E_{\bar\nu_\varphi}\Big[(\eta(0) - R(\varphi))^k\Big] \, , \qquad \omega_k(\varphi) \; = \; E_{\bar\nu_\varphi}\Big[\big|\eta(0) - R(\varphi)\big|^k\Big] \, ,$$

$$\gamma_k(\varphi) \; = \; k! \sum (-1)^{m_1+\cdots+m_k-1} (m_1 + \cdots + m_k - 1)! \prod_{j=1}^k \frac{1}{m_j!} \Big(\frac{\alpha_j}{j!}\Big)^{m_j} \, .$$

In this last formula summation is carried over all nonnegative integer solutions of $m_1 + \cdots + km_k = k$ and α_j stands for the j-th moment of $\eta(0)$: $\alpha_j = E_{\bar\nu_\varphi}[\eta(0)^j]$. We shall often write $\sigma(\varphi)^2$ for $\gamma_2(\varphi)$. Notice also that $\gamma_3(\varphi) = \mu_3(\varphi)$.

Denote by $v_\varphi(t)$ the normalized characteristic function associated to the distribution p_φ:

$$v_\varphi(t) \; = \; E_{\bar\nu_\varphi}\Big[\exp\Big\{\frac{it[\eta(0) - R(\varphi)]}{\sigma(\varphi)}\Big\}\Big] \, .$$

For each $0 \le \varphi_0 < \varphi_1 \le \varphi^*$ and k^* in $\mathbb{N}$ consider the following hypotheses:

(CL1) There exists a finite constant A_0 such that

$$\sup_{\varphi_0\le\varphi<\varphi_1} \omega_k(\varphi)/\sigma(\varphi)^k \; \le \; A_0$$

for all $2 \le k \le k^* + 1$.

(CL2) For every $\delta > 0$, there exists $C(\delta) < 1$ such that

$$\sup_{\varphi_0\le\varphi<\varphi_1} \; \sup_{\delta\le|t|\le\pi\sigma(\varphi)} \big|v_\varphi(t)\big| \le C(\delta) \, .$$

(CL3) There exists $\gamma > 0$ so that

$$\sup_{\varphi_0\le\varphi<\varphi_1} \int_{|t|\le\pi\sigma(\varphi)} \big|v_\varphi(t)\big|^\gamma \, dt \; \le \; C \; < \; \infty \, .$$

Remark 1.2 Assumptions (CL1)–(CL3) are verified on any compact subset of $(0, \varphi^*)$ for any positive integer k^*.

To prove this claim fix an interval $[\varphi_0, \varphi_1]$ contained in $(0, \varphi^*)$ and a positive integer k^*. The proof of Lemma 2.3.5 can easily be adapted to show that $\bar{\nu}_\varphi$ is an increasing family of probability measures. By assumption (1.1), $E_{\bar{\nu}_{\varphi_1}}[\exp\{\theta_0\eta(0)\}]$ is finite for some positive θ_0. Therefore, the expectation $E_{\bar{\nu}_\varphi}[\exp\{\theta_0\eta(0)\}]$ is finite for all $\varphi_0 \le \varphi \le \varphi_1$ and a fortiori $\omega_k(\varphi)$ is finite on $[\varphi_0, \varphi_1]$ for each $2 \le k \le k^*$. Since $\sigma(\varphi)$ is a smooth function strictly positive on $(0, \varphi)$, assumption (CL1) holds on $[\varphi_0, \varphi_1]$. On the other hand, $v_\varphi(t)$ is a continuous function of the pair (φ, t). Assumptions (CL2) and (CL3) follow therefore from the fact that $\sigma(\varphi)$ is a smooth strictly positive function on $(0, \varphi^*)$ and that $|v_\varphi(t\sigma(\varphi))| < 1$ for each fixed (t, φ) in $(0, \pi] \times (0, \varphi^*)$.

We may now state the local central limit theorem uniform over compact sets on the parameter φ. For each $0 \le \varphi < \varphi^*$, let $\{X_j, j \ge 1\}$ be a family of independent identically distributed random variables on a probability space $(\Omega, \mathcal{A}, P_\varphi)$ with common distribution p_φ. For $m \ge 0$, denote by $H_m(x)$ the Hermite polynomial of degree m:

$$H_m(x) = (-1)^m e^{x^2/2} \frac{d^m}{dx^m} e^{-x^2/2} = m! \sum_{k=0}^{[m/2]} (-1)^k \frac{x^{m-2k}}{k!(m-2k)!2^k} \cdot$$

Here $[m/2]$ stands for the integer part of $m/2$. Let $q_0(x)$ denote the density of the normalized Gaussian distribution and, for $j \ge 1$, let

$$q_j(x) = \frac{1}{\sqrt{2\pi}} e^{-x^2/2} \sum H_{j+2a}(x) \prod_{m=1}^{j} \frac{1}{k_m!} \left(\frac{\gamma_{m+2}(\varphi)}{(m+2)!\sigma(\varphi)^{m+2}} \right)^{k_m} ,$$

where the summation is carried out over all non negative integer solutions of $k_1 + 2k_2 + \cdots + jk_j = j$ and $k_1 + k_2 + \cdots + k_j = a$.

Theorem 1.3 *Assume hypotheses (CL1)–(CL3) for some* $0 \le \varphi_0 < \varphi_1 \le \varphi^*$ *and some* $k^* \ge 2$. *There exist constants* $E_0 = E_0(\varphi_0, \varphi_1, k^*)$ *and* $n_0 = n_0(\varphi_0, \varphi_1, k^*)$ *such that*

$$\sup_{\varphi_0 \le \varphi \le \varphi_1} \sup_{K \ge 0} \left| \sqrt{n\sigma(\varphi)^2} P_\varphi \Big[\sum_{i=1}^{n} X_i = K \Big] - \sum_{j=0}^{k^*-2} \frac{1}{n^{j/2}} q_j(x) \right| \le \frac{E_0}{n^{(k^*-1)/2}}$$

for all $n \ge n_0$. *In this formula* $x = (K - nR(\varphi))/\sigma(\varphi)\sqrt{n}$.

The proof of this theorem is omitted since it follows closely the classical arguments given for instance in Petrov (1975) (Theorem VII.12). There is only a slight problem to control the integral I_3 in the proof of this theorem but hypotheses (CL2) and (CL3) are built to estimate this integral.

Theorem 1.1 under assumptions (CL1)–(CL3) is easily deduced from this result.

Corollary 1.4 *Assume hypotheses (CL1)–(CL3) for $\varphi_0 < \varphi_1$ and $k^* = 4$. There exists a constant $C(E_0, A_0)$ depending only on A_0 and E_0, the constants appearing in assumption (CL1) and in the statement of Theorem 1.3, such that for every positive integer ℓ and every cylinder function $f : \mathbb{N}^{\Lambda_\ell} \to \mathbb{R}$ with finite second moments with respect to $\bar\nu_\varphi$, $\varphi_0 \le \varphi \le \varphi_1$,*

$$\Big| E_{\nu_{N,K}}[f] \;-\; E_{\bar\nu_{\Phi(K/N_\star^d)}}[f] \Big| \\ \le\; C(E_0, A_0)\Big(\frac{\ell_\star}{N_\star}\Big)^d \sqrt{E_{\bar\nu_{\Phi(K/N_\star^d)}}\Big[\Big(f - E_{\bar\nu_{\Phi(K/N_\star^d)}}[f]\Big)^2\Big]}$$

for all $N \ge 2\ell$ and all $R(\varphi_0) \le K/N_\star^d \le R(\varphi_1)$. In particular, for all bounded cylinder function f,

$$\limsup_{N\to\infty} \sup_{R(\varphi_0)\le K/N_\star^d \le R(\varphi_1)} N^d \Big| E_{\nu_{N,K}}[f] \;-\; E_{\bar\nu_{\Phi(K/N_\star^d)}}[f] \Big| \;<\; \infty\,.$$

Proof. The difference $|E_{\nu_{N,K}}[f] - E_{\bar\nu_{\Phi(K/N_\star^d)}}[f]|$ may be written as

$$\Big| \sum_{\xi\in\mathbb{N}^{\Lambda_\ell}} \bar\nu_\varphi^\ell(\xi)\Big(f(\xi) - E_{\bar\nu_\varphi}[f]\Big)\Big\{\frac{\bar\nu_\varphi^N\Big(\sum_{x\in\Lambda_N - \Lambda_\ell}\eta(x) = K - M(\xi)\Big)}{\bar\nu_\varphi^N\Big(\sum_{x\in\Lambda_N}\eta(x) = K\Big)} - 1\Big\}\Big| \tag{1.4}$$

where $M(\xi)$ stands for $\sum_{x\in\Lambda_\ell}\xi(x)$ and $\varphi = \Phi(K/N_\star^d)$. Theorem 1.3 and elementary estimates give that the absolute value of the expression inside braces is bounded above by

$$C(A_0, E_0)\Big(\frac{\ell_\star}{N_\star}\Big)^d \Big\{1 \;+\; \frac{|\bar\xi - R(\varphi)|}{\sigma(\varphi)} \;+\; \ell_\star^d\,\frac{\big(\bar\xi - R(\varphi)\big)^2}{\sigma(\varphi)^2}\Big\}$$

for all N large enough. In this formula, $\bar\xi$ stands for the density of particles in Λ_ℓ: $\bar\xi = |\Lambda_\ell|^{-1}\sum_{x\in\Lambda_\ell}\xi(x)$ and C is a constant depending only on A_0 and E_0, the constants appearing in hypothesis (CL1) and in Theorem 1.3. By Schwarz inequality and assumption (CL1), expression (1.4) is therefore bounded above by

$$C'(A_0, E_0)\Big(\frac{\ell_\star}{N_\star}\Big)^d \Big\{1 + \ell_\star^{-d/2}\Big\}\sqrt{E_{\bar\nu_\varphi}\Big[\Big(f - E_{\bar\nu_\varphi}[f]\Big)^2\Big]}\,. \qquad \square$$

To prove Theorem 1.1 it remains therefore to check assumptions (CL1)–(CL3) for $\varphi_0 = 0$, $\varphi_1 < \varphi^*$ and $k^* = 4$. We start with (CL1). By assumption (1.1), $\omega_k(\varphi)/\sigma(\varphi)^k$ is a smooth function on $(0, \varphi^*)$ and is therefore bounded on any compact subset of $(0, \varphi^*)$. However, for φ close to the origin, a straightforward expansion shows that

$$\mu_k(\varphi) \;=\; \frac{p(1)}{p(0)}\varphi \;+\; O(\varphi^2) \quad \text{for all } k \geq 2\,.$$

Assumption (CL1) fails therefore close to the origin because $\mu_k(\varphi)/\sigma(\varphi)^k = O(\varphi^{1-(k/2)})$ and we are forced to consider separately the case of small densities. The analysis relies on the following alternative version of the local central limit theorem.

Theorem 1.5 *For all* $0 < \varphi_0 < \varphi^*$ *and* $k_0 \geq 2$, *there exist finite constants* $E_1 = E_1(\varphi_0, k_0)$ *and* $E_2 = E_2(\varphi_0, k_0)$ *such that*

$$\sup_{K\geq 0} \left| \sqrt{n\sigma(\varphi)^2}\, P_\varphi\Big[\sum_{i=1}^n X_i = K\Big] \;-\; \sum_{j=0}^{k_0-2} \frac{1}{n^{j/2}} q_j(x) \right| \;\leq\; \frac{E_1}{(\sigma(\varphi)^2 n)^{(k_0-1)/2}}$$

for all $\varphi \leq \varphi_0$ *such that* $\sigma(\varphi)^2 n \geq E_2$. *Here again* $x = (K - nR(\varphi))/\sigma(\varphi)\sqrt{n}$.

We omit the proof of this theorem because it follows closely the classical proof (Theorem VII.12 of Petrov (1975)). There are only two modifications needed. Firstly, the integral I_1 must be redefined as the integral of $|f_n(t) - u_{k,n}(t)|$ over the region $|t| < (n\sigma(\varphi)^2)^\chi$ for some $\chi < 1/6$. Secondly, to estimate the integral I_3 we may proceed as follows. Denote by $\tilde v_\varphi(t)$ the characteristic function of $X_1 - R(\varphi)$ under P_φ: $\tilde v_\varphi(t) = E_\varphi[\exp\{it(X_1 - R(\varphi))\}]$. It is easy to see that

$$\left|\tilde v_\varphi(t)\right|^2 - 1 \;\leq\; P_\varphi[X_1 = 0]P_\varphi[X_1 = 1]\{\cos t - 1\} \;\leq\; C\varphi\{\cos t - 1\}$$

for some constant C that depends only on φ_0 and on the original distribution $p(\cdot)$ because

$$P_\varphi[X_1 = 0] \;=\; 1 + O(\varphi) \quad \text{and} \quad P_\varphi[X_1 = 1] \;=\; \frac{p(1)}{p(0)}\varphi \;+\; O(\varphi^2)$$

for φ close to the origin. These estimates permit to bound the integral I_3.

Recall that $\sigma(\varphi)^2 = p(1)p(0)^{-1}\varphi + O(\varphi^2)$ and $R(\varphi) = p(1)p(0)^{-1}\varphi + O(\varphi^2)$. Therefore, for all $\varphi_0 < \varphi^*$, there exists finite positive constants $C_1 = C_1(p, \varphi_0)$ and $C_2 = C_2(p, \varphi_0)$ such that

$$0 \;<\; C_1 \;\leq\; \frac{\sigma(\varphi)^2}{R(\varphi)} \;\leq\; C_2 \;<\infty \tag{1.5}$$

for $0 \leq \varphi \leq \varphi_0$. In particular, taking in Theorem 1.5 $\varphi = \Phi(K/n)$, we have that $n\sigma(\varphi)^2 \sim nR(\varphi) = K$. The remainder in Theorem 1.5 is thus of the order of K^{-1}, the inverse of the total number of particles.

With the same arguments presented in the proof of Corollary 1.4, we obtain the following estimate for the expectation of cylinder functions with respect to canonical measures.

Corollary 1.6 *Fix $0 < \varphi_0 < \varphi^*$, a positive integer ℓ and a cylinder function $f: \mathbb{N}^{\Lambda_\ell} \to \mathbb{R}$ with finite second moment with respect to $\bar\nu_\varphi$ for all $\varphi \le \varphi_0$. There exist finite constants $E_1 = E_1(\varphi_0)$ and $E_2 = E_2(\varphi_0)$ such that*

$$\begin{aligned} &\Big| E_{\nu_{N,K}}[f] - E_{\bar\nu_{\Phi(K/N_\star^d)}}[f] \Big| \\ &\quad \le E_1 \Big(\frac{\ell_\star}{N_\star}\Big)^d \Big\{ \frac{1}{\sigma(\varphi)^2} E_{\bar\nu_\varphi}\Big[\Big| f - E_{\bar\nu_\varphi}[f] \Big| \Big] \\ &\qquad + \frac{1}{\sigma(\varphi)} \sqrt{E_{\bar\nu_\varphi}\Big[\Big(f - E_{\bar\nu_\varphi}[f] \Big)^2 \Big]} \Big\} \end{aligned} \tag{1.6}$$

for all $N \ge 2\ell$ and all K such that $\Phi(K/N_\star^d) \le \varphi_0$, $\sigma(\Phi(K/N_\star^d))^2 N_\star^d \ge E_2$. On the right hand side of the inequality $\varphi = \Phi(K/N_\star^d)$.

Theorem 1.5 also provides a simple proof of the equivalence of ensembles.

Corollary 1.7 *Fix $0 < \varphi_0 < \varphi^*$, a positive integer ℓ and a cylinder function $f : \mathbb{N}^{\Lambda_\ell} \to \mathbb{R}$ with finite second moment:*

$$\sup_{0 \le \varphi \le \varphi_0} E_{\bar\nu_\varphi}\Big[|f|^2 \Big] < \infty .$$

There exists a finite constant $E_1 = E_1(\varphi_0, f)$ such that

$$\Big| E_{\nu_{N,K}}[f] - E_{\bar\nu_{\Phi(K/N_\star^d)}}[f] \Big| \le \frac{E_1}{N_\star^d}$$

for all large enough N and all $0 \le K/N_\star^d \le R(\varphi_0)$.

Proof. Fix $\varphi_0 > 0$ and a cylinder functions $f : \mathbb{N}^{\Lambda_\ell} \to \mathbb{R}$ with finite second moment. Denote by E_1 and E_2 the constants introduced in Corollary 1.6. To prove this lemma for N large enough and K so that $\sigma(\Phi(K/N_\star^d))^2 N_\star^d \ge E_2$, we just need to show that the expression inside braces in (1.6) is bounded. This expression is of course bounded for φ on any compact subset $(0, \varphi_0]$ because f has finite second moment and $\sigma(\varphi)^2$ is strictly positive. We just need therefore to investigate the behavior of this expression for φ close to the origin. Since by assumption $E_{\bar\nu_\varphi}[|f|^2]$ is finite, we have

$$E_{\bar\nu_\varphi}[f] = f(\underline{0}) + \sum_{x \in \Lambda_\ell} \Big[f(\mathfrak{d}_x) - f(\underline{0}) \Big] \frac{p(1)}{p(0)} \varphi + O_f(\varphi^2) ,$$

$$E_{\bar\nu_\varphi}\Big[\big(f - E_{\bar\nu_\varphi}[f]\big)^2 \Big] = \sum_{x \in \Lambda_\ell} \Big[f(\mathfrak{d}_x) - f(\underline{0}) \Big]^2 \frac{p(1)}{p(0)} \varphi + O_f(\varphi^2)$$

and

$$E_{\bar\nu_\varphi}\Big[\big|f - E_{\bar\nu_\varphi}[f]\big|\Big]$$
$$= \Big\{\Big|\sum_{x\in\Lambda_\ell}\big[f(\mathfrak{d}_x) - f(\underline{0})\big]\Big| + \sum_{x\in\Lambda_\ell}\Big|f(\mathfrak{d}_x) - f(\underline{0})\Big|\Big\}\frac{p(1)}{p(0)}\varphi + O_f(\varphi^2)\,.$$

In these formulas $\underline{0}$ stands for the configuration of $\mathbb{N}^{\Lambda_\ell}$ with no particles; for x in Λ_ℓ, $\mathfrak{d}_x$ stands for the configuration with no particles but one at site x and $O_f(\varphi^2)$ indicates a constant bounded in absolute value by $C(f)\varphi^2$. The derivation of these expansions is based on the following simple estimate:

$$\bar\nu_\varphi(\eta) \;\le\; \bar\nu_{\varphi_0}(\eta)\Big(\frac{Z(\varphi_0)}{p_0}\Big)^{|\Lambda_\ell|}\Big(\frac{\varphi}{\varphi_0}\Big)^{M(\eta)} \tag{1.7}$$

that holds for every $\varphi \le \varphi_0$ and configuration η of $\mathbb{N}^{\Lambda_\ell}$ provided $M(\eta) = \sum_{x\in\Lambda_\ell}\eta(x)$.

Replacing f by $\xi(0)$ we obtain that $\sigma(\varphi)^2 = \{p(1)/p(0\}\varphi + O(\varphi^2)$. A straightforward computation shows then that the right hand side of (1.6) is bounded above by $C(E_1, f)N_\star^{-d}$. To conclude the proof of the corollary it remains to consider the cases not covered by Corollary 1.6, i.e., densities $K/N_\star^d$ such that $0 \le \sigma(\Phi(K/N_\star^d))^2 N_\star^d \le E_2$.

Fix such value of K. By inequality (1.5) it follows that K is less than or equal to $C_1(\varphi_0, p)^{-1}E_2$. By estimate (1.7), the difference $|E_{\nu_{N,K}}[f] - E_{\bar\nu_{\Phi(K/N_\star^d)}}[f]|$ is bounded above by

$$\Big|f(\underline{0})\Big\{\nu_{N,K}\Big[\sum_{x\in\Lambda_\ell}\xi(x) = 0\Big] - \bar\nu_{\Phi(K/N_\star^d)}\Big[\sum_{x\in\Lambda_\ell}\xi(x) = 0\Big]\Big\}\Big|$$
$$+\; C(f,K)\nu_{N,K}\Big[\sum_{x\in\Lambda_\ell}\xi(x)\ge 1\Big] + C(f,\varphi_0,\ell)E_{\bar\nu_{\varphi_0}}\big[|f|\big]\Phi(K/N_\star^d)$$
$$\le\; 2C(f,K)\Big\{\nu_{N,K}\Big[\sum_{x\in\Lambda_\ell}\xi(x)\ge 1\Big] + \bar\nu_{\Phi(K/N_\star^d)}\Big[\sum_{x\in\Lambda_\ell}\xi(x)\ge 1\Big]\Big\}$$
$$+\; C(f,\varphi_0,K)N^{-d}$$

because $R(\varphi) = p(1)p(0)^{-1}\varphi + O(\varphi^2)$. In these formulas $C(f,K)$ stands for the maximum value of the cylinder function f over all configurations ξ of $\mathbb{N}^{\Lambda_\ell}$ with less than K particles: $C(f,K) = \max_{\xi;\sum_{x\in\Lambda_\ell}\xi(x)\le K}|f(\xi)|$. By Chebychev inequality this last expression is bounded above by $C(f,\varphi_0)N_\star^{-d}$ because $K \le C_1(\varphi_0, p)^{-1}E_2$. This concludes the proof of the corollary. □

We have seen in this section that the equivalence of ensembles, just proved for product measures or independent random variables, relies strongly on the local central limit theorem. This later result has also been proved for Gibbs measures. We refer to DelGrosso (1974), Dobrushin and Tirozzi (1977) and the references therein.

2. On the Local Central Limit Theorem

We proved in the previous section a local central limit theorem, uniform over the density, for exponential families of independent identically distributed random variables. This uniform local central limit theorem provided some estimates on the expectation of local functions with respect to canonical measures. Theorem 1.1 is a typical example of such an estimate. We prove in this section more refined estimates that are needed in the proof of a spectral gap for conservative dynamics.

We start with a second order expansion of the expectation of cylinder functions with respect to canonical measures. In order to deduce it we just need to recall from Theorem 1.5 the second order expansion of the local central limit theorem and the proof of Corollary 1.4.

Lemma 2.1 *Fix $\varphi_0 < \varphi^*$, a positive integer ℓ and a cylinder function $f : \mathbb{N}^{\Lambda_\ell} \to \mathbb{R}$. There exist finite constants $E_1 = E_1(\varphi_0)$ and $E_2 = E_2(\varphi_0)$ such that*

$$\Big| E_{\nu_{N,K}}[f] - E_{\bar\nu_{\Phi(K/N_\star^d)}}[f] - \Big(\frac{\ell_\star}{N_\star}\Big)^d \Big\{ \frac{\gamma_3(\varphi)}{\sigma(\varphi)^4} < f\,;\, \bar\xi >_{\bar\nu_\varphi} - \frac{\ell_\star^d}{\sigma(\varphi)^2} < f\,;\, \{\bar\xi - \alpha\}^2 >_{\bar\nu_\varphi} \Big\} \Big| \le \frac{E_1}{N^{3d/2}\sigma(\varphi)^2} < f\,;\, f >_{\bar\nu_\varphi}^{1/2}$$

for all $N \ge 2\ell$ and all K such that $\Phi(K/N_\star^d) \le \varphi_0$, $\sigma(\Phi(K/N_\star^d))^2 N_\star^d \ge E_2$. In this formula $< f\,;\, g >_{\bar\nu_\varphi}$ represents the covariance of f and g with respect to $\bar\nu_\varphi$, φ stands for $\Phi(K/N_\star^d)$ and $\bar\xi = \ell_\star^{-d} M(\xi)$.

In the remaining of this section we consider generalized exclusion processes introduced in section 2.4. Recall that in this model at most κ particles are allowed per site and that the grand canonical measures $\{\bar\nu_\varphi;\ \varphi \ge 0\}$ are given by

$$\bar\nu_\varphi\{\eta;\ \eta(0) = j\} = \frac{\varphi^j}{1 + \varphi + \cdots + \varphi^\kappa}$$

for $0 \le j \le \kappa$.

This model presents a special feature, known as the particles–holes duality, that simplifies some computations: for each $0 \le j \le \kappa$,

$$\bar\nu_\varphi\{\eta;\ \eta(0) = \kappa - j\} = \bar\nu_{1/\varphi}\{\eta;\ \eta(0) = j\}\,.$$

This identity can be interpreted as follows. Under $\bar\nu_\varphi$, the distribution of holes is equal to the distribution of particles under $\bar\nu_{1/\varphi}$. In particular, to prove any statement concerning the the distribution of ν_φ, it is enough to prove it for $\varphi \le 1$. This property is used below in the proof of Lemmas 2.2 and 2.3 and in the investigation of the spectral gap for generalized symmetric simple exclusion processes. This duality between holes and particles provides also some simple relations among the central moment of order k of different distributions that may be useful when computing expansions around 0 or ∞. It follows immediately from the previous identity that

$$R(\varphi) = \kappa - R(\varphi^{-1}) \quad \text{and} \quad \mu_j(\varphi) = (-1)^j \mu_j(\varphi^{-1})$$

for all $j \geq 2$.

Lemma 2.2 *Fix a positive integer ℓ. For all cylinder functions $f : \Sigma^{\kappa}_{\Lambda_\ell} \to \mathbb{R}$, there exists a constant $E_0 = E_0(f)$ such that*

$$\Big| E_{\nu_{N,K}}[f] - E_{\bar{\nu}_{\Phi(K/N^d_\star)}}[f] \Big| \leq \frac{E_0(f)}{N^d} \tag{2.1}$$

for all $N \geq \ell$ and all $0 \leq K \leq \kappa N^d_\star$.

Proof. Since there is at most a finite number of particles per site, all cylinder functions are bounded and therefore the assumptions of Corollary 1.7 are satisfied. In particular, inequality (2.1) is satisfied for N large enough and $0 \leq K \leq (2\kappa/3)N^d_\star$.

Consider now holes as particles and particles as holes, i.e., consider the probability measure $q(\cdot)$ on $\{0, \dots, \kappa\}$ defined by $q(k) = p(\kappa - k)$. Corollary 1.7 applied to $q(\cdot)$ states that inequality (2.1) is fulfilled for all N large enough and all $(\kappa/3)N^d_\star \leq K \leq \kappa N^d_\star$. This proves inequality (2.1) for all N large enough.

Since f is a cylinder function (and therefore bounded), to extend inequality (2.1) for small values of N, it is enough to modify, if necessary, the value of the constant $E_0(f)$. □

Lemma 2.3 *Denote by $h(\eta)$ the cylinder function $[1+\eta(0)]\mathbf{1}\{\eta(0) < \kappa\}$. For every $\varepsilon > 0$, there exists $N' = N'(\varepsilon)$ and $\varphi' = \varphi'(\varepsilon)$ such that*

$$\Big| E_{\nu_{N,K+1}}[h(\eta(0))] - E_{\nu_{N,K}}[h(\eta(0))] - \frac{1}{K+1} E_{\nu_{N,K+1}}\Big[\eta(0)\Big\{h(\eta(0)) - h(\eta(0)-1)\Big\}\Big] \Big| \leq \frac{\varepsilon}{N^d}$$

for all $N \geq N'$ and all $K/N^d_\star < R(\varphi')$.

Proof. Fix $\varepsilon > 0$. Here is the idea of the proof. In the case where $N^d\sigma(\varphi)^2$ is large, this statement follows from the second order expansion for the marginal probabilities of the canonical measures proved in Lemma 2.1. In the case where $N^d\sigma(\varphi)^2$ (and therefore K) is small we prove the result by inspection.

We start with the case where the total number of particles K is large. Take φ_0 such that $R(\varphi_0) = 1$, denote by E_1, E_2 the two constants given by Lemma 2.1 and by E_0 the constant introduced in Lemma 2.2

Let $K' = K'(\varepsilon)$ be defined as

$$K' = \max\Big\{C_1^{-1}E_1^2\varepsilon^{-2},\, C_1^{-1}E_2,\, E_0(\kappa+1)^2\varepsilon^{-1}\Big\},$$

where C_1 is the lower bound for $\sigma(\varphi)^2/R(\varphi)$ obtained in (1.5) and set $\tilde{\varphi} = \Phi(K/N_\star^d)$. Notice that for $K' \le K \le N_\star^d$ we have $\sigma(\tilde{\varphi})^2 N_\star^d \ge E_2$ because, by inequality (1.5), $\sigma(\tilde{\varphi})^2 N_\star^d = K\sigma(\tilde{\varphi})^2 R(\tilde{\varphi})^{-1} \ge KC_1$ and we chose $K' \ge C_1^{-1}E_2$. We are thus entitled to estimate the expectation $E_{\nu_{N,K}}[h(\eta(0))]$ with Lemma 2.1 for $K' \le K \le N_\star^d$:

$$
\begin{aligned}
E_{\nu_{N,K}}[h] &= E_{\bar{\nu}_{\tilde{\varphi}}}[h] \\
&+ \frac{1}{N_\star^d}\left\{\frac{\gamma_3(\tilde{\varphi})}{\sigma(\tilde{\varphi})^4}E_{\bar{\nu}_{\tilde{\varphi}}}\Big[h;\eta(0)\Big] - \frac{1}{\sigma(\tilde{\varphi})^2}E_{\bar{\nu}_{\tilde{\varphi}}}\Big[h\,;\,\{\eta(0)-\alpha\}^2\Big]\right\} \\
&\pm \frac{E_1}{N_\star^{3d/2}\sigma(\tilde{\varphi})^2}E_{\bar{\nu}_{\tilde{\varphi}}}[h\,;\,h]^{1/2}\,.
\end{aligned}
\tag{2.2}
$$

We claim that the expression inside braces in this formula vanishes as $\tilde{\varphi}\downarrow 0$. Indeed, performing a change of variables $\xi = \eta + \mathfrak{d}_0$, we obtain that

$$
E_{\bar{\nu}_{\tilde{\varphi}}}\Big[h(\eta(0))f(\eta)\Big] = \frac{1}{\tilde{\varphi}}E_{\bar{\nu}_{\tilde{\varphi}}}\Big[\eta(0)f(\eta-\mathfrak{d}_0)\Big]
$$

for every cylinder function f. Let $\alpha = R(\tilde{\varphi})$. Applying this identity to the cylinder functions 1, $\eta(0)-\alpha$ and $(\eta(0)-\alpha)^2$ we obtain that the expression inside braces in (2.2) is equal to

$$
\begin{aligned}
&\frac{\gamma_3}{\tilde{\varphi}\sigma^4}E_{\bar{\nu}_{\tilde{\varphi}}}\Big[\eta(0)\{\eta(0)-1-\alpha\}\Big] - \frac{1}{\tilde{\varphi}\sigma^2}E_{\bar{\nu}_{\tilde{\varphi}}}\Big[\eta(0)\{\eta(0)-1-\alpha\}^2\Big] \\
&\quad + \frac{1}{\tilde{\varphi}}E_{\bar{\nu}_{\tilde{\varphi}}}[\eta(0)] = \frac{1}{\tilde{\varphi}\sigma^4}\Big\{2\sigma^4 - \gamma_3\alpha - \sigma^2\alpha\Big\}\,.
\end{aligned}
$$

Expanding the functions $R(\varphi)$, $\sigma(\varphi)^2$ and $\gamma_3(\varphi)$ around their value at the origin we obtain that

$$
\begin{aligned}
R(\varphi) = \varphi + \varphi^2 + O(\varphi^3)\,,\quad \sigma(\varphi)^2 = \varphi + 2\varphi^2 + O(\varphi^3) \\
\text{and}\quad \gamma_3(\varphi) = \varphi + 4\varphi^2 + O(\varphi^3)\,.
\end{aligned}
$$

It is easy easy to prove from these estimates that $\varphi^{-1}\sigma(\varphi)^{-4}\{2\sigma^4 - \gamma_3\alpha - \sigma^2\alpha\}$ is of order $O(\varphi)$ for φ close to the origin.

This proves that the expression inside braces in formula (2.2) is of order $O(\varphi)$ for φ close to the origin. A similar argument proves that $E_{\bar{\nu}_\varphi}[h;h]/\sigma(\varphi)^2$ is bounded on compact intervals of $\mathbb{R}_+$. Therefore, for all $N \ge 1$, $K' \le K \le N_\star^d$

$$
E_{\nu_{N,K}}[h] = E_{\bar{\nu}_{\tilde{\varphi}}}[h] + \frac{1}{N^d}\left\{O(\tilde{\varphi}) \pm \sqrt{\frac{R(\tilde{\varphi})E_1^2}{\sigma(\tilde{\varphi})^2 K}}\right\}
$$

because $K/N_\star^d = R(\tilde{\varphi})$. By inequality (1.5) and since $K \ge K' \ge C_1^{-1}E_1^2\varepsilon^{-2}$ the absolute value of the remainder is bounded by $N^{-d}\{O(\tilde{\varphi})+\varepsilon\}$. From this estimate it follows that

$$
E_{\nu_{N,K+1}}[h] - E_{\nu_{N,K}}[h] = \tilde{h}\Big(\frac{K+1}{N_\star^d}\Big) - \tilde{h}\Big(\frac{K}{N_\star^d}\Big) + R_{\varepsilon,N,\tilde{\varphi}} \tag{2.3}
$$

provided $K' \le K \le N_\star^d$. In this formula $\tilde h(\alpha)$ stands for the expectation of h with respect to the grand canonical measure with density α: $\tilde h(\alpha) = E_{\bar\nu_{\Phi(\alpha)}}[h]$ and $R_{\varepsilon,N,\tilde\varphi}$ is a remainder, bounded in absolute value by $N^{-d}\{O(\tilde\varphi)+\varepsilon\}$.

We estimate now $E_{\nu_{N,K}}[\eta(0)\{h(\eta(0)) - h(\eta(0)-1)\}]$. This cylinder function may be rewritten as $\eta(0) - \kappa(\kappa+1)\mathbf{1}\{\eta(0)=\kappa\}$. By Lemma 2.2,

$$E_{\nu_{N,K}}\Big[\mathbf{1}\{\eta(0)=\kappa\}\Big] \;=\; E_{\bar\nu_{\Phi(K/N_\star^d)}}\Big[\mathbf{1}\{\eta(0)=\kappa\}\Big] \;\pm\; \frac{E_0}{N^d}$$

for all $N \ge 1$ and $0 \le K \le \kappa N_\star^d$. Notice that the expectation on the right hand side is of order $O(\tilde\varphi^\kappa)$. Therefore,

$$\begin{aligned}&\frac{1}{K+1} E_{\nu_{N,K+1}}\Big[\eta(0)\{h(\eta(0)) - h(\eta(0)-1)\}\Big]\\ &\qquad = \frac{1}{N_\star^d} + \frac{C(\kappa)O(\tilde\varphi^\kappa)}{N_\star^d R(\tilde\varphi)} \pm \frac{E_0(\kappa+1)^2}{N^d(K+1)}\,.\end{aligned}$$

Since $\kappa \ge 2$, $R(\varphi) = \varphi + O(\varphi^2)$ and $K' \le K \le N_\star^d$, the last two terms are bounded in absolute value by $N^{-d}\{O(\tilde\varphi)+\varepsilon\}$ because by definition $K \ge K' \ge (\kappa+1)^2 E_0 \varepsilon^{-1}$.

In view of this estimate and (2.3), to conclude the proof of the lemma in the case $K' \le K \le N_\star^d$, it remains to show that

$$\tilde h\Big(\frac{K+1}{N_\star^d}\Big) - \tilde h\Big(\frac{K}{N_\star^d}\Big) - \frac{1}{N_\star^d}$$

is bounded by ε/N^d for $\tilde\varphi$ small enough. Denote by $\tilde h'(\alpha)$ the derivative of $\tilde h$ with respect to the density α. A straightforward computation shows that $\tilde h'(\alpha) = 1+O(\alpha)$. In particular, by Taylor expansion, $\tilde h((K+1)/N_\star^d) - \tilde h(K/N_\star^d) - 1/N_\star^d$ is bounded above by $O(\tilde\varphi)N_\star^{-d}$. This concludes the proof of the lemma in the case $K' \le K \le N_\star^d$.

It remains to consider the case $K \le K'$. For a finite subset Λ of $\mathbb{Z}^d$, denote by $Z_\kappa(\Lambda, K)$ the total number of configurations of $\{0,\dots,\kappa\}^\Lambda$ with K particles. We shall expand this expression in Λ since the total number of particles K is bounded by K'. The main contribution to $Z_\kappa(\Lambda,K)$ is the set of all configuration with no site occupied by more than one particle. There are $O(|\Lambda|^K)$ of such configurations. The second main contribution to $Z_\kappa(\Lambda,K)$ is the set of all configurations with one site occupied by two particles and all other sites occupied by at most one particle. There are $O(|\Lambda|^{K-1})$ of such configurations. We have therefore that

$$\begin{aligned}Z_\kappa(\Lambda,K) &= \binom{|\Lambda|}{K} + |\Lambda|\binom{|\Lambda|-1}{K-2} + C(K)O(|\Lambda|^{K-2})\\ &= \frac{|\Lambda|!}{K!(|\Lambda|-K)!}\Big\{1 + \frac{K(K-1)}{|\Lambda|-K+1} \pm \frac{C(K)}{|\Lambda|^2}\Big\}\,.\end{aligned}$$

Since the canonical measure $\nu_{N,K}$ is the uniform measure on $\Sigma_{N,K}^\kappa$, $\nu_{N,K}\{\eta;\ \eta(0)=a\}$ is equal to the proportion of configurations with a particles at the origin:

$$\nu_{N,K}\{\eta; \eta(0) = a\} = \frac{Z_\kappa(\Lambda_N - \{0\}, K - a)}{Z_\kappa(\Lambda_N, K)} . \tag{2.4}$$

This identity and the explicit formula for $Z_\kappa(\Lambda, K)$ shows that the probability $\nu_{N,K}\{\eta; \eta(0) \geq 2\}$ is bounded by $C(K)N_\star^{-2d}$. In particular, since h is bounded, to prove the lemma in the case $K \leq K'$, we have only to show that

$$\begin{aligned} h(0)\Big[\nu_{N,K+1}\{\eta; \eta(0) = 0\} &- \nu_{N,K}\{\eta; \eta(0) = 0\}\Big] \\ + h(1)\Big[\nu_{N,K+1}\{\eta; \eta(0) = 1\} &- \nu_{N,K}\{\eta; \eta(0) = 1\}\Big] \\ + \frac{1}{K+1}[h(0) - h(1)]\nu_{N,K+1}&\{\eta; \eta(0) = 1\} \end{aligned}$$

is bounded by εN^{-d} for all N large enough. Since $\nu_{N,K}\{\eta; \eta(0) = 0\}$ is equal to $1 - \nu_{N,K}\{\eta; \eta(0) \geq 1\}$ and since $h(1) - h(0) = 1$, it is enough to prove that

$$\frac{K}{K+1}\nu_{N,K+1}\{\eta; \eta(0) = 1\} - \nu_{N,K}\{\eta; \eta(0) = 1\} \tag{2.5}$$

is bounded by εN^{-d} for all N large enough. From the explicit formula for $Z_\kappa(\Lambda, K)$, we obtain that

$$\nu_{N,K}\{\eta; \eta(0) = 1\} = \frac{K}{L}\Big\{1 - \frac{2(K-1)}{N_\star^d + 1 - K} \pm C(K)N_\star^{-2d}\Big\} .$$

This identity shows that the difference (2.5) is equal to $-2K/(N_\star^d - K)(N_\star^d + 1 - K) \pm C(K)N_\star^{-2d}$, what concludes the proof of the lemma. □

Lemma 2.4 *There exist a universal constant $B_2 = B_2(\kappa)$ so that*

$$\sup \frac{\nu_{N,K}\{\eta; \eta(0) = a\}}{\nu_{N,K}\{\eta; \eta(0) = b\}} \leq B_2$$

for all $a \geq b$. In this formula the supremum is taken over all N and K such that that $K/N_\star^d \leq \kappa/2$.

Proof. Fix $\varphi_0 > 0$ so that $R(\varphi_0) = \kappa/2$. Denote by E_1 and E_2 the constants introduced in the statement of Theorem 1.5 and by C_1 the lower bound of $\inf_{0 \leq \alpha \leq \kappa/2} \sigma(\alpha)^2/\alpha$ provided by inequality (1.5). Set $K' = E_2 C_1^{-1}$. Notice that for $K' \leq K \leq (\kappa/2)N_\star^d$,

$$\sigma(\Phi(K/N_\star^d))^2 N_\star^d = \sigma(\Phi(K/N_\star^d))^2 K R(\Phi(K/N_\star^d))^{-1} \geq C_1 K \geq E_2$$

by definition of K'. We are thus entitled to apply Theorem 1.5.

By definition of the canonical measure,

$$\frac{\nu_{N,K}\{\eta; \eta(0) = a\}}{\nu_{N,K}\{\eta; \eta(0) = b\}} = \varphi^{a-b} \frac{\sqrt{2\pi(|\Lambda_N| - 1)}\bar{\nu}_\varphi\Big\{\sum_{x \in \Lambda_N - \{0\}} \xi(x) = K - a\Big\}}{\sqrt{2\pi(|\Lambda_N| - 1)}\bar{\nu}_\varphi\Big\{\sum_{x \in \Lambda_N - \{0\}} \xi(x) = K - b\Big\}} ,$$

where $\varphi = \Phi(K/N_\star^d)$. Theorem 1.5 and straightforward computations show that the last ratio is bounded by some constant C depending only on κ for all N large enough and all $K' \le K \le (\kappa/2)N_\star^d$.

It remains to examine the asymptotic behavior of $\nu_{N,K}\{\eta; \eta(0) = a\}/\nu_{N,K}\{\eta; \eta(0) = b\}$ for $K \le K'$. Recall the definition of $Z_\kappa(\Lambda, K)$ given in the proof of Lemma 2.3. By formula (2.4), we have

$$\frac{\nu_{N,K}\{\eta; \eta(0) = a\}}{\nu_{N,K}\{\eta; \eta(0) = b\}} = \frac{Z_\kappa(\Lambda_N - \{0\}, K - a)}{Z_\kappa(\Lambda_N - \{0\}, K - b)} = C(K) O(|\Lambda_N|^{b-a}) .$$

This concludes the proof of the lemma. □

3. Remarks on Large Deviations

We prove in this section two general results on large deviations needed in Chapter 10. The first result is due to Varadhan (1966) and the second one to Donsker and Varadhan (1975c).

Consider a sequence of random variables $\{X_N, N \ge 1\}$ taking values in some metric space $\mathcal{E}$. We say that the sequence $\{X_N, N \ge 1\}$ satisfies a large deviation principle with rate function I and decay rate a_N if for every closed set $\mathcal{F}$ of $\mathcal{E}$

$$\limsup_{N\to\infty} \frac{1}{a_N} \log P[X_N \in \mathcal{F}] \le - \inf_{u\in\mathcal{F}} I(u)$$

and for every open set $\mathcal{G}$ of $\mathcal{E}$

$$\liminf_{N\to\infty} \frac{1}{a_N} \log P[X_N \in \mathcal{G}] \ge - \inf_{u\in\mathcal{G}} I(u)$$

for some positive function $I: \mathcal{E} \to \mathbb{R}_+$ and some increasing sequence $a_N \uparrow \infty$.

Notice that we didn't require the rate function to be lower semicontinuous or the levels sets of I to be compact. The reason is that these assumptions are not needed in the next theorem.

Theorem 3.1 (Laplace–Varadhan) *Let $\{X_N, N \ge 1\}$ be a sequence of random variables satisfying a large deviation principle with rate function I and decay rate a_N. Let $F: \mathcal{E} \to \mathbb{R}$ be a bounded continuous function. Then,*

$$\lim_{N\to\infty} \frac{1}{a_N} \log E\left[e^{a_N F(X_N)}\right] = \sup_{u\in\mathcal{E}} \{F(u) - I(u)\} . \tag{3.1}$$

Proof. We start with the lower bound which is easier. Fix $\varepsilon > 0$. There exists u_0 in $\mathcal{E}$ such that $\sup_{u\in\mathcal{E}}\{F(u) - I(u)\} \le F(u_0) - I(u_0) + \varepsilon$. Fix a neighborhood $\mathcal{V}$ of u_0 such that $F(u) \ge F(u_0) - \varepsilon$ for all u in $\mathcal{V}$. In particular,

$$\frac{1}{a_N} \log E\left[e^{a_N F(X_N)}\right] \geq \frac{1}{a_N} \log E\left[e^{a_N F(X_N)} \mathbf{1}\{X_N \in \mathcal{V}\}\right]$$
$$\geq F(u_0) - \varepsilon + \frac{1}{a_N} \log P\left[X_N \in \mathcal{V}\right] .$$

By the lower bound of the large deviation principle and since u_0 belongs to $\mathcal{V}$, the limit, as $N \uparrow \infty$, of the previous expression is bounded below by

$$F(u_0) - \varepsilon - \inf_{u \in \mathcal{V}} I(u) \geq F(u_0) - I(u_0) - \varepsilon .$$

Since $\sup_{u \in \mathcal{E}}\{F(u) - I(u)\} \leq F(u_0) - I(u_0) + \varepsilon$, we proved that the left hand side of (3.1) is bounded below by the right hand side.

We now turn to the upper bound. Let $B = \|F\|_\infty$. Fix a positive integer K and divide the interval $[-B, B]$ in K contiguous closed intervals of length $2B/K$. Denote these intervals by $I_j = [r_j, r_{j+1}]$, $1 \leq j \leq K$, and denote by $\mathcal{F}_j$ the pre-image by F of I_j: $\mathcal{F}_j = \{u \in \mathcal{E};\ F(u) \in I_j\}$. The subsets $\mathcal{F}_j$ are closed because F is continuous. On the other hand, by definition,

$$E\left[e^{a_N F(X_N)}\right] \leq \sum_{j=1}^{K} e^{r_{j+1} a_N} P\left[X_N \in \mathcal{F}_j\right] .$$

Since $a_N \uparrow \infty$,

$$\limsup_{N\to\infty} a_N^{-1} \log\{d_N + b_N\} = \max\left\{ \limsup_{N\to\infty} a_N^{-1} \log d_N , \limsup_{N\to\infty} a_N^{-1} \log b_N \right\} . \tag{3.2}$$

By the large deviations principle, the left hand side of (3.1) is bounded above by

$$\max_{1\leq j\leq K} \left\{ r_{j+1} + \sup_{u \in \mathcal{F}_j} -I(u) \right\}$$
$$\leq \max_{1\leq j\leq K} \left\{ r_{j+1} + \sup_{u \in \mathcal{F}_j} \{F(u) - I(u)\} + \sup_{u \in \mathcal{F}_j} -F(u) \right\} .$$

Since on $\mathcal{F}_j$ F takes value in the interval $[r_j, r_{j+1}]$ and $r_{j+1} - r_j = 2B/K$, this last expression is bounded above by $\sup_{u\in\mathcal{E}}\{F(u) - I(u)\} + 2B/K$. This shows that the left hand side of (3.1) is bounded above by the right hand side. □

Lemma 3.2 (Minimax lemma) *Let $\mathcal{K}$ be a compact set of a polish space $\mathcal{E}$, let $\mathfrak{A}$ be some parameter set and let $\{J_\beta : \mathcal{E} \to \mathbb{R};\ \beta \in \mathfrak{A}\}$ be a family of upper semi–continuous functions. Then,*

$$\inf_{\mathcal{O}_1,\dots,\mathcal{O}_n} \max_{1\leq j\leq n} \inf_{\beta\in\mathfrak{A}} \sup_{\nu\in\mathcal{O}_j} J_\beta(\nu) \leq \sup_{\mu\in\mathcal{K}} \inf_{\beta\in\mathfrak{A}} J_\beta(\mu) .$$

In the left hand side, the first infimum is carried over all finite open covers $\{\mathcal{O}_1, \dots, \mathcal{O}_n\}$ of $\mathcal{K}$.

Proof. Denote the right hand side by R and fix $\epsilon > 0$. Since for each μ in $\mathcal{K}$, $\inf_{\beta \in \mathfrak{A}} J_\beta(\mu) \leq R$, there exists $\beta(\mu)$ such that $J_{\beta(\mu)}(\mu) \leq R + \epsilon$. Moreover, since each function J_β is assumed to be upper semi–continuous, there exists a neighborhood $\mathcal{O}_\mu$ of μ so that

$$\sup_{\nu \in \mathcal{O}_\mu} J_{\beta(\mu)}(\nu) \leq R + 2\epsilon .$$

The family $\{\mathcal{O}_\mu, \ \mu \in \mathcal{K}\}$ constitutes an open cover of the compact set $\mathcal{K}$ and has a finite subcover that we denote $\mathcal{O}_{\mu_1}, \dots, \mathcal{O}_{\mu_n}$. Since

$$\sup_{\nu \in \mathcal{O}_{\mu_j}} J_{\beta(\mu_j)}(\nu) \leq R + 2\epsilon \quad \text{for} \quad 1 \leq j \leq n ,$$

we have that

$$\inf_{\beta \in \mathfrak{A}} \sup_{\nu \in \mathcal{O}_{\mu_j}} J_\beta(\nu) \leq R + 2\epsilon \quad \text{for} \quad 1 \leq j \leq n .$$

It is now easy to conclude the proof of the lemma. □

This result must be understood as follows. Under very general conditions since we do not impose any restriction on the index set $\mathfrak{A}$ and require only the upper semi–continuity of each function J_β, it allows the replacement of inf sup by sup inf. This exchange constitutes one of the main technical difficulties in the proof of the large deviations upper bound as can be seen in Chapter 10. The next result illustrates how to obtain a large deviations upper bound for compact sets from the minimax lemma and some upper bounds for open sets.

Lemma 3.3 *Consider a polish space $\mathcal{E}$ and a sequence of probabilities P_N on $\mathcal{E}$. Let $\{J_\beta : \mathcal{E} \to \mathbb{R}, \ \beta \in \mathfrak{A}\}$ be a family of upper–semi continuous functions indexed by some set $\mathfrak{A}$. Assume that we are able to prove upper bounds for every open subset $\mathcal{O}$ of $\mathcal{E}$:*

$$\limsup_{N \to \infty} \frac{1}{a_N} \log P_N[\mathcal{O}] \ \leq \ \inf_{\beta \in \mathfrak{A}} \sup_{\mu \in \mathcal{O}} J_\beta(\mu) . \tag{3.3}$$

Then, for every compact set $\mathcal{K}$,

$$\limsup_{N \to \infty} \frac{1}{a_N} \log P_N[\mathcal{K}] \ \leq \ \sup_{\mu \in \mathcal{K}} \inf_{\beta \in \mathfrak{A}} J_\beta(\mu) .$$

Proof. Let $\mathcal{O}_1, \dots, \mathcal{O}_n$ denote a finite open cover of $\mathcal{K}$. By remark (3.2) and assumption (3.3),

$$\limsup_{N \to \infty} \frac{1}{a_N} \log P_N[\mathcal{K}] \ \leq \ \max_{1 \leq j \leq n} \inf_{\beta \in \mathfrak{A}} \sup_{\mu \in \mathcal{O}_j} J_\beta(\mu) .$$

To conclude the proof it remains to minimize over all finite open covers of $\mathcal{K}$ and recall the statement of the minimax lemma. □

4. Weak Solutions of Nonlinear Parabolic Equations

We briefly fix in this section the terminology of weak solutions of non linear parabolic equations and we present the main existence, uniqueness and regularity results of such equations. The reader is referred to Ladyženskaja et al. (1968) for a complete treatment of the question. Hereafter Φ is a smooth strictly increasing function such that $\|\Phi'\|_\infty \le g^* < \infty$ and $\sigma = \{\sigma_{i,j},\ 1 \le i, j \le d\}$ is a symmetric positive definite matrix: $\sigma_{i,j} = \sigma_{j,i}$ and there exists $\delta > 0$ such that $v^* \sigma v \ge \delta |v|^2$ for all v in $\mathbb{R}^d$.

We start considering the problem of the existence and uniqueness of classical solutions.

Theorem 4.1 *Fix $\varepsilon > 0$. For each initial profile $\rho_0 \colon \mathbb{T}^d \to \mathbb{R}$ of class $C^{2+\varepsilon}(\mathbb{T}^d)$, there exists a unique classical solution of class $C^{1+\varepsilon, 2+\varepsilon}(\mathbb{R}_+ \times \mathbb{T}^d)$ of the Cauchy problem*

$$\begin{cases} \partial_t \rho = \displaystyle\sum_{1 \le i,j \le d} \sigma_{i,j} \partial^2_{u_i, u_j} \Phi(\rho) \\ \rho(0, \cdot) = \rho_0(\cdot) \, . \end{cases} \tag{4.1}$$

Moreover, a maximal principle holds:

$$\inf_{u \in \mathbb{T}^d} \rho_0(u) \le \inf_{(t,u) \in \mathbb{R}_+ \times \mathbb{T}^d} \rho(t,u) \le \sup_{(t,u) \in \mathbb{R}_+ \times \mathbb{T}^d} \rho(t,u) \le \sup_{u \in \mathbb{T}^d} \rho_0(u) \, . \tag{4.2}$$

We now turn to weak solutions

Definition 4.2 Fix a bounded initial profile $\rho_0 \colon \mathbb{T}^d \to \mathbb{R}$. A measurable function $\rho \colon \mathbb{R}_+ \times \mathbb{T}^d \to \mathbb{R}$ is a weak solution of the Cauchy problem (4.1) if for every function $G \colon \mathbb{R}_+ \times \mathbb{T}^d \to \mathbb{R}$ of class $C_K^{1,2}(\mathbb{R}_+ \times \mathbb{T}^d)$

$$\int_0^\infty dt \int_{\mathbb{T}^d} du \Big\{ \rho(t,u) \, \partial_t G + \Phi(\rho(t,u)) \sum_{1 \le i,j \le d} \sigma_{i,j} \partial^2_{u_i, u_j} G \Big\} + \int_{\mathbb{T}^d} du \, G(0,u) \rho_0(u) = 0 \, .$$

The following a priori estimate (cf. Oleinik and Kružkov (1961)) on bounded weak solutions of quasi–linear parabolic equations due to Nash plays a central role in the investigation of the existence and uniqueness of weak solutions. It states that bounded weak solutions are uniformly Hölder continuous on each compact subset of $(0, \infty) \times \mathbb{T}^d$.

Theorem 4.3 *Fix a bounded profile ρ_0 and a bounded weak solution $\rho(t,u)$ of (4.1). There exist constants a and A depending only on the dimension d, on σ and on Φ such that for every $0 < s \le t$,*

$$|\rho(t,u) - \rho(s,v)| \;\le\; A\,\|\rho\|_\infty \Big\{ \Big(\frac{|u-v|}{\sqrt{s}}\Big)^a + \Big(\frac{t-s}{\sqrt{s}}\Big)^{a/2(1+a)} \Big\}\,.$$

It follows from this estimate and Theorem 4.1 that there exists a bounded weak solution of the Cauchy problem (4.1) for bounded initial profiles ρ_0. In fact, it is not difficult to prove the existence and uniqueness of weak solutions in the class of measurable functions in $L^2([0,T]\times\mathbb{T}^d)$.

Theorem 4.4 *Fix a bounded profile ρ_0. There exists a unique weak solution of the quasi–linear parabolic equation (4.1) that belongs to $L^2([0,T]\times\mathbb{T}^d)$. Moreover, the solution is uniformly Hölder continuous on each compact subset of $(0,\infty)\times\mathbb{T}^d$ and satisfies the maximum principle (4.2).*

Proof. Fix a bounded profile ρ_0 and consider a sequence of smooth profiles $\{\rho_0^\varepsilon,\ \varepsilon>0\}$ bounded by $\|\rho_0\|_\infty$ and converging weakly to ρ_0 in the sense that

$$\lim_{\varepsilon\to 0}\int du\, H(u)\rho_0^\varepsilon(u) \;=\; \int du\, H(u)\rho_0(u) \tag{4.3}$$

for all continuous function $H\colon\mathbb{T}^d\to\mathbb{R}$. For each $\varepsilon>0$, denote by $\rho^\varepsilon(t,u)$ the unique classical solution of equation (4.1) with initial data ρ_0^ε. By the maximum principle (4.2) the sequence ρ^ε is uniformly bounded. On the other hand, by Theorem 4.3, on each compact subset of $(0,\infty)\times\mathbb{T}^d$, the sequence $\{\rho^\varepsilon,\ \varepsilon>0\}$ is uniformly Hölder continuous. The sequence is therefore relatively compact (for the uniform topology) on each compact set of $(0,\infty)\times\mathbb{T}^d$ and we may obtain a subsequence ρ^{ε_k} that converges uniformly on each compact subset of $(0,\infty)\times\mathbb{T}^d$ to a bounded function ρ. It is very easy to show that ρ is a weak solution of (4.1), what proves the existence of a bounded weak solution. It satisfies, moreover, the maximum principle (4.2) and is uniformly Hölder continuous on each compact subset of $(0,\infty)\times\mathbb{T}^d$.

To prove uniqueness in the class $L^2([0,T]\times\mathbb{T}^d)$, we need to introduce some notation. For each z in $\mathbb{Z}^d$, denote by $\psi_z\colon\mathbb{T}^d\to\mathbb{C}$ the $L^2(\mathbb{T}^d)$ function defined by $\psi(u)=\exp\{(2\pi i)z\cdot u\}$. Here $z\cdot u$ stands for the inner product in $\mathbb{R}^d$. It is well known that $\{\psi_z,\ z\in\mathbb{Z}^d\}$ forms an orthonormal basis of $L^2(\mathbb{T}^d)$. In particular, any function f in $L^2(\mathbb{T}^d)$ can be written as

$$f \;=\; \sum_{z\in\mathbb{Z}^d} <\psi_z, f>\psi_z\ ,$$

where $<\cdot,\cdot>$ stands for the inner product in $L^2(\mathbb{T}^d)$. Since $\{\psi_z,\ z\in\mathbb{Z}^d\}$ is an orthonormal basis, for f, g in $L^2(\mathbb{T}^d)$,

$$\int_{\mathbb{T}^d} du\, f(u)g(u) \;=\; \sum_{z\in\mathbb{Z}^d} <\psi_z, f><\psi_z, g>\ . \tag{4.4}$$

Moreover, an integration by parts shows that

$$< \psi_z, \partial_{u_j} f > = -2\pi i z_j < \psi_z, f >$$

for every function f in $C^1(\mathbb{T}^d)$ and every $z \in \mathbb{Z}^d$, $1 \le j \le d$. Finally, for f, g in $L^2(\mathbb{T}^d)$, denote by $f * g$ the convolution of f and g:

$$(f * g)(u) = \int_{\mathbb{T}^d} dv\, f(v) g(u - v) . \tag{4.5}$$

It is easy to deduce from this definition that

$$< \psi_z, f * g > = < \psi_z, f >< \psi_z, g > . \tag{4.6}$$

To keep notation simple, assume that σ is the identity matrix. Fix a positive integer M, $a > d$ and define the function $F_M: \mathbb{T}^d \to \mathbb{C}$ by the series

$$F_M(u) = \sum_{z \in \mathbb{Z}^d} \frac{M}{(1 + |z|^2)(M + |z|^a)} \psi_z(u) .$$

F_M is a well defined twice continuously differentiable real function because $a > d$.

Consider two weak solutions ρ^1, ρ^2 of (4.1) such that

$$\int_0^T dt \int_{\mathbb{T}^d} du\, |\rho^j(t, u)|^2 < \infty$$

for $j = 1, 2$. Denote the difference $\rho^1 - \rho^2$ (resp. $\Phi(\rho^1) - \Phi(\rho^2)$) by $\bar\rho$ (resp. $\bar\Phi$) and let $R_M: [0, T] \to \mathbb{R}$ be the function defined by

$$R_M(t) = \int_{\mathbb{T}^d} du\, \bar\rho_t(u) (F_M * \bar\rho_t)(u) .$$

R_M is well defined because ρ_t^j, $j = 1, 2$, belong to $L^2(\mathbb{T}^d)$ and F_M is bounded.

By properties (4.5) and (4.6),

$$R_M(t) = \sum_{z \in \mathbb{Z}^d} \frac{M}{(1 + |z|^2)(M + |z|^a)} < \psi_z, \bar\rho_t >^2 .$$

Moreover, since ρ^1, ρ^2 are in $L^2(\mathbb{T}^d)$,

$$\lim_{M \to \infty} R_M(t) = \sum_{z \in \mathbb{Z}^d} \frac{1}{(1 + |z|^2)} < \psi_z, \bar\rho_t >^2 .$$

Denote the right hand side by $R(t)$. Since $a > d$ and ρ^1, ρ^2 are weak solutions, R_M is time differentiable and

$$(\frac{d}{dt} R_M)(t) = -8\pi^2 \sum_{z \in \mathbb{Z}^d} \frac{M |z|^2}{(1 + |z|^2)(M + |z|^a)} < \psi_z, \bar\rho_t >< \psi_z, \bar\Phi_t > .$$

We may rewrite the right hand side as

$$
\begin{aligned}
&- 8\pi^2 \sum_{z\in\mathbb{Z}^d} < \psi_z, \bar\rho_t >< \psi_z, \bar\Phi_t > \\
&\quad + 8\pi^2 \sum_{z\in\mathbb{Z}^d} \frac{|z|^a}{M+|z|^a} < \psi_z, \bar\rho_t >< \psi_z, \bar\Phi_t > \\
&\quad\quad + 8\pi^2 \sum_{z\in\mathbb{Z}^d} \frac{M}{(1+|z|^2)(M+|z|^a)} < \psi_z, \bar\rho_t >< \psi_z, \bar\Phi_t > .
\end{aligned}
$$

By Schwarz inequality the third expression is less than or equal to

$$
\begin{aligned}
&8A\pi^2 \sum_{z\in\mathbb{Z}^d} \frac{M}{(1+|z|^2)(M+|z|^a)} < \psi_z, \bar\rho_t >^2 \\
&\quad + \frac{2\pi^2}{A} \sum_{z\in\mathbb{Z}^d} \frac{M}{(1+|z|^2)(M+|z|^a)} < \psi_z, \bar\Phi_t >^2
\end{aligned}
$$

for every $A > 0$. In virtue of (4.4), the second term of this sum is bounded above by

$$
\begin{aligned}
\frac{2\pi^2}{A} \sum_{z\in\mathbb{Z}^d} < \psi_z, \bar\Phi_t >^2 &= \frac{2\pi^2}{A} \int_{\mathbb{T}^d} du\, \bar\Phi_t(u)^2 \\
&\le \frac{2\pi^2 g^*}{A} \int_{\mathbb{T}^d} du\, \bar\Phi_t(u)\bar\rho_t(u) \;=\; \frac{2\pi^2 g^*}{A} \sum_{z\in\mathbb{Z}^d} < \psi_z, \bar\Phi_t >< \psi_z, \bar\rho_t >
\end{aligned}
$$

because $\bar\Phi_t(u) = \Phi(\rho^1(t,u)) - \Phi(\rho^2(t,u))$, $\Phi(\cdot)$ is strictly increasing and Φ' is bounded in absolute value by g^*. Therefore, setting $A = g^*$, integrating in time and applying identity (4.4), we obtain that $R_M(t)$ is bounded above by

$$
R_M(0) + B_M - 6\pi^2 \int_0^t ds \int_{\mathbb{T}^d} du\, \bar\rho(s,u)\bar\Phi(s,u) + 8\pi^2 g^* \int_0^t ds\, R_M(s) ,
$$

where

$$
B_M = 8\pi^2 \int_0^T dt \sum_{z\in\mathbb{Z}^d} \frac{|z|^a}{M+|z|^a} \Big| < \psi_z, \bar\rho_t >< \psi_z, \bar\Phi_t > \Big| .
$$

By Gronwall inequality,

$$
R_M(t) + 6\pi^2 \int_0^t ds \int_{\mathbb{T}^d} du\, \bar\rho(s,u)\bar\Phi(s,u) \le \Big\{ R_M(0) + B_M \Big\} \exp\Big\{ 8\pi^2 g^* t \Big\}
$$

for every $t \le T$. Since ρ^1, ρ^2 (and therefore $\bar\Phi$) belong to $L^2([0,T]\times\mathbb{T}^d)$, by (4.4) $\lim_{M\to\infty} B_M = 0$. Therefore, letting $M \uparrow \infty$, from the definition of $R(t)$,

$$
R(t) + 6\pi^2 \int_0^t ds \int_{\mathbb{T}^d} du\, \bar\rho(s,u)\bar\Phi(s,u) \le R(0) \exp\Big\{ 8\pi^2 g^* t \Big\} ,
$$

what concludes the proof of the theorem. □

It follows from the uniqueness of bounded weak solutions and from the first part of the proof of this theorem that solutions of quasi–linear parabolic equations depend continuously on the initial data.

Theorem 4.5 *Fix a bounded profile ρ_0 and a sequence of bounded profiles ρ_0^ε converging weakly to ρ_0 in the sense (4.3). For each $\varepsilon > 0$, denote by $\rho^\varepsilon(t,u)$ (resp. $\rho(t,u)$) the unique bounded weak solution of equation (4.1) with initial data ρ_0^ε (resp. ρ_0). The sequence ρ^ε converges uniformly on each compact set of $(0,\infty)\times\mathbb{T}^d$ to ρ.*

5. Entropy Solutions of Quasi–Linear Hyperbolic Equations

We review in this section some properties of weak and entropy solutions of the conservation law $\partial_t\rho + m\cdot\nabla\Phi(\rho) = 0$. Proofs, examples and further details can be found in Lax (1957), Kružkov (1970) and Smoller (1983).

Definition 5.1 Fix a bounded initial profile $\rho_0\colon\mathbb{R}^d\to\mathbb{R}$ and a vector $m = (m_1,\dots,m_d)$ in $\mathbb{R}^d$. A bounded function $\rho\colon\mathbb{R}_+\times\mathbb{R}^d\to\mathbb{R}$ is a weak solution of the Cauchy problem

$$\begin{cases} \partial_t\rho + \sum_{j=1}^d m_j\partial_{u_j}\Phi(\rho) = 0 \\ \rho(0,\cdot) = \rho_0(\cdot) \end{cases} \tag{5.1}$$

if for every function $G\colon\mathbb{R}_+\times\mathbb{R}^d\to\mathbb{R}$ of class $C_K^{1,1}(\mathbb{R}_+\times\mathbb{R}^d)$

$$\int_0^\infty dt\int_{\mathbb{R}^d} du\,\Big\{\rho(t,u)\,\partial_t G \;+\; \Phi(\rho(t,u))\sum_{i=1}^d m_i\partial_{u_i}G\Big\} \\ +\int_{\mathbb{R}^d} du\,G(0,u)\rho_0(u) \;=\; 0\,.$$

It turns out that weak solutions are not determined uniquely by their initial value. An additional criterion is therefore needed to select among the weak solutions the physically relevant. The entropy condition presented below is due to Kružkov (1970). In dimension 1, if Φ is convex, this condition is equivalent to require the weak solution to have only decreasing (resp. increasing) shocks in the case $m_1 > 0$ (resp. $m_1 < 0$), what explains the terminology.

Definition 5.2 A bounded function $\rho\colon\mathbb{R}_+\times\mathbb{R}^d\to\mathbb{R}$ is an entropy solution of the Cauchy problem (5.1) if for every *positive* function G of class $C_K^{1,1}((0,\infty)\times\mathbb{R}^d)$ and for every a in $\mathbb{R}_+$,

$$\int_0^{\infty} dt \int_{\mathbb{R}^d} du \Big\{ |\rho(t,u) - a|\, \partial_t G \,+\, |\Phi(\rho(t,u)) - \Phi(a)| \sum_{i=1}^{d} m_i \partial_{u_i} G \Big\} \;\ge\; 0 \quad (5.2)$$

and

$$\lim_{t\to 0} \int_K du\, |\rho(t,u) - \rho_0(u)| \;=\; 0 \qquad (5.3)$$

for all compact sets K of $\mathbb{R}^d$.

Kružkov (1970) proved the existence and uniqueness of entropy solutions:

Theorem 5.3 *For every bounded profile $\rho_0 \colon \mathbb{R}^d \to \mathbb{R}$, there exists a unique entropy solution of equation (5.1).*

The first condition imposes $\partial_t|\rho(t,u) - a| + \sum_{1\le i\le d} m_i \partial_{u_i} |\Phi(\rho(t,u)) - \Phi(a)|$ to be negative in the weak sense on $(0,\infty)\times\mathbb{R}^d$ for all a in $\mathbb{R}$. It is only the second condition that connects the solution to the initial data since in the first G is taken with compact support on the open set $(0,\infty)\times\mathbb{R}^d$.

Kružkov also proved that the entropy solutions are monotone and stable in the L^1 norm:

Theorem 5.4 *Consider two bounded profiles ρ_0^1, $\rho_0^2 \colon \mathbb{R}^d \to \mathbb{R}$ such that $\rho_0^1 \le \rho_0^2$ and denote by $\rho^i(t,u)$ the entropy solutions of (5.1) with initial data ρ_0^i, $i = 1, 2$. For every $t \ge 0$, $\rho^1(t,u) \le \rho^2(t,u)$.*

Theorem 5.5 *Consider two bounded functions ρ^1, ρ^2 defined on $(0,\infty)\times\mathbb{R}^d$ that satisfy inequality (5.2). Let $n = \max_{i=1,2} \|\rho^i\|_\infty$. For every positive real R and every $0 < s \le t < \infty$,*

$$\int_{|u|\le R} du\, |\rho^1(t,u) - \rho^2(t,u)| \;\le\; \int_{|u|\le R+\sigma_n(t-s)} du\, |\rho^1(s,u) - \rho^2(s,u)|\,,$$

where, for each positive integer n, $\sigma_n = |m| \sup_{|a|\le n} |\Phi'(a)|$.

Since constants are entropy solutions of (5.1), it follows from Theorem 5.4 that $\|\rho(t,\cdot)\|_\infty \le \|\rho_0\|_\infty$ for every $t \ge 0$. On the other hand, Theorem 5.5 concerns functions defined in the open set $(0,\infty)\times\mathbb{R}^d$ and no assumption nor statement is made on the behavior at time 0. Uniqueness of entropy solutions follows from this stability result since by assumption (5.3) there is L^1 local convergence at time 0:

Corollary 5.6 *Consider two bounded profiles ρ_0^1, $\rho_0^2 \colon \mathbb{R}^d \to \mathbb{R}$. Let $n = \max_{i=1,2} \|\rho_0^i\|_\infty$ and denote by $\rho^i(t,u)$ the entropy solutions of (5.1) with initial data ρ_0^i, $i = 1, 2$. For every positive real R and every $0 \le s \le t < \infty$,*

$$\int_{|u|\le R} du\, |\rho^1(t,u) - \rho^2(t,u)| \;\le\; \int_{|u|\le R+\sigma_n(t-s)} du\, |\rho^1(s,u) - \rho^2(s,u)|\,.$$

Moreover, the entropy solutions are continuous in L^1_{loc}:

Corollary 5.7 *Consider the entropy solution $\rho(t,u)$ of equation (5.1) with initial data ρ_0. For every positive real R and every $0 \le t < \infty$,*

$$\lim_{\varepsilon\to 0} \int_{|u|\le R} du\, |\rho(t+\varepsilon,u) - \rho(t,u)| \;=\; 0\,.$$

Proof. For $t = 0$ the result follows from assumption (5.3). Fix thus $t > 0$. For each $\varepsilon > 0$, the bounded function $\rho^\varepsilon(t,u)$ defined as $\rho(t+\varepsilon,u)$ satisfies inequality (5.2). Therefore, by Theorem 5.5, for every $R > 0$,

$$\int_{|u|\le R} du\, |\rho(t+\varepsilon,u) - \rho(t,u)| \;\le\; \int_{|u|\le R+\sigma_n t} du\, |\rho(2\varepsilon,u) - \rho(\varepsilon,u)|$$

that vanishes as $\varepsilon \downarrow 0$ by assumption (5.3). □

We conclude this section with a result on the continuous dependence on the initial data of entropy solutions of one-dimensional equations due to P. Lax (1957). Unless otherwise stated, up to the end of this section, we consider the one-dimensional differential equation

$$\begin{cases} \partial_t \rho \;+\; \partial_u \Phi(\rho) \;=\; 0 \\ \rho(0,\cdot) = \rho_0(\cdot) \end{cases} \tag{5.4}$$

and assume Φ to be strictly concave or convex in the range of ρ_0. In this context an explicit formula for the entropy solution was derived by E. Hopf (1950) for a quadratic equation and by P. Lax (1957) for the general case.

To fix ideas denote by $[a_1,a_2]$ the range of ρ_0 and assume that Φ is strictly convex in $[a_1,a_2]$. If necessary we redefine Φ in $[a_1,a_2]^c$ in order for Φ to be strictly convex on $\mathbb{R}$. Denote by Υ the Legendre transform of Φ:

$$\Upsilon(u) \;=\; \sup_{v\in\mathbb{R}}\{uv - \Phi(v)\}$$

and define R as the integral of ρ_0:

$$R(u) \;=\; \int_0^u dv\, \rho_0(v)\,.$$

The following three results are taken from Lax (1957).

Lemma 5.8 *Consider the expression*

$$B(v) \;=\; R(v) \;+\; t\Upsilon\Big(\frac{u-v}{t}\Big)\,.$$

For a fixed $t > 0$, with the exception of a countable set of values of u, B assumes its minimum at a single point denoted by $v_0(t,u)$.

We are now in a position to write an explicit formula for the entropy solution of equation (5.4):

Theorem 5.9 *The function* $\rho(t,u)$ *defined by*

$$\rho(t,u) \;=\; (\Phi')^{-1}\Big(\frac{u - v_0(t,u)}{t}\Big)$$

is the entropy solution of (5.4).

This explicit formula permits to prove the continuous dependence on the initial data of entropy solutions of one-dimensional equations:

Theorem 5.10 *Let* ρ_0^ε *a sequence of bounded functions converging weakly to a bounded limit* ρ_0 *as* $\varepsilon \downarrow 0$. *Denote by* ρ^ε *(resp.* $\rho(t,u)$*) the entropy solution of equation (5.4) with initial data* ρ_0^ε *(resp.* ρ_0*). The sequence* $\rho^\varepsilon(t,u)$ *converges to* $\rho(t,u)$ *at all continuity points* u *of* $\rho(t,\cdot)$.

All analysis carried out above can be extended to weak or entropy solutions defined on the torus $\mathbb{T}^d$ since each bounded function ρ on $\mathbb{T}^d$ can be interpreted as a periodic bounded function on $\mathbb{R}^d$ with period $\mathbb{T}^d$ and equal to ρ on $\mathbb{T}^d$. In the periodic case, Corollaries 5.6 and 5.7 state that the L^1 norm of the difference of two entropy solutions decreases in time and that the entropy solution is L^1 continuous in time:

Theorem 5.11 *Consider two bounded profiles* ρ_0^1, ρ_0^2: $\mathbb{T}^d \to \mathbb{R}$ *and denote by* $\rho^i(t,u)$ *the entropy solutions of (5.1) with initial data* ρ_0^i, $i = 1, 2$. *For every* $0 \le s \le t < \infty$,

$$\int_{\mathbb{T}^d} du\, |\rho^1(t,u) - \rho^2(t,u)| \;\le\; \int_{\mathbb{T}^d} du\, |\rho^1(s,u) - \rho^2(s,u)|\;.$$

Moreover, for every $0 \le t < \infty$,

$$\lim_{\varepsilon\to 0} \int_{\mathbb{T}^d} du\, |\rho(t+\varepsilon,u) - \rho(t,u)| \;=\; 0\;.$$

Appendix 3. Nongradient Tools: Spectral Gaps and Closed Forms

We present in this chapter the main tools used in the proof of the hydrodynamic behavior of reversible nongradient systems: estimates on the rate of convergence to equilibrium of reversible Markov processes and closed and exact forms in the context of interacting particle systems. The chapter is organized as follows. In section 1 we prove a second order expansion for the largest eigenvalue of a small perturbation of a reversible generator. In the hydrodynamic setting this expansion reduces the proof of a two block estimates to the computation of some central limit variances (cf. section 7.3). In sections 2 and 3 we prove that the spectral gap of the generator of a symmetric generalized exclusion process on a cube of length N shrinks as N^{-2} in any dimension and uniformly over the density. Finally, in section 4 we investigate the closed and exact forms in the setting of interacting particle systems.

In order to motivate the derivation of bounds on the largest negative eigenvalue of the generator of a reversible process, we conclude this section with a brief investigation of the rate of convergence to equilibrium in L^2 of reversible Markov processes. Consider an irreducible Markov process X_t on some countable state space $\mathcal{E}$. Denote by $\{P_t,\, t \geq 0\}$ the semigroup of the process and by L its generator:

$$(Lf)(x) \;=\; \sum_{y\in\mathcal{E}} \lambda(x)p(x,y)[f(y) - f(x)]\;.$$

Assume that the process is reversible with respect to an invariant probability measure ν. Since L is a self adjoint operator on $L^2(\nu)$ all its eigenvalues are real and nonpositive. Moreover, 0 is an eigenvalue associated at least to the constant functions. We claim that 0 has multiplicity 1. To prove this statement, consider an eigenfunction f in $L^2(\nu)$ associated to the eigenvalue 0: $Lf = 0$. Multiply both sides of this equation by f and integrate with respect to ν to get that the Dirichlet form of f vanishes:

$$0 \;=\; \mathfrak{D}(f) \;=<-Lf, f>_\nu \;=\; (1/2) \sum_{x,y\in\mathcal{E}} \nu(x)\lambda(x)p(x,y)[f(y) - f(x)]^2\;.$$

Since we assumed the process to be irreducible, this in turn implies that f is constant.

Recall that the Dirichlet form is well defined in $L^2(\nu)$. Denote by λ_1 the lower bound of the strictly positive part of the spectrum of the generator $-L$:

$$\lambda_1 := \inf_{\substack{f \in L^2(\nu) \\ <f,1>_\nu = 0}} \frac{< -Lf, f >_\nu}{< f, f >_\nu} = \inf_{f \in L^2(\nu)} \frac{\mathfrak{D}(f)}{\mathbf{Var}(\nu, f)} \cdot$$

In this formula $\mathbf{Var}(\nu, f)$ stands for the variance of f with respect to ν and the first infimum is taken over all $L^2(\nu)$ functions f which are orthogonal to the constants ($< f >_\nu = < f, 1 >_\nu = 0$). We shall refer to λ_1 as the spectral gap of the generator L. Notice that λ_1 is not necessarily an eigenvalue of L and that λ_1 may vanish because $L^2(\nu)$ is an infinite dimensional space. The following result establishes that λ_1 is closely related to the exponential rate of convergence to equilibrium in $L^2(\nu)$.

Theorem 0.1 *Denote by λ_0 the largest real λ such that for every function f in $L^2(\nu)$*

$$\Big\| P_t f - < f >_\nu \Big\|_2 \le C(f) e^{-\lambda t}$$

for all $t > 0$ and some finite constant $C(f)$ depending only on f. λ_0 coincides with the spectral gap: $\lambda_0 = \lambda_1$.

Proof. Denote by $\int_{-\infty}^0 \lambda dE_\lambda$ the spectral decomposition of the self adjoint non-positive operator L. For each function f in $L^2(\nu)$, denote by μ_f the spectral measure on $\mathbb{R}_-$ associated to f: $\mu_f(d\lambda) = d < E_\lambda f, f >$. It follows from the definition of λ_1 that $\mu_f((-\lambda_1, 0]) = 0$ for each function f orthogonal to the constants. In particular,

$$\Big\| P_t f - < f >_\nu \Big\|_2 = \int_{-\infty}^0 e^{\lambda t} \mu_{f - <f>_\nu}(d\lambda) \le e^{-\lambda_1 t} \| f - < f > \|_2$$

for every f in $L^2(\nu)$. This shows that $\lambda_0 \ge \lambda_1$.

We turn to the inverse inequality. Since for every $L^2(\nu)$ function f and for every $t \ge 0$,

$$\int_{-\infty}^0 e^{\lambda t} \mu_{f - <f>_\nu}(d\lambda) = \Big\| P_t f - < f >_\nu \Big\|_2 \le C(f) e^{-\lambda_0 t}$$

for some finite constant $C(f)$, $\mu_{f-<f>_\nu}((-\lambda_0, 0]) = 0$ for every f in $L^2(\nu)$. Therefore, for each function f orthogonal to the constants, $\mu_f((-\lambda_0, 0]) = 0$. In particular, $\lambda_1 \ge \lambda_0$. □

We have just shown that the spectral gap λ_1 is intimately connected to the rate of convergence to equilibrium in L^2. It is therefore natural to try to prove lower bounds for λ_1. The next result provides such an estimate in a general context.

Proposition 0.2 *Suppose $f, g : \mathcal{E} \to R_+$ are functions satisfying $Lg + fg = 0$. If $f(x) \ge \lambda > 0$ and $g(x)$ is bounded below by some strictly positive constant, the spectral gap is bounded below by λ.*

Proof. By assumption, the function $f(x) = -(Lg)(x)/g(x)$ is bounded below by λ. In particular, for every $L^2(\nu)$ function h with norm $\|h\|_2 = 1$,

$$\int \frac{-(Lg)(x)}{g(x)} h^2(x)\nu(dx) \;\geq\; \lambda\,.$$

If g was bounded above, by Theorem A1.10.2, the Dirichlet form of h would be bounded below by λ for every h with L^2 norm equal to 1. The proposition would therefore follow from the definition of λ_1. Hence, to conclude the proof it remains to approximate g by bounded functions.

For each positive integer M, denote by g_M the function defined by $g_M(x) = g(x) \wedge M$. By assumption g_M is bounded below by a strictly positive constant and bounded above by M. In particular, by Theorem A1.10.2,

$$\mathfrak{D}(h) \;\geq\; \int \frac{-(Lg_M)(x)}{g_M(x)} h^2(x)\nu(dx)$$

for every M. A straightforward analysis shows that $-(Lg_M)(x)/g_M(x)$ is bounded below by $-(Lg)(x)/g(x)$ if $g(x) \leq M$ and that $-(Lg_M)(x)/g_M(x)$ is positive if $g(x) \geq M$. In particular, the right hand side of the last inequality is bounded below by

$$\int \frac{-(Lg)(x)}{g(x)} \mathbf{1}\{g(x) \leq M\} h^2(x)\nu(dx)\,.$$

Since by assumption $-(Lg)(x)/g(x)$ is a positive function bounded below by λ, by the monotone convergence theorem, as $M \uparrow \infty$, this expression converges to

$$\int \frac{-(Lg)(x)}{g(x)} h^2(x)\nu(dx) \;\geq\; \lambda\,.$$

This conclude the proof of the proposition. □

1. On the Spectrum of Reversible Markov Processes

We consider in this section a reversible Markov process on a countable state space $\mathcal{E}$ and keep the notation of the beginning of the chapter.

Theorem 1.1 *Assume that the generator L has a spectral gap of magnitude γ^{-1}:*

$$\mathbf{Var}(\nu, f) \;\leq\; \gamma\,\mathfrak{D}(f)$$

for every f in $L^2(\nu)$. Let V be a mean-zero bounded function such that $<(-L)^{-1}V, V>_\nu < \infty$. Denote by λ_ε the upper bound of the spectrum of $L + \varepsilon V$:

$$\lambda_\varepsilon \;=\; \sup_{f;\, \|f\|_2=1} \Big\{ < f, (L+\varepsilon V) f >_\nu \Big\} \;=\; \sup_{f;\, \|f\|_2=1} \Big\{ \varepsilon < f, V f >_\nu \;-\; \mathfrak{D}(f) \Big\}\,.$$

Then,

$$0 \le \lambda_\varepsilon \le \frac{\varepsilon^2}{1 - 2\|V\|_\infty \varepsilon\gamma} < (-L)^{-1}V, V >_\nu \cdot$$

Proof. Fix $\varepsilon > 0$. To keep notation simple, denote by L_ε the operator $L + \varepsilon V$. Let $\{G_{\varepsilon,n},\ n \ge 1\}$ be a sequence of $L^2(\nu)$ functions that approaches the supremum:

$$\|G_{\varepsilon,n}\|_2 = 1 \quad \text{and} \quad \lim_{n\to\infty} < G_{\varepsilon,n}, L_\varepsilon G_{\varepsilon,n} >_\nu = \lambda_\varepsilon .$$

Notice that $< G_{\varepsilon,n}, (\lambda_\varepsilon - L_\varepsilon) G_{\varepsilon,n} >_\nu$ vanishes as $n \uparrow \infty$ by definition of $G_{\varepsilon,n}$. We may assume without loss of generality that $G_{\varepsilon,n}$ has positive expectation:

$$< G_{\varepsilon,n} >_\nu \ge 0 .$$

The eigenfunctions associated to the largest eigenvalue of the generator L are the constants. Since we are considering a small perturbation of the generator and $G_{\varepsilon,n}$ is normalized to have positive mean, we expect $G_{\varepsilon,n}$ to be close to 1. The idea of the proof is therefore to expand $G_{\varepsilon,n}$ around 1. By definition of $G_{\varepsilon,n}$, we have

$$\lambda_\varepsilon = < G_{\varepsilon,n}, (\lambda_\varepsilon - L_\varepsilon) G_{\varepsilon,n} >_\nu + < G_{\varepsilon,n}, L_\varepsilon G_{\varepsilon,n} >_\nu .$$

Since $L_\varepsilon = L + \varepsilon V$, we may rewrite the second term on the right hand side as

$$\varepsilon \Big\{ < V >_\nu + 2 < V[G_{\varepsilon,n} - 1] >_\nu + < V[G_{\varepsilon,n} - 1]^2 >_\nu \Big\} - \mathfrak{D}(G_{\varepsilon,n}) .$$

Recall that V has mean 0. By Schwarz inequality the expression inside braces is bounded above by

$$\begin{aligned} &\varepsilon \Big\{ \frac{\varepsilon}{A} < (-L)^{-1}V, V >_\nu + \frac{A}{\varepsilon} < (-L)[G_{\varepsilon,n} - 1], [G_{\varepsilon,n} - 1] >_\nu \\ &\quad + < V[G_{\varepsilon,n} - 1]^2 >_\nu \Big\} \\ &\quad \le \frac{\varepsilon^2}{A} < (-L)^{-1}V, V >_\nu + 2\varepsilon \|V\|_\infty \Big(1 - < G_{\varepsilon,n} >_\nu \Big) + A\,\mathfrak{D}(G_{\varepsilon,n}) \end{aligned}$$

for every positive A.

To bound the second term of the last expression, notice that by Schwarz inequality and our choice of $G_{\varepsilon,n}$,

$$0 \le < G_{\varepsilon,n} >_\nu \le < G_{\varepsilon,n}^2 >_\nu^{1/2} = 1$$

so that

$$\begin{aligned} 0 &\le 1 - < G_{\varepsilon,n} >_\nu \le 1 - < G_{\varepsilon,n} >_\nu^2 \\ &= < G_{\varepsilon,n}^2 >_\nu - < G_{\varepsilon,n} >_\nu^2 \le \gamma \mathfrak{D}(G_{\varepsilon,n}) \end{aligned}$$

provided we have a spectral gap of magnitude γ^{-1}. Recollecting all previous estimates we obtain that λ_ε is bounded above by

$$< G_{\varepsilon,n}, (\lambda_\varepsilon - L_\varepsilon) G_{\varepsilon,n} >_\nu + \frac{\varepsilon^2}{A} < (-L)^{-1} V, V >_\nu$$
$$- (1 - A - 2\|V\|_\infty \varepsilon\gamma) \mathfrak{D}(G_{\varepsilon,n}) .$$

We conclude the proof of the upper bound by choosing A to be equal to $1 - 2\|V\|_\infty \varepsilon\gamma$ and letting $n \uparrow \infty$ because $< G_{\varepsilon,n}, (\lambda_\varepsilon - L_\varepsilon) G_{\varepsilon,n} >_\nu$ vanishes as $n \uparrow \infty$.

The lower bound follows from the fact that V has mean 0. We have just to set $f = 1$ in the variational formula for λ_ε. □

The following result is a simple consequence of the previous Theorem.

Corollary 1.2 *Assume that the generator L has a strictly positive spectral gap of magnitude γ^{-1}. Let V be a bounded function. For every sufficiently small ε,*

$$\sup_f \left\{ \varepsilon < V f^2 >_\nu - < (-L) f, f >_\nu \right\}$$
$$\le \varepsilon < V >_\nu + \frac{\varepsilon^2}{1 - 2\|V\|_\infty \varepsilon\gamma} < (-L)^{-1} V, V >_\nu ,$$

where the supremum is taken over all functions f in $L^2(\nu)$ such that $\|f\|_2 = 1$.

We conclude this section presenting an alternative variational formula for the largest eigenvalue of a perturbation of a generator which is reversible with respect to some measure ν. This alternative version is constantly used in the proof of large deviations principles for Markov processes.

Consider a continuous time Markov process X_t on a countable space $\mathcal{E}$ with generator L reversible with respect to an invariant state ν. Let $V : \mathcal{E} \to \mathbb{R}$ be a bounded function and denote by λ_V the largest eigenvalue of the symmetric operator $L + V$ in $L^2(\nu)$. We claim that

$$\lambda_V = \sup_f \left\{ < V, f >_\nu - \mathfrak{D}(\sqrt{f}) \right\} , \tag{1.1}$$

where the supremum is taken over all densities with respect to ν: $f \ge 0$ and $\int f d\nu = 1$.

To prove this statement recall the variational formula for the largest eigenvalue of a symmetric operator:

$$\lambda_V = \sup_f \left\{ < V, f^2 >_\nu - \mathfrak{D}(f) \right\} .$$

In this formula the supremum is carried over all functions f in $L^2(\nu)$ such that $\|f\|_2 = 1$. We have proved in Appendix 1 that $\mathfrak{D}(V(f)) \le \mathfrak{D}(f)$ for every function $V : \mathbb{R} \to \mathbb{R}$ such that $|V(b) - V(a)| \le |b - a|$. In particular, $\mathfrak{D}(|f|) \le \mathfrak{D}(f)$ and the previous supremum is equal to

$$\sup_{f \geq 0} \left\{ < V, f^2 >_\nu - \mathfrak{D}(f) \right\} ,$$

where the supremum now is carried over all positive functions with L^2 norm equal to 1. To obtain (1.1) it remains to replace f by $\sqrt{f}$.

2. Spectral Gap for Generalized Exclusion Processes

We investigate in this section the spectral gap of symmetric generalized exclusion dynamics on finite d-dimensional cubes. For each positive integer N, denote by Ω_N a cube of linear size N:

$$\Omega_N = \{1, \ldots, N\}^d$$

and by L_{Ω_N} the generator of the symmetric generalized exclusion process on Ω_N:

$$(L_{\Omega_N} f)(\eta) = (1/2) \sum_{\substack{x,y \in \Omega_N \\ |x-y|=1}} r_{x,y}(\eta)[f(\eta^{x,y}) - f(\eta)] .$$

In this formula $r_{x,y}(\eta)$ is equal to 1 whenever a jump is possible from x to y:

$$r_{x,y}(\eta) = \mathbf{1}\{\eta(x) > 0, \eta(y) < \kappa\} . \tag{2.1}$$

For each fixed total number of particles $0 \leq K \leq \kappa N^d$, the Markov process with generator L_{Ω_N} and state space $\Sigma^\kappa_{\Omega_N,K}$ is a finite state irreducible Markov process. In particular, it has a unique ergodic invariant probability measure that we denote by $\nu_{\Omega_N,K}$ or $\nu_{N,K}$. Since the transition probability is symmetric ($p(y) = (1/2)$ if $|y| = 1$ and 0 otherwise), this measure is in fact reversible. Moreover, a simple computation shows that $\nu_{N,K}$ is the uniform probability measure on $\Sigma^\kappa_{\Omega_N,K}$, the space of all configurations of $\{0, \ldots, \kappa\}^{\Omega_N}$ with K particles:

$$\Sigma^\kappa_{\Omega_N,K} = \{\eta \in \{0, \ldots, \kappa\}^{\Omega_N}; \sum_{x \in \Omega_N} \eta(x) = K\} .$$

Keep in mind that $L^2(\nu_{N,K})$ is a finite dimensional space. Since L_{Ω_N} is a self adjoint generator all its eigenvalues are real and nonpositive. Denote them by $0 \geq -\lambda_0 > -\lambda_1 > \cdots > -\lambda_R$. To keep notation simple, we omitted the dependence of λ_i on N and K. Since L_{Ω_N} is a generator, 0 is an eigenvalue associated at least to the constant functions. We claim that 0 has multiplicity 1: let f be a eigenfunction in $L^2(\nu_{N,K})$ associated to the eigenvalue 0: $L_{\Omega_N} f = 0$. Multiply both sides of this equation by f and integrate with respect to $\nu_{N,K}$ to show that f is constant. In particular, for each $N \geq 1$ and $0 \leq K \leq \kappa N^d$, L_{Ω_N} has a positive spectral gap denoted by $\lambda_1 = \lambda_1(K, N)$. We may express λ_1 by a variational formula:

$$\lambda_1(N,K) = \inf \frac{< -L_{\Omega_N} f, f >_{\nu_{N,K}}}{< f, f >_{\nu_{N,K}}} ,$$

where the infimum is carried over all functions in $L^2(\nu_{N,K})$ that are orthogonal to the eigenspace associated to λ_0, i.e., that are orthogonal to the constants: $< f, 1 >_{\nu_{N,K}} = 0$. Thus, if we denote by $W(N,K)$ the inverse of the spectral gap, $W(N,K) = \lambda_1(N,K)^{-1}$,

$$W(N,K) = \sup_{f \in L^2(\nu_{N,K})} \left\{ \frac{\mathbf{Var}\,(\nu_{N,K}, f)}{\mathfrak{D}(\nu_{N,K}, f)} \right\} .$$

In this formula, for a finite subset Λ of $\mathbb{Z}^d$ and function f in $L^2(\nu_{\Lambda,K})$ $\mathbf{Var}\,(\nu_{\Lambda,K}, f)$ and $\mathfrak{D}(\nu_{\Lambda,K}, f)$ denote respectively the variance and the Dirichlet form of f with respect to $\nu_{\Lambda,K}$:

$$\mathbf{Var}\,(\nu_{\Lambda,K}, f) = E_{\nu_{\Lambda,K}}\left[\left(f - E_{\nu_{\Lambda,K}}[f] \right)^2 \right]$$

$$\mathfrak{D}(\nu_{\Lambda,K}, f) = (1/4) \sum_{x \in \Lambda} \sum_{\substack{y \in \Lambda \\ |y-x|=1}} \int r_{x,y}(\eta) \left\{ f(\eta^{x,y}) - f(\eta) \right\}^2 \nu_{\Lambda,K}(d\eta) .$$

We investigate in this section the asymptotic behavior, as $N \uparrow \infty$, of the spectral gap of the generator of a generalized symmetric exclusion process restricted to a d-dimensional cube of linear size N. We prove that the spectral gap shrinks as N^{-2} in all dimensions:

Theorem 2.1 *There exists a universal constant C_0 such that*

$$\mathbf{Var}\,(\nu_{N,K}, f) \le C_0 N^2 \mathfrak{D}(\nu_{N,K}, f)$$

for all $N \ge 1$, $0 \le K \le \kappa N^d$ and all functions f in $L^2(\nu_{N,K})$.

Fix a finite subset Λ of $\mathbb{Z}^d$, a subset Λ_1 of Λ and a configuration η of $\{0, \dots, \kappa\}^\Lambda$. Recall that $M(\eta) = M_\Lambda(\eta)$ stands for the total number of particles for the configuration η:

$$M(\eta) = \sum_{x \in \Lambda} \eta(x) .$$

For a function $f \colon \Sigma_{\Lambda,K} \to \mathbb{R}$ and a configuration ζ of $\{0, \dots, \kappa\}^{\Lambda_1}$ with at most K particles, denote by f_ζ the function on $\Sigma_{\Lambda - \Lambda_1, K - M(\zeta)}$ whose value at a configuration ξ is equal to $f(\zeta, \xi)$. Here (ζ, ξ) is the configuration of $\Sigma_{\Lambda,K}$ defined by

$$(\zeta,\xi)(x) = \begin{cases} \zeta(x) & \text{if } x \in \Lambda_1 \\ \xi(x) & \text{if } x \notin \Lambda_1 . \end{cases}$$

This notation permits to express in a simple form the conditional expectation of $f \colon \Sigma_{\Lambda,K} \to \mathbb{R}$ given a configuration ζ of $\Lambda_1 \subset \Lambda$. An elementary computation shows that it is equal to the expectation of f_ς with respect to the canonical measure $\nu_{\Lambda-\Lambda_1,K-M(\varsigma)}$:

$$E_{\nu_{\Lambda,K}}\Big[f(\eta)\,\big|\,\eta(x)=\zeta(x) \text{ for } x\in\Lambda_1\Big] \;=\; E_{\nu_{\Lambda-\Lambda_1,K-M(\varsigma)}}[f_\varsigma]\,. \tag{2.2}$$

Denote by $W(N)$ the maximum over all densities of the inverse of the spectral gaps $W(N,K)$:

$$W(N) \;=\; \max_{0\le K\le \kappa N^d} W(N,K)\,.$$

With this notation the statement of the theorem reduces to the existence of a universal constant C_0 such that

$$W(N) \;\le\; C_0 N^2\,.$$

for all positive integers N. This statement is proved by induction on N.

It follows from the discussion preceding Theorem 2.1 that for each fixed N and $0\le K\le \kappa N$, L_{Ω_N} has a strictly positive spectral gap: $W(N,K)<\infty$. In particular, since there are only a finite number of cases, for any N', there exists a finite constant d_0 so that

$$\mathbf{Var}\,(\nu_{N,K}, f) \;\le\; d_0\, N^2\, \mathfrak{D}(\nu_{N,K}, f) \tag{2.3}$$

for all $N\le N'$, all $0\le K\le \kappa(N')^d$ and all f in $L^2(\nu_{N,K})$. We may restate this inequality saying that for any N_0 there exists a finite constant d_0 so that

$$W(N) \;\le\; d_0 N^2 \quad \text{for all } N\le N'.$$

Proof of Theorem 2.1. To avoid unnecessary heavy notation and to detach the main ideas of the proof, we first consider the problem in one dimension. We indicate in the next section the ingredients needed to extend this proof to higher dimensions.

For generalized exclusion processes with jump rate given by (2.1), there is a duality between particles and holes, i.e., the holes evolve with the same dynamics as particles do. We may therefore assume that the density of particles $K/|\Omega_N|$ is bounded above by $\kappa/2$. We shall do so without further comment.

The idea of the proof consists in using the induction hypothesis to set up a recursive equation for $W(N)$. With this purpose in mind we write the identity

$$\begin{aligned} f(\eta) \;-\; E_{N,K}[f] \;&=\; \Big\{f(\eta) \;-\; E_{N,K}\Big[f\,\big|\,\eta(N)\Big]\Big\} \\ &\quad+\; \Big\{E_{N,K}\Big[f\,\big|\,\eta(N)\Big] \;-\; E_{N,K}[f]\Big\}\,. \end{aligned}$$

Hereafter $E_{N,K}$ stands for the expectation with respect to the measure $\nu_{N,K}$. Through this decomposition and since the two terms on the right hand side of the last identity are orthogonal in $L^2(\nu_{N,K})$, we may express the variance of f as

$$\begin{aligned}\mathbf{Var}(\nu_{N,K}, f) &= E_{N,K}\Big[\Big(f - E_{N,K}[f \,|\, \eta(N)]\Big)^2\Big] \\ &\quad + E_{N,K}\Big[\Big(E_{N,K}\big[f \,|\, \eta(N)\big] - E_{N,K}[f]\Big)^2\Big]\,.\end{aligned} \tag{2.4}$$

The first term on the right hand side is easily analyzed through the induction assumption. Indeed, from identity (2.2) we have that $E_{N,K}[f \,|\, \eta(N)] = E_{N-1,K-\eta(N)}[f_{\eta(N)}]$. To keep notation simple, we abbreviate $N-1$ by N_1 and $K-\eta(N)$ by $K_{\eta(N)}$ and denote configurations of $\Sigma_{N_1,K_{\eta(N)}}$ by the Greek letter ξ. Taking conditional expectation with respect to $\eta(N)$ and applying once more identity (2.2), the first term on the right hand side of (2.4) becomes

$$\begin{aligned}&E_{N,K}\Big[\Big(f - E_{N_1,K_{\eta(N)}}[f_{\eta(N)}]\Big)^2\Big] \\ &\quad = E_{N,K}\Big[E_{N_1,K_{\eta(N)}}\Big[\Big(f_{\eta(N)} - E_{N_1,K_{\eta(N)}}[f_{\eta(N)}]\Big)^2\Big]\Big]\,.\end{aligned}$$

Notice that $E_{N_1,K_{\eta(N)}}\Big[\Big(f_{\eta(N)} - E_{N_1,K_{\eta(N)}}[f_{\eta(N)}]\Big)^2\Big]$ is the variance, with respect to $\nu_{N_1,K_{\eta(N)}}$, of $f_{\eta(N)}$, a function of $N-1$ variables. By the induction hypothesis, this expression is bounded by $W(N-1)\mathfrak{D}(\nu_{N_1,K_{\eta(N)}}, f_{\eta(N)})$. In particular, the first term on the right hand side of (2.4) is bounded by

$$\begin{aligned}&W(N-1)E_{N,K}\Big[\mathfrak{D}(\nu_{N_1,K_{\eta(N)}}, f_{\eta(N)})\Big] = \\ &\frac{W(N-1)}{4}\sum_{\substack{x,y\in\Omega_{N_1}\\ |x-y|=1}} E_{N,K}\Big[E_{N_1,K_{\eta(N)}}\Big[r_{x,y}(\xi)\Big\{f_{\eta(N)}(\xi^{x,y}) - f_{\eta(N)}(\xi)\Big\}^2\Big]\Big].\end{aligned}$$

Since expectation with respect to $\nu_{N_1,K_{\eta(N)}}$ corresponds to conditional expectation with respect to $\eta(N)$, the last expression is equal to

$$\begin{aligned}&(1/4)W(N-1)\sum_{\substack{x,y\in\Omega_{N_1}\\ |x-y|=1}} E_{N,K}\Big[E_{N,K}\Big[r_{x,y}(\eta)\Big\{f(\eta^{x,y}) - f(\eta)\Big\}^2 \,\Big|\, \eta(N)\Big]\Big] \\ &\quad \le W(N-1)\mathfrak{D}(\nu_{N,K}, f)\,.\end{aligned}$$

This last relation misses to be an equality only because in the Dirichlet form $\mathfrak{D}(\nu_{N,K}, f)$ there is a piece that measures the dependence of f on jumps over the bond $\{N-1, N\}$ that does not appear at the left hand side of the inequality.

Up to his point we showed by means of the induction hypothesis that the first term on the right hand side of (2.4) is bounded by $W(N-1)\mathfrak{D}(\nu_{N,K}, f)$:

$$E_{N,K}\Big[\Big(f - E_{N,K}[f \,|\, \eta(N)]\Big)^2\Big] \le W(N-1)\mathfrak{D}(\nu_{N,K}, f)\,. \tag{2.5}$$

We turn now to the second term on the right hand side of equation (2.4). Since the expected value, with respect to $\nu_{N,K}$, of $E_{N,K}[f|\eta(N)]$ is equal to $E_{N,K}\big[f\big]$, this expression is the variance of $E_{N,K}[f \mid \eta(N)]$, a function of one variable. The estimation of this variance constitutes the second step in our route to build a recurrent formula for $W(N)$. We summarize it in the following proposition.

Proposition 2.2 *There exists a finite constant $C = C(\kappa)$ depending only on κ such that for each $\varepsilon > 0$, there exists $\ell = \ell(\varepsilon)$ and $N' = N'(\varepsilon)$ for which*

$$\begin{aligned} &\mathbf{Var}\Big(\nu_{N,K}, E_{N,K}\big[f \mid \eta(N)\big]\Big) \\ &\quad\le\; C(\kappa)\Big\{N + \frac{\ell W(\ell+1)}{N}\Big\}\mathfrak{D}(\nu_{N,K}, f) + \frac{\varepsilon}{N}\mathbf{Var}(\nu_{N,K}, f) \end{aligned} \tag{2.6}$$

for all $N \ge N'(\varepsilon)$ and all $0 \le K \le (\kappa/2)N$.

We conclude now the proof of Theorem 2.1 assuming Proposition 2.2. Take $\varepsilon < 2$ in Proposition 2.2. From identity (2.4), estimate (2.5) and Proposition 2.2, for N large enough,

$$\Big(1 - \frac{\varepsilon}{N}\Big)\mathbf{Var}\Big(\nu_{N,K}, f\Big) \;\le\; \Big\{W(N-1) + C(\kappa)\,N\Big\}\mathfrak{D}(\nu_{N,K}, f)$$

for some constant $C(\kappa)$ that depends only on κ. There exists, therefore, a constant $C(\kappa)$ such that

$$W(N) \;\le\; \Big(1 - \frac{\varepsilon}{N}\Big)^{-1}\Big\{W(N-1) + C(\kappa)N\Big\}$$

for all N large enough. It is elementary to prove from this recursive inequality and estimate (2.3) the existence of a constant C_0 for which $W(N) \le C_0 N^2$. for all $N \ge 1$. □

We turn now to the proof of Proposition 2.2. The strategy may roughly be described as follows. We shall first prove a spectral gap, uniform over the parameters K and N, for functions depending only on one site (hereafter called one site functions). More precisely, we shall prove that there exists a universal constant $B_1 = B_1(\kappa)$ such that for every $H : \{0, \dots, \kappa\} \to \mathbb{R}$,

$$\mathbf{Var}\Big(\nu_{N,K}, H(\eta(N))\Big) \;\le\; B_1(\kappa)\mathcal{D}\Big(\nu_{N,K}, H(\eta(N))\Big)\,,$$

where $\mathcal{D}$ is a slight modification of the Dirichlet form $\mathfrak{D}$. Applying this result to the one site function $E_{N,K}[f|\eta(N)]$, we shall reduce the proof of Proposition 2.2 to the proof that the Dirichlet form of $E_{N,K}[f|\eta(N)]$ is bounded by the right hand side of (2.6).

The statement of the uniform spectral gap for one site functions requires some notation. Denote by $\nu^1_{N,K}$ the one site marginal of $\nu_{N,K}$:

$$\nu^1_{N,K}(a) \;=\; \nu_{N,K}\{\eta;\ \eta(N) = a\} \qquad \text{for } 0 \le a \le \kappa\,.$$

Consider the one site Dirichlet form $\mathcal{D}(\nu^1_{N,K}, \cdot)$ defined on $L^2(\nu^1_{N,K})$ by

$$\mathcal{D}(\nu^1_{N,K}, H) = \sum_{a=1}^{\kappa} [H(a-1) - H(a)]^2 \nu^1_{N,K}(a) .$$

We have the following one–coordinate Poincaré inequality:

Lemma 2.3 *There exists a constant $B_1 = B_1(\kappa)$, so that for all functions $H: \{0, \dots, \kappa\} \to \mathbb{R}$,*

$$E_{\nu_{N,K}}\Big[\Big(H(\eta(N)) - E_{\nu_{N,K}}[H(\eta(N))]\Big)^2\Big] \le B_1\, \mathcal{D}(\nu^1_{N,K}, H) . \tag{2.7}$$

Proof. Since $H(\eta(N))$ depends only on $\eta(N)$, we may write the variance of H as

$$\sum_{a=0}^{\kappa} \nu^1_{N,K}(a) \Big\{ H(a) - E_{\nu^1_{N,K}}[H] \Big\}^2 \le \sum_{a=0}^{\kappa} \nu^1_{N,K}(a) \Big\{ H(a) - H(0) \Big\}^2 .$$

By Schwarz inequality and a summation by parts, the right hand side of this last formula is bounded above by

$$\kappa \sum_{b=1}^{\kappa} \nu^1_{N,K}(b) \Big\{ H(b-1) - H(b) \Big\}^2 \sum_{a=b}^{\kappa} \frac{\nu^1_{N,K}(a)}{\nu^1_{N,K}(b)} .$$

Recall that we assumed the density K/N to be bounded above by $\kappa/2$. By Lemma A2.2.4, uniformly over densities bounded by $\kappa/2$, the ratio $\nu^1_{N,K}(a)/\nu^1_{N,K}(b)$ for $a \ge b$ is bounded by some constant depending only on κ. The last expression is therefore dominated by

$$B_1(\kappa) \sum_{b=1}^{\kappa} \nu^1_{N,K}(b) \Big\{ H(b-1) - H(b) \Big\}^2 .$$

This concludes the proof of the lemma. □

Applying this lemma to the one site function $E_{N,K}[f \mid \eta(N)]$ we obtain an estimate for the variance of $E_{N,K}[f \mid \eta(N)]$:

$$\begin{aligned} &\mathbf{Var}\Big(\nu_{N,K}, E_{N,K}\Big[f \,\Big|\, \eta(N)\Big]\Big) \\ &\quad \le B_1(\kappa) \sum_{b=1}^{\kappa} \nu^1_{N,K}(b) \Big\{ E_{N_1,K-b+1}[f_{b-1}] - E_{N_1,K-b}[f_b] \Big\}^2 . \end{aligned} \tag{2.8}$$

In this formula, for each $0 \le b \le \kappa$, $f_b : \Sigma_{N_1,K-b} \to \mathbb{R}$ stands for the function defined by $f_b(\xi) = f(\xi, b)$.

We have to estimate the difference $E_{N_1,K_b+1}[f_{b-1}] - E_{N_1,K_b}[f_b]$. Since the measure ν_{N,K_b+1} is concentrated on configurations with K_b+1 particles, we have that

$$E_{N_1,K_b+1}[f_{b-1}] \;=\; \frac{1}{K_b+1} \sum_{x\in\Omega_{N_1}} E_{N_1,K_b+1}\Big[f_{b-1}(\xi)\,\xi(x)\Big]\,. \tag{2.9}$$

Performing a change of variables $\xi' = \xi - \mathfrak{d}_x$, where $\mathfrak{d}_x$ denotes the configuration with no particles but one at site x, we obtain that the last expression is equal to

$$\frac{R_{N_1,K_b}}{K_b+1} \sum_{x\in\Omega_{N_1}} E_{N_1,K_b}\Big[f_{b-1}(\xi+\mathfrak{d}_x)\,[1+\xi(x)]\,\mathbf{1}\{\xi(x)<\kappa\}\Big]\,, \tag{2.10}$$

where R_{N_1,K_b} is the factor $\nu_{N_1,K_b+1}(\xi)/\nu_{N_1,K_b}(\xi)$ coming from the change of measures. It is given by

$$R_{N,K} \;=\; \frac{Z_\kappa(\Omega_N,K)}{Z_\kappa(\Omega_N,K+1)}\,,$$

provided $Z_\kappa(N,K)$ stands for the total number of configurations of $\Sigma^\kappa_{\Omega_N,K}$:

$$Z_\kappa(\Omega_N,K) \;=\; \sum_{\substack{J_i\ge 0,\,0\le i\le\kappa\\ J_1+\cdots+\kappa J_\kappa=K\\ J_0+\cdots+J_\kappa=N}} \frac{N!}{J_0!\cdots J_\kappa!}\,.$$

Denote the cylinder function $[1+\xi(x)]\,\mathbf{1}\{\xi(x)<\kappa\}$ by $h(\xi(x))$. Replacing f_{b-1} by 1 in formulas (2.9) and (2.10), we get that the expected value, with respect to ν_{N_1,K_b}, of $\{N_1 R_{N_1,K_b}/(K_b+1)\}h(\xi(1))$ is equal to 1. To keep notation simple we shall abbreviate $(NR_{N,K}/K+1)$ by $S_{N,K}$ and denote by $g_{N_1,K_b}(\xi(x))$ the cylinder function $S_{N_1,K_b}h(\xi(x))$. We may now write

$$\begin{aligned} E_{N_1,K_b+1}[f_{b-1}] \;&-\; E_{N_1,K_b}[f_b]\\ &= \frac{1}{N_1}\sum_{x\in\Omega_{N_1}} E_{N_1,K_b}\Big[\Big\{f_{b-1}(\xi+\mathfrak{d}_x)-f_b(\xi)\Big\}g_{N_1,K_b}(\xi(x))\Big]\\ &\quad+\frac{1}{N_1}\sum_{x\in\Omega_{N_1}} E_{N_1,K_b}\Big[f_b(\xi)\Big\{g_{N_1,K_b}(\xi(x))-1\Big\}\Big]\,. \end{aligned} \tag{2.11}$$

Notice that the second term is the covariance of f_b and $g_{N_1,K_b}(\xi(x))$ because $g_{N_1,K_b}(\xi(x))$ has mean one. We shall estimate each term of (2.11) separately. We start with the first one which is simpler.

Lemma 2.4 *There exists a constant $C(\kappa)$ depending only on κ such that*

$$\begin{aligned} &\sum_{b=1}^{\kappa} \nu^1_{N,K}(b)\Big\{\frac{1}{N_1}\sum_{x\in\Omega_{N_1}} E_{N_1,K_b}\Big[\Big\{f_{b-1}(\xi+\mathfrak{d}_x)-f_b(\xi)\Big\}g_{N_1,K_b}(\xi(x))\Big]\Big\}^2\\ &\quad\le\; C(\kappa)\,N\,\mathfrak{D}(\nu_{N,K},f)\,. \end{aligned} \tag{2.12}$$

Proof. By Schwarz inequality and since g_{N_1,K_b} has mean 1 with respect to the measure ν_{N_1,K_b}, the left hand side of (2.12) is bounded above by

$$\sum_{b=1}^{\kappa} \nu^1_{N,K}(b) \frac{1}{N_1} \sum_{x \in \Omega_{N_1}} E_{N_1,K_b}\Big[\Big\{ f_{b-1}(\xi + \mathfrak{d}_x) - f_b(\xi) \Big\}^2 g_{N_1,K_b}(\xi(x)) \Big] .$$

We shall prove at the end of this lemma that $S_{N,K}$ is uniformly bounded over all densities less than $(2/3)\kappa$:

$$\sup_{\substack{N \ge 1 \\ 0 \le K \le (2/3)\kappa N}} S_{N,K} \;\le\; C(\kappa) \tag{2.13}$$

for some constant C depending only on κ. From this estimate it follows that the function $g_{N_1,K_b}(\xi(x))$ is bounded above by $C(\kappa)\mathbf{1}\{\xi(x) < \kappa\}$ because there are at most κ particles per site. In particular, the left hand side of (2.12) is bounded above by

$$\begin{aligned} \frac{C(\kappa)}{N_1} \sum_{x \in \Omega_{N_1}} \sum_{b=1}^{\kappa} \nu^1_{N,K}(b) E_{N_1,K_b}\Big[\Big\{ f_{b-1}(\xi + \mathfrak{d}_x) - f_b(\xi) \Big\}^2 \mathbf{1}\{\xi(x) < \kappa\} \Big] \\ = \frac{C(\kappa)}{N_1} \sum_{x \in \Omega_{N_1}} E_{N,K}\Big[E_{N_1,K_{\eta(N)}}\Big[\Big\{ f_{\eta(N)-1}(\xi + \mathfrak{d}_x) - f_{\eta(N)}(\xi) \Big\}^2 \times \\ \times \mathbf{1}\{\eta(N) > 0, \xi(x) < \kappa\} \Big]\Big] . \end{aligned}$$

In these formulas $C(\kappa)$ is a constant depending only on κ that may change from line to line. Recall that $f_{\eta(N)}(\xi) = f(\xi, \eta(N))$ and that $\nu_{N,K}$–conditional expectation with respect to $\eta(N)$ corresponds to expectation with respect to $\nu_{N_1,K-\eta(N)}$. Therefore, the last sum is equal to

$$\frac{C(\kappa)}{N_1} \sum_{x \in \Omega_{N_1}} E_{N,K}\Big[r_{N,x}(\eta) \Big\{ f(\eta^{N,x}) - f(\eta) \Big\}^2 \Big] . \tag{2.14}$$

It remains to estimate this expression by the Dirichlet form. The difficulty here is to evaluate the effect of a long range jump from N to x with a Dirichlet form that measures only modifications due to nearest neighbor jumps. We already faced this problem when proving the two block estimates for zero range processes in section 5.5. The indicator function $r_{N,x}(\eta)$ adds here a minor difficulty. In sake of completeness we shall prove at the end of this section that for each fixed sites x and y,

$$\begin{aligned} & E_{N,K}\Big[r_{x,y}(\eta) \Big\{ f(\eta^{x,y}) - f(\eta) \Big\}^2 \Big] \\ & \quad \le C(\kappa) |x-y| \sum_{z = x \wedge y}^{x \vee y - 1} E_{N,K}\Big[r_{z,z+1}(\eta) \Big\{ f(\eta^{z,z+1}) - f(\eta) \Big\}^2 \Big] \end{aligned}$$

for some finite constant $C(\kappa)$ depending only on κ. It follows from this estimate that (2.14) is bounded above by

$$\begin{aligned}
&\frac{C(\kappa)}{N_1} \sum_{x\in\Omega_{N_1}} (N-x) \sum_{y=x}^{N-1} E_{N,K}\Big[r_{y,y+1}(\eta)\Big\{f(\eta^{y,y+1}) - f(\eta)\Big\}^2\Big] \\
&\quad\le C(\kappa) N \sum_{y\in\Omega_{N-1}} E_{N,K}\Big[r_{y,y+1}(\eta)\Big\{f(\eta^{y,y+1}) - f(\eta)\Big\}^2\Big] \\
&\quad= C(\kappa)\, N\, \mathfrak{D}(\nu_{N,K}, f)\,.
\end{aligned}$$

To conclude the proof of the lemma it remains to show that $S_{N,K}$ is bounded. Recall that this expression is equal to $(E_{N,K}[h(\xi(1))])^{-1}$, because the expectation of $g_{N,K}(\xi(1))$ with respect to $\nu_{N,K}$ is equal to 1. For each fixed N and K this expression is of course bounded because $h(0) = 1$. To prove the statement we have therefore to investigate the asymptotic behavior as $N \uparrow \infty$. By the equivalence of ensembles (Lemma A2.2.2), there exists a finite constant $B(\kappa)$ depending only on κ such that

$$\Big| E_{N,K}[h(\xi(1))] - E_{K/N}[h(\xi(1)]\Big| \le B(\kappa)/N$$

for all $0 \le K/N \le 2\kappa/3$. In this formula and below E_α indicates expectation with respect to the grand canonical invariant measure with density α, that we denoted by ν_α. Changing variables we get that $E_\alpha[h(\xi(0))] = \alpha/\Phi(\alpha)$. A straightforward expansion around the origin shows that this function is bounded below by a strictly positive constant on any compact subset of $[0, 2)$. In particular, $E_{N,K}[h(\xi(0))]$ is bounded below by a positive constant for N large enough and $0 \le K < (2/3)\kappa N$. □

We turn now to the second line in decomposition (2.11). Since $g_{N_1,K_b}(\xi(x))$ has mean 1 with respect to ν_{N_1,K_b} the second line reduces to the covariance

$$E_{N_1,K_b}\Big[f_b(\xi)\,;\, \frac{1}{N_1}\sum_{x\in\Omega_{N_1}} g_{N_1,K_b}(\xi(x))\Big]\,.$$

Proposition 2.5 *For each fixed $\varepsilon > 0$, there exists $\ell = \ell(\varepsilon)$ and $N' = N'(\varepsilon)$ such that,*

$$\begin{aligned}
&\sum_{b=1}^{\kappa} \nu^1_{N,K}(b)\Big\{E_{N_1,K_b}\Big[f_b(\xi)\,;\, \frac{1}{N_1}\sum_{x\in\Omega_{N_1}} g_{N_1,K_b}(\xi(x))\Big]\Big\}^2 \\
&\quad\le \frac{C(\kappa)\ell W(\ell+1)}{N}\mathfrak{D}(\nu_{N,K}, f) + \frac{\varepsilon}{N}\mathbf{Var}(\nu_{N,K}, f)
\end{aligned} \tag{2.15}$$

for all $N \ge N'(\varepsilon)$ and all $0 \le K \le (\kappa/2)N$.

Before proving this statement, notice that Proposition 2.2 follows from inequality (2.8), formula (2.11), Lemma 2.4 and Proposition 2.5. We turn now to the proof of Proposition 2.5. We first single out the case of small densities.

Lemma 2.6 *For every $\varepsilon > 0$, there exists a positive integer N' and a density α_0 such that*

$$\mathbf{Var}\Big(\nu_{N,K}, \frac{1}{N_1} \sum_{x\in\Omega_{N_1}} g_{N_1,K_b}(\xi(x))\Big) \le \frac{\varepsilon}{N}\cdot$$

for all $N \ge N'$ and all $K/N \le \alpha_0$.

Proof. The variance may be written as

$$\frac{1}{N_1} E_{N_1,K_b}\Big[g_{N_1,K_b}(\xi(1))\,;\, g_{N_1,K_b}(\xi(1))\Big] + \Big(1 - \frac{1}{N_1}\Big) E_{N_1,K_b}\Big[g_{N_1,K_b}(\xi(1))\,;\, g_{N_1,K_b}(\xi(2))\Big]\,.$$

Since $g_{N_1,K_b}(\xi(1))$ has mean 1, a change of variables $\xi' = \xi + \mathfrak{d}_1$ permits to rewrite the variance of $g_{N_1,K_b}(\xi(1))$ as

$$\frac{N_1}{K_b+1} E_{N_1,K_b+1}\Big[g_{N_1,K_b}(\xi(1)-1)\xi(1)\Big] - E_{N_1,K_b}\Big[g_{N_1,K_b}(\xi(1))\Big]\,.$$

By similar reasons the covariance of $g_{N_1,K_b}(\xi(1))$ and $g_{N_1,K_b}(\xi(2))$ is equal to

$$\frac{N_1}{K_b+1} E_{N_1,K_b+1}\Big[g_{N_1,K_b}(\xi(1))\xi(2)\Big] - E_{N_1,K_b}\Big[g_{N_1,K_b}(\xi(1))\Big]\,.$$

In order to obtain a simpler expression for this covariance observe that the first expectation may be rewritten as

$$\frac{N-1}{(K_b+1)(N-2)} \sum_{x=2}^{N-1} E_{N_1,K_b+1}\Big[g_{N_1,K_b}(\xi(1))\xi(x)\Big]$$

because $\nu_{N,K}$ is the uniform measure over all configurations of $\Sigma^{\kappa}_{N,K}$. Since $\nu_{N,K}$ is concentrated on configurations with K particles, $\sum_{x=2}^{N-1} \xi(x) = K_b + 1 - \xi(1)$. Therefore, last expression is equal to

$$\frac{N-1}{(N-2)} E_{N_1,K_b+1}\Big[g_{N_1,K_b}(\xi(1))\Big] - \frac{N-1}{(K_b+1)(N-2)} E_{N_1,K_b+1}\Big[g_{N_1,K_b}(\xi(1))\xi(1)\Big]\,.$$

Up to this point we showed that the variance of $\sum_{x\in\Omega_{N_1}} g_{N_1,K_b}(\xi(x))$ with respect to ν_{N_1,K_b} is equal to

$$S_{N_1,K_b}\Big\{ E_{N_1,K_b+1}[h(\xi(1))] - E_{N_1,K_b}[h(\xi(1))] - \frac{1}{K_b+1} E_{N_1,K_b+1}\Big[\xi(1)\Big\{h(\xi(1)) - h(\xi(1)-1)\Big\}\Big]\Big\}\,,$$

where $h(\xi(1))$ is the cylinder function $[1 + \xi(1)]\mathbf{1}\{\xi(1) < \kappa\}$. To conclude the proof of the lemma it remains to recall that $S_{N,K}$ is bounded and apply Lemma A2.2.3, □

Lemma 2.7 *For each fixed $\varepsilon > 0$ and $\alpha_0 > 0$, there exists $\ell = \ell(\alpha_0, \varepsilon)$ and $N = N(\alpha_0, \varepsilon)$ such that*

$$\sum_{b=1}^{\kappa} \nu_{N,K}^1(b)\Big\{ E_{N_1,K_b}\Big[f_b(\xi)\,;\, \frac{1}{N_1}\sum_{x\in\Omega_{N_1}} g_{N_1,K_b}(\xi(x))\Big]\Big\}^2 \le \frac{C(\kappa)\ell W(\ell+1)}{N}\mathfrak{D}(\nu_{N,K}, f) + \frac{\varepsilon}{N}\mathbf{Var}(\nu_{N,K}, f)\,. \tag{2.16}$$

for all $N \ge N(\alpha_0, \varepsilon)$ and all $\alpha_0 N \le K \le (\kappa/2)N$.

Proof. Fix a positive integer $\ell \le \sqrt{N}$ and divide the set $\{1, \dots, N_1\}$ in non overlapping intervals $\{B_a,\ 1 \le a \le p\}$ of length ℓ or $\ell + 1$:

$$\Omega_{N_1} = \bigcup_{a=1}^{p} B_a \qquad \text{and} \quad B_{a_1} \cap B_{a_2} = \phi \quad \text{for} \quad a_1 \neq a_2\,.$$

Denote by $M_a = M_a(\xi)$ the total number of particles in the interval B_a for the configuration ξ: $M_a = \sum_{x\in B_a} \xi(x)$, by $|B_a|$ the cardinality of B_a, by ξ_a^ℓ the density of particles in B_a: $\xi_a^\ell = |B_a|^{-1}M_a$ and by $\tilde{g}_a(\xi_a^\ell)$ the expected value of the cylinder function g_{N_1,K_b} with respect to the measure ν_{B_a,M_a}:

$$\tilde{g}_a(\xi_a^\ell) = E_{\nu_{B_a,M_a}}\Big[g_{N_1,K_b}(\xi(x))\Big] \qquad \text{for any} \quad x \in B_a\,.$$

Since the covariance is bilinear, the left hand side of (2.16) is bounded by

$$2\sum_{b=1}^{\kappa} \nu_{N,K}^1(b)\Big\{ \frac{1}{N_1}\sum_{a=1}^{p} |B_a| E_{N_1,K_b}\Big[f_b(\xi); \frac{1}{|B_a|}\sum_{x\in B_a} g_{N_1,K_b}(\xi(x)) - \tilde{g}_a(\xi_a^\ell)\Big]\Big\}^2 + 2\sum_{b=1}^{\kappa} \nu_{N,K}^1(b)\Big\{ E_{N_1,K_b}\Big[f_b(\xi)\,;\, \frac{1}{N_1}\sum_{a=1}^{p} |B_a|\, \tilde{g}_a(\xi_a^\ell)\Big]\Big\}^2 . \tag{2.17}$$

We consider these two expressions separately.

To estimate the first line, apply Schwarz inequality and take conditional expectation with respect to M_a. Since to take conditional expectation with respect to the total number of particles M_a corresponds to integrate with respect to ν_{B_a,M_a}, the first expression in (2.17) is equal to

$$2\sum_{b=1}^{\kappa} \nu_{N,K}^1(b)\frac{1}{N_1}\sum_{a=1}^{p} |B_a| \times E_{N_1,K_b}\Big[\Big\{ E_{B_a,M_a}\Big[f_b(\xi)\,;\, \frac{1}{|B_a|}\sum_{x\in B_a} g_{N_1,K_b}(\xi(x))\Big]\Big\}^2\Big]$$

because, by definition, $\tilde{g}_a(\xi_a^\ell)$ is the expected value of $g_{N_1,K_b}(\cdot)$ with respect to ν_{B_a,M_a}. In this formula f_b should be understood as a function of $\{\xi(x);\ x \in B_a\}$ with all remaining coordinates frozen.

By (2.13), the function $g_{N_1,K_b}(\cdot)$ is bounded by a constant depending only on κ. Therefore, by Schwarz inequality and by the induction hypothesis, for each a we have that

$$\Big\{ E_{B_a,M_a}\Big[f_b(\xi)\, ;\, \frac{1}{|B_a|}\sum_{x\in B_a} g_{N_1,K_b}(\xi(x)) \Big]\Big\}^2 \le C(\kappa)\mathbf{Var}(\nu_{B_a,M_a}, f_b)$$
$$\le C(\kappa) W(|B_a|)\mathfrak{D}(\nu_{B_a,M_a}, f_b)\,.$$

Moreover,

$$\sum_{a=1}^{p} E_{N_1,K_b}\Big[\mathfrak{D}(\nu_{B_a,M_a}, f_b)\Big] \le \mathfrak{D}(\nu_{N_1,K_b}, f_b)\,.$$

This inequality misses to be an equality because the Dirichlet form on the left hand side does not take into account modifications due to jumps from one cube B_a to another cube B_b for $a \neq b$. Therefore, the first expression in (2.17) is bounded above by

$$C(\kappa)N^{-1}(\ell+1)W(\ell+1)\sum_{b=1}^{\kappa} \nu^1_{N,K}(b)\mathfrak{D}(\nu_{N_1,K_b}, f_b)$$
$$\le \frac{C(\kappa)\ell W(\ell+1)}{N}\mathfrak{D}(\nu_{N,K}, f)\,.$$

because all cubes B_a have length at most $(\ell+1)$.

We turn now to the second expression of formula (2.17). Denote by $\tilde{h}(\alpha)$ the expectation, with respect to ν_α, of the cylinder function $h(\xi) = [1+\xi(0)]\mathbf{1}\{\xi(0) < \kappa\}$, by $\tilde{g}(\alpha)$ the expectation of g_{N_1,K_b} with respect to ν_α and by $\tilde{g}'(\alpha)$ the derivative of the smooth function $\tilde{g}$ calculated at α:

$$\tilde{h}(\alpha) = E_{\nu_\alpha}\Big[[1+\xi(0)]\mathbf{1}\{\xi(0)<\kappa\}\Big]\,, \quad \tilde{g}(\alpha) = S_{N_1,K_b}\tilde{h}(\alpha)$$
$$\text{and}\quad \tilde{g}'(\alpha_0) = \frac{d}{d\alpha}\tilde{g}(\alpha)\Big|_{\alpha=\alpha_0}\,.$$

Consider now the second term in formula (2.17). Since $\sum_a |B_a|\xi_a^\ell$ is equal to the total number of particles on Ω_{N_1}, we may add $\tilde{g}(K_b/N_1) - \tilde{g}'(K_b/N_1)[\xi_a^\ell - K_b/N_1]$ to the second term of the covariance without modifying the value of the covariance. The second expression in (2.17) is therefore equal to

$$2\sum_{b=1}^{\kappa}\nu^1_{N,K}(b)\Big\{E_{N_1,K_b}\Big[f_b(\xi)\,;\,\frac{1}{N_1}\sum_{a=1}^{p}|B_a|F_a(\xi_a^\ell)\Big]\Big\}^2$$
$$\le 2\sum_{b=1}^{\kappa}\nu^1_{N,K}(b)\mathbf{Var}\Big(\nu_{N_1,K_b}, f_b\Big)E_{N_1,K_b}\Big[\Big(\frac{1}{N_1}\sum_{a=1}^{p}|B_a|F_a(\xi_a^\ell)\Big)^2\Big]\,, \tag{2.18}$$

where

$$F_a(\xi_a^\ell) = \tilde g_a(\xi_a^\ell) - \tilde g(K_b/N_1) - \tilde g'(K_b/N_1)[\xi_a^\ell - K_b/N_1] . \tag{2.19}$$

To conclude the proof of the lemma it remains to show that the second variance is bounded by εN^{-1} for some $\ell = \ell(\alpha_0, \varepsilon)$, all $N \geq N(\alpha_0, \varepsilon)$ and all $\alpha_0 N \leq K \leq (\kappa/2)N$. This variance is equal to

$$\sum_a \frac{|B_a|^2}{N_1^2} E_{N_1,K_b}\Big[F_a(\xi_a^\ell)^2\Big] + \sum_{a\neq b} \frac{|B_a||B_b|}{N_1^2} E_{N_1,K_b}\Big[F_a(\xi_a^\ell)F_b(\xi_b^\ell)\Big] .$$

Consider a total number of particles K so that the density K_b/N_1 is bounded below by α_0 and above by $\kappa/2$. By Remark A2.1.2, Corollary A2.1.4 and since the grand canonical measures ν_α are product, for N large enough,

$$E_{N_1,K_b}[F_a(\xi_a^\ell)^2] \leq E_{K_b/N_1}\Big[F_a(\xi_a^\ell)^2\Big] + \frac{E(\alpha_0)\ell}{N}\Big\{E_{K_b/N_1}\Big[F_a(\xi_a^\ell)^4\Big]\Big\}^{1/2}$$

and, for $a \neq b$,

$$E_{N_1,K_b}[F_a(\xi_a^\ell)F_b(\xi_b^\ell)] \leq E_{K_b/N_1}\Big[F_a(\xi_a^\ell)\Big] E_{K_b/N_1}\Big[F_b(\xi_b^\ell)\Big] + \frac{E(\alpha_0)\ell}{N}\Big\{E_{K_b/N_1}\Big[F_a(\xi_a^\ell)^2\Big] E_{K_b/N_1}\Big[F_b(\xi_b^\ell)^2\Big]\Big\}^{1/2} .$$

By definition of $\tilde g_a(\cdot)$ and of $\tilde g(\cdot)$, the expectation of $F_a(\xi_a^\ell)$ with respect to the grand canonical measure ν_{K_b/N_1} is equal to 0. In particular, applying the elementary inequality $2ab \leq a^2 + b^2$, we obtain that the second variance in (2.18) is bounded by

$$\frac{E(\alpha_0)\ell}{N}\sum_a \frac{|B_a|}{N} E_{K_b/N_1}\Big[F_a(\xi_a^\ell)^2\Big] + \frac{E(\alpha_0)\ell^2}{N^2}\sum_a \frac{|B_a|}{N}\Big\{E_{K_b/N_1}\Big[F_a(\xi_a^\ell)^4\Big]\Big\}^{1/2} . \tag{2.20}$$

We shall estimate these two terms separately. Notice, however, that we reduced the original problem involving canonical measures to a simpler estimate concerning only product measures.

Recall that $\tilde g_a(\xi_a)$ (resp. $\tilde g(\alpha)$) denotes the expectation of $S_{N_1,K_b}h$ with respect to ν_{B_a,M_a} (resp. ν_α). In particular, by (2.13), $\tilde g_a$ and $\tilde g$ are bounded by a constant depending only on κ. Moreover, we get an explicit formula for $\tilde h$ changing variables: $\tilde h(\alpha) = \alpha/\Phi(\alpha)$. This formula permits to compute the derivative of $\tilde g(\cdot)$:

$$\tilde g'(\alpha) = S_{N_1,K_b}\tilde h'(\alpha) \quad\text{and}\quad \tilde h'(\alpha) = \frac{\sigma(\Phi(\alpha))^2 - \alpha}{\Phi(\alpha)\sigma(\Phi(\alpha))^2} .$$

Hence, $\tilde{h}'(\cdot)$ is a smooth function on $(0, \kappa)$. A straightforward expansion permits to describe its behavior at the boundary:

$$\lim_{\alpha \to 0} \tilde{h}'(\alpha) = 1 , \quad \lim_{\alpha \to \kappa} \tilde{h}'(\alpha) = -\kappa .$$

In particular, $\tilde{h}'(\cdot)$ may be extended as a continuous function on $[0, \kappa]$. This estimate together with the previous bounds on $\tilde{g}_a(\xi_a)$ and $\tilde{g}(\alpha)$ show that F_a is bounded by a constant depending only on κ. In particular, the second expression in (2.20) is bounded above by $E(\alpha_0, \kappa)\ell^2/N^2$.

On the other hand, by Lemma A2.2.2,

$$\left| \tilde{g}_a(\xi_a^\ell) - \tilde{g}(\xi_a^\ell) \right| \le S_{N_1, K_b} \frac{C(\kappa)}{\ell}$$

for some constant $C(\kappa)$ depending only on κ. In particular, by (2.13) and Schwarz inequality, the first expression in (2.20) is bounded above by

$$\frac{E(\alpha_0, \kappa)}{\ell N} + \frac{E(\alpha_0)\ell}{N} \sum_a \frac{|B_a|}{N} E_{K_b/N_1} \left[G_a(\xi_a^\ell)^2 \right] ,$$

where $G_a(\xi_a^\ell) = \tilde{g}(\xi_a^\ell) - \tilde{g}(K_b/N_1) - \tilde{g}'(K_b/N_1)(\xi_a^\ell - K_b/N_1)$. By Taylor formula, the absolute value of G_a is bounded above by $S_{N_1, K_b} \|\tilde{h}''\|_\infty (\xi_a^\ell - K_b/N_1)^2$. The first expression of (2.20) is thus bounded above by

$$S_{N_1, K_b} \|\tilde{h}''\|_\infty^2 \frac{E(\alpha_0, \kappa)\ell}{N\ell^2} \left\{ \|m_4\|_\infty + \|\sigma^2\|_\infty^2 \right\} .$$

In this formula $\|\cdot\|_\infty$ stands for the sup norm of functions defined on $(0, \kappa)$. To conclude the proof it remains to show that $\|\tilde{h}''\|_\infty$, $\|m_4\|_\infty$ and $\|\sigma^2\|_\infty$ are bounded. As α approaches 0 (resp. κ), the one site marginal of ν_α converges to the Dirac measure concentrated on the configuration with 0 (resp. κ) particles. Therefore, $m_4(\Phi(\alpha))$ and $\sigma(\Phi(\alpha))^2$ must vanish as α converges to 0 or κ. This behavior can also be checked expanding m_4 and σ^2 around the origin and around κ. This proves that m_4 and σ^2 are bounded functions because they are smooth on $(0, \kappa)$.

On the other hand, straightforward computations show that

$$\tilde{h}''(\alpha) = \frac{\alpha\sigma(\Phi(\alpha))^2 + \alpha m_3(\Phi(\alpha)) - 2\sigma(\Phi(\alpha))^4}{\Phi(\alpha)\sigma(\Phi(\alpha))^6} .$$

From this formula, one can show that $\lim_{\alpha\to 0} \tilde{h}''(\alpha) = 0$ and $\lim_{\alpha\to 0} \tilde{h}''(\alpha) = -2(\kappa + 1)$. Therefore $\tilde{h}''$ can be extend as a continuous function on $[0, \kappa]$ and the lemma is proved □

We conclude this section with an estimate on the Dirichlet form of generalized symmetric simple exclusion processes with long jumps that was used in the proof of Lemma 2.4. Since this estimate reappears in the sequel for processes evolving in higher dimensions, we prove it in this more general setup.

Denote by $\Omega_{N,K}$ the cube $\{1,\dots,N\}^d$ and by $\nu_{N,K}$ the uniform measure on $\Sigma^{\kappa}_{\Omega_{N,K}}$. For each pair of sites $x = (x_1,\dots,x_d)$ and $y = (y_1,\dots y_d)$, let $n = n(x,y)$ stand for the distance from x to y: $n = \|x - y\| = \sum_{1\le i\le d} |x_i - y_i|$ and denote by $\Gamma(x,y)$ a path from x to y, i.e., a sequence $x = x_0,\dots,x_n = y$ of sites such that $\|x_{i+1} - x_i\| = 1$ for $0 \le i \le n-1$. Note that we used the same symbol x_i to represent a sequence of sites in $\mathbb{Z}^d$ and the i-th coordinate of a site x. In the following, to avoid confusions and whenever necessary, we will clarify the meaning of x_i.

Lemma 2.8 *For each fixed pair of sites x, y in $\Omega_{N,K}$ and each path $\Gamma(x,y) = (x = x_0,\dots,x_n = y)$,*

$$E_{\nu_{N,K}}\Big[r_{x,y}(\xi)\Big\{f(\xi^{x,y}) - f(\xi)\Big\}^2\Big]$$
$$\le\ 4\kappa^2\|x-y\|\sum_{k=0}^{n-1} E_{\nu_{N,K}}\Big[r_{x_k,x_{k+1}}(\xi)\Big\{f(\xi^{x_k,x_{k+1}}) - f(\xi)\Big\}^2\Big]\,.$$

Proof. The proof of this lemma relies on two observations. On the one hand, the Dirichlet form associated to exchange of occupation variables (rather than jumps of particles) is bounded above by the Dirichlet form associated to exclusion jumps. This is clear because to exchange the occupation variables of two distinct sites z_0 and z_1, one may just perform few jumps from one site to the other. On the other hand, to displace one particle from site x to site y, we may simply exchange the occupation variables of sites $x = x_0$ and x_1, than exchange the occupation variables of sites x_1 and x_2 and repeat this procedure up to the point where particles originally at x sit at site x_{n-1}. Then, one may let one particle jump from site x_{n-1} to site $x_n = y$ and repeat back the exchange procedure to retrieve the original configuration with one particle less at site x and one additional particle at site y.

A rigorous proof requires some notation. For each pair of neighbor sites z_0, z_1 and each configuration η, denote by T_{z_0,z_1} the configuration obtained from η changing the occupation variables $\eta(z_0)$ and $\eta(z_1)$:

$$(T_{z_0,z_1}\eta)(z) = \begin{cases} \eta(z) & \text{if } z \neq z_0, z_1\,, \\ \eta(z_1) & \text{if } z = z_0 \text{ and} \\ \eta(z_0) & \text{if } z = z_1\,. \end{cases}$$

We first claim that

$$\int \Big[f(T_{z_0,z_1}\eta) - f(\eta)\Big]^2 \nu_{N,K}(d\eta)\ \le\ 4\kappa^2 < -L_{z_0,z_1} f, f >_{\nu_{N,K}} \tag{2.21}$$

for every f in $L^2(\nu_{N,K})$ and every bond $\{z_0, z_1\}$.

To prove this statement, for fixed $0 \le a < b \le \kappa$, consider the set of configurations such that $\eta(z_0) = a$, $\eta(z_1) = b$. For $0 \le k \le b-a$, let $\xi_k = \xi_k(\eta) =$

$\eta + k\mathfrak{d}_{z_0} - k\mathfrak{d}_{z_1}$ and notice that $\xi_0 = \eta$, $\xi_{b-a} = T_{z_0,z_1}\eta$ and $\xi_{k+1} = (\xi_k)^{z_1,z_0}$. We may thus rewrite the difference $f(T_{z_0,z_1}\eta) - f(\eta)$ as $\sum_{0\le k\le b-a-1}[f((\xi_k)^{z_1,z_0}) - f(\xi_k)]$. In particular, by Schwarz inequality,

$$\int \mathbf{1}\{\eta(z_0) = a, \eta(z_1) = b\}\Big[f(T_{z_0,z_1}\eta) - f(\eta)\Big]^2 \nu_{N,K}(d\eta)$$

$$\le (b-a)\sum_{k=0}^{b-a-1} \int \mathbf{1}\{\eta(z_0) = a, \eta(z_1) = b\}\Big[f((\xi_k)^{z_1,z_0}) - f(\xi_k)\Big]^2 \nu_{N,K}(d\eta)\,.$$

Performing a change of variables $\zeta = \xi_k$ we obtain that the last expression is equal to

$$(b-a)\sum_{k=0}^{b-a-1} \int \mathbf{1}\{\zeta(z_0) = a+k, \zeta(z_1) = b-k\}\Big[f(\zeta^{z_1,z_0}) - f(\zeta)\Big]^2 \nu_{N,K}(d\zeta)$$

because $\nu_{N,K}$ is the uniform measure. Summing over all $0 \le a < b \le \kappa$ gives that

$$\int \mathbf{1}\{\eta(z_0) < \eta(z_1)\}\Big[f(T_{z_0,z_1}\eta) - f(\eta)\Big]^2 \nu_{N,K}(d\eta)$$

$$\le \kappa^2 \int r_{z_1,z_0}(\eta)\Big[f(\eta^{z_1,z_0}) - f(\eta)\Big]^2 \nu_{N,K}(d\eta)\,.$$

To conclude the proof of the claim it remains to repeat the same argument in the case $\eta(z_0) > \eta(z_1)$ and recall the explicit expression for the Dirichlet form $< -L_{z_0,z_1}f, f >_{\nu_{N,K}}$.

Recall now the definition of the path $\Gamma(x,y) = (x_0,\dots,x_n)$ from x to y. For a fixed configuration ξ and for $0 \le k \le n-1$, denote by $\zeta_k = \zeta_k(\xi)$ the configuration $T_{x_{k-1},x_k}\cdots T_{x_0,x_1}\xi$, $\zeta_0 = \xi$. Let $\zeta_n = \zeta_n(\xi)$ be the configuration obtained from ζ_{n-1} letting one particle jump from x_{n-1} to y: $\zeta_n = (\zeta_{n-1})^{x_{n-1},y}$. For $1 \le j \le n-1$, let $\zeta_{n+j} = T_{x_{n-1-j},x_{n-j}}\cdots T_{x_{n-1},x_{n-2}}\zeta_n$. Notice that $\zeta_0 = \xi$ and that $\zeta_{2n-1} = \xi^{x,y}$. We may therefore rewrite the difference $f(\xi^{x,y}) - f(\xi)$ as $\sum_{0\le k\le 2n-2}[f(\zeta_{k+1}) - f(\zeta_k)]$. By Schwarz inequality,

$$\Big\{f(\xi^{x,y}) - f(\xi)\Big\}^2 \le (2n-1)\sum_{k=0}^{2n-2}\Big\{f(\zeta_{k+1}) - f(\zeta_k)\Big\}^2\,.$$

Notice that $\zeta_{k+1} = T_{x_k,x_{k+1}}\zeta_k$ for $0 \le k < n-1$ and $\zeta_{k+1} = T_{x_{2n-k-2},x_{2n-k-1}}\zeta_k$ for $n \le k < 2n-1$. Therefore, if we define z_k as x_k for $0 \le k \le n-1$ and as x_{2n-k-2} for $n \le k < 2n-1$, we have that

$$\zeta_{k+1} = T_{z_k,z_{k+1}}\zeta_k \quad \text{for } 0 \le k < 2n-1,\ k \ne n-1\,.$$

With this notation and since the jump rate $r_{x,y}$ is bounded by 1,

$$\sum_{\substack{0\le k\le 2n-2\\ k\ne n-1}} E_{\nu_{N,K}}\Big[r_{x,y}(\xi)\Big\{f(\zeta_{k+1})-f(\zeta_k)\Big\}^2\Big]$$

$$\le \sum_{\substack{0\le k\le 2n-2\\ k\ne n-1}} E_{\nu_{N,K}}\Big[\Big\{f(T_{z_k,z_{k+1}}\zeta_k)-f(\zeta_k)\Big\}^2\Big]\,.$$

Performing the change of variables $\eta=\zeta_k$, applying inequality (2.21) and recalling the definition of z_k, we obtain that the last expression is bounded above by

$$4\kappa^2\sum_{\substack{0\le k\le n-2\\ k\ne n-1}} E_{\nu_{N,K}}\Big[r_{x_k,x_{k+1}}(\xi)\Big\{f(\xi^{x_k,x_{k+1}})-f(\xi)\Big\}^2\Big]\,.$$

On the other hand, $\zeta_n=(\zeta_{n-1})^{x_{n-1},y}$ and $r_{x,y}(\xi)=r_{x_{n-1},y}(\zeta_{n-1})$ because $\zeta_{n-1}(x_{n-1})=\xi(x)$. Therefore, performing the change of variables $\eta=\zeta_{n-1}$, we get that

$$E_{\nu_{N,K}}\Big[r_{x,y}(\xi)\Big\{f(\zeta_{n-1})-f(\zeta_n)\Big\}^2\Big]$$

$$=E_{\nu_{N,K}}\Big[r_{x_{n-1},y}(\zeta_{n-1})\Big\{f((\zeta_{n-1})^{x_{n-1},y})-f(\zeta_{n-1})\Big\}^2\Big]$$

$$=E_{\nu_{N,K}}\Big[r_{x_{n-1},y}(\eta)\Big\{f(\eta^{x_{n-1},y})-f(\eta)\Big\}^2\Big]$$

This estimate together with the previous one concludes the proof of the lemma. □

In dimension one the previous lemma states that for each fixed $x>0$,

$$E_{\nu_{N,K}}\Big[r_{0,x}(\xi)\Big\{f(\xi^{0,x})-f(\xi)\Big\}^2\Big]$$

$$\le 4\kappa^2 x\sum_{y=0}^{x-1}E_{\nu_{N,K}}\Big[r_{y,y+1}(\xi)\Big\{f(\xi^{y,y+1})-f(\xi)\Big\}^2\Big]\,.$$

3. Spectral Gap in Dimension $d\ge 2$

We indicate in this section the main modifications required to extend the proof of the spectral gap in dimension 1 to higher dimensions. To fix ideas we investigate the 2-dimensional case.

To set up a recursive equation for the inverse of the spectral gap $W(N)$ we need to introduce some notation. For each fixed N consider the sequence of sites $\{x_{N,k},\ 1\le k\le 2N-1\}$ defined by

$$x_{N,k}=\begin{cases}(k,N) & \text{for } 1\le k\le N\\ (N,2N-k) & \text{for } N+1\le k\le 2N-1\end{cases}$$

so that $\Omega_N = \Omega_{N-1} \cup \{x_{N,k}, 1 \leq k \leq 2N-1\}$. To keep notation simple, we shall denote $x_{N,k}$ simply by x_k when no confusion arises.

For each $1 \leq k \leq 2N-1$, denote by $\mathcal{F}_k = \mathcal{F}_{N,k}$ the σ-algebra generated by $\{\eta(x_j), 1 \leq j \leq k\}$, by ζ_k the configuration $(\eta(x_1), \dots, \eta(x_k))$, by $\tilde{\Omega}_k$ the set $\Omega_N - \{x_1, \dots, x_k\}$ and by f_k the conditional expectation $E_{\nu_{N,K}}[f|\mathcal{F}_k]$. Recall the definition of the function f_{ζ_k} introduced just before (2.2). Equation (2.2) states that f_k is equal to $E_{\nu_{\tilde{\Omega}_k, K-M(\zeta_k)}}[f_{\zeta_k}]$.

Since $\mathcal{F}_k$ is an increasing sequence of σ-algebras, with the convention that $\mathcal{F}_0$ is the trivial σ-algebra and that $f_0 = E_{\nu_{N,K}}[f]$, $\{f_k, 0 \leq k \leq 2N-1\}$ is a martingale. We may thus express the variance of f as

$$\mathbf{Var}(\nu_{N,K}, f) = E_{\nu_{N,K}}\Big[(f - f_{2N-1})^2\Big] + \sum_{k=0}^{2N-2} E_{\nu_{N,K}}\Big[(f_{k+1} - f_k)^2\Big]. \tag{3.1}$$

The very same arguments presented in the previous section to derive inequality (2.5) and the induction assumption permit to estimate the first term on the right hand side of this identity:

$$E_{\nu_{N,K}}\Big[(f - f_{2N-1})^2\Big] \leq W(N-1)\mathfrak{D}(\nu_{N,K}, f).$$

To estimate the second term on the right hand side of (3.1), recall that $\{f_k, 0 \leq k \leq 2N-1\}$ is a martingale. Therefore, for each $0 \leq k \leq 2N-2$, we may rewrite the expectation $E_{\nu_{N,K}}[(f_{k+1} - f_k)^2]$ as

$$E_{\nu_{N,K}}\Big[E_{\nu_{N,K}}\Big[\Big(f_{k+1} - E_{\nu_{N,K}}[f_{k+1} \,|\, \mathcal{F}_k]\Big)^2 \Big| \mathcal{F}_k\Big]\Big]$$
$$= E_{\nu_{N,K}}\Big[E_{\nu_{\tilde{\Omega}_k, K-M(\zeta_k)}}\Big[\Big(f_{k+1,\zeta_k}(\eta(x_{k+1})) - E_{\nu_{\tilde{\Omega}_k, K-M(\zeta_k)}}[f_{k+1,\zeta_k}]\Big)^2\Big]\Big].$$

In this formula $f_{k+1,\zeta_k}(\eta(x_{k+1}))$ is the function f_{k+1} evaluated at the configuration $(\zeta_k, \eta(x_{k+1}))$. We write it in this way to indicate that $\eta(x_k{+}1)$ only is integrated with respect to $\nu_{\tilde{\Omega}_k, K-M(\zeta_k)}$. In particular, the term $E_{\nu_{\tilde{\Omega}_k, K-M(\zeta_k)}}[(f_{k+1,\zeta_k}(\eta(x_{k+1})) - E_{\nu_{\tilde{\Omega}_k, K-M(\zeta_k)}}[f_{k+1}])^2]$ is the variance of a function depending only on one site. Applying Lemma 2.3 we obtain that this variance is bounded above by

$$B_1(\kappa)\sum_{b=1}^{\kappa} \nu^1_{\tilde{\Omega}_k, K-M(\zeta_k)}(b)\Big\{f_{k+1,\zeta_k}(b-1) - f_{k+1,\zeta_k}(b)\Big\}^2$$
$$= B_1(\kappa)\sum_{b=1}^{\kappa} \nu^1_{\tilde{\Omega}_k, K-M(\zeta_k)}(b)\Big\{f_{k+1}(\zeta_k, b-1) - f_{k+1}(\zeta_k, b)\Big\}^2,$$

where $\nu^1_{\tilde{\Omega}_k, K-M(\zeta_k)}$ stands for the one site marginal of $\nu_{\tilde{\Omega}_k, K-M(\zeta_k)}$.

Since f_{k+1} is the conditional expectation of f given $\mathcal{F}_{k+1}$, by equation (2.2),

$$f_{k+1}(\zeta_{k+1}) = E_{\nu_{\tilde{\Omega}_{k+1}, K-M(\zeta_{k+1})}}[f_{\zeta_{k+1}}].$$

In particular,

$$
\begin{aligned}
& f_{k+1,\zeta_k}(b-1) - f_{k+1,\zeta_k}(b) \\
& \quad = E_{\nu_{\tilde\Omega_{k+1}, K-M(\zeta_k)-b+1}}[f_{\zeta_k,b-1}] - E_{\nu_{\tilde\Omega_{k+1}, K-M(\zeta_k)-b}}[f_{\zeta_k,b}]\;.
\end{aligned}
$$

Adapting the arguments presented between formula (2.9) and (2.11) to the 2-dimensional setting, we show that this difference is equal to

$$
\begin{aligned}
& \frac{1}{|\tilde\Omega_{k+1}|} \sum_{x\in\tilde\Omega_{k+1}} E_{\nu_{\tilde\Omega_{k+1}, K-M(\zeta_k)-b}} \Big[\Big\{ f_{\zeta_k,b-1}(\xi+\mathfrak{d}_x) - f_{\zeta_k,b}(\xi) \Big\} g_{k,K,\zeta_k,b}(\xi(x)) \Big] \\
& \quad + \frac{1}{|\tilde\Omega_{k+1}|} \sum_{x\in\tilde\Omega_{k+1}} E_{\nu_{\tilde\Omega_{k+1}, K-M(\zeta_k)-b}} \Big[f_{\zeta_k,b}(\xi) \Big\{ g_{k,K,\zeta_k,b}(\xi(x)) - 1 \Big\} \Big]\;.
\end{aligned}
$$

In this formula $g_{k,K,\zeta_k,b}$ stands for the mean-one cylinder function defined by $g_{k,K,\zeta_k,b}(\xi(x)) = S_{|\tilde\Omega_{k+1}|, K-M(\zeta_k)-b}[1+\xi(x)]\mathbf{1}\{\xi(x) < \kappa\}$. Notice that the second term is the covariance of $f_{\zeta_k,b}$ and $g_{k,K,\zeta_k,b}(\xi(x))$ because the function $g_{k,K,\zeta_k,b}(\xi(x))$ has mean one.

The following result permits to estimate the first term in the above decomposition. This lemma is the main difference between the proof in dimension one and the proof in higher dimension.

Lemma 3.1 *There exists a constant $C(\kappa)$ depending only on κ such that*

$$
\begin{aligned}
& \sum_{k=0}^{2N-2} \sum_{b=1}^{\kappa} \nu^1_{\tilde\Omega_k, K-M(\zeta_k)}(b) \Big\{ \frac{1}{|\tilde\Omega_{k+1}|} \sum_{x\in\tilde\Omega_{k+1}} \\
& \qquad E_{\nu_{\tilde\Omega_{k+1}, K-M(\zeta_k)-b}} \Big[\Big\{ f_{\zeta_k,b-1}(\xi+\mathfrak{d}_x) - f_{\zeta_k,b}(\xi) \Big\} g_{k,K,\zeta_k,b}(\xi(x)) \Big] \Big\}^2 \\
& \qquad \le C(\kappa)\, N\, \mathfrak{D}(\nu_{N,K}, f)\;.
\end{aligned}
$$

Proof. With the same arguments presented before formula (2.14) in the proof of Lemma 2.4, we obtain that the expression on the left hand side of the above inequality is bounded by

$$
\begin{aligned}
& C(\kappa) \sum_{k=1}^{2N-1} \frac{1}{|\tilde\Omega_{k+1}|} \sum_{x\in\tilde\Omega_{k+1}} E_{\nu_{N,K}} \Big[r_{x_k,x}(\eta) \Big\{ f(\eta^{x_k,x}) - f(\eta) \Big\}^2 \Big] \\
& \quad \le \frac{C(\kappa)}{N^2} \sum_{k=1}^{2N-1} \sum_{x\in\Omega_N} E_{\nu_{N,K}} \Big[r_{x_k,x}(\eta) \Big\{ f(\eta^{x_k,x}) - f(\eta) \Big\}^2 \Big]
\end{aligned}
$$

for some constant $C(\kappa)$ depending only on κ and that may change from line to line.

It remains to estimate this expression by the Dirichlet form of f. We shall consider the case $1 \leq k \leq N$. By symmetry the arguments apply to $N < k \leq 2N - 1$.

Recall the terminology and the notation introduced just before Lemma 2.8. To each site x in Ω_N and each $1 \leq k \leq N$, consider a path $\Gamma(x_k, x)$ from x_k to x moving first along the ordinate and than along the abscissa. More precisely, let $\Gamma(x_k, x) = (x = z_0, z_1, \ldots, z_n = x)$ for $n = \|x_k - x\|$. Among all possible paths from x_k to x we choose the one with the following property: there exists $0 \leq n_0 < n$ so that $z_{j+1} - z_j = -e_2$ for $0 \leq j \leq n_0 - 1$ and $y_{j+1} - y_j = \pm e_1$ for $n_0 \leq j \leq n - 1$.

By Lemma 2.8, the summation over $1 \leq k \leq N$ in last expression is bounded above by

$$\frac{C(\kappa)}{N} \sum_{k=1}^{N} \sum_{x \in \Omega_N} \sum_{j=0}^{\|x_k - x\| - 1} E_{\nu_{N,K}} \Big[r_{z_j, z_{j+1}}(\eta) \Big\{ f(\eta^{z_j, z_{j+1}}) - f(\eta) \Big\}^2 \Big]$$

because all paths have length at most $2N$.

Fix a bond $b = (b_1, b_2)$. From the way we construct the paths, there are at most N^2 different paths using the bond b. In particular, changing the order of summation, we rewrite last formula as

$$\begin{aligned} &\frac{C(\kappa)}{N} \sum_{b=(b_1,b_2)} E_{\nu_{N,K}} \Big[r_{b_1,b_2}(\eta) \Big\{ f(\eta^{b_1,b_2}) - f(\eta) \Big\}^2 \Big] \sum_{k,x;\ \{b_1,b_2\} \subset \Gamma(x_k,x)} \\ &\quad \leq C(\kappa) N \sum_{b=(b_1,b_2)} E_{\nu_{N,K}} \Big[r_{b_1,b_2}(\eta) \Big\{ f(\eta^{b_1,b_2}) - f(\eta) \Big\}^2 \Big] \\ &\quad = C(\kappa) N \mathfrak{D}(\nu_{N,K}, f) \,. \end{aligned}$$

In the first formula, summation over b is carried over all oriented bonds b in Ω_N and the second summation is carried over all sites x and x_k whose path $\Gamma(x_k, x)$ contains b_1 and b_2. Since by symmetry the same argument applies to $N < k \leq 2N - 1$, the lemma is thus proved $\square$

The proof of Proposition 2.5 is easily adapted to higher dimensional processes because the geometry is almost irrelevant. In the present context this proposition states that for each fixed $\varepsilon > 0$, there exists $\ell = \ell(\varepsilon)$ and $N' = N'(\varepsilon)$ such that,

$$\begin{aligned} &\sum_{b=1}^{\kappa} \nu^1_{\tilde{\Omega}_k, K - M(\zeta_k)}(b) \Big\{ \frac{1}{|\tilde{\Omega}_{k+1}|} \sum_{x \in \tilde{\Omega}_{k+1}} \\ &\qquad E_{\nu_{\tilde{\Omega}_{k+1}, K - M(\zeta_k) - b}} \Big[f_{\zeta_k, b}(\xi) \Big\{ g_{k,K,\zeta_k,b}(\xi(x)) - 1 \Big\} \Big] \Big\}^2 \\ &\quad \leq \frac{C(\kappa) \ell W(\ell + 1)}{N^2} \mathfrak{D}(\nu_{N,K}, f) + \frac{\varepsilon}{N^2} \mathbf{Var}(\nu_{N,K}, f) \end{aligned}$$

for all $N \geq N'(\varepsilon)$ and all $0 \leq K \leq (\kappa/2) N^2$.

Summing over $1 \le k \le 2N-1$, we get from this result and Lemma 3.1 that for every $\varepsilon > 0$, the second term on the right hand side of (3.1) is bounded above by

$$C(\kappa)\Big\{N + \frac{\ell W(\ell+1)}{N}\Big\}\mathfrak{D}(\nu_{N,K}, f) + \frac{\varepsilon}{N}\mathbf{Var}(\nu_{N,K}, f)$$

for some $\ell = \ell(\varepsilon)$ and all $N \ge N'(\varepsilon)$, $0 \le K \le (\kappa/2)N^2$. To conclude the proof of the spectral gap in dimension 2 it remains to argues as we did in the previous section just after the statement of Proposition 2.2.

4. Closed and Exact Forms

We investigate in this section the closed and exact forms on Σ^κ, the space of all configuration with at most κ particles per site. To justify some definitions and to clarify the ideas, we briefly overview in the beginning of the section the closed and exact forms on $\mathbb{Z}^d$.

Consider $\mathbb{Z}^d$ endowed with the distance $|||(x_1, \dots, x_d)||| = \sum_{1\le i\le d} |x_i|$ and recall from section 4 that a path $\Gamma(x, y) = (x = x_0, \dots, x_n = y)$ from x to y is a sequence of sites $\{z_i,\ 0 \le i \le n\}$ such that $z_0 = x$, $z_n = y$ and $|||z_{i+1} - z_i||| = 1$ for $0 \le i \le n-1$. The integer n is called the length of the path and x, y the end points. A closed path is a path whose end points are equal and a n-step path is a path of length n. By analogy with the continuous case we define a closed form as

Definition 4.1 A collection $\mathrm{u} = \{\mathrm{u}_x = (\mathrm{u}_x^1, \dots, \mathrm{u}_x^d),\ x \in \mathbb{Z}^d\}$ is a closed form on $\mathbb{Z}^d$ if

$$\mathrm{u}_x^i + \mathrm{u}_{x+e_i}^j = \mathrm{u}_x^j + \mathrm{u}_{x+e_j}^i \tag{4.1}$$

for every $1 \le i, j \le d$ and $x \in \mathbb{Z}^d$.

u_x^i should be interpreted as a i-th partial discrete derivative at x. It represents therefore the price to jump from x to $x+e_i$. With this interpretation condition (4.1) means that the price to jump from x to $x+e_i$ and then from $x+e_i$ to $x+e_i+e_j$ is the same as the one to jump from x to $x+e_j$ and then from $x+e_j$ to $x+e_i+e_j$. In other words, condition (4.1) imposes that the price of a 2-step path depends on the path only through its end points.

Since u_x^i stands for the price to jump from x to $x+e_i$, $-\mathrm{u}_x^i$ is the price to jump from $x+e_i$ to x. Moreover, condition (4.1) considers only "increasing" paths (paths from x to $x+e_i+e_j$ for some x in $\mathbb{Z}^d$ and $1 \le i, j \le d$). Nevertheless, it follows from (4.1) that the price of any 2-step path depends on the path only through its end points. Fix, for instance, x in $\mathbb{Z}^d$, $1 \le i \ne j \le d$ and consider the paths $\Gamma_1 = (x, x+e_i, x+e_i-e_j)$ and $\Gamma_2 = (x, x-e_j, x+e_i-e_j)$. By definition, the price of Γ_1 is $\mathrm{u}_x^i - \mathrm{u}_{x+e_i-e_j}^j$ and the price of Γ_2 is $-\mathrm{u}_{x-e_j}^j + \mathrm{u}_{x-e_j}^i$. These prices are the same if $\mathrm{u}_{x-e_j}^j + \mathrm{u}_x^i = \mathrm{u}_{x-e_j}^i + \mathrm{u}_{x+e_i-e_j}^j$. This equality follows from (4.1) setting $y = x - e_j$.

Like in the continuous case each function $F: \mathbb{Z}^d \to \mathbb{R}$ gives rise to a closed form. These closed forms play a particular role and deserve a special terminology.

Definition 4.2 An exact form is a closed form for which there exists $F: \mathbb{Z}^d \to \mathbb{R}$ such that

$$\mathrm{u}_x^i \;=\; F(x+e_i) \;-\; F(x) \tag{4.2}$$

for x in $\mathbb{Z}^d$ and $1 \le i \le d$. Such form is denoted by $\mathrm{u}^F = (\mathrm{u}^{F,1}, \dots, \mathrm{u}^{F,d})$.

Notice that condition (4.1) follows from identity (4.2), for $\mathrm{u}_x^{F,i} + \mathrm{u}_{x+e_i}^{F,j} = F(x+e_i+e_j) - F(x) = \mathrm{u}_x^{F,j} + \mathrm{u}_{x+e_j}^{F,i}$. Notice, furthermore, that the exact form associated to $F+C$ coincides with the one associated to F for any function F and constant C.

At this point we need to elucidate whether all closed forms are exact forms. To prove that a closed form is an exact form we need to exhibit a function $F: \mathbb{Z}^d \to \mathbb{R}$ for which $\mathrm{u} = \mathrm{u}^F$. In this case, by (4.2), u_x^i would be equal to the i-th partial discrete derivative of F at x. The strategy to build such integral F of the closed form u is thus clear. We fix a site, say the origin, and assign an arbitrary value to F at this site since we have seen that the closed forms associated to F and $F+C$ are equal. Then, for a site x we find a path from 0 to x and define $F(x)$ as the path integral of u along this path. More precisely, let $a \in \mathbb{R}$ and set $F(0) = a$. Fix a site x and select a path $\Gamma(0,x) = (0 = x_0, x_1, \dots, x_n = x)$ from the origin to x. Write $F(x) = a + F(x) - F(0) = a + \sum_{0 \le j \le n-1} F(x_{j+1}) - F(x_j)$. Since, by definition of a path, sites x_j, x_{j+1} are neighbors and since u should be the discrete derivative of F, the difference $F(x_{j+1}) - F(x_j)$ can be expressed with the form u. In the case where $x_{k+1} = x_k + e_i$ for some $1 \le i \le d$, $F(x_{k+1}) - F(x_k) = \mathrm{u}_{x_k}^i = < \mathrm{u}_{x_k}, x_{k+1} - x_k >$, provided $< \cdot, \cdot >$ stand for the inner product in $\mathbb{R}^d$. In the case where $x_{k+1} = x_k - e_i$ for some $1 \le i \le d$, $F(x_{k+1}) - F(x_k) = -\mathrm{u}_{x_{k+1}}^i = < \mathrm{u}_{x_{k+1}}, x_{k+1} - x_k >$. Therefore, if for two sites x, y, we denote by $x \wedge y$ the largest site smaller than x and y for the natural partial order of $\mathbb{Z}^d$ ($(x_1, \dots, x_d) \le (y_1, \dots, y_d)$ if $x_i \le y_i$ for $1 \le i \le d$), $F(x_{k+1}) - F(x_k) = < \mathrm{u}_{x_k \wedge x_{k+1}}, x_{k+1} - x_k >$ and

$$F(x) \;=\; a \;+\; \sum_{k=0}^{n-1} < \mathrm{u}_{x_k \wedge x_{k+1}}, x_{k+1} - x_k > \;\cdot \tag{4.3}$$

The second term on the right hand side of this expression is the path integral of the closed form u along the path $\Gamma(0,x) = (0 = x_0, x_1, \dots, x_n = x)$ that we shall denote by $I_{\Gamma(0,x)}(\mathrm{u})$. Therefore, if the closed form u is the exact form u^F associated to F, F should satisfy the above relation. In principle nothing prevent $I_{\Gamma(0,x)}(\mathrm{u})$ to depend on the particular path $\Gamma(0,x)$ chosen. If we can prove, however, that the path integral $I_{\Gamma(0,x)}(\mathrm{u})$ does not depend on the particular path chosen, formula (4.3) defines a function $F: \mathbb{Z}^d \to \mathbb{R}$ and a straightforward computation shows that $F(x+e_i) - F(x) = \mathrm{u}_x^i$ for all $x \in \mathbb{Z}^d$ and $1 \le i \le d$, i.e., that $\mathrm{u} = \mathrm{u}^F$.

The previous argument shows that if the state space is countable and simply connected, the statement that all closed forms are exact forms follows from the statement that the path integral $I_{\Gamma(0,x)}(\mathrm{u})$ does not depend on the particular path chosen but only on the end points 0, x.

Of course, we may extend the definition of a path integral for any pair of sites x, y in $\mathbb{Z}^d$: for a closed form u and a path $\Gamma(x,y) = (x = x_0, \dots, x_n = y)$ from x to y, denote by $I_{\Gamma(x,y)}(\mathrm{u})$ the path integral of u along $\Gamma(x,y)$:

$$I_{\Gamma(x,y)}(\mathrm{u}) = \sum_{k=0}^{n-1} < \mathrm{u}_{x_k \wedge x_{k+1}}, x_{k+1} - x_k > .$$

We claim that on $\mathbb{Z}^d$ the integral $I_{\Gamma(x,y)}(\mathrm{u})$ depend on the path $\Gamma(x,y)$ only through its end points x and y:

Lemma 4.3 *Consider a closed form* u *on* $\mathbb{Z}^d$ *and two paths* $\Gamma_1(x,y)$, $\Gamma_2(x,y)$ *from* x *to* y. *Then,*

$$I_{\Gamma_1(x,y)}(\mathrm{u}) = I_{\Gamma_2(x,y)}(\mathrm{u}) .$$

Proof. A straightforward induction argument, left to the reader, shows that it is enough to prove the statement for the elementary paths $\Gamma_1 = (x, x \pm e_i, x \pm e_i \pm e_j)$, $\Gamma_2 = (x, x \pm e_j, x \pm e_i \pm e_j)$ for $1 \le i, j \le d$. In this case the identity $I_{\Gamma_1}(\mathrm{u}) = I_{\Gamma_2}(\mathrm{u})$ follows immediately from property (4.1) of the closed form u. We shall prove in Lemma 4.9 below this statement in greater generality. □

Corollary 4.4 *On* $\mathbb{Z}^d$ *all closed forms are exact forms.*

Proof. Fix a closed form u. We have just proved that we may define unambiguously the path integral of u. In particular, the function $F: \mathbb{Z}^d \to \mathbb{R}$ given by $F(x) = I_{\Gamma(0,x)}(\mathrm{u})$ is well defined. Moreover, by definition of the path integral, $F(x + e_i) - F(x) =< \mathrm{u}_x, e_i >= \mathrm{u}_x^i$ for every x in $\mathbb{Z}^d$ and $1 \le i \le d$. Therefore u is the exact form u^F. □

We conclude the examination of closed and exact forms on a countable simply connected space with an example of space that admits closed forms that are not exact. For a positive integer N, consider $\mathbb{T}_N$ the one-dimensional torus with N points and the closed form u identically equal to 1. u is clearly not an exact form. In fact the path integral now depends on the path chosen since $I_{(0,1,\dots x)}(\mathrm{u}) = x$ and $I_{(0,-1,\dots x-N)}(\mathrm{u}) = x - N$ for every $0 < x \le N$.

To investigate closed and exact forms in the context of infinite particle systems, we summarize the concepts and the main ideas introduced up to this point through another perspective.

We started with a topological space $(\mathbb{Z}^d, \|\!| \cdot |\!\|)$ endowed with a discrete metric. This discrete metric permitted to define paths between two sites. We then introduced the concept of a closed form. Condition (4.1) can be interpreted as requiring the path integral of the closed form u along any 2-step path $\Gamma(x,y)$ to depend only on the end points x, y. In Lemma 4.3 we extended this property to a finite length path $\Gamma(x,y)$. This result permitted to integrate $\mathfrak{u}$ unambiguously and to prove that all closed forms are exact forms.

Consider now the state space Σ^κ. For two configurations η and ξ, denote by $D(\eta,\xi)$ the minimum number of nearest neighbor jumps in order to obtain ξ from η. For example, η is at distance 1 from a configuration ξ if ξ is obtained from η by a jump of a particle to a nearest neighbor site: $\xi = \eta^{x,x\pm e_i}$ for some site x in $\mathbb{Z}^d$ and some $1 \le i \le d$.

A path $\Gamma(\eta,\xi) = (\eta = \eta_0, \dots, \eta_n = \xi)$ from a configuration η to ξ is a sequence of configurations η_k such that every two consecutive configurations are at distance 1:

$$\eta_0 = \eta\,, \quad \eta_n = \xi \quad \text{and} \quad D(\eta_k,\eta_{k+1}) = 1 \quad \text{for} \quad 0 \le k \le n-1\,.$$

To avoid confusion, we should point out that we consider always Σ^κ endowed with the product topology and not the discrete topology generated by the distance D. In particular, when referring to continuous functions, we mean continuous functions with respect to the product topology.

To keep notation simple, for two sites x, y, denote by $\mathcal{H}_x^0$ (resp. $\mathcal{H}_x^\kappa$) the set of configurations with at least one (resp. at most $\kappa - 1$) particles at site x: $\mathcal{H}_x^0 = \{\eta \in \Sigma^\kappa,\ \eta(x) > 0\}$ (resp. $\mathcal{H}_x^\kappa = \{\eta \in \Sigma^\kappa,\ \eta(x) < \kappa\}$) and by $\mathcal{H}_{x,y}$ the set $\mathcal{H}_x^0 \cap \mathcal{H}_y^\kappa$. Let $\sigma^{x,y}\colon \mathcal{H}_{x,y} \to \mathcal{H}_{y,x}$ be the operator that moves a particle from x to y:

$$(\sigma^{x,y}\eta)(z) = \begin{cases} \eta(z) & z \ne x\,,y\,; \\ \eta(x) - 1 & z = x\,; \\ \eta(y) + 1 & z = y\,. \end{cases}$$

We may now introduce the closed forms. Consider a family $\mathfrak{u} = \{(\mathfrak{u}_x^1, \dots, \mathfrak{u}_x^d);\ x \in \mathbb{Z}^d\}$ of continuous functions $\mathfrak{u}_x^i\colon \mathcal{H}_{x,x+e_i} \to \mathbb{R}$ and interpret $\mathfrak{u}_x^i(\eta)$ as the price to move a particle from site x to site $x+e_i$ when the configuration is η. In particular, the price to move a particle from x to $x - e_i$ when the configuration is η is equal to $-\mathfrak{u}_{x-e_i}^i(\sigma^{x,x-e_i}\eta)$. We have seen in the first part of this section that a closed form gives the same price for any 2-step path with equal end points. In the present context of particle systems with the distance adopted above, there are two types of 2-step paths. We may either move a particle two times or move two particles one time each.

Fix a site x, $1 \le i,j \le d$ and a configuration η in $\mathcal{H}_{x,x+e_i+e_j}$. There are four possible different 2-step paths from η to $\eta - \mathfrak{d}_x + \mathfrak{d}_{x+e_i+e_j}$. The first one is obtained letting a particle jump from x to $x+e_i$ and then from $x+e_i$ to $x+e_i+e_j$. Formally this becomes $\Gamma_1 = (\eta, \sigma^{x,x+e_i}\eta, \sigma^{x+e_i,x+e_i+e_j}\sigma^{x,x+e_i}\eta)$. This path is possible only if η belongs to $\mathcal{H}_1 = \mathcal{H}_{x+e_i}^\kappa$ and its price, denoted by $I_{\Gamma_1}(\mathfrak{u})$, is $\mathfrak{u}_x^i(\eta) + \mathfrak{u}_{x+e_i}^j(\sigma^{x,x+e_i}\eta)$. The second path is obtained letting a particle jump from x to $x+e_j$ and then from $x+e_j$ to $x+e_i+e_j$: $\Gamma_2 = (\eta, \sigma^{x,x+e_j}\eta, \sigma^{x+e_j,x+e_i+e_j}\sigma^{x,x+e_j}\eta)$. This path is defined on $\mathcal{H}_2 = \mathcal{H}_{x+e_j}^\kappa$ and its price is $I_{\Gamma_2}(\mathfrak{u}) = \mathfrak{u}_x^j(\eta) + \mathfrak{u}_{x+e_j}^i(\sigma^{x,x+e_j}\eta)$. We may also let first a particle jump from $x+e_i$ to $x+e_i+e_j$ and then let a particle jump from x to $x+e_i$. We obtain in this way $\Gamma_3 = (\eta, \sigma^{x+e_i,x+e_i+e_j}\eta, \sigma^{x,x+e_i}\sigma^{x+e_i,x+e_i+e_j}\eta)$ defined on $\mathcal{H}_3 = \mathcal{H}_{x+e_i}^0$ with price $I_{\Gamma_3}(\mathfrak{u}) = \mathfrak{u}_{x+e_i}^j(\eta) + \mathfrak{u}_x^i(\sigma^{x+e_i,x+e_i+e_j}\eta)$. Finally, we may let first a particle jump from $x+e_j$ to $x+e_j+e_i$ and then let a particle jump from x to $x+e_j$: $\Gamma_4 = (\eta, \sigma^{x+e_j,x+e_i+e_j}\eta, \sigma^{x,x+e_j}\sigma^{x+e_j,x+e_i+e_j}\eta)$. This path is defined on the set $\mathcal{H}_4 = \mathcal{H}^0_{x+e_j}$ and its price is $I_{\Gamma_4}(\mathrm{u}) = \mathrm{u}^i_{x+e_j}(\eta) + \mathrm{u}^j_x(\sigma^{x+e_j,x+e_i+e_j}\eta)$.

In the spirit of the beginning of this section, a closed form u has to assign the same price for all different paths constructed above: for all fixed $1 \le i, j \le d$,

$$I_{\Gamma_k}(\mathrm{u}) = I_{\Gamma_l}(\mathrm{u}) \quad \text{for all } \ 1 \le k, l \le 4 \ \text{ and } \ \eta \in \mathcal{H}_k \cap \mathcal{H}_l \cap \mathcal{H}_{x,x+e_i+e_j} . \tag{4.4}$$

Furthermore, for any $1 \le i, j \le d$, any two sites x, y such that $x + e_i \ne y$ and $y + e_j \ne x$, and any configuration η in $\mathcal{H}_{x,x+e_i} \cap \mathcal{H}_{y,y+e_j}$, there are two ways to move a particle from site x to $x + e_i$ and from site y to site $y + e_j$. We imposed $x + e_i$ (resp. x) to be different from y (resp. $y + e_j$) because these cases belong to the first type of 2-step paths where a particle moves twice. We may first move the particle at x and then the particle sitting at y or in the other way around. In the first case the path is $\tilde{\Gamma}_1(x,y) = (\eta, \sigma^{x,x+e_i}\eta, \sigma^{y,y+e_j}\sigma^{x,x+e_i}\eta)$. This path is possible only if $\eta - \delta_{x,y}\mathfrak{d}_x + \delta_{x+e_i,y+e_j}\mathfrak{d}_{x+e_i}$ belongs to $\mathcal{H}_{x,x+e_i} \cap \mathcal{H}_{y,y+e_j}$. This additional restriction must be imposed because, in the case where $x = y$ for instance, two particles leave site x. This path is thus possible only if η has at least two particle at x or, equivalently, if $\eta - \mathfrak{d}_x$ belongs to $\mathcal{H}^0_x$. The price of this path, denoted by $I_{\tilde{\Gamma}_1(x,y)}(\mathrm{u})$, is equal to $\mathrm{u}^i_x(\eta) + \mathrm{u}^j_y(\sigma^{x,x+e_i}\eta)$. In the second case, the path is $\tilde{\Gamma}_2(x,y) = (\eta, \sigma^{y,y+e_j}\eta, \sigma^{x,x+e_i}\sigma^{y,y+e_j}\eta)$. This path is possible under the same restrictions and its price is $I_{\tilde{\Gamma}_2(x,y)}(\mathrm{u}) = \mathrm{u}^j_y(\eta) + \mathrm{u}^i_x(\sigma^{y,y+e_j}\eta)$. Once more, for u to be a closed form it must assign the same price for these two paths:

$$\mathrm{u}^i_x(\eta) + \mathrm{u}^j_y(\sigma^{x,x+e_i}\eta) = \mathrm{u}^j_y(\eta) + \mathrm{u}^i_x(\sigma^{y,y+e_j}\eta) \tag{4.5}$$

for every $1 \le i, j \le d$, every sites x, y such that $x+e_i \ne y$ and $x \ne y+e_j$, and every configuration η such that $\eta - \delta_{x,y}\mathfrak{d}_x + \delta_{x+e_i,y+e_j}\mathfrak{d}_{x+e_i}$ belong to $\mathcal{H}_{x,x+e_i} \cap \mathcal{H}_{y,y+e_j}$.

Notice that in conditions (4.4) and (4.5) we considered only increasing paths. We leave to the reader to check that it follows from (4.4) and (4.5) that the price of any 2-step path depends on the path only through its end points. We give just an example to illustrate. Fix x, $1 \le i \ne j \le d$ and assume that η is a configuration such that $\eta(x) > 0$, $\eta(x + e_i) < \kappa$, $\eta(x - e_j) < \kappa$, $\eta(x + e_i - e_j) < \kappa$. We want to show that the price for moving a particle from x to $x - e_j$ and then from $x - e_j$ to $x - e_j + e_i$ is the same as the price for moving a particle from x to $x + e_i$ and then from $x + e_i$ to $x + e_i - e_j$ when the configuration is η. The price of the first path is $-\mathrm{u}^j_{x-e_j}(\sigma^{x,x-e_j}\eta) + \mathrm{u}^i_{x-e_j}(\sigma^{x,x-e_j}\eta)$ and the second is $\mathrm{u}^i_x(\eta) - \mathrm{u}^j_{x+e_i-e_j}(\sigma^{x,x+e_i-e_j}\eta)$. These two expressions are equal if and only if $\mathrm{u}^i_x(\eta) + \mathrm{u}^j_{x-e_j}(\sigma^{x,x-e_j}\eta) = \mathrm{u}^i_{x-e_j}(\sigma^{x,x-e_j}\eta) + \mathrm{u}^j_{x+e_i-e_j}(\sigma^{x,x+e_i-e_j}\eta)$. Setting $\xi = \sigma^{x,x-e_j}\eta$ and $y = x - e_j$, we see that $\xi(y) > 0$, $\xi(y + e_j) < \kappa$, $\xi(y + e_i) < \kappa$, $\xi(y + e_i + e_j) < \kappa$. Moreover, the last equality holds if and only if

$$\mathrm{u}^j_y(\xi) + \mathrm{u}^i_{y+e_j}(\sigma^{y,y+e_j}\xi) = \mathrm{u}^i_y(\xi) + \mathrm{u}^j_{y+e_i}(\sigma^{y,y+e_i}\xi) .$$

This last identity follows from (4.5).

We are now ready to define closed forms in Σ^κ:

Definition 4.5 A collection $\mathfrak{u} = \{(\mathfrak{u}_x^1, \dots, \mathfrak{u}_x^d);\ x \in \mathbb{Z}^d\}$ of continuous functions $\mathfrak{u}_x^i : \mathcal{H}_{x,x+e_i} \to \mathbb{R}$ is a closed form if it satisfies equations (4.4) and (4.5).

Here again $\mathfrak{u}_x^i(\eta)$ has to be interpreted as a discrete derivative at η for jumps from x to $x+e_i$, i.e., as the price to move a particle from x to $x+e_i$ when the configuration is η. For this reason we defined $\mathfrak{u}_x^i$ only for configurations with at least one particle at x and less than κ particles at $x+e_i$, otherwise the jump would not be allowed.

Like in the previous setting, to each continuous function $W: \Sigma^\kappa \to \mathbb{R}$ is associated a closed form:

Definition 4.6 A closed form $\mathfrak{u}$ is said to be an exact form if there exists a continuous function $W: \Sigma^\kappa \to \mathbb{R}$ such that

$$\mathfrak{u}_x^i(\eta) \;=\; W(\sigma^{x,x+e_i}\eta) \;-\; W(\eta) \tag{4.6}$$

for every $x \in \mathbb{Z}^d$, $1 \le i \le d$ and configuration η in $\mathcal{H}_{x,x+e_i}$. This closed form is denoted by $\mathfrak{u}^W = (\mathfrak{u}^{W,1}, \dots, \mathfrak{u}^{W,d})$.

Conditions (4.4) and (4.5) follow from relation (4.6). On the one hand, for every site x, $1 \le i,j \le d$, $1 \le k \le 4$ and configuration η in $\mathcal{H}_k \cap \mathcal{H}_{x,x+e_i+e_j}$, $I_{\Gamma_k}(\mathfrak{u}) = W(\eta - \mathfrak{d}_x + \mathfrak{d}_{x+e_i+e_j}) - W(\eta)$. This proves (4.4). On the other hand, for every site x, y, $1 \le i,j \le d$ and configuration η such that $\eta - \delta_{x,y}\mathfrak{d}_x + \delta_{x+e_i,y+e_j}\mathfrak{d}_{x+e_i}$ belongs to $\mathcal{H}_{x,x+e_i} \cap \mathcal{H}_{y,y+e_j}$, $I_{\Gamma_1(x,y)}(\mathfrak{u}) = I_{\Gamma_2(x,y)}(\mathfrak{u}) = W(\eta - \mathfrak{d}_x - \mathfrak{d}_y + \mathfrak{d}_{x+e_i} + \mathfrak{d}_{y+e_j}) - W(\eta)$.

We now present two examples of closed forms that will play a central role in the sequel.

Example 4.7 Fix $1 \le i \le d$ and denote by $\mathfrak{a}^i$ the closed form defined by

$$(\mathfrak{a}^i)_x^j(\eta) \;=\; \mathfrak{a}_x^{i,j}(\eta) \;=\; \delta_{i,j}$$

for $1 \le j \le d$, x in $\mathbb{Z}^d$ and configurations η in $\mathcal{H}_{x,x+e_j}$. $\mathfrak{a}^i$ is a closed form since it adds 1 whenever a particle jumps from some site x to $x+e_i$. This closed form corresponds to the formal function $W_i(\eta)$ defined by $W_i(\eta) = \sum_x < x, e_i > \eta(x)$ since $W_i(\sigma^{x,x+e_j}\eta) - W_i(\eta) = \delta_{i,j}$. This last observation indicates that $\mathfrak{a}^i$ is not an exact form. To prove this statement, assume by contradiction that $\mathfrak{a}^i$ is the closed form associated to a continuous function W. Denote by $\underline{1}$ the configuration with all sites occupied by 1 particle and by $\underline{1}'$ the configuration with no particles at the origin and all other sites occupied by one particle. For a positive integer k, set $x_k = ke_i$ and η_k as the configuration such that $\eta_k(x) = 1$ for $x \ne 0$, x_k; $\eta_k(0) = 0$ and $\eta_k(x_k) = 2$. Since W is continuous and η_k converges to $\underline{1}'$, $W(\underline{1}') = \lim_k W(\eta_k)$. On the other hand, since η_k is obtained from $\underline{1}$ moving k times a particle in the i-th direction, $W(\eta_k) = k + W(\underline{1})$. This leads to the contradiction $W(\underline{1}') = \infty$.

Example 4.8 Let h be a cylinder function. Recall from Chapter 7 that we denote by Γ_h the formal sum $\sum_x \tau_x h$. Define $\mathfrak{v}^h$ as

$$(\mathfrak{v}^h)^i_x(\eta) = \mathfrak{v}^{h,i}_x(\eta) = \Gamma_h(\sigma^{x,x+e_i}\eta) - \Gamma_h(\eta)$$

for x in $\mathbb{Z}^d$, $1 \le i \le d$ and configurations η in $\mathcal{H}_{x,x+e_i}$. Though $\sum_x \tau_x h$ is a formal sum, the difference $\Gamma_h(\eta + \mathfrak{d}_{x+e_i}) - \Gamma_h(\eta + \mathfrak{d}_x)$ is well defined. We leave to the reader to check that $\mathfrak{v}^h$ is a closed form that is not exact, unless h is constant.

We have just presented two examples of closed forms on Σ^κ that are not exact. On finite cubes, however, all closed forms are exact forms. To prove this claim, consider a finite cube Λ in $\mathbb{Z}^d$ and recall that Σ^κ_Λ designates the configuration space $\{0, \dots, \kappa\}^\Lambda$. We have seen in the beginning of this section that for countable simply connected state spaces, the proof that all closed forms are exact forms reduces to the proof that the path integrals of the closed form depend on the path chosen only through its end points.

In order to define a path integral in our particle system setting, consider two configurations η and ξ at distance 1. Denote by $b_+(\eta, \xi)$ the site at which the configuration ξ has one particle more than the configuration η and by $b_-(\eta, \xi)$ the site at which ξ has one particle less than η. With this notation,

$$\xi = \sigma^{b_-, b_+}\eta$$

where $b_+ = b_+(\eta, \xi)$ and $b_- = b_-(\eta, \xi)$. Denote by $\beta = \beta(\eta, \xi)$ the configuration defined by

$$\beta(\eta, \xi) = \begin{cases} \eta & \text{if } b_+ \ge b_- \,, \\ \xi & \text{otherwise} \,. \end{cases}$$

In this formula we adopted the order introduced in the first part of this section. In one dimension, β is equal to η if ξ is obtained from η letting one particle jump to the right and β is equal to ξ if in order to obtain ξ from η we need to move a particle to the left.

Consider the exact form associated to a continuous function $W \colon \Sigma^\kappa \to \mathbb{R}$. Fix two configurations η, ξ at finite distance and a path $\Gamma(\eta, \xi) = (\eta = \eta_0, \dots, \eta_n = \xi)$. We have that $W(\xi) - W(\eta)$ is equal to $W(\eta_n) - W(\eta_0) = \sum_{0 \le k \le n-1} [W(\eta_{k+1}) - W(\eta_k)]$. Fix $0 \le k \le n-1$ and suppose that $\eta_{k+1} = \sigma^{x,x+e_j}\eta_k$ for some site x and some $1 \le j \le d$. In this case, with the notation just introduced, $\beta_k = \beta(\eta_k, \eta_{k+1}) = \eta_k$, $b_{k,-} = b_-(\eta_k, \eta_{k+1}) = x$, $b_{k,+} = b_+(\eta_k, \eta_{k+1}) = x + e_j$. On the other hand, by (4.6), $W(\eta_{k+1}) - W(\eta_k) = \mathfrak{u}^{W,j}_x(\eta_k)$. We may rewrite this expression as $< \mathfrak{u}^W_{b_{k,-} \wedge b_{k,+}}(\beta_k), b_{k,+} - b_{k,-} >$. In this formula $< \cdot, \cdot >$ stands for the inner product on $\mathbb{R}^d$ and $\mathfrak{u}^W_{b_{k,-} \wedge b_{k,+}}$ for the vector $(\mathfrak{u}^{W,1}_{b_{k,-} \wedge b_{k,+}}, \dots, \mathfrak{u}^{W,d}_{b_{k,-} \wedge b_{k,+}})$. In contrast, in the case where $\eta_{k+1} = \sigma^{x,x-e_j}\eta_k$ for some site x and some $1 \le j \le d$, $\beta_k = \beta(\eta_k, \eta_{k+1}) = \eta_{k+1}$, $b_{k,-} = b_-(\eta_k, \eta_{k+1}) = x$, $b_{k,+} = b_+(\eta_k, \eta_{k+1}) = x - e_j$. Since, by (4.6), $W(\eta_k) - W(\eta_{k+1}) = \mathfrak{u}^{W,j}_{x-e_j}(\eta_{k+1})$, we have that $W(\eta_{k+1}) - W(\eta_k) =< \mathfrak{u}^W_{b_{k,-} \wedge b_{k,+}}(\beta_k), b_{k,+} - b_{k,-} >$. In conclusion, we have that

$$W(\eta) - W(\xi) = \sum_{k=0}^{n-1} < u^W_{b_{k,-}\wedge b_{k,+}}(\beta_k), b_{k,+} - b_{k,-} > \cdot$$

By extension, if u is a closed form on Σ^κ and $\Gamma(\eta, \xi) = (\eta = \eta_0, \dots, \eta_n = \xi)$ is a path from η to ξ, we denote by $I_{\Gamma(\eta,\xi)}(u)$ the path integral of u along $\Gamma(\eta, \xi)$ defined by

$$I_{\Gamma(\eta,\xi)}(u) = \sum_{k=0}^{n-1} < u_{b_{k,-}\wedge b_{k,+}}(\beta_k), b_{k,+} - b_{k,-} > \cdot \tag{4.7}$$

Lemma 4.9 *On Σ^κ the path integral of a closed form depends on the path chosen only through its end points.*

Proof. We shall prove this lemma in dimension 2. Fix a closed form u. To prove the lemma, we have to show that the path integral of u along any closed path vanishes. Consider a configuration η and a closed path $\Gamma(\eta, \eta) = (\eta_0, \dots, \eta_n)$. For each $0 \le k \le n - 1$, let $x_k = b_-(\eta_k, \eta_{k+1})$, $y_k = b_+(\eta_k, \eta_{k+1})$ so that $\eta_{k+1} = \sigma^{x_k, y_k}\eta_k$. There exists ℓ large enough so that x_k, y_k belong to the cube Λ_ℓ for every $0 \le k \le n - 1$. In order to enumerate all sites of Λ_ℓ, define $J\colon \Lambda_\ell \to \{0, \dots, (2\ell + 1)^2 - 1\}$ by

$$J(x^1, x^2) = \begin{cases} (2\ell + 1)(x^2 + \ell) + (x^1 + \ell) & \text{if } x_2 + \ell \text{ is even}, \\ (2\ell + 1)(x^2 + \ell) + (-x^1 + \ell) & \text{if } x_2 + \ell \text{ is odd} \end{cases}$$

and define implicitly the sequence $\{z_k,\, 0 \le k \le (2\ell + 1)^2 - 1\}$ by $z_{J(x)} = x$ so that $\Lambda_\ell = \{z_0, \dots, z_{(2\ell+1)^2-1}\}$.

The proof is divided in two steps. We first reduce the problem to a one-dimensional problem by constructing a new closed path $\tilde\Gamma(\eta, \eta) = (\xi_0, \dots, \xi_m)$ that uses only bonds $(z_j, z_{j\pm1})$ and whose path integral $I_{\tilde\Gamma(\eta,\eta)}$ is equal to the original path integral $I_{\Gamma(\eta,\eta)}$:

$$I_{\tilde\Gamma(\eta,\eta)} = I_{\Gamma(\eta,\eta)} \quad \text{and} \quad \xi_{k+1} = \xi_k^{z_j, z_{j\pm1}} \tag{4.8}$$

for some $0 \le j \le (2\ell + 1)^2 - 1$ and all $0 \le k \le m - 1$.

In order to construct such new path, fix $0 \le k \le n - 1$ and recall that $\eta_{k+1} = \sigma^{x_k, y_k}\eta_k$. It is enough to show that there exists a path $\tilde\Gamma_k(\eta_k, \eta_{k+1}) = (\eta_k = \xi_0^k, \dots, \xi_m^k = \eta_{k+1})$ fulfilling the second requirement of (4.8) and such that $I_{\Gamma_k(\eta_k,\eta_{k+1})} = I_{\tilde\Gamma_k(\eta_k,\eta_{k+1})}$, where $\Gamma_k(\eta_k, \eta_{k+1})$ stands for the one step path leading from η_k to η_{k+1}. By definition of the sequence $\{z_j,\, 0 \le j \le (2\ell + 1)^2 - 1\}$, if $y_k = x_k \pm e_1$, there exists j such that $x_k = z_j$ and $y_k = z_{j\pm1}$. In this case nothing has to be done and we may take $\tilde\Gamma_k(\eta_k, \eta_{k+1}) = \Gamma_k(\eta_k, \eta_{k+1})$.

Assume now that $y_k = x_k + e_2$. The case where $y_k = x_k - e_2$ is treated in an analogous way. Let $x_k = (x_k^1, x_k^2)$ and assume without loss of generality that $x_k^2 + \ell$ is an odd number. In this case the path $(\eta_k, \sigma^{(-\ell, x_k^2),(-\ell, x_k^2+1)}\eta_k)$ fulfills the second property in (4.8) because $J(-\ell, x_k^2 + 1) = 1 + J(-\ell, x_k^2)$. There is therefore nothing to prove if $x_k^1 = -\ell$ since we may set $\tilde\Gamma_k(\eta_k, \eta_{k+1}) = \Gamma_k(\eta_k, \eta_{k+1})$.

Fix now $-\ell < a \le \ell$ and assume that $x^1_k = a$ so that $\eta_{k+1} = \sigma^{(a,x^2_k),(a,x^2_k+1)}\eta_k$. We shall construct a new 3-step path $\tilde\Gamma_{k,a}(\eta_k, \eta_{k+1})$ that uses the bonds $(x_k, x_k - e_1)$, $(x_k - e_1, y_k - e_1)$ and $(y_k, y_k - e_1)$ and whose path integral coincides with $I_{\Gamma_k(\eta_k,\eta_{k+1})}(\mathfrak{u})$. Notice that this new path does not use the bond (x_k, y_k) but the bond $(x_k - e_1, y_k - e_1)$ instead.

There are four possible cases that must be treated separately. According to the values of $\eta_k(x_k - e_1)$ and $\eta_k(y_k - e_1)$, η_k belongs to at least one of the following sets: $\mathcal{H}^0_{x_k-e_1} \cap \mathcal{H}^0_{y_k-e_1}$, $\mathcal{H}^0_{x_k-e_1} \cap \mathcal{H}^\kappa_{y_k-e_1}$, $\mathcal{H}^\kappa_{x_k-e_1} \cap \mathcal{H}^0_{y_k-e_1}$ and $\mathcal{H}^\kappa_{x_k-e_1} \cap \mathcal{H}^\kappa_{y_k-e_1}$. We consider the first one and leave the other cases to the reader. Assume that η_k belongs to $\mathcal{H}^0_{x_k-e_1} \cap \mathcal{H}^0_{y_k-e_1}$. Consider the path $\tilde\Gamma_{k,a}(\eta_k, \eta_{k+1})$ given by

$$(\eta_k, \sigma^{y_k-e_1,y_k}\eta_k, \sigma^{x_k-e_1,y_k-e_1}\sigma^{y_k-e_1,y_k}\eta_k,$$
$$\sigma^{x_k,x_k-e_1}\sigma^{x_k-e_1,y_k-e_1}\sigma^{y_k-e_1,y_k}\eta_k = \eta_{k+1}) \, .$$

We claim that the path integral of the closed form $\mathfrak{u}$ along this path is equal to $I_{\Gamma_k(\eta_k,\eta_{k+1})}(\mathfrak{u})$. To prove this claim, notice that the path integral along $\tilde\Gamma_{k,a}(\eta_k, \eta_{k+1})$ is equal to

$$\mathfrak{u}^1_{y_k-e_1}(\eta_k) \; + \; \mathfrak{u}^2_{x_k-e_1}(\sigma^{y_k-e_1,y_k}\eta_k) \; - \; \mathfrak{u}^1_{x_k-e_1}(\eta_{k+1})$$

because $\eta_{k+1} = \sigma^{x_k,x_k-e_1}\sigma^{x_k-e_1,y_k-e_1}\sigma^{y_k-e_1,y_k}\eta_k$. By (4.4), this expression is equal to $\mathfrak{u}^2_{x_k}(\eta_k) = I_{\Gamma_k(\eta_k,\eta_{k+1})}(\mathfrak{u})$. Therefore both path integrals are the same. On the other hand, the new path $\Gamma_{k,a}(\eta_k, \eta_{k+1})$ does not use the bond (x_k, y_k) but the bond $(x_k - e_1, y_k - e_1)$ instead. Repeating this procedure $x^1_k + \ell$ times we obtain a path $\tilde\Gamma_k(\eta_k, \eta_{k+1})$ fulfilling the second requirement of (4.8) and such that $I_{\tilde\Gamma_k(\eta_k,\eta_{k+1})}(\mathfrak{u}) = I_{\Gamma_k(\eta_k,\eta_{k+1})}(\mathfrak{u})$. Juxtaposing these paths we prove (4.8) and conclude the first step.

Consider now a closed path $\Gamma(\eta, \eta) = (\eta_0, \dots, \eta_n)$ such that for all $0 \le k \le n-1$, $\eta_{k+1} = \eta_k^{z_j,z_{j\pm1}}$ for some j. We want to prove that the path integral along this path vanishes. Notice that this is a one-dimensional problem since only jumps from sites z_j to $z_{j\pm1}$ are allowed.

The strategy consists in constructing a new path with length $n-2$ and same path integral. The idea to construct such a new path is simple. Suppose that a particle jumps from some site x to y and immediately after from y to x. A new closed path of length $n-2$ can be constructed suppressing these two jumps and it is easy to show that both path integrals are equal. If there are no such consecutive jumps, a new closed path can still be constructed. We just need to change the order of the jumps, preserving the path integral by property (4.5), up to obtain two consecutive jumps as described before.

Recall that we are now considering a path $(\eta = \eta_0, \dots, \eta_n = \eta)$ such that for each $0 \le k \le n-1$ $\eta_{k+1} = \sigma^{x_k,y_k}\eta_k$ where $x_k = z_j$ and $y_k = z_{j\pm1}$ for some j. Assume without loss of generality that $J(y_0) = J(x_0) + 1$. Since $\Gamma(\eta, \eta)$ can be interpreted as a closed, one-dimensional path on $\{0, \dots, \kappa\}^{\mathbb{Z}}$ and a particle jumped from $z_{J(x_0)}$ to $z_{J(x_0)+1}$, later a particle must jump from $z_{J(x_0)+1}$ to $z_{J(x_0)}$. Let k_0 be the first time when this happen: $k_0 = \min\{k \ge 0, \, x_k = z_{J(x_0)+1} \; , \; y_k = z_{J(x_0)}\}$

and let k_1 be the last time before k_0 that a particle jumps from $z_{J(x_0)}$ to $z_{J(x_0)+1}$: $k_1 = \max\{k \le k_0,\ x_k = z_{J(x_0)}\ ,\ y_k = z_{J(x_0)+1}\}$.

We shall assume without loss of generality that between k_1 and k_0, there are no jumps from some site x_1 to some site y_1 and then a jump from site y_1 to x_1. Otherwise we can repeat the same arguments to x_1, y_1 in place of x_0, y_0.

Consider the path $\Gamma(\eta,\eta)$ between its k_1 and k_0-step. and denote it by $\Gamma_{k_1,k_0+1}(\eta_{k_1},\eta_{k_0+1})$. Keep in mind that $\eta_{k_0+1} = \sigma^{y_0,x_0}\eta_{k_0}$ by definition of k_0. Denote by $\tilde{\Gamma}^1(\eta_{k_1},\eta_{k_0+1})$ the path $(\eta_{k_1},\dots,\eta_{k_0-1},\xi_{k_0},\eta_{k_0+1})$, where $\xi_{k_0} = \sigma^{y_0,x_0}\eta_{k_0-1}$. The difference between this path and the original one is that in the penultimate step instead of moving a particle from x_{k_0-1} to y_{k_0-1} and then a particle from y_0 to x_0, we inverted the order of these jumps. We leave to the reader to check that the assumption made in last paragraph guarantees that we may change the order of the jumps and that property (4.5) ensures that the path integrals along both paths are the same. Repeating this procedure k_0-k_1 times we obtain at the end a new closed path $\tilde{\Gamma}^{k_0-k_1}(\eta,\eta) = (\xi_0,\dots,\xi_n)$ whose path integral is equal to the path integral of $\Gamma(\eta,\eta)$ and such that $\xi_0 = \eta$, $\xi_1 = \sigma^{x_0,y_0}\eta$ and $\xi_2 = \sigma^{y_0,x_0}\xi_1 = \eta$. Consider now the $(n-2)$-step path obtained from $\tilde{\Gamma}^{k_0-k_1}$ suppressing the first two jumps: $\Gamma_1^*(\eta,\eta) = (\eta,\xi_3,\dots,\xi_n)$. By definition of the path integral, the price to move a particle from x to y for the configuration η is equal to the price to move a particle from y to x for the configuration $\sigma^{x,y}\eta$. In particular, the path integral of $\mathfrak{u}$ along $\Gamma_1^*(\eta,\eta)$ is equal to the path integral along $\Gamma(\eta,\eta)$. Repeating this argument $n/2$ times we obtain that the path integral along $\Gamma(\eta,\eta)$ vanishes. □

From this result and the proof of Corollary 4.4, it follows that on finite cubes, all closed forms are exact forms:

Corollary 4.10 *Let Λ be a finite cube in $\mathbb{Z}^d$. All closed forms on Σ_Λ^κ are exact forms.*

Proof. Let $\mathfrak{u}$ be a closed form on Σ_Λ^κ. Fix $0 \le K \le \kappa|\Lambda|$ and a configuration ξ_K in the hyperplane $\Sigma_{\Lambda,K}^\kappa$. Set an arbitrary value $F(\xi_K)$ for F at ξ_K. Since Λ is simply connected, for every configuration η in the hyperplane $\Sigma_{\Lambda,K}^\kappa$ there is a path $(\eta_k)_{0\le k\le n}$ from ξ_K to η. Define $F(\eta)$ by

$$F(\eta) \;=\; F(\xi_K) \;+\; I_{\Gamma(\xi_K,\eta)}(\mathfrak{u})\,.$$

It follows from the previous lemma that F is a well defined function on $\Sigma_{\Lambda,M(\xi_K)}^\kappa$. Moreover, a simple computation shows that $\mathfrak{u}(\eta) = \mathfrak{u}^F(\eta)$ for all configurations η of $\Sigma_{\Lambda,M(\xi_K)}^\kappa$. This shows that $\mathfrak{u}$ is an exact form. □

Notice that in the proof of the previous Corollary, we have a degree of freedom in the definition of F on each hyperplane $\Sigma_{\Lambda,K}^\kappa$ because the value of $F(\xi_K)$ is arbitrary. We may choose, for instance, $F(\xi_K)$ so that F has mean zero with respect to all canonical measures $\nu_{\Lambda,K}$.

Since all closed forms are exact forms on simply connected, finite subsets of $\mathbb{Z}^d$, on the cube $\Lambda_\ell = \{-\ell,\dots,\ell\}^d$, the closed forms $\{\mathfrak{a}^i,\ 1\le i\le d\}$ and $\mathfrak{v}^h$

introduced in Examples 4.7 and 4.8 are exact forms. It is easy to check that the closed form $\mathfrak{a}^i$ is the exact form associated to the continuous function $W_i(\eta) = \sum_{x \in \Lambda_\ell} < x, e_i > \eta(x)$.

To introduce a special class of closed forms, we needed to extend its definition to include collections of $L^2(\nu_\alpha)$ functions.

Definition 4.11 A collection $\mathfrak{u} = \{(\mathfrak{u}_x^1, \dots, \mathfrak{u}_x^d);\ x \in \mathbb{Z}^d\}$ of $L^2(\nu_\alpha)$ functions $\mathfrak{u}_x^i$: $\mathcal{H}_{x,x+e_i} \to \mathbb{R}$ is a closed form if it satisfies equations (4.4) and (4.5) in $L^2(\nu_\alpha)$.

We may now introduce the main object of this section.

Definition 4.12 A class of $L^2(\nu_\alpha)$ functions $\mathfrak{g} = \{\mathfrak{g}^i: \mathcal{H}_{0,e_i} \to \mathbb{R},\ 1 \le i \le d\}$ is a germ of a closed form if the collection $\{(\mathfrak{g}_x^1, \dots, \mathfrak{g}_x^d),\ x \in \mathbb{Z}^d\}$ defined through translations of $\mathfrak{g}$ by the formula

$$\mathfrak{g}_x^i(\eta) = (\tau_x \mathfrak{g}^i)(\eta) \tag{4.9}$$

is a closed form. Notice that $\mathfrak{g}_x^i$ is defined on $\mathcal{H}_{x,x+e_i}$ as required. The space of germs of closed forms is denoted by $\mathfrak{G}$.

For a cylinder function h and a bond (x, y), denote by $(\nabla_{x,y} h): \mathcal{H}_{x,y} \to \mathbb{R}$ the cylinder function defined by $(\nabla_{x,y} h)(\eta) = h(\sigma^{x,y}\eta) - h(\eta)$. We extend the definition of $\nabla_{x,y} h$ to Σ^κ setting $(\nabla_{x,y} h)(\eta) = 0$ if $\eta \notin \mathcal{H}_{x,y}$: $(\nabla_{x,y} h)(\eta) = \mathbf{1}\{\eta(x) > 0, \eta(y) < \kappa\}[h(\sigma^{x,y}\eta) - h(\eta)]$. We sometimes abbreviate $\nabla_{0,e_i} h$ by $\nabla_i h$ and we denote by ∇h the vector $(\nabla_1 h, \dots, \nabla_d h)$.

Examples 4.7 and 4.8 provide two types of germs of closed forms. For $1 \le i \le d$, consider the class of cylinder function $\mathfrak{A}^i = \{\mathfrak{A}^{i,1}, \dots, \mathfrak{A}^{i,d}\}$ with $\mathfrak{A}^{i,j} = \delta_{i,j}$, $1 \le j \le d$. The collection $\{(\mathfrak{A}_x^{i,1}, \dots, \mathfrak{A}_x^{i,d}),\ x \in \mathbb{Z}^d\}$ obtained from $\mathfrak{A}^i$ through formula (4.9) is the closed form $\mathfrak{a}^i$ of Example 4.7.

For a cylinder function h, the collection $\{(\nabla_{x,x+e_1}\Gamma_h, \dots, \nabla_{x,x+e_d}\Gamma_h),\ x \in \mathbb{Z}^d\}$ obtained from formula (4.9) through the cylinder function $(\nabla_1\Gamma_h, \dots, \nabla_d\Gamma_h)$ is the closed form $\mathfrak{v}^h$ of Example 4.8. These germs of closed forms associated to translations of cylinder functions are called germs of exact forms:

Definition 4.13 The germ $\mathfrak{g}$ of a closed form is said to be the germ of an exact form if there exists a cylinder function h such that $\mathfrak{g} = \nabla\Gamma_h$. The space of germs of exact forms is denoted by $\mathfrak{E}$:

$$\mathfrak{E} = \{\nabla\Gamma_h;\ h \text{ is a cylinder function}\}.$$

The goal of the remaining of this section is to prove that in $L^2(\nu_\alpha)$ the germs of closed forms are generated by the germs of exact forms and by the germs $\{\mathfrak{A}^i,\ 1 \le i \le d\}$ introduced above. Let $\mathfrak{G}_\alpha$ and $\mathfrak{E}_\alpha$ denote, respectively, the closure in $L^2(\nu_\alpha)$ of germs of closed forms and of germs of exact forms.

$$\mathfrak{G}_\alpha = \overline{\mathfrak{G}} \text{ in } L^2(\nu_\alpha) \qquad \text{and} \qquad \mathfrak{E}_\alpha = \overline{\mathfrak{E}} \text{ in } L^2(\nu_\alpha).$$

Theorem 4.14 *In $L^2(\nu_\alpha)$ the space of germs of closed forms is a direct sum of the linear space generated by the germs $\{\mathfrak{A}^i,\ 1 \le i \le d\}$ and the closure of germs of exact forms:*

$$\mathfrak{G}_\alpha \;=\; \{\mathfrak{A}^i,\ 1 \le i \le d\} \,\oplus\, \mathfrak{E}_\alpha\,.$$

Proof. $\mathfrak{G}_\alpha$, $\mathfrak{E}_\alpha$ are closed linear subspaces of $L^2(\nu_\alpha)$. We have already seen that for any $1 \le i \le d$ and any cylinder function h, the germs $\mathfrak{A}^i$, $\nabla\Gamma_h$ belong to $\mathfrak{G}$. In particular, $\{\mathfrak{A}^i,\ 1 \le i \le d\} + \mathfrak{E}_\alpha$ is a subset of $\mathfrak{G}_\alpha$. To conclude the proof of the theorem it remain to show that the sum is direct and that

$$\mathfrak{G} \subset \{\mathfrak{A}^i,\ 1 \le i \le d\} \;+\; \mathfrak{E}_\alpha.$$

We first prove the inclusion. The set $\mathfrak{E} = \{\nabla\Gamma_h;\ h \text{ cylinder function}\}$ is a linear subspace of $L^2(\nu_\alpha)$ and so is $\{\mathfrak{A}^i,\ 1 \le i \le d\} + \mathfrak{E}$. It is well known that in a Hilbert space the strong closure of a linear subspace coincides with its weak closure (cf. Theorem V.1.11 of Yosida (1997)). Therefore, we have just to prove that for every germ $\mathfrak{g}$ of a closed form there is a sequence of cylinder functions (ψ_n) and constants $C_1, \dots, C_d$ such that $\nabla\Gamma_{\psi_n}$ converges weakly in $L^2(\nu_\alpha)$ to $\mathfrak{g} + \sum_i C_i \mathfrak{A}^i$.

The strategy of the proof of the Theorem is easy to understand. Consider a germ $\mathfrak{g}$ of a closed form $\{(\mathfrak{g}^1_x, \dots, \mathfrak{g}^d_x),\ x \in \mathbb{Z}^d\}$ and recall that $\mathfrak{g}^i_x = \tau_x \mathfrak{g}^i$. Assume, to fix ideas, that $\mathfrak{g}^1, \dots \mathfrak{g}^d$ are cylinder functions. We proved in Corollary 4.10 that all closed forms on Σ^κ_Λ are exact forms if Λ is a cube of finite length. We are thus tempted to project the closed form $\{(\mathfrak{g}^1_x, \dots, \mathfrak{g}^d_x),\ x \in \mathbb{Z}^d\}$ on Σ^κ_Λ to recover a cylinder function ψ_n whose discrete partial derivatives are $\mathfrak{g}^i_x$. More precisely, fix a positive integer n and denote by $\mathcal{F}_n$ the σ-algebra generated by $\{\eta(x),\ x \in \Lambda_n\}$. For $1 \le i \le d$ and a site x in $\Lambda_{n,i} = \{y \in \Lambda_n;\ y + e_i \in \Lambda_n\}$, define $\mathfrak{g}^{i,n}_x$ as the conditional expectation of $\mathfrak{g}^i_x$ given $\mathcal{F}_n$:

$$\mathfrak{g}^{i,n}_x \;=\; E_\alpha\Big[\mathfrak{g}^i_x \,\big|\, \mathcal{F}_n\Big]\,.$$

It is easy to check that $\{\mathfrak{g}^{i,n}_x;\ x \in \Lambda_{n,i}\}$ is a closed form on $\Sigma^\kappa_n = \Sigma^\kappa_{\Lambda_n}$. By Corollary 4.10, there exists a function $\widetilde{\psi}_n$ $\mathcal{F}_n$-measurable such that

$$\nabla_{x,x+e_i}\widetilde{\psi}_n \;=\; \mathfrak{g}^{i,n}_x \quad \text{for all sites } x \text{ in } \Lambda_{i,n} \text{ and}$$
$$E\Big[\widetilde{\psi}_n \Big| \sum_{x\in\Lambda_n} \eta(x) \;=\; K\Big] \;=\; 0 \qquad \text{for all} \quad 0 \le K \le \kappa|\Lambda_n|\,.$$

This last requirement picks a function among the one-parameter family of integrals constructed in Corollary 4.10 on each hyperplane $\Sigma^\kappa_{n,K}$.

Since $\mathfrak{g}^i$ is a cylinder function, $\mathfrak{g}^i_x = \tau_x\mathfrak{g}^i$ is $\mathcal{F}_n$-measurable for x not too close from the boundary of Λ_n. In particular, there exists $n_0 = n_0(\mathfrak{g})$ such that $\mathfrak{g}^{i,n}_x = \tau_x\mathfrak{g}^i$ for all x in Λ_{n-n_0} and

$$\tau_x \nabla_{0,e_i} \tau_{-x}\widetilde{\psi}_n \;=\; \nabla_{x,x+e_i}\widetilde{\psi}_n \;=\; \mathfrak{g}^{i,n}_x \;=\; \tau_x\mathfrak{g}^i$$

for x in Λ_{n-n_0}. The first equality follows from the identity $\tau_z \nabla_{y,y+e_j} = \nabla_{y+z,y+z+e_j} \tau_z$ for z, y in $\mathbb{Z}^d$ and $1 \le j \le d$, to be used repeatedly hereafter. Averaging over x in Λ_{n-n_0}, this identity shows that

$$\mathfrak{g}^i = \nabla_{0,e_i} \frac{1}{[2(n-n_0)+1]^d} \sum_{x \in \Lambda_{n-n_0}} \tau_{-x} \widetilde{\psi}_n .$$

The difference between this last average and $\nabla_{0,e_i} \Gamma_{[2(n-n_0)+1]^{-d}\widetilde{\psi}_n} = [2(n-n_0)+1]^{-d} \nabla_{0,e_i} \Gamma_{\widetilde{\psi}_n}$ is equal to

$$\tilde{\sigma}_n^i = \nabla_{0,e_i} [2(n-n_0)+1]^{-d} \sum_{x \in \Lambda_{n+1} - \Lambda_{n-n_0}} \tau_{-x} \widetilde{\psi}_n$$

because $\widetilde{\psi}_n$ is $\mathcal{F}_n$-measurable and thus $\nabla_{0,e_i} \tau_{-x} \widetilde{\psi}_n$ vanishes for $x \notin \Lambda_{n+1}$. In particular, the inclusion $\mathfrak{G} \subset \{\mathfrak{A}^1, \dots, \mathfrak{A}^d\} + \mathfrak{E}_\alpha$ will bc proved as soon as we show that the boundary term $\tilde{\sigma}_n^i$ is a weakly relatively compact sequence whose limit points belong to the linear space generated by $\mathfrak{A}^i$. This is the strategy of the proof.

The proof is divided in several steps to detach the main ideas. We first project the closed form $\{(\mathfrak{g}_x^1, \dots, \mathfrak{g}_x^d),\ x \in \mathbb{Z}^d\}$ on finite cubes and apply Corollary 4.10 to obtain the cylinder function $\widetilde{\psi}_n$. In steps 2 and 3 we estimate the $L^2(\nu_\alpha)$ norm of $\widetilde{\psi}_n$ and show that a modification R_n^i of the boundary term $\tilde{\sigma}_n^i$ is a weakly relatively compact sequence. In step 4 we prove that the non boundary term in $(2n)^{-d} \nabla_{0,e_i} \Gamma_{\widetilde{\psi}_n}$ converges strongly in $L^2(\nu_\alpha)$ to the germ $\mathfrak{g}^i$. Finally, in steps 5 and 6 we show that all limit points of the sequence R_n^i belong to the linear space generated by $\mathfrak{A}^i$.

Step 1. Projection to finite boxes. Fix a germ $(\mathfrak{g}^1, \dots, \mathfrak{g}^d)$ of a closed form and recall the definition of $\mathcal{F}_n$, $\mathfrak{g}_x^{i,n}$ and $\widetilde{\psi}_n \in \mathcal{F}_n$ given in the beginning of the proof. For $1 \le i \le d$, consider the expression $(2n)^{-d} \nabla_{0,e_i} \Gamma_{\widetilde{\psi}_n}$. Since $\widetilde{\psi}_n$ is $\mathcal{F}_n$-measurable, this sum is

$$\frac{1}{(2n)^d} \nabla_{0,e_i} \Gamma_{\widetilde{\psi}_n} = \frac{1}{(2n)^d} \nabla_{0,e_i} \sum_{\substack{-n \le x_i \le n+1 \\ |x_j| \le n,\, j \ne i}} \tau_x \widetilde{\psi}_n .$$

It is not difficult to see that $(2n)^{-d} \nabla_{0,e_i} \sum_{x \in \Lambda_n, -n < x_i \le n} \tau_x \widetilde{\psi}_n$ converges in $L^2(\nu_\alpha)$ to $\mathfrak{g}^i$ (the proof given in Step 4 for a slight modification of $\widetilde{\psi}_n$ applies also to $\widetilde{\psi}_n$). The problem comes from the boundary terms

$$\sigma_n^i = \frac{1}{(2n)^d} \nabla_{0,e_i} \sum_{y \in \Lambda_n;\, y_i = -n} \tau_y \widetilde{\psi}_n + \frac{1}{(2n)^d} \nabla_{0,e_i} \sum_{y \in e_i + \Lambda_n;\, y_i = n+1} \tau_y \widetilde{\psi}_n . \quad (4.10)$$

The proof of the existence of a subsequence $\sigma_{n_k}^i$ that converges weakly to a multiple of $\mathfrak{A}^i$ is derived in two steps. We first show that the sequence $\{\sigma_n^i,\ n \ge 1\}$

is weakly relatively compact and then prove that all limit points of this set are in the linear subspace generated by $\mathfrak{A}^i$.

By Alaoglu's theorem, the unit ball for the strong topology is weakly relatively compact (cf. Theorem V.2.1 of Yosida (1997)). Therefore, to prove that the sequence $\{\sigma_n^i,\ n \geq 1\}$ is weakly relatively compact we just need a bound on the $L^2(\nu_\alpha)$ norm of σ_n^i. To prove such a bound we first estimate the $L^2(\nu_\alpha)$ norm of $\widetilde{\psi}_n$.

Step 2. Bound on the $L^2(\nu_\alpha)$ norm of $\widetilde{\psi}_n$. Since by construction, for each $0 \leq K \leq \kappa|\Lambda_n|$, $\widetilde{\psi}_n$ has mean 0 with respect to $\nu_{\Lambda_n,K}$, by the spectral gap proved in sections 2 and 3, there exists a universal constant C_1 such that

$$\left\langle (\widetilde{\psi}_n)^2 \right\rangle_\alpha \leq C_1 n^2 \mathfrak{D}\left(\nu_\alpha^{\Lambda_n}, \widetilde{\psi}_n\right),$$

where $\mathfrak{D}(\nu_\alpha^{\Lambda_n}, \cdot)$ is the Dirichlet form with respect to $\nu_\alpha^{\Lambda_n}$. This is the unique point in the proof of hydrodynamic limit of nongradient systems where a sharp estimate on the spectral gap for the generator restricted to finite boxes is needed.

An estimate of the Dirichlet form $\widetilde{\psi}_n$ is easy to derive. By definition of $\widetilde{\psi}_n$, for x in $\Lambda_{n,i}$, $\nabla_{x,x+e_i}\widetilde{\psi}_n = \mathfrak{g}_x^{i,n}$. By Schwarz inequality and by translation invariance of ν_α, $< (\mathfrak{g}_x^{i,n})^2 >_\alpha$ is bounded above by $< (\mathfrak{g}_x^i)^2 >_\alpha = < (\mathfrak{g}^i)^2 >_\alpha$. Therefore,

$$\begin{aligned} \mathfrak{D}(\nu_\alpha^{\Lambda_n}, \widetilde{\psi}_n) &= (1/2)\sum_{i=1}^d \sum_{x\in\Lambda_{n,i}} \left\langle \left(\nabla_{x,x+e_i}\widetilde{\psi}_n\right)^2 \right\rangle_\alpha \\ &\leq C n^d \sum_{1\leq i\leq d} < (\mathfrak{g}^i)^2 >_\alpha . \end{aligned}$$

Henceforth we use the shorthand $||\mathfrak{g}||^2$ for $\sum_{1\leq i\leq d}(\mathfrak{g}^i)^2$. The two previous estimates show that

$$\mathfrak{D}(\nu_\alpha^{\Lambda_n}, \widetilde{\psi}_n) \leq C n^d \left\langle ||\mathfrak{g}||^2 \right\rangle_\alpha, \qquad \left\langle (\widetilde{\psi}_n)^2 \right\rangle_\alpha \leq C n^{d+2} \left\langle ||\mathfrak{g}||^2 \right\rangle_\alpha$$

for some universal constant C.

Step 3. Bound on the L^2 norm of the boundary terms. We start with an estimate of the $L^2(\nu_\alpha)$ norm of σ_n^i in terms of the Dirichlet form of $\widetilde{\psi}_n$. To understand the difficulty, fix a site x in $e_i + \Lambda_n$ such that $x_i = n + 1$. $\tau_x\widetilde{\psi}_n$ is therefore a function of $\{\eta(x),\ x_i \geq 1\}$ and $\nabla_{0,e_i}\tau_x\widetilde{\psi}_n(\eta) = (\tau_x\widetilde{\psi}_n)(\eta + \mathfrak{d}_{e_i}) - (\tau_x\widetilde{\psi}_n)(\eta)$. We have thus to estimate the modification caused on $\widetilde{\psi}_n$ by the creation of a particle with a Dirichlet form that measures only displacement of particles. Next lemma shows that such an estimation is possible provided we smooth out $\widetilde{\psi}_n$ by taking conditional expectations.

Lemma 4.15 *Consider a $\mathcal{F}_{3n}$-measurable function f. Denote by h the conditional expectation of f with respect to $\mathcal{F}_n$: $h(\xi) = E_\alpha[f \mid \eta(x),\ x \in \Lambda_n]$. For each $1 \leq i \leq d$, denote by $\partial_{i,+}\Lambda_n$ the positive boundary of Λ_n in the i-th direction: $\partial_{i,+}\Lambda_n = \{y \in \Lambda_n;\ y_i = n\}$. There exist finite constants $C_1(\alpha)$, $C_2(\alpha)$ such that*

$$\sum_{y\in\partial_{i,+}\Lambda_n} E_\alpha\Big[\Big\{h(\eta-\mathfrak{d}_y)-h(\eta)\Big\}^2\mathbf{1}\{\eta(y)>0\}\Big]$$
$$\le C_1(\alpha)n^{-1}E_\alpha[f^2] + C_2(\alpha)n\,\mathfrak{D}(\nu_\alpha^{\Lambda_{3n}},f)\,.$$

A similar statement holds if the positive boundary $\partial_{i,+}\Lambda_n$, $\eta-\mathfrak{d}_y$ and $\mathbf{1}\{\eta(y)>0\}$ are replaced by the negative boundary $\partial_{i,-}\Lambda_n=\{y\in\Lambda_n;\,y_i=-n\}$, $\eta+\mathfrak{d}_y$ and $\mathbf{1}\{\eta(y)<\kappa\}$.

Proof. Set $i=1$ and fix a site y in $\partial_{1,+}\Lambda_n$. To keep notation simple, we shall denote the configurations of Λ_{3n} (resp. Λ_n, $\Lambda_{3n}-\Lambda_n$) by the symbol η (resp. ζ, ξ). Recall from section 4 that for each function f on Σ_{3n}^κ and each configuration ζ of Σ_n^κ, $f_\zeta\colon\Lambda_{3n}-\Lambda_n\to\mathbb{R}$ stands for the function defined by $f_\zeta(\xi)=f(\zeta,\xi)$ for all configurations ξ of $\Lambda_{3n}-\Lambda_n$.

Since h is defined as the conditional expectation of f, we have that $h(\zeta)=E_\alpha[f\,|\,\zeta]=E_\alpha[f_\zeta(\xi)]$. In this last expectation it should be understood that integration is made with respect to the variable ξ, while ζ is kept frozen. In particular, if we denote $\zeta-\mathfrak{d}_y$ by ζ^y, the difference $h(\zeta-\mathfrak{d}_y)-h(\zeta)$ can be rewritten as $E_\alpha[f_{\zeta^y}(\xi)]-E_\alpha[f_\zeta(\xi)]$.

Denote by $\Lambda_{y,n}$ the hypercube Λ_{n-1} translated by $y+(n+1)e_1$: $\Lambda_{y,n}=y+(n+1)e_1+\Lambda_{n-1}$. Notice that $\Lambda_{y,n}\subset\Lambda_{3n}$ for all y in $\partial_{1,+}\Lambda_n$. We may rewrite $E_\alpha[f_{\zeta^y}(\xi)]$ as

$$\frac{1}{\alpha|\Lambda_{y,n}|}\sum_{z\in\Lambda_{y,n}}E_\alpha\Big[f_{\zeta^y}(\xi)\xi(z)\Big] + \frac{1}{\alpha|\Lambda_{y,n}|}\sum_{z\in\Lambda_{y,n}}E_\alpha\Big[f_{\zeta^y}(\xi)[\alpha-\xi(z)]\Big]\,.$$

In the first term, for each z in $\Lambda_{y,n}$, we perform a change of variables $\tilde\xi=\xi-\mathfrak{d}_z$ to rewrite it as

$$\frac{\Phi(\alpha)}{\alpha|\Lambda_{y,n}|}\sum_{z\in\Lambda_{y,n}}E_\alpha\Big[f_{\zeta^y}(\xi+\mathfrak{d}_z)\{1+\xi(z)\}\mathbf{1}\{\xi(z)<\kappa\}\Big]$$
$$=\frac{1}{|\Lambda_{y,n}|}\sum_{z\in\Lambda_{y,n}}E_\alpha\Big[\Big\{f_{\zeta^y}(\xi+\mathfrak{d}_z)-f_\zeta(\xi)\Big\}g(\xi(z))\Big]$$
$$+\frac{1}{|\Lambda_{y,n}|}\sum_{z\in\Lambda_{y,n}}E_\alpha\Big[f_\zeta(\xi)g(\xi(z))\Big]\,.$$

In this last formula $g(\xi(z))$ stands for $(\Phi(\alpha)/\alpha)[1+\xi(z)]\mathbf{1}\{\xi(z)<\kappa\}$. Subtracting $h(\zeta)=E_\alpha[f_\zeta]$ we obtain that the difference $h(\zeta-\mathfrak{d}_y)-h(\zeta)$ is equal to

$$
\begin{aligned}
&\frac{1}{|\Lambda_{y,n}|} \sum_{z\in\Lambda_{y,n}} E_\alpha\Big[\Big\{ f_{\zeta^y}(\xi+\mathfrak{d}_z) - f_\zeta(\xi)\Big\} g(\xi(z))\Big] \\
&\qquad + E_\alpha\Big[f_\zeta(\xi) \frac{1}{|\Lambda_{y,n}|} \sum_{z\in\Lambda_{y,n}} \{ g(\xi(z)) - 1\}\Big] \\
&\qquad\qquad + E_\alpha\Big[f_{\zeta^y}(\xi) \frac{1}{|\Lambda_{y,n}|} \sum_{z\in\Lambda_{y,n}} [1 - (\xi(z)/\alpha)]\Big] .
\end{aligned}
$$

The expected value of $g(\xi(z))$ and of $\xi(z)/\alpha$ with respect to ν_α is equal to 1. Therefore, the second and third terms are covariances of f_ζ and f_{ζ^y} with averages of cylinder functions. In particular, by Schwarz inequality, the square of the second and third term in the above decomposition are bounded by $C(\alpha)|\Lambda_{y,n}|^{-1}\{E_\alpha[f_\zeta^2]+ E_\alpha[f_{\zeta^y}^2]\}$ for some finite constant $C(\alpha)$ depending only on α. On the other hand, by Schwarz inequality, the square of the first term is bounded by

$$
C(\alpha)\frac{1}{|\Lambda_{y,n}|} \sum_{z\in\Lambda_{y,n}} E_\alpha\Big[\Big\{ f_{\zeta^y}(\xi+\mathfrak{d}_z) - f_\zeta(\xi)\Big\}^2 \mathbf{1}\{\xi(z)<\kappa\}\Big]
$$

for some finite constant $C(\alpha)$ because $g(\cdot)$ is bounded. Hereafter the constant $C(\alpha)$ may change from line to line.

Recall that $E_\alpha[f_\zeta(\xi)] = E_\alpha[f\,|\,\zeta]$. From the previous bounds and Schwarz inequality, we obtain that the expected value, with respect to ν_α, of $[h(\zeta-\mathfrak{d}_y) - h(\zeta)]^2\mathbf{1}\{\zeta(y)>0\}$ is bounded by

$$
\begin{aligned}
&C(\alpha)|\Lambda_{y,n}|^{-1}\Big\{E_\alpha[f^2] + E_\alpha[f^2(\eta-\mathfrak{d}_y)\mathbf{1}\{\eta(y)>0\}]\Big\} \\
&\qquad + C(\alpha)\frac{1}{|\Lambda_{y,n}|}\sum_{z\in\Lambda_{y,n}} E_\alpha\Big[r_{y,z}(\eta)\Big\{ f(\eta^{y,z}) - f(\eta)\Big\}^2\Big] .
\end{aligned}
$$

Performing a change of variables $\eta' = \eta - \mathfrak{d}_y$ in the second expectation and summing over y in the positive boundary of Λ_n, we obtain that the first two terms are bounded by $C(\alpha)n^{-1}E_\alpha[f^2]$. On the other hand, by Lemma 2.8 and like in the proof of Lemma 3.1, the third expression is bounded by $C(\alpha)n\mathfrak{D}(\nu_\alpha^{\Lambda_{3n}}, f)$ for some constant $C(\alpha)$ depending only on α. □

We have now all elements to prove that a slight modification of the sequence σ_n^i defined in (4.10) is bounded in $L^2(\nu_\alpha)$. In view of Lemma 4.15, for each positive integer n, let

$$
\psi_n = E\Big[\widetilde{\psi}_{3n}|\mathcal{F}_n\Big] .
$$

By Schwarz inequality and the translation invariance of ν_α, for each $1\le i\le d$,

$$
E_\alpha\Big[\Big(\frac{1}{(2n)^d}\nabla_{0,e_i}\sum_{y\in\partial_{i,-}\Lambda_n}\tau_y\psi_n\Big)^2\Big] \le \frac{C}{n^{d+1}}\sum_{y\in\partial_{i,+}\Lambda_n} E_\alpha\Big[\Big(\nabla_{y,y+e_i}\psi_n\Big)^2\Big]
$$

because $\tau_y \nabla_{0,e_i} = \nabla_{y,y+e_i}\tau_y$ for all y in $\mathbb{Z}^d$. By Step 2, $< (\widetilde{\psi}_{3n})^2 >_\alpha$ is bounded above by $Cn^{d+2} < \|\mathfrak{g}\|^2 >_\alpha$ and the Dirichlet form $\mathfrak{D}(\nu_\alpha^{\Lambda_n}, \widetilde{\psi}_{3n})$ is bounded by $Cn^d < \|\mathfrak{g}\|^2 >_\alpha$. Therefore, Lemma 4.15 applied to the function ψ_n shows that the right hand side of the last formula is bounded by $C(\alpha) < \|\mathfrak{g}\|^2 >_\alpha$. The same argument applies to $\sum_{y\in e_i+\Lambda_n,\, y_i=n+1} \tau_y\psi_n$.

In view of this estimate, for $1 \le i \le d$, define $R_n = (R_n^1, \dots, R_n^d)$ by

$$R_n^i = \frac{1}{(2n)^d}\nabla_{0,e_i}\Gamma_{\psi_n} = \frac{1}{(2n)^d}\nabla_{0,e_i}\sum_x \tau_x\psi_n .$$

R_n^i can be decomposed as

$$\frac{1}{(2n)^d}\nabla_{0,e_i}\sum_{\substack{y\in\Lambda_n\\ -n<y_i\le n}} \tau_y\psi_n + \frac{1}{(2n)^d}\nabla_{0,e_i}\sum_{\substack{y\in\Lambda_n\\ y_i=-n}} \tau_y\psi_n$$
$$+ \frac{1}{(2n)^d}\nabla_{0,e_i}\sum_{\substack{y\in e_i+\Lambda_n\\ y_i=n+1}} \tau_y\psi_n .$$

We shall refer to the first sum as the non boundary term, to the second one, denoted by $\sigma_n^{i,-}$, as the negative boundary term and to the third sum, denoted by $\sigma_n^{i,+}$, as the positive boundary term. We have just proved that the sequences $\sigma_n^{i,+}$, $\sigma_n^{i,-}$ are bounded in $L^2(\nu_\alpha)$ and are thus weakly relatively compact. In the next step we show that the non boundary term converges to $\mathfrak{g}^i$ in $L^2(\nu_\alpha)$ and in the Steps 5 and 6 we prove that all limit points of the sequence $\sigma_n^{i,-} + \sigma_n^{i,+}$ belong to the subspace generated by $\mathfrak{A}^i$.

Step 4. Convergence of the non boundary term to $\mathfrak{g}^i$. For a fixed y in Λ_n such that $-n < y_i \le n$ we have

$$\nabla_{0,e_i}\tau_y\psi_n = \tau_y\nabla_{-y,-y+e_i}\psi_n = \tau_y\nabla_{-y,-y+e_i}E_\alpha\left[\widetilde{\psi}_{3n}\,|\,\mathcal{F}_n\right] .$$

Since $-n < y_i \le n$ we may move the gradient $\nabla_{-y,-y+e_i}$ inside the expectation. By definition, $\nabla_{-y,-y+e_i}\widetilde{\psi}_{3n} = \mathfrak{g}_{-y}^{i,3n} = E[\mathfrak{g}_{-y}^i\,|\,\mathcal{F}_{3n}]$. Therefore,

$$\nabla_{0,e_i}\tau_y\psi_n = \tau_y E\Big[E\big[\mathfrak{g}_{-y}^i\,|\,\mathcal{F}_{3n}\big]\,|\,\mathcal{F}_n\Big] = \tau_y\mathfrak{g}_{-y}^{i,n}$$

and the first term in the decomposition of R_n^i is

$$\frac{1}{(2n)^d}\sum_{\substack{y\in\Lambda_n\\ -n<y_i\le n}} \tau_y\mathfrak{g}_{-y}^{i,n} .$$

By the martingale convergence theorem $\mathfrak{g}_0^{i,n} \to \mathfrak{g}_0^i = \mathfrak{g}^i$ in $L^2(\nu_\alpha)$, as $n\uparrow\infty$. Therefore, for every $\varepsilon > 0$, there exists $n_0 \in \mathbb{N}$ such that $< (\mathfrak{g}_0^{i,n} - \mathfrak{g}^i)^2 >_\alpha \le \varepsilon$ for every $n \ge n_0$. In particular, since ν_α is translation invariant, for every $n \ge n_0$ and every y in Λ_{n-n_0},

$$
\begin{aligned}
\Big\langle \big(\mathfrak{g}^{i,n}_y - \mathfrak{g}^i_y\big)^2 \Big\rangle_\alpha &= \Big\langle \Big(E\big[\tau_y \mathfrak{g}^i \,|\, \mathcal{F}_n\big] - \tau_y \mathfrak{g}^i\Big)^2 \Big\rangle_\alpha \\
&\le \Big\langle \Big(E\big[\tau_y \mathfrak{g}^i \,|\, \mathcal{F}_{y+\Lambda_{n_0}}\big] - \tau_y \mathfrak{g}^i\Big)^2 \Big\rangle_\alpha \\
&= \Big\langle \Big(\tau_y E\big[\mathfrak{g}^i \,|\, \mathcal{F}_{n_0}\big] - \tau_y \mathfrak{g}^i\Big)^2 \Big\rangle_\alpha \le \varepsilon\,.
\end{aligned}
$$

Thus, for $n \ge n_0$, we have that

$$
\begin{aligned}
&\Big\langle \Big(\frac{1}{(2n)^d} \sum_{y\in\Lambda_n;\, -n<y_i\le n} \tau_y \mathfrak{g}^{i,n}_{-y} - \mathfrak{g}^i \Big)^2 \Big\rangle_\alpha \\
&\quad \le \frac{C(d) n_0}{n} \big\langle (\mathfrak{g}^i)^2 \big\rangle_\alpha + \frac{1}{(2n)^d} \sum_{y\in\Lambda_{n-n_0}} \Big\langle \big(\mathfrak{g}^{i,n}_y - \mathfrak{g}^i_y\big)^2 \Big\rangle_\alpha \\
&\quad \le \frac{C(d) n_0}{n} \big\langle (\mathfrak{g}^i)^2 \big\rangle_\alpha + \varepsilon\,.
\end{aligned}
$$

This estimate shows that

$$
\frac{1}{(2n)^d} \nabla_{0,e_i} \sum_{\substack{y\in\Lambda_n \\ -n<y_i\le n}} \tau_y \psi_n = (2n)^{-d} \sum_{y\in\Lambda_n;\, -n<y_i\le n} \tau_y \mathfrak{g}^{i,n}_{-y}
$$

converges to $\mathfrak{g}^i$ in $L^2(\nu_\alpha)$.

We turn now to the boundary terms. By Step 3 the sequences $(\sigma^{1,-}_n, \dots, \sigma^{d,-}_n)$ and $(\sigma^{1,+}_n, \dots, \sigma^{d,+}_n)$ are weakly relatively compact. To prove that all limit points belong to the linear space generated by $\{\mathfrak{A}^i,\ 1 \le i \le d\}$, consider a weakly convergent subsequence. To keep notation simple assume that the sequences converge weakly and denote by $\mathfrak{b}_+$ (resp. $\mathfrak{b}_-$) the weak limit of the positive (resp. negative) boundary term.

We start proving that for each $1 \le i \le d$, $\mathfrak{b}^i_+$ and $\mathfrak{b}^i_-$ depend on η only through $\eta(0)$ and $\eta(e_i)$.

Step 5. For $1 \le i \le d$, $\mathfrak{b}^i_\pm(\eta) = \mathfrak{b}^i_\pm(\eta(0), \eta(e_i))$. We shall prove this statement for $\mathfrak{b}^i_-$; the same arguments apply to $\mathfrak{b}^i_+$. Notice that the negative boundary term $\sigma^{i,-}_n$ is measurable with respect to the σ-algebra generated by $\{\eta(e_i),\ \eta(x),\ x \in \mathbb{Z}^d, x_i \le 0\}$. Of course the weak limit $\mathfrak{b}^i_-$ inherits this property.

We claim that $\nabla_{z,z+e_j} \mathfrak{b}^i_- = 0$ for all bonds $\{z, z+e_j\}$ in the half space $\mathbb{Z}^d_{i,-} = \{x \in \mathbb{Z}^d, x_i \le 0\}$ that do not intersect the origin, i.e., such that z, $z + e_j \ne 0$. We prove at the end of this step that this property implies that $\mathfrak{b}^i_-$ depends on η only through $\eta(0)$, $\eta(e_i)$. Notice that we do not require here j to be different from i.

Fix such a bond and assume that n is large enough so that z, $z+e_j$ belong to Λ_n. Since $\nabla_{z,z+e_j}$ is continuous with respect to the weak topology, it is enough to show that the sequence $\nabla_{z,z+e_j}\sigma^{i,-}_n$ converges to 0 in $L^2(\nu_\alpha)$. Since $\{z, z+e_j\}$ does not intersect $\{0, e_i\}$, $\nabla_{z,z+e_j}$ and ∇_{0,e_i} commute. In particular, for each y in Λ_n,

$$\nabla_{z,z+e_j}\nabla_{0,e_i}\tau_y\psi_n \;=\; \nabla_{0,e_i}\nabla_{z,z+e_j}\tau_y E\Big[\widetilde{\psi}_{3n}\,|\,\mathcal{F}_n\Big]\,.$$

Therefore, by Schwarz inequality, $< (\nabla_{z,z+e_j}\sigma_n^{i,-})^2 >_\alpha$, which is equal to

$$\Big\langle\Big((2n)^{-d}\sum_{\substack{y\in\Lambda_n\\ y_i=-n}} \nabla_{0,e_i}\nabla_{z,z+e_j}\tau_y\psi_n\Big)^2\Big\rangle_\alpha$$
$$\le\; \frac{C(d)}{n^{d+1}}\sum_{\substack{y\in\Lambda_n\\ y_i=-n}}\Big\langle(\nabla_{0,e_i}\nabla_{z,z+e_j}\tau_y\psi_n)^2\Big\rangle_\alpha$$

for some finite constant $C(d)$ that hereafter may change from line to line.

To estimate the right hand side, we have to consider three different cases. Either the bond z, $z+e_j$ belong to $y+\Lambda_n$ or z, $z+e_j$ belong to $(y+\Lambda_n)^c$ or $\{z, z+e_j\}$ links $y+\Lambda_n$ to its complement. In the first case, since z, $z+e_j$ belong to $y+\Lambda_n$, $z-y$, $z-y+e_j$ belong to Λ_n. In particular, $\nabla_{z-y,z+e_j-y}$ and the conditional expectation with respect to $\mathcal{F}_n$ commute. Thus,

$$\nabla_{z,z+e_j}\tau_y\psi_n \;=\; \nabla_{z,z+e_j}\tau_y E\Big[\widetilde{\psi}_{3n}\,|\,\mathcal{F}_n\Big]$$
$$=\; \tau_y E\Big[\nabla_{z-y,z+e_j-y}\widetilde{\psi}_{3n}\,|\,\mathcal{F}_n\Big] \;=\; \tau_y E\Big[\mathfrak{g}^{j,3n}_{z-y}\,|\,\mathcal{F}_n\Big]$$

because $\nabla_{z,z+e_j}\tau_y \;=\; \tau_y\nabla_{z-y,z-y+e_j}$. The L^2 norm of the last expression is bounded above by $< \|\mathfrak{g}\|^2 >_\alpha^{1/2}$. This proves that

$$\frac{C(d)}{n^{d+1}}\sum_{\substack{y\in\Lambda_n,\, y_i=-n\\ y;z,z+e_j\in y+\Lambda_n}}\Big\langle(\nabla_{0,e_i}\nabla_{z,z+e_j}\tau_y\psi_n)^2\Big\rangle_\alpha$$

vanishes as $n\uparrow\infty$.

On the other hand, if z, $z+e_j$ belong to $(y+\Lambda_n)^c$, $\nabla_{z,z+e_j}\tau_y E[\widetilde{\psi}_{3n}\,|\,\mathcal{F}_n]$ vanishes because $\tau_y E[\widetilde{\psi}_{3n}\,|\,\mathcal{F}_n]$ is measurable with respect to $\{\eta(x),\ x\in y+\Lambda_n\}$. Therefore,

$$\frac{C(d)}{n^{d+1}}\sum_{\substack{y\in\Lambda_n,\, y_i=-n\\ y;z,z+e_j\in(y+\Lambda_n)^c}}\Big\langle(\nabla_{0,e_i}\nabla_{z,z+e_j}\tau_y\psi_n)^2\Big\rangle_\alpha \;=\; 0\,.$$

It remains to consider the case where $\{z, z+e_j\}$ links $y+\Lambda_n$ to $(y+\Lambda_n)^c$. Assume that $z\in y+\Lambda_n$ and $z+e_j\in(y+\Lambda_n)^c$, the other possibility is handled in a similar way. In this case, $y_j = z_j - n$ and we are reduced to estimate

$$\frac{C(d)}{n^{d+1}}\sum_{\substack{y\in\Lambda_n,\, y_i=-n\\ y_j=z_j-n}}\Big\langle(\nabla_{0,e_i}\tau_y\nabla_{z-y,z-y+e_j}\psi_n)^2\Big\rangle_\alpha\,.$$

Notice that $j \neq i$ because $\{z, z+e_j\}$ and $\{0, e_i\}$ are disjoints. Since $< (\nabla_{0,e_i} f)^2 >_\alpha \le 4 < f^2 >_\alpha$ and ν_α is translation invariant, setting $x = z - y$, we have that the last expression is bounded above by

$$\frac{C(d)}{n^{d+1}} \sum_{\substack{x \in z+\Lambda_n \\ x_j = n,\, x_i = z_i + n}} \Big\langle (\nabla_{x,x+e_j} \psi_n)^2 \Big\rangle_\alpha .$$

It is not hard to adapt the proof of Lemma 4.15 to show that this expression vanishes as $n \uparrow \infty$.

Up to this point we proved that $\nabla_{z,z+e_j} \mathfrak{b}^i_-$ vanishes for all bonds $\{z, z+e_j\}$ in the half space $\mathbb{Z}^d_{i,-}$ that do not intersect the origin. We claim that this property in addition to the fact that $\mathfrak{b}^i_-$ is measurable with respect to $\{\eta(e_i), \eta(x),\ x \in \mathbb{Z}^d_{i,-}\}$ implies that $\mathfrak{b}^i_-$ is a function of $\eta(0)$ and $\eta(e_i)$ only: $\mathfrak{b}^i_- = E_\alpha[\mathfrak{b}^i_- \,|\, \eta(0),\, \eta(e_i)]$.

To prove this statement, consider a sequence $\mathfrak{b}^{i,(n)}_-$ of finite approximations of $\mathfrak{b}^i_-$: $\mathfrak{b}^{i,(n)}_- = E[\mathfrak{b}^i_- \,|\, \mathcal{F}_n]$. It is easy to check that sequence $\mathfrak{b}^{i,(n)}_-$ inherits from $\mathfrak{b}^i_-$ both properties: $\mathfrak{b}^{i,(n)}_-$ is measurable with respect to $\{\eta(e_i), \eta(x),\ x \in \Lambda_n \cap \mathbb{Z}^d_{i,-}\}$ and $\nabla_{z,z+e_j} \mathfrak{b}^{i,(n)}_- = 0$ for all bonds $\{z, z+e_j\}$ in $\Lambda_n \cap \mathbb{Z}^d_{i,-}$ that do not intersect the origin. In particular,

$$\mathfrak{b}^{i,(n)}_- = E_\alpha \Big[\mathfrak{b}^{i,(n)}_- \,|\, \eta(0), \eta(e_i), \sum_{x \in \Lambda^{i,-}_n} \eta(x) \Big] , \tag{4.11}$$

where $\Lambda^{i,-}_n = \Lambda_n \cap \mathbb{Z}^d_{i,-}$.

Fix now a cylinder function g. Denote by g_- the conditional expectation of g with respect to the σ-algebra generated by $\{\eta(e_i),\ \eta(x),\ x \in \mathbb{Z}^d_{i,-}\}$. Since $\mathfrak{b}^i_-$ is measurable with respect to this σ-algebra, $< g, \mathfrak{b}^i_- >_\alpha = < g_- , \mathfrak{b}^i_- >_\alpha$. Moreover, since g is a cylinder function, for n sufficiently large, $< g_- , \mathfrak{b}^i_- >_\alpha = < g_- , \mathfrak{b}^{i,(n)}_- >_\alpha$. Therefore, by (4.11),

$$\begin{aligned} < g\,, \mathfrak{b}^i_- >_\alpha &= \Big\langle E_\alpha \Big[\mathfrak{b}^{i,(n)}_- \,|\, \eta(0), \eta(e_i), \sum_{x \in \Lambda^{i,-}_n} \eta(x) \Big] , g_- \Big\rangle_\alpha \\ &= \Big\langle \mathfrak{b}^{i,(n)}_- , E_\alpha \Big[g_- \,|\, \eta(0), \eta(e_i), \sum_{x \in \Lambda^{i,-}_n} \eta(x) \Big] \Big\rangle_\alpha \\ &= \Big\langle \mathfrak{b}^i_- , E_\alpha \Big[g_- \,|\, \eta(0), \eta(e_i), \sum_{x \in \Lambda^{i,-}_n} \eta(x) \Big] \Big\rangle_\alpha . \end{aligned}$$

Since g is a cylinder function, as $n \uparrow \infty$, $E_\alpha \Big[g_- \,|\, \eta(0), \eta(e_i), \sum_{x \in \Lambda^{i,-}_n} \eta(x) \Big]$ converges in $L^2(\nu_\alpha)$ to $E_\alpha[g_- \,|\, \eta(0), \eta(e_i)]$. Therefore,

$$< g\,, \mathfrak{b}^i_- >_\alpha = \Big\langle E_\alpha[g_- \,|\, \eta(0), \eta(e_i)]\,, \mathfrak{b}^i_- \Big\rangle_\alpha = \Big\langle g\,, E_\alpha[\mathfrak{b}^i_- \,|\, \eta(0), \eta(e_i)] \Big\rangle_\alpha .$$

This concludes the proof of Step 5.

It remains to prove that $\mathfrak{b}_- + \mathfrak{b}_+$ belongs to the linear space spanned by $\{\mathfrak{A}^i, 1 \le i \le d\}$. This identity is a simple consequence of the following two relations.

Step 6. Conclusion. For $1 \le i \le d$, we have that

$$\begin{aligned} \nabla_{2e_i,e_i} \mathfrak{b}^i_{\pm} &= -\mathbf{1}\{\eta(e_i) = \kappa - 1\}\mathbf{1}\{\eta(2e_i) > 0\}\mathfrak{b}^i_{\pm} \\ \nabla_{0,-e_i} \mathfrak{b}^i_{\pm} &= -\mathbf{1}\{\eta(-e_i) < \kappa\}\mathbf{1}\{\eta(0) = 1\}\mathfrak{b}^i_{\pm} \,. \end{aligned} \tag{4.12}$$

We leave to the reader to check that these relations implies that $\mathfrak{b}^i_{\pm} = C_{\pm}\mathfrak{A}^i$.

We shall prove these relations for $\mathfrak{b}^i_-$, all arguments apply to $\mathfrak{b}^i_+$. Since ∇_{2e_i,e_i} is continuous with respect to the weak topology and $\mathfrak{b}^i_-$ is the weak limit of the negative boundary term, we have that

$$\nabla_{2e_i,e_i} \mathfrak{b}^i_- = \lim_{n\to\infty} (2n)^{-d} \sum_{\substack{y\in\Lambda_n \\ y_i=-n}} \nabla_{2e_i,e_i} \nabla_{0,e_i} \tau_y \psi_n \,. \tag{4.13}$$

Since $y_i = -n$ and ψ_n is $\mathcal{F}_n$-measurable, we have that

$$\begin{aligned} &\nabla_{2e_i,e_i} \nabla_{0,e_i} \tau_y \psi_n \\ &= \nabla_{2e_i,e_i} \Big\{ \mathbf{1}\{\eta(0) > 0\}\mathbf{1}\{\eta(e_i) < \kappa\} \Big[(\tau_y\psi_n)(\eta - \mathfrak{d}_0) - (\tau_y\psi_n)(\eta) \Big] \Big\} \\ &\quad = \mathbf{1}\{\eta(0) > 0\} \Big[(\tau_y\psi_n)(\eta - \mathfrak{d}_0) - (\tau_y\psi_n)(\eta) \Big] \nabla_{2e_i,e_i} \mathbf{1}\{\eta(e_i) < \kappa\} \,. \end{aligned}$$

A trivial computation gives that $\nabla_{2e_i,e_i}\mathbf{1}\{\eta(e_i) < \kappa\} = -\mathbf{1}\{\eta(2e_i) > 0\}\mathbf{1}\{\eta(e_i) = \kappa - 1\}$. In particular,

$$\nabla_{2e_i,e_i} \nabla_{0,e_i} \tau_y \psi_n = -\mathbf{1}\{\eta(2e_i) > 0\}\mathbf{1}\{\eta(e_i) = \kappa - 1\} \nabla_{0,e_i} \tau_y \psi_n \,.$$

Replacing in equation (4.13) $\nabla_{2e_i,e_i} \nabla_{0,e_i} \tau_y \psi_n$ by the right hand side of last identity and recalling that $(2n)^{-d} \sum_{y\in\Lambda_n,\, y_i=-n} \nabla_{0,e_i}\tau_y\psi_n$ weakly converges to $\mathfrak{b}^i_-$, we obtain the first relation in (4.12).

We turn now to the second identity in (4.12). We need to compute $\nabla_{0,-e_i}\nabla_{0,e_i}\tau_y\psi_n$ for y in Λ_n with $y_i = -n$. The difference, with respect to the proof of the first identity, is that $\tau_y\psi_n$ depends on the occupation variables $\eta(0)$, $\eta(-e_i)$. For this reason, computations are slightly more troublesome. It is easy to show that $\nabla_{0,-e_i}\nabla_{0,e_i}\tau_y\psi_n$ is equal to

$$\begin{aligned} &- \mathbf{1}\{\eta(-e_i) < \kappa\}\mathbf{1}\{\eta(0) = 1\}\nabla_{0,e_i}\tau_y\psi_n \\ &\qquad + \mathbf{1}\{\eta(e_i) < \kappa,\ \eta(0) > 1\}(\nabla_{0,-e_i}\tau_y\psi_n)(\eta - \mathfrak{d}_0) \\ &\qquad\quad - \mathbf{1}\{\eta(e_i) < \kappa,\ \eta(0) > 1\}(\nabla_{0,-e_i}\tau_y\psi_n)(\eta) \,. \end{aligned}$$

On the one hand, since $(2n)^{-d} \sum_{y\in\Lambda_n,\, y_i=-n} \nabla_{0,e_i}\tau_y\psi_n$ weakly converges to $\mathfrak{b}^i_-$, we have that

$$\begin{aligned} &\lim_{n\to\infty} -(2n)^{-d} \sum_{\substack{y\in\Lambda_n \\ y_i=-n}} \mathbf{1}\{\eta(-e_i) < \kappa\}\mathbf{1}\{\eta(0) = 1\}\nabla_{0,e_i}\tau_y\psi_n \\ &\quad = -\mathbf{1}\{\eta(-e_i) < \kappa\}\mathbf{1}\{\eta(0) = 1\}\mathfrak{b}^i_- \,. \end{aligned}$$

On the other hand, by definition of ψ_n, $(\nabla_{0,-e_i}\tau_y\psi_n)(\eta)$ is equal to

$$\begin{aligned}\tau_y\nabla_{-y,-y-e_i}\psi_n(\eta) &= \tau_y E_\alpha\Big[\nabla_{-y,-y-e_i}\widetilde{\psi}_{3n} \,|\, \mathcal{F}_n\Big] \\ &= -\tau_y \mathfrak{g}^{i,n}_{-y-e_i}(\sigma^{-y,-y-e_i}\eta)\,.\end{aligned}$$

Since the closed forms $\{\mathfrak{g}^{i,n}_x\}$ are uniformly bounded in $L^2(\nu_\alpha)$, applying Schwarz inequality, it is easy to show that

$$(2n)^{-d}\sum_{\substack{y\in\Lambda_n\\ y_i=-n}} \mathbf{1}\{\eta(0) > 1\}\Big\{(\nabla_{0,-e_i}\tau_y\psi_n)(\eta-\mathfrak{d}_0) \;-\; (\nabla_{0,-e_i}\tau_y\psi_n)(\eta)\Big\}$$

vanishes as $n\uparrow\infty$. This concludes the proof of step 6 and shows that $\mathfrak{G}$ is included in $\mathfrak{E}_\alpha + \{\mathfrak{A}^i,\ 1\le i\le d\}$.

To conclude the proof of Theorem 4.14 we have to show that the intersection of $\mathfrak{E}$ and $\{\mathfrak{A}^i,\ 1\le i\le d\}$ in $L^2(\nu_\alpha)$ is trivial. Assume that there exists a cylinder function h and a vector C in $\mathbb{R}^d$ such that $\sum_{1\le i\le d} C_i\mathfrak{A}^i = \nabla\Gamma_h$. In particular, $C_i\mathfrak{A}^i = \nabla_{0,e_i}\Gamma_h$. taking inner product in $L^2(\nu_\alpha)$ with respect to $\eta(e_i)-\eta(0)$, we obtain that $C_i = 0$ because $<\eta(e_i)-\eta(0)\,,\, r_{0,e_i}>_\alpha \;=\; -<r_{0,e_i}>_\alpha\neq 0$ and $<\eta(e_i)-\eta(0)\,,\,\nabla_{0,e_i}\Gamma_h>_\alpha = 0$. This proves that the sum is direct and concludes the proof of the theorem. □

5. Comments and References

The estimate of the spectral gap presented in Proposition 0.2 is taken from Landim, Sethuraman and Varadhan (1996). The second order expansion for the largest eigenvalue of a small perturbation of a reversible generator stated in Theorem 1.1 is due to Varadhan (1994a). The proof that the spectral gap for the generator of the generalized symmetric simple exclusion process restricted to a cube of length ℓ is of order ℓ^{-2} is based on the martingale approach introduced by Lu and Yau (1993). The proof presented here is taken from Landim, Sethuraman and Varadhan (1996). The characterization of the closed and exact forms in the context of interacting particle systems is due in dimension 1 to Quastel (1992). It is extended to higher dimensions by Funaki, Uchiyama and Yau (1995).

Spectral gap of conservative dynamics. Lu and Yau (1993) proved that the generator of the symmetric exclusion process with speed change, the so called Kawasaki dynamics, restricted to a cube of length ℓ as a gap of order ℓ^{-2} in any dimension and for any boundary condition provided some mixing conditions are satisfied. Landim, Sethuraman and Varadhan (1996) applied the martingale approach introduced by Lu and Yau to prove a spectral gap for zero range dynamics for which the jump rate $g(\cdot)$ satisfies the Lipschitz condition $\sup_{k\ge 1}|g(k+1) - g(k)| < \infty$ and for which there exists $\delta > 0$ and k_0 in $\mathbb{N}$ such that $g(k+j)-g(k)\ge\delta$

for $j \geq k_0$. Posta (1997) obtains a sharp estimate for the spectral gap of a two-dimensional Kawasaki dynamics of unbounded discrete spins on finite cubes.

Nash inequalities, L^2 decay to equilibrium. We have seen that the spectral gap of the generator is closely related to the rate of convergence to equilibrium in L^2. In fact, it is easy to see that for reversible systems the existence of a gap is equivalent to an exponential rate of convergence to equilibrium in L^2. Consider a Markov process with reversible invariant measure ν. Denote by γ the gap of the spectrum:

$$\gamma = \inf_f \frac{< (-L)f, f >_\nu}{< f, f >_\nu} ,$$

where the infimum is carried out over all $L^2(\nu)$ functions orthogonal to the constants and where $< \cdot, \cdot >_\nu$ stands for the inner product of $L^2(\nu)$. Then,

$$\gamma = \inf_{t>0} \frac{1}{t} \inf_f \Big\{ -\log \frac{\|S_t f\|_2^2}{< f, f >_\nu} \Big\} ,$$

where the infimum is carried out over all $L^2(\nu)$ functions orthogonal to the constants. Here $\{S_t,\, t \geq 0\}$ stands for the semigroup of the process and $\| \cdot \|_2$ for the $L^2(\nu)$ norm. We refer to Liggett (1989) for a proof of this statement.

As we have seen in this chapter, there is no spectral gap for conservative dynamics in infinite volume. In this case a polynomial decay to equilibrium is expected:

$$\big\| S_t f - < f >_\nu \big\|_2^2 \leq \frac{V(f)}{t^a}$$

for some exponent a and some function V on $L^2(\nu)$. Here $< f >_\nu$ stands for the expected value of f with respect to ν. As already noticed by Deuschel (1989), the exponent a might depend on the class of functions considered, which introduces an additional difficulty in the analysis.

A simple computation on the symmetric simple exclusion process using duality shows that for conservative, infinite volume interacting particle systems the expected algebraic decay for cylinder functions is $t^{-d/2}$.

There is a general method to prove polynomial decay to equilibrium in L^2 based on Nash inequalities. This type of inequality consists in estimating the variance of a L^2 function by the Dirichlet form and a norm V defined in L^2:

$$\mathbf{Var}(\nu, f) \leq C\,\mathfrak{D}(f)^{1/p} V(f)^{1/q} .$$

for some universal constant C. In this formula $\mathfrak{D}$ is the Dirichlet form ($\mathfrak{D}(f) = < (-L)f, f >_\nu$), $1 < p, q < \infty$ are conjugates $p^{-1} + q^{-1} = 1$ and V is a function in $L^2(\nu)$ satisfying $0 \leq V(f) \leq \infty$, $V(rf + s) = r^2 V(f)$ for all r, s in $\mathbb{R}$. It is not difficult to prove (cf. Liggett (1991)) that if the semigroup is a contraction for V, i.e., if $V(S_t f) \leq V(f)$ for all f in $L^2(\nu)$ and $t > 0$, then there is polynomial convergence to equilibrium in $L^2(\nu)$ with exponent $a = q - 1$:

$$\big\| S_t f - < f > \big\|_2^2 \leq C \frac{V(f)}{t^{q-1}}$$

for all f in L^2, $t > 0$. The converse of this statement is also true because we assumed the process to be reversible.

Nash inequalities have been proved by Bertini and Zegarlinski (1996a,b) for symmetric simple exclusion processes and for symmetric exclusion processes with speed change, the so called Kawasaki dynamics, with V given by the square of a triple norm:

$$V(f) = \Big(\sum_{x \in \mathbb{Z}^d} \Big\| \frac{\partial f}{\partial \eta(x)} \Big\|_\infty \Big)^2 ,$$

where $\partial f / \partial \eta(x)$ measures the dependence of the function f on the site x: $\partial f / \partial \eta(x) = f(\sigma^x \eta) - f(\eta)$ and $\sigma^x \eta$ is the configuration η with the occupation variable at x flipped: $(\sigma^x \eta)(y) = \delta_{x,y}(1 - \eta(y)) + (1 - \delta_{x,y})\eta(y)$. Janvresse, Landim, Quastel and Yau (1997) proved Nash inequality for zero range dynamics with the triple norm $V(f)$ given by

$$V(f) = \Big(\sum_{x \in \mathbb{Z}^d} \Big\| \frac{\partial f}{\partial \eta(x)} \Big\|_2 \Big)^2 ,$$

where $\partial f / \partial \eta(x)$ is now $f(\eta + \mathfrak{d}_x) - f(\eta)$.

The proof that the semigroup is a contraction for the triple norms defined above has only been achieved for the symmetric simple exclusion process and seems to be a very difficult problem for conservative particle systems. Janvresse, Landim, Quastel and Yau (1997) presented a direct method to prove the algebraic decay to equilibrium for zero range processes. Cancrini and Galves (1995) obtained an upper bound for the rate of convergence to equilibrium for d-dimensional symmetric exclusion processes starting either form a periodic configuration or a stationary mixing distribution. They proved the existence of a finite constant $C(d, N)$ such that

$$\Big| \mathbb{P}_\mu[\eta(x) = 1,\ 1 \le x \le N] - \alpha^N \Big| \le C(d, N) \Big(\frac{\log t}{\sqrt{t}} \Big)^d ,$$

where μ is either a periodic configuration or a stationary mixing distribution with density α. Galves and Guiol (1997) extended this result to the one-dimensional nearest neighbor symmetric zero range process with jump rate given by $g(k) = \mathbf{1}\{k \ge 1\}$. Deuschel (1994) develops a general theory to prove polynomial L^2 decay to equilibrium for attractive critical processes constructed by means of transient random walks on $\mathbb{Z}^d$.

The question of polynomial decay to equilibrium of infinite volume conservative dynamics is far to be well understood and constitutes one of the main open problems in the field.

Non conservative dynamics. L^2 decay to equilibrium has also been considered for non conservative dynamics. Aizenman and Holley (1987) proved exponential L^2 convergence for infinite volume Glauber dynamics satisfying the Dobrushin–Shlosman uniqueness condition (cf. Dobrushin and Shlosman (1987)). Thomas

(1989) proved that at low enough temperature the generator of reversible stochastic Ising models restricted to a cube of length ℓ with free boundary condition is less than $C(\beta)\exp\{-\beta C_0 \ell^{d-1}\}$ for some universal constant C_0. Here β is the inverse of the temperature and $C(\beta)$ is a function of β only. Liggett (1989) proved exponential L^2 convergence to equilibrium for supercritical reversible nearest particle systems and proved in Liggett (1991) an algebraic L^2 decay to equilibrium in the critical case. Neuhauser (1990) proves the exponential decay to equilibrium of a stochastic Ising model superposed with a stirring process multiplied by a small constant.

Logarithmic Sobolev inequalities and convergence to equilibrium in L^∞. Closely related to the spectral gap inequalities are the logarithmic Sobolev inequalities. It consists in proving that the entropy of a density f with respect to a canonical measure $\nu_{N,K}$ is bounded by the Dirichlet form of $\sqrt{f}$:

$$\int f \log f \, d\nu_{N,K} \;\le\; C_0 N^2 < \sqrt{f}, (-L_N)\sqrt{f} >_{N,K}$$

for some universal constant C_0, all $0 \le K \le N^d$ and all $N \ge 1$. We refer to Deuschel and Stroock (1989), Davies (1989) Davies et al. (1992) for the relationship between spectral gaps, logarithmic Sobolev inequalities and the hypercontractivity of the semigoup.

Logarithmic Sobolev inequalities have been introduced by Gross (1976). He showed the connection between this type of inequality and the hypercontractivity of the semigroup. By the same time Holley and Stroock (1976a,b) started to investigate exponential L^2 and L^∞ convergence to equilibrium of stochastic Ising model. Holley (1985a,b) proved exponential L^∞ convergence of finite range, translation invariant, attractive Glauber dynamics. Holley and Stroock (1987b) extended this result to non attractive, one-dimensional dynamics.

In the context of interacting particle systems on a lattice, logarithmic Sobolev inequalities were first derived by Holley and Stroock (1987a) for continuous spin Ising models. Deuschel and Stroock (1990) investigated the relations between spectral gap, logarithmic Sobolev inequalities and hypercontractivity for translation invariant, finite range, continuous spins Glauber dynamics. Zegarlinski (1990a,b), (1992) presented a general method to prove logarithmic Sobolev inequalities for Glauber dynamics based on the Gibbs structure and on Dobrushin uniqueness condition (cf. Dobrushin (1968), (1970)). He applied this approach to one-dimensional Gibbs measures with bounded, finite range interactions and to continuous or finite spin systems satisfying the Dobrushin uniqueness condition. Stroock and Zegarlinski (1992a,b,c) proved the equivalence of a uniform logarithmic Sobolev inequality and the Dobrushin–Shlosman mixing conditions (cf. Dobrushin and Shlosman (1987)) for finite range lattice gases with compact and continuous or finite spin spaces. From this result they deduced exponential L^∞ convergence to equilibrium for stochastic Ising models satisfying the Dobrushin–Shlosman mixing conditions.

Martinelli, Olivieri and Scopolla (1990) proved exponential convergence to equilibrium of ferromagnetic stochastic Ising model in two dimensions at low temperature and with a small external field, uniformly on the volume and on the boundary conditions. Mountford (1995) proves exponential convergence to equilibrium of attractive reversible subcritical nearest particle systems

Lu and Yau (1993) proved a logarithmic Sobolev inequality for Glauber dynamics by the martingale method, which is the only one that has been successfully adapted to the conservative case by Yau (1997) for generalized symmetric exclusion processes and Yau (1996) for Kawasaki dynamics. Martinelli and Olivieri (1994a) analyze the connections between several mixing conditions and prove uniform exponential convergence for finite range attractive stochastic Ising models. Martinelli and Olivieri (1994b) prove a logarithmic Sobolev inequality for Glauber dynamics satisfying strong mixing conditions and deduce the hypercontractivity of the semigroup and exponential L^∞ convergence to equilibrium in infinite volume.

Logarithmic Sobolev inequalities has not yet been proved for conservative dynamics with unrestricted number of particles per site. A version of L^∞ convergence to equilibrium for conservative dynamics is a major problem in the field.

References

Aizenman, M., Holley R. (1987): Rapid convergence to equilibrium of stochastic Ising models in the Dobrushin–Shlosman regime. In H. Kesten, editor, *Percolation Theory and Ergodic Theory of Infinite Particle Systems*, volume 8 of IMA Volumes in Mathematics, pages 1–11, Springer-Verlag, Berlin.

Aldous, D.J. (1978): Stopping times and tightness. Ann. Probab. **6**, 335–340

Aldous, D.J. (1982): Markov chains with almost exponential hitting times. Stoch. Proc. Appl. **13**, 305–310

Aldous, D.J. (1989): *Probability Approximations via the Poisson Clumping Heuristics*. Volume 77 of Applied Mathematical Sciences. Springer-Verlag, New York

Aldous, D.J., Brown, M. (1992): Inequalities for rare events in time reversible Markov chains I. In M. Shaked and Y. L. Tong, editors, *Stochastic Inequalities*, volume 22 of IMS Lecture Notes, pages 1–16. IMS

Aldous, D.J., Brown, M. (1993): Inequalities for rare events in time reversible Markov chains II. Stoch. Proc. Appl. **44**, 15–25

Aldous, D.J., Diaconis, P. (1995): Hammersley's interacting particle process and longest increasing subsequences. Probab. Th. Rel. Fields **103**, 199-213

Alexander, F.J., Cheng, Z., Janowsky, S.A., Lebowitz, J.L. (1992): Shock fluctuations in the two–dimensional asymmetric simple exclusion process. J. Stat. Phys. **68**, 761–785

Andjel, E.D. (1982): Invariant measures for the zero-range process. Ann. Probab. **10**, 525–547

Andjel, E.D. (1986): Convergence to a non extremal equilibrium measure in the exclusion process. Probab. Th. Rel. Fields **73**, 127–134

Andjel, E.D., Bramson, M., Liggett, T.M. (1988): Shocks in the asymmetric exclusion process. Probab. Th. Rel. Fields **78**, 231–247

Andjel, E.D., Kipnis, C. (1984): Derivation of the hydrodynamical equation for the zero-range interaction process. Ann. Probab. **12**, 325–334

Andjel, E.D., Vares, M.E. (1987): Hydrodynamic equations for attractive particle systems on Z. J. Stat. Phys. **47**, 265–288

Arratia, R. (1983): The motion of a tagged particle in the simple symmetric exclusion system on Z. Ann. Probab. **11**, 362–373

Asselah, A., Brito, R., Lebowitz, J.L. (1997): Self diffusion in simple models: systems with long range jumps. J. Stat. Phys. **87** 1131–1144

Asselah, A., Dai Pra, P. (1997): Sharp estimates for the occurrence of rare events for symmetric simple exclusion. Stoch. Proc. Appl. **71**, 259–273

Bahadoran, C. (1996a): Hydrodynamical limit for non-homogeneous zero range processes, preprint

Bahadoran, C. (1996b): Hydrodynamical limit for spatially heterogeneous simple exclusion processes, preprint

Bahadoran, C. (1997): Hydrodynamics of asymmetric misanthrope processes with general initial configurations, preprint

Bellman, R., Harris, T.E. (1951): Recurrence times for the Ehrenfest model. Pacific J. Math. **1** 179–193

Benassi, A., Fouque, J.P. (1987): Hydrodynamical limit for the asymmetric exclusion process. Ann. Probab. **15**, 546–560

Benassi, A., Fouque, J.P. (1988): Hydrodynamical limit for the asymmetric zero-range process. Ann. Inst. H. Poincaré, Probabilités **24**, 189–200

Benassi, A., Fouque, J.P. (1991): Fluctuation field for the asymmetric simple exclusion process. In U. Hornung, P. Kotelenez and G. Papanicolaou, editors, *Random Partial Differential Equations*, volume 102 of International Series of Numerical Mathematics, pages 33–43, Birhäuser, Boston

Benassi, A., Fouque, J.P., Saada, E., Vares, M.E. (1991): Asymmetric attractive particle systems on Z: hydrodynamical limit for monotone initial profiles. J. Stat. Phys. **63**, 719–735

Benjamini, I., Ferrari, P.A., Landim, C. (1996): Asymmetric processes with random rate. Stoch. Proc. Appl. **61**, 181–204

Benois, O. (1996): Large deviations for the occupation times of independent particle systems. Ann. Appl. Probab. **6** 269–296

Benois, O., Kipnis, C., Landim, C. (1995): Large deviations for mean zero asymmetric zero-range processes in infinite volume. Stoch. Proc. Appl. **55**, 65–89

Benois, O., Koukkous, A., Landim, C. (1997): Diffusive behavior of asymmetric zero range processes. J. Stat. Phys. **87**, 577–591

Bertini, L., Landim, C., Olla, S. (1997): Derivation of Cahn–Hilliard equations from Ginzburg–Landau models. J. Stat. Phys. **88**, 365–381

Bertini, L., Presutti, E., Rüdiger, B., Saada, E. (1994): Dynamical fluctuations at the critical point: convergence to a non linear stochastic PDE. Theory Probab. Appl. **38**, 586–629

Bertini, L., Zegarlinski, B. (1996a): Coercive inequalities for Kawasaki dynamics: the product case. University of Texas Mathematical Physics archive preprint 96-561

Bertini, L., Zegarlinski, B. (1996b): Coercive inequalities for Gibss measures. University of Texas Mathematical Physics archive preprint 96-562.

Billingsley, P. (1968): *Convergence of Probability Measures*, John Wiley & Sons, New York

Boldrighini, C., De Masi, A., Pellegrinotti, A. (1991): Nonequilibrium fluctuations in particle system modeling reaction–diffusion equations. Stoch. Proc. Appl. **42**, 1–30

Boldrighini, C., De Masi, A., Pellegrinotti, A., Presutti, E. (1987): Collective phenomena in interacting particle systems. Stoch. Proc. Appl. **25**, 137–152

Boldrighini, C., Dobrushin, R.L., Suhov, Yu.M. (1983): One–dimensional hard rod caricature of hydrodynamics. J. Stat. Phys. **31**, 577–616

Bonaventura, L. (1995): Interface dynamics in an interacting spin system. Nonlinear Anal. **25**, 799–819

Bramson, M. (1988): Front propagation in certain one dimensional exclusion models. J. Stat. Phys. **51**, 863–870

Bramson, M., Calderoni, P., De Masi, A., Ferrari, P.A., Lebowitz, J.L., Schonmann, R.H. (1989): Microscopic selection principle for a diffusion–reaction equation. J. Stat. Phys. **45**, 905–920

Bramson, M., Cox, J.T., Griffeath, D. (1988): Occupation time large deviations of the voter model. Probab. Th. Rel. Fields **73**, 613-625

Bramson, M., Lebowitz, J.L. (1991): Spatial structure in diffusion–limited two–particle reactions. J. Stat. Phys. **65**, 941–951

Breiman, L. (1968): *Probability*. Addison–Wesley, New York

Brezis, H., Crandall, M.G. (1979): Uniqueness of solutions of the initial value problem for $u_t - \Delta\phi(u) = 0$. J. Math. Pure Appl. **58**, 153–163

Brox, T., Rost, H. (1984): Equilibrium fluctuations of stochastic particle systems: the role of conserved quantities. Ann. Probab. **12**, 742–759

Buttà, P. (1993): On the validity of an Einstein relation in models of interface dynamics. J. Stat. Phys. **72** 1401–1406

Buttà, P. (1994): Motion by mean curvature by scaling a nonlocal equation: convergence at all times in the two–dimensional case. Lett. Math. Phys. **31**, 41–55

Calderoni, P., Pellegrinotti, A., Presutti, E., Vares, M.E. (1989): Transient bimodality in interacting particle systems. J. Stat. Phys. **55**, 523–577

Cancrini, N., Galves, A. (1995): Approach to equilibrium in the symmetric simple exclusion process. Markov Proc. Rel. Fields **2**, 175–184

Caprino, S., De Masi, A., Presutti, E., Pulvirenti, M. (1989): A stochastic particle system modeling the Carleman equation. J. Stat. Phys. **55**, 625–638

Caprino, S., De Masi, A., Presutti, E., Pulvirenti, M. (1990): A stochastic particle system modeling the Carleman equation: Addendum. J. Stat. Phys. **59**, 535–537

Caprino, S., De Masi, A., Presutti, E., Pulvirenti, M. (1991): A derivation of the Broadwell equation. Commun. Math. Phys. **135**, 443–465

Caprino, S., Pulvirenti, M. (1995): A cluster expansion approach to a one–dimensional Boltzmann equation: a validity result. Commun. Math. Phys. **166**, 603–631

Caprino, S., Pulvirenti, M. (1991): The Boltzmann–Grad limit for a one–dimensional Boltzmann equation in a stationary state. Commun. Math. Phys. **177**, 63–81

Carlson, J.M., Grannan, E.R., Swindle, G.H. (1993): A limit theorem for tagged particles in a class of self–organizing particle systems. Stoch. Proc. Appl. **47**, 1–16

Carlson, J.M., Grannan, E.R., Swindle, G.H., Tour, J. (1993) Singular diffusion limits of a class of reversible self–organizing particle systems. Ann. Probab. **21**, 1372–1393

Carmona, R.A., Xu, L. (1997): Diffusive hydrodynamic limit for systems of interacting diffusions with finite range random interaction. Commun. Math. Phys. **188**, 565–584

Cassandro, M., Galves, A., Olivieri, E. Vares, M. E. (1984): Metastable behavior of stochastic dynamics: a pathwise approach. J. Stat. Phys. **35**, 603–628

Chang, C.C. (1994): Equilibrium fluctuations of gradient reversible particle systems. Probab. Th. Rel. Fields **100**, 269–283

Chang, C.C. (1995): Equilibrium fluctuations of nongradient reversible particle systems. In T. Funaki and W. Woyczinky, editor, *Proc. On Stochastic Method for Nonlinear P.D.E.*, volume 77 of IMA vol. in Mathematics, pages 41–51, Springer-Verlag, New York

Chang, C.C., Yau, H.T. (1992): Fluctuations of one dimensional Ginzburg–Landau models in nonequilibrium. Commun. Math. Phys. **145**, 209–234

Chayes, L., Swindle, G. (1996): Hydrodynamic limits for one dimensional particle systems with moving boundaries. Ann. Probab. **24**, 559–598

Chow, Y.S., Teicher, H. (1988): *Probability Theory*, Springer-Verlag, New York

Chung, K.L. (1974): *A Course in Probability Theory*, 2nd edition, Academic Press, New York.

Cocozza, C. (1985): Processus des misanthropes. Z. Wahrsch. Verw. Gebiete **70**, 509–523

Cogburn, R. (1985): On the distribution of first passage and return times for small sets. Ann. Probab. **13**, 1219–1223

Comets, F. (1987): Nucleation for a long-range magnetic model. Ann. Inst. H. Poincaré, Probabilités **23**, 135–178

Comets, F., Eisele, Th. (1988): Asymptotic dynamics, noncritical and critical fluctuations for a geometric long-range interacting model. Commun. Math. Phys. **118**, 531–567

Covert, P., Rezakhanlou, F. (1996): Hydrodynamic limit for particle systems with non–constant speed parameter. J. Stat. Phys. **88**, 383–426

Cox, J.T., Griffeath, D. (1984): Large deviation for Poisson systems of independent random walks. Z. Wahrsch. Verw. Gebiete **66**, 543–558

Davies, E.B. (1989): *Heat kernels and spectral theory*, Cambridge University Press, Cambridge

Davies, E.B., Gross, L., Simon, B. (1992): Hypercontractivity: a bibliographical review. In S. Albeverio, J.E. Fenstand, H. Holden and T. Lindstrom, editors, *In memoriam of Raphael Hoegh-Krohn*. Ideas and Methods of Mathematic and Physics, Cambridge University Press, Cambridge

De Masi, A., Ferrari, P. A. (1984): A remark on the hydrodynamics of the zero range processes. J. Stat. Phys. **36**, 81–87

De Masi, A., Ferrari, P. A. (1985): Self-diffusion in one-dimensional lattice gases in the presence of an external field, J. Stat. Phys. **38**, 603–613

De Masi, A., Ferrari, P. A., Goldstein, S., Wick, W. D. (1985): Invariance principle for reversible Markov processes with application to diffusion in the percolation regime. Contemp. Math. **41**, 71–85

De Masi, A., Ferrari, P. A., Goldstein, S., Wick, W. D. (1989): An invariance principle for reversible Markov processes. Applications to random motions in random environments. J. Stat. Phys. **55**, 787–855

De Masi, A., Ferrari, P. A., Ianiro, N., Presutti, E. (1982): Small deviations from local equilibrium for a process which exhibits hydrodynamical behavior II. J. Stat. Phys. **29**, 81–93

De Masi, A., Ferrari, P. A., Lebowitz, J.L. (1986): Reaction-diffusion equations for interacting particle systems, J. Stat. Phys. **44**, 589–644

De Masi, A., Ferrari, P. A., Vares, M.E. (1989): A microscopic model of interface related to the Burger's equation. J. Stat. Phys. **55**, 601–609

De Masi, A., Ianiro, N., Pellegrinotti, A., Presutti, E. (1984): A survey of the hydrodynamical behavior of many-particle systems. In J. L. Lebowitz and E. W. Montroll, editors, *Nonequilibrium phenomena II: From stochastics to hydrodynamics* Volume 11 of Studies in Statistical Mechanics, pages 123–294, North-Holland, Amsterdam

De Masi, A., Ianiro, N., Presutti, E. (1982): Small deviations from local equilibrium for a process which exhibits hydrodynamical behavior. I. J. Stat. Phys. **29**, 57–79

De Masi, A., Kipnis, C., Presutti, E., Saada, E. (1989): Microscopic structure at the shock in the asymmetric simple exclusion. Stochastics **27**, 151–165

De Masi, A., Orlandi, E., Presutti, E., Triolo, L. (1994): Glauber evolution with Kac potentials: I. Mesoscopic and macroscopic limits, interface dynamics. Nonlinearity **7**, 633–696

De Masi, A., Orlandi, E., Presutti, E., Triolo, L. (1996a): Glauber evolution with Kac potentials: II. Fluctuations. Nonlinearity **9**, 27–51

De Masi, A., Orlandi, E., Presutti, E., Triolo, L. (1996b): Glauber evolution with Kac potentials: III. Spinodal decompositions. Nonlinearity **9**, 53–114

De Masi, A., Pellegrinotti, A., Presutti, E., Vares, E. (1994): Spatial patterns when phases separate in an interacting particle system. Ann. Probab. **22**, 334–371

De Masi, A., Presutti, E. (1991): *Mathematical methods for hydrodynamic limits*, volume 1501 of Lecture Notes in Mathematics, Springer-Verlag, New York

De Masi, A., Presutti, E., Scacciatelli, E. (1989): The weakly asymmetric simple exclusion process. Ann. Inst. H. Poincaré, Probabilités **25**, 1–38

De Masi, A., Presutti, E., Spohn, H., Wick, D. (1986): Asymptotic equivalence of fluctuation fields for reversible exclusion processes with speed change. Ann. Probab. **14**, 409–423

De Masi, A., Presutti, E., Vares, M.E. (1986): Escape from the unstable equilibrium in a random process with infinitely many interacting particles. J. Stat. Phys. **44**, 645–696

DelGrosso, G. (1974): On the local central limit theorem for Gibbs processes. Commun. Math. Phys. **37**, 141–160

Derrida, B., Domany, E., Mukamel, D. (1992): An exact solution of a one–dimensional asymmetric exclusion model with open boundaries. J. Stat. Phys. **69**, 667–687

Derrida, B., Evans, M.T., Mallick, K. (1995): Exact diffusion constant for a one–dimensional asymmetric exclusion model with open boundaries. J. Stat. Phys. **79**, 833–874

Derrida, B., Janowsky, S.A., Lebowitz, J.L., Speer, E.R. (1993): Exact solutions of the totally asymmetric simple exclusion process: shock profiles. J. Stat. Phys. **73**, 813–842

Deuschel, J.D. (1989): Invariance principle and empirical mean large deviations of the critical Ornstein–Uhlenbeck process. Ann. Probab. **17**, 74–90
Deuschel, J.D. (1994): Algebraic L^2 decay of attractive critical processes on the lattice. Ann. Probab. **22**, 264–283
Deuschel, J.D., Stroock, D.W. (1989): *Large Deviations*. Academic Press, New York
Deuschel, J.D., Stroock, D.W. (1990): Hypercontractivity and spectral gap of symmetric diffusions with applications to the stochastic Ising model. J. Funct. Anal. **92**, 30–48
DiPerna, R.J. (1985): Measure-valued solutions to conservation laws. Arch. Rational Mech. Anal. **88**, 223–270
Dittrich, P. (1987): Limit theorems for branching diffusions in hydrodynamical rescaling. Math. Nachr. **131**, 59–72
Dittrich, P. (1988a): A stochastic model of a chemical reaction with diffusion. Probab. Th. Rel. Fields **79**, 115–128
Dittrich, P. (1988b): A stochastic particle system: fluctuations around a non-linear reaction-diffusion equation. Stoch. Proc. Appl. **30**, 149–164
Dittrich, P. (1990): Travelling waves and long-time behaviour of the weakly asymmetric exclusion process. Probab. Th. Rel. Fields **86**, 443–455
Dittrich, P. (1992): Long-time behaviour of the weakly asymmetric exclusion process and the Burgers equation without viscosity. Math. Nachr. **155**, 279–287
Dittrich, P., Gartner, J. (1991): A central limit theorem for the weakly asymmetric simple exclusion process. Math. Nachr. **151**, 75–93
Dobrushin, R.L. (1968): The problem of uniqueness of a Gibbs random field and problem of phase transition. Funct. Anal. Appl. **2**, 302–312
Dobrushin, R.L. (1970): Prescribing a system of random variables by conditional distributions. Theory Probab. Appl. **15**, 453–486
Dobrushin, R.L. (1989): Caricatures of hydrodynamics. In B. Simon and A. Truman and I. M. Davies, editors, *IXth International Congress on Mathematical Physics*, pages 117–132, Adam Hilger, Bristol
Dobrushin, R.L., Pellegrinotti, A., Suhov, Yu.M. (1990): One dimensional harmonic lattice caricature of hydrodynamics: a higher correction. J. Stat. Phys. **61**, 387–402
Dobrushin, R.L., Pellegrinotti, A., Suhov, Yu.M., Triolo, L. (1986): One dimensional harmonic lattice caricature of hydrodynamics. J. Stat. Phys. **43**, 571–607
Dobrushin, R.L., Pellegrinotti, A., Suhov, Yu.M., Triolo, L. (1988): One dimensional harmonic lattice caricature of hydrodynamics: second approximation. J. Stat. Phys. **52**, 423–439
Dobrushin, R.L., Shlosman, S.B. (1987): Completely analytical interactions constructive descriptions. J. Stat. Phys. **56**, 983–1014
Dobrushin, R.L., Siegmund-Schultze, R. (1982): The hydrodynamic limit for systems of particles with independent evolution. Math. Nachr. **105**, 199–224
Dobrushin, R.L., Tirozzi, B. (1977): The central limit theorem and the problem of equivalence of ensembles. Commun. Math. Phys. **54**, 173–192
Donsker, M.D., Varadhan, S.R.S. (1975a): Asymptotic evaluation of certain Markov process expectations for large time I. Comm. Pure Appl. Math. **XXVIII**, 1–47
Donsker, M.D., Varadhan, S.R.S. (1975b): Asymptotic evaluation of certain Markov process expectations for large time II. Comm. Pure Appl. Math. **XXVIII**, 279–301
Donsker, M.D., Varadhan, S.R.S. (1975c): Asymptotic evaluation of certain Wiener integrals for large time. In A. M. Arthurs, editor, *Functional Integration and Its Applications, Proceedings of the International Conference held at the Cumberland Lodge, Windsor Great Park, London 1974*, pages 15–33, Clarendon Press, Oxford
Donsker, M.D., Varadhan, S.R.S. (1976): Asymptotic evaluation of certain Markov process expectations for large time III. Comm. Pure Appl. Math. **XXIX**, 389–461
Donsker, M.D., Varadhan, S.R.S. (1989): Large deviations from a hydrodynamic scaling limit. Comm. Pure Appl. Math. **XLII**, 243–270

Doob, J.L. (1953): *Stochastic Processes*, John Wiley & Sons, New York

Durrett, R. (1991): *Probability: Theory and Examples*, Wadsworth, Belmont

Durrett, R., Neuhauser, C. (1994): Particle systems and reaction–diffusion equations. Ann. Probab. **22**, 289–333

Ekhaus, M., Seppäläinen, T. (1996): Stochastic dynamics macroscopically governed by the porous medium equation for isothermal flow. Ann. Acad. Sci. Fenn. Math. **21**, 309–352

Esposito, R., Marra, R. (1994): On the derivation of the incompressible Navier–Stokes equation for Hamiltonian particle systems. J. Stat. Phys. **74**, 981–1004

Esposito, R., Marra, R., Yau, H.T. (1994): Diffusive limit of asymmetric simple exclusion. Rev. Math. Phys. **6**, 1233–1267

Esposito, R., Marra, R., Yau, H.T. (1996): Navier–Stokes equations for stochastic lattice gases. Comm. Math. Phys. **182**, 395–456

Ethier, S.N., Kurtz, T. G. (1986): *Markov Processes, Characterization and Convergence*. John Wiley & Sons, New York

Evans, L.C., Rezakhanlou, F. (1997): A stochastic model for growing sandpiles and its continuum limit, preprint

Evans, L.C., Soner, H.M., Souganidis, P.E. (1992): Phase transition and generalized motion by mean curvature. Comm. Pure Appl. Math. **XLV**, 1097–1123

Evans, L.C., Spruck, J. (1991): Motion of level sets by mean curvature. I. J. Diff. Geom. **33**, 635–681

Evans, M.R., Foster, D.P., Godrèche, C., Mukamel, D. (1995): Asymmetric exclusion model with two species: spontaneous symmetry breaking. J. Stat. Phys. **80**, 69–102

Eyink, G.L. (1990): Dissipation and large thermodynamic fluctuations. J. Stat. Phys. **61**, 533–572

Eyink, G.L., Lebowitz, J.L., Spohn, H. (1990): Hydrodynamics of stationary non-equilibrium states for some stochastic lattice gas models. Commun. Math. Phys. **132**, 253–283

Eyink, G.L., Lebowitz, J.L., Spohn, H. (1991): Lattice gas models in contact with stochastic reservoirs: local equilibrium and relaxation to the steady state. Commun. Math. Phys. **140**, 119–131

Eyink, G.L., Lebowitz, J.L., Spohn, H. (1996): Hydrodynamics and fluctuations outside of local equilibrium: driven diffusive systems. J. Stat. Phys. **83**, 385–472

Farmer, J., Goldstein, S., Speer, E.R. (1984): Invariant states of a thermally conducting barrier. J. Stat. Phys. **34**, 263–277

Feller, W. (1966): *An Introduction to Probability Theory and its Applications*, John Wiley & Sons, New York

Feng, S., Iscoe, I., Seppäläinen, T. (1997): A microscopic mechanism for the porous medium equation. Stoch. Proc. Appl. **66**, 147–182

Ferrari, P.A. (1986): The simple exclusion process as seen from a tagged particle. Ann. Probab. **14**, 1277–1290

Ferrari, P.A. (1992): Shock fluctuations in asymmetric simple exclusion. Probab. Th. Rel. Fields **91**, 81–101

Ferrari, P.A. (1996): Limit theorems for tagged particles. Markov Proc. Rel. Fields **2**, 17–40

Ferrari, P.A., Fontes, L.R.G. (1994a): Current fluctuations for the asymmetric simple exclusion process. Ann. Probab. **22**, 820–832

Ferrari, P.A., Fontes, L.R.G. (1994b): Shock fluctuations in the asymmetric simple exclusion process. Probab. Th. Rel. Fields **99**, 305–319

Ferrari, P.A., Fontes, L.R.G. (1996): Poissonian approximation for the tagged particle in asymmetric simple exclusion. J. Appl. Prob. **33**, 411–419

Ferrari, P.A., Fontes, L.R.G., Kohayakawa, Y. (1994): Invariant measures for a two species asymmetric process. J. Stat. Phys. **76**, 1153–1177

Ferrari, P.A., Galves, A., Landim, C. (1994): Exponential waiting times for a big gap in a one-dimensional zero range process. Ann. Probab. **22**, 284–288

Ferrari, P.A., Galves, A., Liggett, T.M. (1995): Exponential waiting time for filling a large interval in the symmetric simple exclusion process. Ann. Inst. H. Poincaré, Probabilités **31**, 155–175

Ferrari, P.A., Goldstein, S. (1988): Microscopic stationary states for stochastic systems with particle flux. Probab. Th. Rel. Fields **78**, 455–471

Ferrari, P.A., Goldstein, S., Lebowitz, J.L. (1985): Diffusion, mobility and the Einstein relation. In, J. Fritz, A. Jaffe, D. Száz, editors, *Statistical Physics and Dynamical Systems*, pages 405–441, Birkhäuser, Boston

Ferrari, P.A., Kipnis, C. (1995): Second class particles in the rarefaction fan. Ann. Inst. H. Poincaré, Probabilités **31**, 143–154

Ferrari, P.A., Kipnis, C., Saada, E. (1991): Microscopic structure of traveling waves in the asymmetric simple exclusion process. Ann. Probab. **19**, 226–244

Ferrari, P.A., Presutti, E., Vares, M.E. (1987): Local equilibrium for a one dimensional zero range proces. Stoch. Proc. Appl. **26**, 31–45

Ferrari, P.A., Presutti, E., Vares, M.E. (1988): Nonequilibrium fluctuations for a zero range process, Ann. Inst. H. Poincaré, Probabilités **24**, 237–268

Filippov, A.F. (1960): Differential equations with discontinuous right–hand side. Mat. Sb. (N.S.) **51** (**93**), 99–128

Foster, D.P., Godrèche, C. (1994): Finite–size effects for phase segregation in a two–dimensional asymmetric exclusion model with two species. J. Stat. Phys. **76**, 1129–1151

Fouque, J.P., Saada, E. (1994): Totally asymmetric attractive particle systems on Z: hydrodynamical limit for general initial profiles. Stoch. Proc. Appl. **51**, 9–23

Fritz, J. (1982): Local stability and hydrodynamical limit of Spitzer's one–dimensional lattice model. Commun. Math. Phys. **86**, 363–373

Fritz, J. (1985): On the asymptotic behavior of Spitzer's model for evolution of one–dimensional point systems. J. Stat. Phys. **38**, 615–645

Fritz, J. (1987a): On the hydrodynamic limit of a one dimensional Ginzburg–Landau lattice model: the a priori bounds. J. Stat. Phys. **47**, 551–572

Fritz, J. (1987b): On the hydrodynamic limit of a scalar Ginzburg–Landau Lattice model: the resolvent approach. In, G. Papanicolaou, editor, *Hydrodynamic behaviour and interacting particle systems*, volume 9 of IMA volumes in in Mathematics, pages 75–97. Springer-Verlag, New York

Fritz, J. (1989a): On the hydrodynamic limit of a Ginzburg–Landau model. The law of large numbers in arbitrary dimensions. Probab. Th. Rel. Fields **81**, 291–318

Fritz, J. (1989b): Hydrodynamics in a symmetric random medium. Commun. Math. Phys. **125**, 13–25

Fritz, J. (1990): On the diffusive nature of entropy flow in infinite systems: remarks to a paper by Guo, Papanicolaou and Varadhan. Commun. Math. Phys. **133**, 331–352

Fritz, J., Funaki, T., Lebowitz, J.L. (1994): Stationary states of random Hamiltonian systems. Probab. Th. Rel. Fields **99**, 211–236

Fritz, J., Liverani, C., Olla, S. (1997): Reversibility in infinite Hamiltonian systems with conservative noise. Commun. Math. Phys. **189**, 481–496

Fritz, J., Maes, C. (1988): Derivation of a hydrodynamic equation for Ginzburg–Landau models in an external field. J. Stat. Phys. **53**, 1179–1206

Fritz, J., Rüdiger, B. (1995): Time dependent critical fluctuations of a one dimensional local mean field model. Probab. Th. Rel. Fields **103**, 381–407

Funaki, T. (1989a): The hydrodynamical limit for a scalar Ginzburg–Landau model on R. In M. Métivier and S. Watanabe, editors, *Stochastic Analysis*, Lecture Notes in Mathematics **1322**, pages 28–36, Springer-Verlag, Berlin

Funaki, T. (1989b): Derivation of the hydrodynamical equation for a one–dimensional Ginzburg–Landau model. Probab. Th. Rel. Fields **82**, 39–93

Funaki, T., Handa, K, Uchiyama, K. (1991): Hydrodynamic limit of one dimensional exclusion processes with speed change. Ann. Probab. **19**, 245–265

Funaki, T., Spohn, H. (1997): Motion by mean curvature from the Ginzburg–Landau $\nabla\phi$ interface model. Commun. Math. Phys. **185**, 1–36

Funaki, T., Uchiyama, K., Yau, H.T. (1995): Hydrodynamic limit for lattice gas reversible under Bernoulli measures. In T. Funaki and W. Woyczinky, editors, *Proceedings On Stochastic Method for Nonlinear P.D.E.*. IMA volumes in Mathematics **77**, pages 1–40, Springer-Verlag, New York

Gabrielli, D., Jona–Lasinio, G., Landim, C. (1996): Onsager reciprocity relations without microscopic reversibility. Phys. Rev. Lett. **77**, 1202–1205

Gabrielli, D., Jona–Lasinio, G., Landim, C. (1998): Onsager symmetry from microscopic TP invariance, preprint

Gabrielli, D., Jona–Lasinio, G., Landim, C., Vares, M.E. (1997): Microscopic reversibility and thermodynamic fluctuations. In C. Cercignani, G. Jona–Lasinio, G. Parisi and L. A. Radicati di Brozolo, editors, *Boltzmann's Legacy 150 Years After His Birth.* volume 131 of Atti dei Convegni Licei, pages 79–88, Accademia Nazionale dei Lincei, Roma

Galves, A., Guiol, H. (1997): Relaxation time of the one dimensional zero range process with constant rate. Markov Proc. Rel. Fields **3**, 323–332

Galves, A., Kipnis, C., Marchioro, C., Presutti, E. (1981): Nonequilibrium measures which exhibit a temperature gradient: study of a model. Commun. Math. Phys. **81**, 121–147

Galves, A., Martinelli, F., Olivieri, E. (1989): Large density fluctuations for the one dimensional supercritical contact process. J. Stat. Phys. **55**, 639–648

Gärtner, J. (1988): Convergence towards Burger's equation and propagation of chaos for weakly asymmetric exclusion processes. Stoch. Proc. Appl. **27**, 233–260

Gärtner, J., Presutti, E. (1990): Shock fluctuations in a particle system. Ann. Inst. H. Poincaré, Physique Théorique **53**, 1–14

Giacomin, G. (1991): Van der Waals limit and phase separation in a particle model with Kawasaki dynamics. J. Stat. Phys. **65**, 217–234

Giacomin, G. (1994): Phase separation and random domain patterns in a stochastic particle model. Stoch. Proc. Appl. **51**, 25–62

Giacomin, G. (1995): Onset and structure of interfaces in a Kawasaki + Glauber interacting particle system. Probab. Th. Rel. Fields **103**, 1–24

Giacomin, G., Lebowitz, J.L. (1997): Phase segregation dynamics in particle systems with long range interactions I. Macroscopic limits. J. Stat. Phys. **87**, 37–61

Giacomin, G., Lebowitz, J.L. (1998): Phase segregation dynamics in particle systems with long range interactions II. Interface motion. preprint.

Giacomin, G., Olla, S., Spohn, H. (1998): Equilibrium fluctutation for $\nabla\phi$ interface model, preprint

Gielis, G., Koukkous, A., Landim, C. (1997): Equilibrium fluctuations for zero range processes in random environment. To appear in *Stoch. Proc. Appl.*

Gikhman, I.I., Skorohod, A.V. (1969): *Introduction to the Theory of Random Processes*. W. B. Saunders Company, Philadelphia

Goldstein, S. (1995): Antisymmetric functionals of reversible Markov processes. Ann. Inst. H. Poincaré, Probabilités **31**, 177–190

Goldstein, S., Kipnis, C., Ianiro, N. (1985): Stationary states for a mechanical system with stochastic boundary conditions, J. Stat. Phys. **41**, 915–939

Goldstein, S., Lebowitz, J.L., Presutti, E. (1981): Mechanical systems with stochastic boundary conditions. In J. Fritz, J.L. Lebowitz and D. Szász, editors, *Random Fields*, volume 27 of Colloquia Mathematica Societatis Janos Bolyai, pages 421–461, North-Holland, Amsterdam

Goldstein, S., Lebowitz, J.L., Ravishankar, K. (1982): Ergodic properties of a system in contact with a heat bath: a one-dimensional model. Commun. Math. Phys. **85**, 419–427

Gravner, J., Quastel, J. (1998): Internal DLA and the Stefan problem. Preprint

Greven, A. (1984): The hydrodynamical behavior of the coupled branching process, Ann. Probab. **12**, 760–767

Grigorescu, I. (1997a): Self–diffusion for Brownian motion with local interaction. Preprint.

Grigorescu, I. (1997b): Uniqueness of the tagged particle process in a system with local interactions. Preprint.

Gross, L. (1976): Logarithmic Sobolev inequalities. Am. J. Math. **97**, 1061–1083

Guo, M.Z., Papanicolaou, G.C., Varadhan, S.R.S. (1988): Nonlinear diffusion limit for a system with nearest neighbor interactions. Commun. Math. Phys. **118**, 31–59

Hammersley, J.M. (1972): A few seedlings of research. *Proceedings of the 6th Berkeley symposium in Mathematics, Statistics and Probability*, pages 345–394, University of California Press.

Harris, T.E. (1952): First passage and recurrence distributions. Trans. Amer. Math. Soc. **73**, 471–486

Harris, T.E. (1965): Diffusions with collisions between particles. J. Appl. Prob. **2**, 323–338

Harris, T.E. (1967): Random measures and motions of point processes. Z. Wahrsch. Verw. Gebiete **9**, 36–58

Helland, I. (1982): Central limit theorems for martingales with discrete or continuous time. Scand. J. Stat. **9**, 79–94

Holley, R. (1985a): Rapid convergence to equilibrium in one-dimensional stochastic Ising models. Ann. Probab. **13**, 72–89

Holley, R. (1985b): Possible rates of convergence in finite range, attractive spin systems. In *Particle Systems, Random Media and Large Deviations*, Contemporary Mathematics **41**, pages 215–234

Holley, R., Stroock, D.W. (1976a): L^2-Theory for the stochastic Ising model. Z. Wahrsch. Verw. Gebiete **35**, 87–101

Holley, R., Stroock, D.W. (1976b): Applications of the stochastic Ising model to the Gibbs states. Commun. Math. Phys. **48**, 249–265

Holley, R., Stroock, D.W. (1978): Generalized Ornstein-Uhlenbeck processes and infinite particle branching Brownian motions. RIMSA Kyoto Publ. **14**, 741–788

Holley, R., Stroock, D.W. (1979a): Central limit phenomena of various interacting systems. Ann. Math. **110**, 333–393

Holley, R., Stroock, D.W. (1979b): Rescaling short–range interacting stochastic processes in higher dimensions. In J. Fritz, J.L. Lebowitz and D. Szász, editors, *Random Fields*, volume 27 of Colloquia Mathematica Societatis Janos Bolyai, pages 535–550, North-Holland, Amsterdam

Holley, R., Stroock, D.W. (1987a): Logarithmic Sobolev inequalities and stochastic Ising models. J. Stat. Phys. **46**, 1159–1194

Holley, R., Stroock, D.W. (1987b): Uniform and L^2 convergence in one dimensional stochastic Ising models. Commun. Math. Phys. **123**, 85–93

Hopf, E. (1950): The partial differential equation $u_t + uu_x = \mu u_{xx}$. Comm. Pure Appl. Math. **III**, 201–230

Janowski, S.A., Lebowitz, J.L. (1994): Exact results for the asymmetric simple exclusion proces with a blockage. J. Stat. Phys. **77**, 35–51

Janvresse, E. (1997): First order correction for the hydrodynamic limit of symmetric simple exclusion processes with speed change in dimension $d \geq 3$, preprint

Janvresse, E., Landim, C., Quastel, J., Yau, H.T. (1997): Relaxation to equilibrium of conservative dynamics I: zero range dynamics. Preprint

Jensen, L., Yau, H.T. (1997): *Hydrodynamical Scaling Limits of Simple Exclusion Models.* To appear in Park City-IAS Summer School Lecture Series.

Jockusch, W., Propp, J., Shor, P. (1995): Random domino tillings and the artic circle theorem, preprint

Jona–Lasinio, G. (1991): Stochastic reaction diffusion equations and interacting particle systems. Ann. Inst. H. Poincaré, Physique Théorique **55**, 751–758

Jona–Lasinio, G. (1992): Structure of hydrodynamic fluctutations in interacting particle systems. In F. Guerra, M. I. Loffredo and C. Marchioro, editors, *Probabilistic Methods in Mathematical Physics*, pages 262–263, World Scientific, Singapore

Jona–Lasinio, G., Landim, C., Vares, M.E. (1993): Large deviations for a reaction-diffusion model. Probab. Th. Rel. Fields **97**, 339–361

Katsoulakis, M.A., Souganidis, P.E. (1994): Interacting particle systems and generalized evolution fronts. Arch. Rational Mech. Anal. **127**, 133–159

Katsoulakis, M.A., Souganidis, P.E. (1995): Generalized motion by mean curvature as a macroscopic limit of stochastic Ising models with long range interaction and Glauber dynamics. Commun. Math. Phys. **169**, 61–97

Kemeny, J.G., Snell, J.L., Knapp, A.W. (1966): *Denumerable Markov Chains*, Van Nostrand Company, Princeton

Kipnis, C. (1985): Recent results on the movement of a tagged particle in simple exclusion. Contemp. Math. **41**, 259–265

Kipnis, C. (1986): Central limit theorems for infinite series of queues and applications to simple exclusion. Ann. Probab. **14**, 397–408

Kipnis, C., Landim, C., Olla, S. (1994): Hydrodynamical limit for a nongradient system: the generalized symmetric simple exclusion process. Comm. Pure Appl. Math. **XLVII**, 1475–1545

Kipnis, C., Landim, C., Olla, S. (1995): Macroscopic properties of a stationary non-equilibrium distribution for a non-gradient interacting particle system. Ann. Inst. H. Poincaré, Probabilités **31**, 191–221

Kipnis, C., Léonard, C. (1995): Grandes Déviations pour un système hydrodynamique asymétrique de particules indépendantes. Ann. Inst. H. Poincaré, Probabilités **31**, 223–248

Kipnis, C., Marchioro, C., Presutti, E. (1982): Heat flow in an exactly solvable model J. Stat. Phys. **27**, 65–74

Kipnis, C., Olla, S., Varadhan, S.R.S. (1989): Hydrodynamics and large deviations for simple exclusion processes, Comm. Pure Appl. Math. **XLII**, 115–137

Kipnis, C., Varadhan, S.R.S. (1986): Central limit theorem for additive functionals of reversible Markov processes and applications to simple exclusion, Commun. Math. Phys. **106**, 1–19

Komoriya, K. (1997): Hydrodynamic limit for asymetric mean zero exclusion processes with speed change. preprint

Koukkous, A. (1997): Comportement hydrodynamique de differents processus de zero range. Thèse de doctorat de l'Université de Rouen.

Korolyuk, D.V., Sil'vestrov, D.S. (1984): Entry times into asymptotically receding domains for ergodic Markov chains. Theory Probab. Appl. **28**, 432–442

Kružkov, S.N. (1970): First order quasilinear equations in several independent variables. Math. USSR Sbornik **10**, 217–243

Ladyženskaja, O.A., Solonnikov, U.A., Ural'ceva, N.N. (1968): *Linear and Quasi–Linear Equations of Parabolic Type*. Volume 23 of Translations of mathematical monographs, A. M. S., Providence, Rhode Island

Landim, C. (1991a): Hydrodynamical equations for attractive particle systems on Z^d. Ann. Probab. **19**, 1537–1558

Landim, C. (1991b): Hydrodynamical limit for asymmetric attractive particle systems on Z^d. Ann. Inst. H. Poincaré, Probabilités **27**, 559–581

Landim, C. (1991c): An overview on large deviations of interacting particle systems. Ann. Inst. H. Poincaré, Physique Théorique **55**, 615–635

Landim, C. (1992): Occupation time large deviations for the symmetric simple exclusion process. Ann. Probab. **20**, 206–231

Landim, C. (1993): Conservation of local equilibrium for attractive particle systems on Z^d. Ann. Probab. **21**, 1782–1808

Landim, C. (1996): Hydrodynamical limit for space inhomogenuous one dimensional totally asymmetric zero range processes. Ann. Probab. **24**, 599–638

Landim, C., Mourragui, M. (1997): Hydrodynamic limit of mean zero asymmetric zero range processes in infinite volume. Ann. Inst. H. Poincaré, Probabilités **33**, 65–82

Landim, C., Olla, S., Volchan, S.B. (1997): Driven tracer particle and Einstein relation in one dimensional symmetric simple exclusion process. Resenhas IME–USP **3**, 173–209

Landim, C., Olla, S., Volchan, S.B. (1998): Driven tracer particle in one dimensional symmetric simple exclusion. To appear in Commun. Math. Phys.

Landim, C., Olla, S., Yau, H.T. (1996): Some properties of the diffusion coefficient for asymmetric simple exclusion processes. Ann. Probab. **24**, 1779–1807

Landim, C., Olla, S., Yau, H.T. (1997): First order correction for the hydrodynamic limit of asymmetric simple exclusion processes in dimension $d \geq 3$, Comm. Pure Appl. Math. **L**, 149–203

Landim, C., Sethuraman, S., Varadhan, S.R.S. (1996): Spectral gap for zero range dynamics. Ann. Probab. **24**, 1871–1902

Landim, C., Vares, M.E. (1994): Equilibrium fluctuations for exclusion processes with speed change. Stoch. Proc. Appl. **52**, 107–118

Landim, C., Vares M.E. (1996): Exponential estimate for reaction diffusion models. Probab. Th. Rel. Fields **106**, 151–186

Landim, C., Yau, H.T. (1995): Large deviations of interacting particle systems in infinite volume. Comm. Pure Appl. Math. **XLVIII**, 339–379

Landim, C., Yau, H.T. (1997): Fluctuation–dissipation equation of asymmetric simple exclusion processes. Probab. Th. Rel. Fields **108**, 321–356

Lax, P.D. (1957): Hyperbolic systems of conservation laws II. Comm. Pure Appl. Math. **X**, 537–566

Lebowitz, J.L., Neuhauser, C., Ravishankar, K. (1996): Dynamics of a spin–exchange model. Stoch. Proc. Appl. **64**, 187–208

Lebowitz, J.L., Orlandi, E., Presutti, E. (1991): A particle model for spinodal decomposition. J. Stat. Phys. **63**, 933–974

Lebowitz, J.L., Penrose, O. (1966): Rigorous treatment of the Van der Waals-Maxwell theory of liquid vapour transition. J. Math. Phys. **7**, 98–113

Lebowitz, J.L., Presutti, E., Spohn, H. (1988): Microscopic models of hydrodynamic behavior. J. Stat. Phys. **51**, 841–862

Lebowitz, J.L., Rost, H. (1994): The Einstein relation for the displacement of a test particle in a random environment. Stoch. Proc. Appl. **54**, 183–196

Lebowitz, J.L., Schonmann, R.H. (1987): On the asymptotics of occurence times of rare events for stochastic spin systems. J. Stat. Phys. **48**, 727–751

Lebowitz, J.L., Spohn, H. (1982a): Microscopic basis for Fick's law of self–diffusion. J. Stat. Phys. **28**, 539–556

Lebowitz, J.L., Spohn, H. (1982b): Steady-state diffusion at low density. J. Stat. Phys. **29**, 39–55

Lebowitz, J.L., Spohn, H. (1983): On the time evolution of macroscopic systems. Comm. Pure Appl. Math. **XXXVI**, 595–613

Lebowitz, J.L., Spohn, H. (1997): Comment on "Onsager reciprocity relations without micrsocpic reversibility". Phys. Rev. Lett. **78**, 394–395

Liggett, T.M. (1973): An infinite particle system with zero range interactions. Ann. Probab. **1**, 240-253

Liggett, T.M. (1975): Ergodic theorems for the asymmetric simple exclusion process. Trans. Amer. Math. Soc. **213**, 237–260

Liggett, T.M. (1985): *Interacting Particle Systems*, Springer-Verlag, New York

Liggett, T.M. (1989): Exponential L^2 convergence of attractive reversible nearest particle systems. Ann. Probab. **17**, 403–432

Liggett, T.M. (1991): L^2 rates of convergence of attractive reversible particle systems: the critical case. Ann. Probab. **19**, 935–959

Liverani, C., Olla, S. (1996): Ergodicity in infinite Hamiltonian systems with conservative noisė. Probab. Th. Rel. Fields **106**, 401–445

Lu, S.L. (1994): Equilibrium fluctuations of a one-dimensional nongradient Ginzburg–Landau model. Ann. Probab. **22**, 1252–1272

Lu, S.L. (1995), Hydrodynamic scaling limits with deterministic initial configurations. Ann. Probab. **23**, 1831–1852

Lu, S.L., Yau, H.T. (1993): Spectral gap and logarithmic Sobolev inequality for Kawasaki and Glauber dynamics. Commun. Math. Phys. **156**, 399–433

Maes, C. (1990): Kinetic limit of a conservative lattice gas dynamics showing long range correlations. J. Stat. Phys. **61**, 667–681

Malyshev, V.A., Manita, A.D., Petrova, E.N., Scacciatelli, E. (1995): Hydrodynamics of the weakly perturbed voter model. Markov Proc. Rel. Fields **1**, 3–56

Marchand, J.P., Martin, Ph. A. (1986): Exclusion process and droplet shape. J. Stat. Phys. **44**, 491–504, J. Stat. Phys. **50**, 469–471 (1988)

Martin-Löf, A. (1976): Limit theorems for the motion of a Poisson system of independent Markovian particles at high density. Z. Wahrsch. Verw. Gebiete **34**, 205–223

Martinelli, F., Olivieri, E. (1994a): Approach to equilibrium of Glauber dynamics in the one phase region I: the attractive case. Commun. Math. Phys. **161**, 447–486

Martinelli, F., Olivieri, E. (1994b): Approach to equilibrium of Glauber dynamics in the one phase region II: the general case. Commun. Math. Phys. **161**, 487–514

Martinelli, F., Olivieri, E., Scoppola, E. (1990): Metastability and exponential approach to equilibrium for low-temperature stochastic Ising models. J. Stat. Phys. **61** 1005–1119

Morrey, C.B. (1955): On the derivation of the equation of hydrodynamics from statistical mechanics. Comm. Pure Appl. Math. **VIII**, 279–326

Mountford, T.S. (1995): Exponential convergence for attractive reversible subcritical nearest particle systems. Stoch. Proc. Appl. **59**, 235–249

Mourragui, M. (1996): Comportement hydrodynamique et entropie relative des processus de sauts, de naissances et de morts. Ann. Inst. H. Poincaré, Probabilités **32**, 361–385

Naddaf, A., Spencer, T. (1997): On homogenization and scaling limit of some grandient perturbations of a massless free field. Commun. Math. Phys. **183**, 55–84

Nappo, G., Orlandi, E. (1988): Limit laws for a coagulation model of interacting random particles. Ann. Inst. H. Poincaré, Probabilités **24**, 319–344

Nappo, G., Orlandi, E., Rost, H. (1989): A reaction–diffusion model for moderately interacting particles. J. Stat. Phys. **55**, 579–600

Neuhauser, C. (1990): One dimensional stochastic Ising models with small migration. Ann. Probab. **18**, 1539–1546

Neveu, J. (1972): *Martingales à Temps Discret*, Dunod, Paris

Noble, C. (1992): Equilibrium behavior of the sexual reproduction process with rapid diffusion. Ann. Probab. **20**, 724–745

Oelschläger, K. (1984): A martingale approach to the law of large numbers for weakly interacting stochastic processses. Ann. Probab. **12**, 458–479

Oelschläger, K. (1985): A law of large numbers for moderately interacting diffusion processes. Z. Wahrsch. Verw. Gebiete **69**, 279–322

Oelschläger, K. (1987): A fluctuation theorem for moderately interacting diffusion processes. Probab. Th. Rel. Fields **74**, 591–616

Oelschläger, K. (1989): On the derivation of reaction-diffusion equations as limit dynamics of systems of moderately interacting stochastic processes. Probab. Th. Rel. Fields **82**, 565–586

Oleinik, O.A., Kružkov, S.N. (1961): Quasi–linear second–order parabolic equations with many independent variables. Russian Math. Surveys **16-5**, 105–146

Olivieri, E., Vares, M.E. (1998): *Large deviations and Metastability*. To be published by Cambridge Universtiy Press, Cambridge

Olla, S., Varadhan, S.R.S. (1991): Scaling limit for interacting Ornstein-Uhlenbeck processes. Commun. Math. Phys. **135**, 355–378

Olla, S., Varadhan, S.R.S., Yau, H.T. (1991): Hydrodynamic limit for a Hamiltonian system with weak noise. Commun. Math. Phys. **155**, 523–560

Onsager, L. (1931): Reciprocal relations in irreversible processes I, II. Phys. Rev. **37**, 405–426; **38**, 2265–2279

Onsager, L., Machlup, S. (1953): Fluctuation and irreversible processes I, II. Phys. Rev. **91**, 1505–1512, 1512–1515

Penrose, O., Lebowitz, J.L. (1987): Towards a rigorous molecular theory of metastability. In E. W. Montroll and J. L. Lebowitz, editors, *Fluctutation Phenomena*, second edition. North-Holland Physics Publishing, Amsterdam

Petrov, V.V. (1975): *Sums of Independent Random Variables*, Springer-Verlag, New York

Port, S.C., Stone, C.J. (1973): Infinite particle systems. Trans. Amer. Math. Soc. **178**, 307–340

Posta, G. (1995): Spectral gap for an unrestricted Kawasaki type dynamics. European Series App. Ind. Math. Prob. Stat. **1**, 145–181

Presutti, E. (1987): Collective phenomena in stochastic particle systems. In S. Albeverio, Ph. Blanchard and L. Streit, editors, *Stochastic Processes – Mathematics and Physics II*. Volume 1250 of Lecture Notes in Mathematics, pages 195–232, Springer-Verlag, Berlin

Presutti, E. (1997): Hydrodynamics in particle systems. In C. Cercignani, G. Jona–Lasinio, G. Parisi and L. A. Radicati di Brozolo, editors, *Boltzmann's Legacy 150 Years After His Birth*. Volume 131 of Atti dei convegni Licei, pages 189–207, Accademia Nazionale dei Lincei, Roma

Presutti, E., Spohn, H. (1983): Hydrodynamics of the voter model. Ann. Probab. **11**, 867–875

Quastel, J. (1992): Diffusion of color in the simple exclusion process. Comm. Pure Appl. Math. **XLV**, 623–679

Quastel, J. (1995a): Large deviations from a hydrodynamical scaling limit for a nongradient system. Ann. Probab. **23**, 724–742

Quastel, J. (1995b): Diffusion in disorder media. In T. Funaki and W. Woyczinky, editors, *Proceedings On Stochastic Method for Nonlinear P.D.E.*, IMA volumes in Mathematics **77**, pages 65–79. Springer-Verlag, New York

Quastel, J., Rezakhanlou, F., Varadhan, S.R.S. (1997): Large deviations for the symmetric simple exclusion process in dimension $d \geq 3$. Preprint

Quastel, J., Yau, H.T. (1997): Lattice gases, large deviations and the incompressible Navier–Stokes equation, preprint

Ravishankar, K. (1992a): Fluctuations from the hydrodynamical limit for the symmetric simple exclusion in Z^d. Stoch. Proc. Appl. **42**, 31–37

Ravishankar, K. (1992b): Interface fluctuations in the two–dimensional weakly asymmetric simple exclusion process. Stoch. Proc. Appl. **43**, 223–247

Reed, M., Simon, B. (1975): *Methods of Modern Mathematical Physics*. Academic Press, New York

Revuz, D., Yor, M. (1991): *Continuous Martingales and Brownian Motion*, Springer-Verlag, Berlin

Rezakhanlou, F. (1990): Hydrodynamic limit for a system with finite range interactions. Commun. Math. Phys. **129**, 445–480

Rezakhanlou, F. (1991): Hydrodynamic limit for attractive particle systems on Z^d. Commun. Math. Phys. **140**, 417–448

Rezakhanlou, F. (1994a): Evolution of tagged particles in non-reversible particle systems. Commun. Math. Phys. **165**, 1–32

Rezakhanlou, F. (1994b): Propagation of chaos for symmetric simple exclusion. Comm. Pure Appl. Math. **XLVII**, 943–957

Rezakhanlou, F. (1995): Microscopic structure of shocks in one conservation laws. Ann. Inst. H. Poincaré, Analyse non Linéaire **12**, 119–153

Rezakhanlou, F. (1996a): Kinetic limits for a class of interacting particle systems. Probab. Th. Rel. Fields **104**, 97–146

Rezakhanlou, F. (1996b): Propagation of chaos for particle systems associated with discret Boltzmann equation. Stoch. Proc. Appl. **64**, 55–72

Rezakhanlou, F. (1997): Equilibrium fluctuations for the discrete Boltzmann equation. To appear in Duke Math. J.

Rezakhanlou, F., Tarver III, J.E. (1997): Boltzmann–Grad limit for a particle system in continuum. Ann. Inst. H. Poincaré, Probabilités **33**, 753–796

Rost, H. (1981): Nonequilibrium behavior of many particle systems: density profile and local equilibria. Z. Wahrsch. Verw. Gebiete **58**, 41–53

Rost, H. (1983): Hydrodynamik gekoppelter diffusionen: fluktuationen im gleichgewicht. In P. Blanchard and L. Streit, editors, *Dynamics and Processes*. Volume 1031 of Lecture Notes in Mathematics, pages 97–107. Springer-Verlag, Berlin

Rost, H. (1984): Diffusion de sphères dures dans la droite réelle: comportement macroscopique et équilibre local. In J. P. Azéma and M. Yor, editors, Séminaire de Probabilités XVIII. Volume 1059 of Lecture Notes in Mathematics, 127–143. Springer-Verlag, Berlin

Rost, H. (1985): A central limit theorem for a system of interacting particles. In L. Arnold and P. Kotelenez, editors, *Stochastic space–time models and limit theorems*, 243–248, D. Reidel Publishing Company

Rost, H., Vares, M.E. (1985): Hydrodynamics of a one dimensional nearest neighbor model. Contemp. Math. **41**, 329–342

Saada, E. (1987a): A limit theorem for the position of a tagged particle in a simple exclusion process. Ann. Probab. **15**, 375–381

Saada, E. (1987b): Processus de zero-range avec particule marquée. Ann. Inst. H. Poincaré, Probabilités **26**, 5–18

Schütz, G. (1993): Generalized Bethe ansatz solution of a one–dimensional asymmetric exclusion process on a ring with blockage. J. Stat. Phys. **71**, 471–505

Schütz, G., Domany, E. (1993): Phase transition in an exactly soluble one–dimensional exclusion model. J. Stat. Phys. **72**, 277–296

Sellami, S. (1998): Equilibrium density fluctuations field of a one–dimensional non gradient reversible model: the generalized exclusion process, preprint

Seppäläinen, T. (1996a): A microscopic model for the Burgers equation and longest increasing subsequences. Electronic J. Probab. **1**, 1–51

Seppäläinen, T. (1996b): Hydrodynamic scaling, convex duality and asymptotic shapes of growth models, preprint

Seppäläinen, T. (1997a): A scaling limit for queues in series. Ann. Appl. Probab. **7**, 855–872

Seppäläinen, T. (1997b): Increasing sequences of independent points on the planar lattice. Ann. Appl. Probab. **7**, 886–898

Shiga, T. (1988): Tagged particle motion in a clustered random walk system. Stoch. Proc. Appl. **30**, 225–252

Siri, P. (1996): Inhomogeneous zero range process. Ph.D. thesis, Politecnico di Torino

Smoller, J. (1983): *Shock Waves and Reaction-Diffusion Equations*, Springer-Verlag, New York

Speer, E. (1994): The two species totally asymmetric simple exclusion process. In M. Fannes, C. Maes and A. Verbeure, editors, *On Three Levels: Micro-, Meso- and Macro-Approaches in Physics*. Volume 324 of Nato ASI series B, pages 91–102.

Spitzer, F. (1970): Interaction of Markov processes. Adv. Math. **5**, 246–290.

Spohn, H. (1985): Equilibrium fluctuations for some stochastic particle systems. In J. Fritz and A. Jaffe and D. Szász, editors, *Statistical Physics and Dynamical Systems*, pages 67–81. Birkhäuser, Boston

Spohn, H. (1986): Equilibrium fluctuations for interacting Brownian particles. Commun. Math. Phys. **103**, 1–33

Spohn, H. (1987): Interacting Brownian particles: a study of Dyson's model. In G. C. Papanicolaou, editor, *Hydrodynamic Behavior and Interacting Particle Systems*, IMA volumes in Mathematics **9**, pages 151–179. Springer-Verlag, New York

Spohn, H. (1990): Tracer diffusion in lattice gases. J. Stat. Phys. **59**, 1227–1239

Spohn, H. (1991): *Large Scale Dynamics of Interacting Particles*, Springer-Verlag, Berlin

Spohn, H. (1993): Interface motion in models with stochastic dynamics. J. Stat. Phys. **71**, 1081–1132

Spohn, H., Yau, H.T. (1995): Bulk diffusivity of lattice gases close to criticality. J. Stat. Phys. **79**, 231–241

Stroock, D.W., Varadhan, S.R.S. (1979): *Multidimensional Diffusion Processes*. Springer-Verlag, New York

Stroock, D.W., Zegarlinski, B. (1992a): The equivalence of the logarithmic Sobolev inequality and the Dobrushin–Shlosman mixing condition. Commun. Math. Phys. **144**, 303–323

Stroock, D.W., Zegarlinski, B. (1992b) The logarithmic Sobolev inequality for continuous spin systems on a lattice. J. Funct. Anal. **104**, 299-326

Stroock, D.W., Zegarlinski, B. (1992c): The logarithmic Sobolev inequality for discrete spin systems on a lattice. Commun. Math. Phys. **149**, 175–193

Suzuki, Y., Uchiyama, K. (1993): Hydrodynamic limit for a spin system on a multidimensional lattice. Probab. Th. Rel. Fields **95**, 47–74

Thomas, L.E. (1989): Bound on the mass gap for finite volume stochastic Ising models at low temperature. Commun. Math. Phys. **126**, 1–11

Uchiyama, K. (1994): Scaling limits of interacting diffusions with arbitrary initial distributions. Probab. Th. Rel. Fields **99**, 97–110

Varadhan, S.R.S. (1966): Asymptotic probabilities and differential equations. Comm. Pure Appl. Math. **XIX**, 261–286

Varadhan, S.R.S. (1991): Scaling limits for interacting diffusions. Commun. Math. Phys. **135**, 313–353

Varadhan, S.R.S. (1994a): Nonlinear diffusion limit for a system with nearest neighbor interactions II. In K. D. Elworthy and N. Ikeda, editors, *Asymptotic Problems in Probability Theory: Stochastic Models and Diffusion on Fractals*. Volume 283 of Pitman Research Notes in Mathematics, pages 75–128. John Wiley & Sons, New York

Varadhan, S.R.S. (1994b): Regularity of self diffusion coefficient. In M. Freidlin, editor, *The Dynkin Festschrift, Markov Processes and their Applications*, pages 387–397, Birkhäuser, Boston

Varadhan, S.R.S. (1995): Self diffusion of a tagged particle in equilibrium for asymmetric mean zero random walk with simple exclusion. Ann. Inst. H. Poincaré, Probabilités **31**, 273–285

Varadhan, S.R.S., Yau, H.T. (1997): Diffusive limit of lattice gases with mixing conditions, preprint

Vares, M.E. (1991): On long time behavior of a class of reaction–diffusion models. Ann. Inst. H. Poincaré, Physique Théorique **55**, 601–613

Venkatraman, R. (1994): Hydrodynamic limit of the asymmetric simple exclusion process with deterministic initial data and the Hammersley process on $S(1)$. Ph.D. Thesis, Courant Institute, New York University.

Wick, W.D. (1985): A dynamical phase transition in an infinite particle system. J. Stat. Phys. **38**, 1015–1025

Wick, W.D. (1989): Hydrodynamic limit of nongradient interacting particle processes. J. Stat. Phys. **89**, 873–892

Xu, L. (1993): Diffusion limit for the lattice gas with short range interactions. Ph.D. thesis, New York University

Yau, H.T. (1991): Relative entropy and hydrodynamics of Ginzburg–Landau models. Lett. Math. Phys. **22**, 63–80

Yau, H.T. (1994): Metastability of Ginzburg–Landau model with a conservation law. J. Stat. Phys. **74**, 705–742

Yau, H.T. (1996): Logarithmic Sobolev inequality for lattice gases with mixing conditions. Commun. Math. Phys. **181**, 367–408

Yau, H.T. (1997): Logarithmic Sobolev inequality for generalized simple exclusion processes. Probab. Th. Rel. Fields **109**, 507–538

Yosida, K. (1995): *Functional Analysis.* Reprint from the sixth edition, Springer-Verlag, Berlin

Zegarlinski, B. (1990a): On logarithmic Sobolev inequalities for infinite lattice systems. Lett. Math. Phys. **20**, 173–182

Zegarlinski, B. (1990b): Logarithmic Sobolev inequalities for infinite one dimensional lattice systems. Commun. Math. Phys. **133**, 147–162

Zegarlinski, B. (1992): Dobrushin uniqueness theorem and logarithmic Sobolev inequalities. J. Funct. Anal. **105**, 77–111

Zhu, M. (1990): Equlibrium fluctuations for one dimensional Ginzburg–Landau lattice model. Nagoya Math. J. **117**, 63–92.

Subject Index

Grundlehren der mathematischen Wissenschaften

A Series of Comprehensive Studies in Mathematics

A Selection

212. Switzer: Algebraic Topology – Homotopy and Homology
215. Schaefer: Banach Lattices and Positive Operators
217. Stenström: Rings of Quotients
218. Gihman/Skorohod: The Theory of Stochastic Processes II
219. Duvaut/Lions: Inequalities in Mechanics and Physics
220. Kirillov: Elements of the Theory of Representations
221. Mumford: Algebraic Geometry I: Complex Projective Varieties
222. Lang: Introduction to Modular Forms
223. Bergh/Löfström: Interpolation Spaces. An Introduction
224. Gilbarg/Trudinger: Elliptic Partial Differential Equations of Second Order
225. Schütte: Proof Theory
226. Karoubi: K-Theory. An Introduction
227. Grauert/Remmert: Theorie der Steinschen Räume
228. Segal/Kunze: Integrals and Operators
229. Hasse: Number Theory
230. Klingenberg: Lectures on Closed Geodesics
231. Lang: Elliptic Curves. Diophantine Analysis
232. Gihman/Skorohod: The Theory of Stochastic Processes III
233. Stroock/Varadhan: Multidimensional Diffusion Processes
234. Aigner: Combinatorial Theory
235. Dynkin/Yushkevich: Controlled Markov Processes
236. Grauert/Remmert: Theory of Stein Spaces
237. Köthe: Topological Vector Spaces II
238. Graham/McGehee: Essays in Commutative Harmonic Analysis
239. Elliott: Probabilistic Number Theory I
240. Elliott: Probabilistic Number Theory II
241. Rudin: Function Theory in the Unit Ball of C^n
242. Huppert/Blackburn: Finite Groups II
243. Huppert/Blackburn: Finite Groups III
244. Kubert/Lang: Modular Units
245. Cornfeld/Fomin/Sinai: Ergodic Theory
246. Naimark/Stern: Theory of Group Representations
247. Suzuki: Group Theory I
248. Suzuki: Group Theory II
249. Chung: Lectures from Markov Processes to Brownian Motion
250. Arnold: Geometrical Methods in the Theory of Ordinary Differential Equations
251. Chow/Hale: Methods of Bifurcation Theory
252. Aubin: Nonlinear Analysis on Manifolds. Monge-Ampère Equations
253. Dwork: Lectures on ρ-adic Differential Equations
254. Freitag: Siegelsche Modulfunktionen
255. Lang: Complex Multiplication
256. Hörmander: The Analysis of Linear Partial Differential Operators I
257. Hörmander: The Analysis of Linear Partial Differential Operators II
258. Smoller: Shock Waves and Reaction-Diffusion Equations
259. Duren: Univalent Functions
260. Freidlin/Wentzell: Random Perturbations of Dynamical Systems
261. Bosch/Güntzer/Remmert: Non Archimedian Analysis – A System Approach to Rigid Analytic Geometry
262. Doob: Classical Potential Theory and Its Probabilistic Counterpart
263. Krasnosel'skiĭ/Zabreĭko: Geometrical Methods of Nonlinear Analysis
264. Aubin/Cellina: Differential Inclusions

265. Grauert/Remmert: Coherent Analytic Sheaves
266. de Rham: Differentiable Manifolds
267. Arbarello/Cornalba/Griffiths/Harris: Geometry of Algebraic Curves, Vol. I
268. Arbarello/Cornalba/Griffiths/Harris: Geometry of Algebraic Curves, Vol. II
269. Schapira: Microdifferential Systems in the Complex Domain
270. Scharlau: Quadratic and Hermitian Forms
271. Ellis: Entropy, Large Deviations, and Statistical Mechanics
272. Elliott: Arithmetic Functions and Integer Products
273. Nikol'skiĭ: Treatise on the Shift Operator
274. Hörmander: The Analysis of Linear Partial Differential Operators III
275. Hörmander: The Analysis of Linear Partial Differential Operators IV
276. Ligget: Interacting Particle Systems
277. Fulton/Lang: Riemann-Roch Algebra
278. Barr/Wells: Toposes, Triples and Theories
279. Bishop/Bridges: Constructive Analysis
280. Neukirch: Class Field Theory
281. Chandrasekharan: Elliptic Functions
282. Lelong/Gruman: Entire Functions of Several Complex Variables
283. Kodaira: Complex Manifolds and Deformation of Complex Structures
284. Finn: Equilibrium Capillary Surfaces
285. Burago/Zalgaller: Geometric Inequalities
286. Andrianaov: Quadratic Forms and Hecke Operators
287. Maskit: Kleinian Groups
288. Jacod/Shiryaev: Limit Theorems for Stochastic Processes
289. Manin: Gauge Field Theory and Complex Geometry
290. Conway/Sloane: Sphere Packings, Lattices and Groups
291. Hahn/O'Meara: The Classical Groups and K-Theory
292. Kashiwara/Schapira: Sheaves on Manifolds
293. Revuz/Yor: Continuous Martingales and Brownian Motion
294. Knus: Quadratic and Hermitian Forms over Rings
295. Dierkes/Hildebrandt/Küster/Wohlrab: Minimal Surfaces I
296. Dierkes/Hildebrandt/Küster/Wohlrab: Minimal Surfaces II
297. Pastur/Figotin: Spectra of Random and Almost-Periodic Operators
298. Berline/Getzler/Vergne: Heat Kernels and Dirac Operators
299. Pommerenke: Boundary Behaviour of Conformal Maps
300. Orlik/Terao: Arrangements of Hyperplanes
301. Loday: Cyclic Homology
302. Lange/Birkenhake: Complex Abelian Varieties
303. DeVore/Lorentz: Constructive Approximation
304. Lorentz/v. Golitschek/Makovoz: Construcitve Approximation. Advanced Problems
305. Hiriart-Urruty/Lemaréchal: Convex Analysis and Minimization Algorithms I. Fundamentals
306. Hiriart-Urruty/Lemaréchal: Convex Analysis and Minimization Algorithms II. Advanced Theory and Bundle Methods
307. Schwarz: Quantum Field Theory and Topology
308. Schwarz: Topology for Physicists
309. Adem/Milgram: Cohomology of Finite Groups
310. Giaquinta/Hildebrandt: Calculus of Variations I: The Lagrangian Formalism
311. Giaquinta/Hildebrandt: Calculus of Variations II: The Hamiltonian Formalism
312. Chung/Zhao: From Brownian Motion to Schrödinger's Equation
313. Malliavin: Stochastic Analysis
314. Adams/Hedberg: Function Spaces and Potential Theory
315. Bürgisser/Clausen/Shokrollahi: Algebraic Complexity Theory
316. Saff/Totik: Logarithmic Potentials with External Fields
317. Rockafellar/Wets: Variational Analysis
318. Kobayashi: Hyperbolic Complex Spaces
319. Bridson/Haefliger: Metric Spaces of Non-Positive Curvature
320. Kipnis/Landim: Scaling Limits of Interacting Particle Systems

Printing: Mercedesdruck, Berlin
Binding: Buchbinderei Lüderitz & Bauer, Berlin